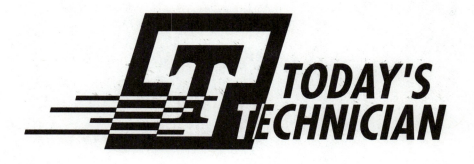

Shop Manual for

Automotive Computer Systems

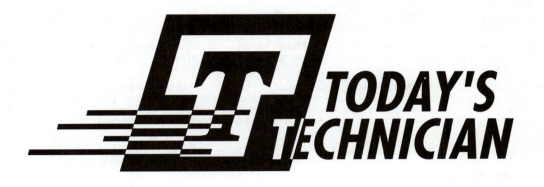

Shop Manual for
Automotive Computer Systems

Don Knowles
Knowles Automotive Training
Moose Jaw, Saskatchewan
Canada

Jack Erjavec
Series Advisor
Columbus State Community College
Columbus, Ohio

Delmar Publishers
I(T)P™ An International Thomson Publishing Company

Albany • Bonn • Boston • Cincinnati • Detroit • London • Madrid • Melbourne
Mexico City • New York • Pacific Grove • Paris • San Francisco • Singapore • Tokyo
Toronto • Washington

NOTICE TO THE READER

Publisher does not warrant or guarantee any of the products described herein or perform any independent analysis in connection with any of the product information contained herein. Publisher does not assume, and expressly disclaims, any obligation to obtain and include information other than that provided to it by the manufacturer.

The reader is expressly warned to consider and adopt all safety precautions that might be indicated by the activities described herein and to avoid all potential hazards. By following the instructions contained herein, the reader willingly assumes all risks in connection with such instructions.

The publisher makes no representation or warranties of any kind, including but not limited to, the warranties of fitness for particular purpose or merchantability, nor are any such representations implied with respect to the material set forth herein, and the publisher takes no responsibility with respect to such material. The publisher shall not be liable for any special, consequential, or exemplary damages resulting, in whole or in part, from the readers' use of, or reliance upon, this material.

Photo sequences: Jeff Hinckley and Rod Dixon Associates

DELMAR STAFF
Publisher: Susan Simpfenderfer
Acquisitions Editor: Vernon Anthony
Developmental Editor: Catherine Eads
Project Editor: Thomas Smith
Production Coordinator: Karen Smith
Art/Design Coordinator: Michael Prinzo

COPYRIGHT © 1996
By Delmar Publishers
an International Thomson Publishing Company
The ITP logo is a trademark under license.

Printed in the United States of America

For information, contact:

Delmar Publishers
3 Columbia Circle, Box 15015
Albany, New York 12212-5015

International Thomson Editores
Campos Eliseos 385, Piso 7
Col Polanco
11560 Mexico DF Mexico

International Thomson Publishing Europe
Berkshire House 168-173
High Holborn
London, WC1V7AA
England

International Thomson Publishing GmbH
Königswinterer Strasse 418
53227 Bonn
Germany

Thomas Nelson Australia
102 Dodds Street
South Melbourne, 3205
Victoria, Australia

International Thomson Publishing Asia
221 Henderson Road
#05-10 Henderson Building
Singapore 0315

Nelson Canada
1120 Birchmont Road
Scarborough, Ontario
Canada M1K 5G4

International Thomson Publishing-Japan
Hirakawacho Kyowa Building, 3F
2-2-1 Hirakawacho
Chiyoda-ku, Tokyo 102
Japan

All rights reserved. No part of this work covered by the copyright hereon may be reproduced or used in any form or by any means—graphic, electronic, or mechanical, including photocopying, recording, taping, or information storage and retrieval systems—without written permission of the publisher.

1 2 3 4 5 6 7 8 9 10 XXX 02 01 00 99 98 97 96

Library of Congress Cataloging-in-Publication Data

Knowles, Don.
    Automotive computer systems / Don Knowles, Jack Erjavec.
      p. cm. — (Today's technician)
    Includes index.
    Contents: [1]. Shop manual — [2]. Classroom manual.
    ISBN 0-8273-6884-4
    1. Automobiles — Motors — Computer control systems. I. Erjavec,
Jack. II. Title. III. Series
TL214.C64K59 1996
629.25'49 — dc20   95-25176
                                                 CIP

CONTENTS

Photo Sequences ix
Preface x

CHAPTER 1 **Shop Practices** 1

Shop Layout 1 • Shop Rules 2 • Vehicle Operation 3 • Housekeeping 3 • Air Quality 4 • Employer and Employee Obligations 6 • Job Responsibilities 7 • National Institute for Automotive Service Excellence (ASE) Certification 8 • Shop Projects to Enhance the Theories in Chapter 1 of the Classroom Manual 8 • Guidelines for Following Proper Shop Practices 9 • Terms to Know 11 • ASE Style Review Questions 11

CHAPTER 2 **Tools, Safety Practices, and Test Equipment** 13

Measuring Systems 13 • Personal Safety 15 • Lifting and Carrying 16 • Hand Tool Safety 17 • Power Tool Safety 17 • Compressed Air Equipment Safety 18 • Hydraulic Pressing and Lifting Equipment 18 • Hydraulic Jack and Jack Stand Safety 22 • Cleaning Equipment Safety and Environmental Considerations 23 • Electronic and Fuel System Diagnostic Equipment 25 • Ignition Test Equipment 32 • Scan Testers 39 • Emissions Analyzer 41 • Engine Analyzers 43 • Safety Training Exercises 44 • Guidelines for Use of Tools, Safety Practices, and the Use of Test Equipment 45 • Terms To Know 47 • ASE Style Review Questions 47

CHAPTER 3 **Input Sensor Diagnosis and Service** 49

Diagnosis of Computer Voltage Supply and Ground Wires 49 • Input Sensor Diagnosis and Service 50 • Guidelines for Input Sensor Diagnosis 67 • Terms To Know 69 • ASE Style Review Questions 70

CHAPTER 4 **Ignition System Service and Diagnosis** 73

Ignition System Diagnosis 73 • No-Start Ignition Diagnosis, Primary Ignition Circuit 74 • No-Start Ignition Diagnosis, Secondary Ignition Circuit 75 • Ignition Module Testing 75 • Pickup Coil Adjustment and Tests 76 • Ignition Coil Inspection and Tests 77 • Inspection of Distributor Cap and Rotor 78 • Testing Secondary Ignition Wires 79 • Ignition Module Removal and Replacement 80 • Distributor Service 80 • Spark Plug Service 87 • Computer-Controlled Ignition System Service and Diagnosis 89 • Electronic Ignition (EI) System Diagnosis and Service 91 • Low Data Rate and High Data Rate EI Service and Diagnosis 95 • General Motors Electronic EI System Service and Diagnosis 98 • Diagnosis of EI Systems with Magnetic Sensors 102 • Engine Misfire Diagnosis 103 • Guidelines for Servicing Distributor and Electronic Ignition Systems 104 • Terms To Know 106 • ASE Style Review Questions 106

CHAPTER 5   **Fuel Tank, Line, Filter, and Pump Diagnosis and Service**   **113**

Alcohol in Fuel Test 113 • Fuel System Pressure Relief 113 • Fuel Tank Service 115 • Fuel Line Service 120 • Fuel Filter Service 122 • Mechanical Fuel Pump Service and Diagnosis 123 • Electric Fuel Pump Testing 126 • Guidelines for Fuel Tank, Line, Filter, and Pump Service and Diagnosis 126 • Terms to Know 127 • ASE Style Review Questions 128

CHAPTER 6   **Computer-Controlled Carburetor Diagnosis and Service**   **131**

Computer-Controlled Carburetor Diagnosis and Service 131 • Computer-Controlled Carburetor System Performance Test 134 • Flash Code Diagnosis 137 • Voltmeter Diagnosis 139 • Scan Tester Diagnosis 143 • Guidelines for Conventional and Computer-Controlled Carburetor System Diagnosis and Service 145 • Terms to Know 146 • ASE Style Review Questions 147

CHAPTER 7   **Electronic Fuel Injection Diagnosis and Service**   **149**

Throttle Body, Multiport, and Sequential Fuel Injection Service and Diagnosis 149 • Injector Service and Diagnosis 157 • Removing and Replacing Fuel Rail, Injectors, and Pressure Regulator 160 • Cold-Start Injector Diagnosis and Service 163 • Minimum Idle Speed Adjustment and Throttle Position Sensor Adjustment 164 • Minimum Idle Speed Adjustment Throttle Body Injection 165 • Throttle Body Service 166 • Fuel Cut RPM Check 168 • Flash Code Diagnosis of TBI, MFI, and SFI Systems 169 • General TBI, MFI, and SFI Diagnosis 176 • Diagnosis of Specific Problems and Necessary Corrections 178 • Guidelines for Servicing TBI, MFI, and SFI Systems 182 • Terms to Know 183 • ASE Style Review Questions 183

CHAPTER 8   **Scan Tester and Digital Storage Oscilloscope Diagnosis of Electronic Fuel Injection and On-Board Diagnostics II**   **189**

Scan Tester Diagnosis 189 • Idle Air Control Motor Service and Diagnosis 203 • Digital Storage Oscilloscope Diagnosis of Electronic Fuel Injection Systems 207 • Diagnosing Repeated Component Failures 217 • Diagnosing Multiple Component Failures 219 • PCM Service, General Motors 220 • OBD II Diagnosis 222 • Guidelines for Scan Tester, DSO and OBD II Diagnosis 230 • Terms to Know 231 • ASE Style Review Questions 231

CHAPTER 9   **Emission Control Systems, Diagnosis and Service**   **237**

Locating Service Information 237 • Preliminary Emission System Inspection 238 • Catalytic Converter Diagnosis and Service 238 • PCV System Service and Diagnosis 239 • Diagnosis of Exhaust Gas Recirculation (EGR) Systems 241 • EGR Vacuum Regulator (EVR) Tests 246 • Exhaust Gas Temperature Sensor Diagnosis 248 • EGR Pressure Transducer (EPT) Diagnosis 248 • Diagnosis of Specific Problems Related to Emission Systems and Necessary Corrections 249 • Guidelines for Servicing and Diagnosing Catalytic Converters, PCV, and EGR Systems 250 • Terms to Know 251 • ASE Style Review Questions 251

CHAPTER 10   Emission System Diagnosis, Part II: Five-Gas Emission Diagnosis and IM240 Failure Corrections   257

Preliminary Emission System Inspection 257 • Pulsed Secondary Air Injection System Diagnosis 258 • Secondary Air Injection System Service and Diagnosis 258 • Evaporative (EVAP) System Diagnosis and Service 260 • EVAP System Thermal Vacuum Valve (TVV) Diagnosis 262 • Diagnosis of Knock Sensor and Knock Sensor Module 262 • Vacuum-Operated Decel Valve Diagnosis 263 • Service and Diagnosis of Combination Throttle Kicker and Idle Stop Solenoid 264 • Computer-Controlled Heat Riser Valve Diagnosis 266 • Heated Air Inlet System Diagnosis 267 • Diagnosis of Repeated Failures 268 • Diagnosis of Multiple Failures 268 • Diagnosis of Specific Emission-Related Problems and Necessary Corrections 269 • Emissions Analyzer Testing 270 • IM240 Emission Testing 277 • Guidelines for Servicing and Diagnosing Emission Systems and Testing Vehicle Emissions 281 • Terms to Know 283 • ASE Style Review Questions 283

CHAPTER 11   Servicing and Diagnosing of Body Computer Systems   293

Preliminary Diagnostic Procedure 293 • Scan Tester Diagnosis of Body Computer Faults 293 • Body Computer Menu Tests 295 • Body Computer State Displays 298 • Individual System and Component Diagnosis 299 • Diagnosis of Central Timer Module (CTM) System 304 • Generic Electronic Module (GEM) System Diagnosis 305 • Guidelines for Diagnosing and Servicing Body Computer Systems 306 • Terms to Know 308 • ASE Style Review Questions 308

CHAPTER 12   Electronic Instrument Cluster and Vehicle Theft Security System Diagnosis and Service   311

Diagnosis of a Typical Electronic Instrument Cluster 311 • Diagnosis of a Typical Import Electronic Instrument Cluster 314 • Head-Up Display (HUD) Diagnosis 318 • Diagnosis of Vehicle Theft Security Systems 320 • Guidelines for Servicing EICs, HUDs, and Vehicle Theft Security Systems 324 • Terms to Know 325 • ASE Style Review Questions 326

CHAPTER 13   Air Bag System Diagnosis and Service   329

Diagnostic System Check 329 • Air Bag System Flash Code and Voltmeter Diagnosis 330 • Scan Tester Diagnosis of Air Bag Systems 334 • Disabling the Air Bag System 335 • Removing and Replacing the Inflator Module 336 • Centering the Clock Spring Electrical Connector 339 • Air Bag System Wiring Repairs 339 • Air Bag Deployment Before Vehicle Scrapping 340 • Guidelines for Diagnosing and Servicing Air Bag Systems 341 • Terms to Know 342 • ASE Style Review Questions 342

CHAPTER 14	**Servicing and Diagnosing Computer-Controlled Transmissions** **345**

General Diagnosis of Computer-Controlled Transaxles and Transmissions 345 • Hydraulic and Mechanical Malfunctions 348 • Electronic Diagnosis 349 • Electronic Diagnosis of Ford Computer-Controlled Transaxles and Transaxles with Electronic Pressure Control Solenoids 353 • Computer-Controlled Transmission and Transaxle Diagnosis with Transmission Tester 356 • Diagnosis of General Motors 4L80-E Transmission 359 • Diagnosis of Computer-Controlled Transaxle with Pattern Select Switch 362 • Guidelines for Diagnosing and Servicing Computer-Controlled Transmissions and Transaxles 365 • Terms to Know 366 • ASE Style Review Questions 366

CHAPTER 15	**Antilock Brake and Traction Control System Diagnosis and Service** **369**

Diagnosis of Rear Wheel Antilock (RWAL) Systems 369 • Diagnosis of Nonintegral ABS with High-Pressure Accumulator 371 • Diagnosis of Integral ABS with High-Pressure Accumulator 379 • Diagnosis of Four Wheel Nonintegral ABS with Low-Pressure Accumulators 382 • Delco Moraine ABS VI Diagnosis 382 • Diagnosis of ABS with Wheel Spin Traction Control System 388 • Diagnosis of ABS with Spark Advance Reduction and Transmission Upshift Traction Control 391 • Diagnosis of ABS with Throttle Control, Wheel Spin Control, and Spark Advance Reduction Traction Control 392 • Guidelines for Diagnosing and Servicing Antilock Brake and Traction Control Systems 395 • Terms to Know 396 • ASE Style Review Questions 396

CHAPTER 16	**Computer-Controlled Suspension System Diagnosis and Service** **401**

Electronic Air Suspension System Diagnosis and Service 401 • Rear Load-Leveling Air Suspension System Service and Diagnosis 406 • Programmed Ride Control System Diagnosis 408 • Automatic Air Suspension Diagnosis and Service 413 • Diagnosis of Air Suspension with Speed Leveling Capabilities 416 • Diagnosis of Road-Sensing Suspension System 416 • Guidelines for Diagnosing and Servicing Computer-Controlled Suspension Systems 419 • Terms to Know 421 • ASE Style Review Questions 421

CHAPTER 17	**Computer-Controlled Air Conditioning Diagnosis and Service** **425**

Preliminary Inspection 425 • A/C Performance Test 426 • Refrigerant System Charge Test 428 • Vehicle Self-Diagnostic Tests 430 • Scan Tester Diagnosis of Computer-Controlled Air Conditioning Systems 435 • Guidelines for Diagnosing and Servicing Computer-Controlled Air Conditioning Systems 437 • Terms to Know 440 • ASE Style Review Questions 440

Appendix A	**441**
Appendix B	**442**
Glossary	**443**
Index	**459**

PHOTO SEQUENCES

1. Typical Procedure for Testing a Ford MAP Sensor **60**
2. Typical Procedure for Timing the Distributor to the Engine **84**
3. Typical Procedure for Relieving Fuel Pressure and Removing a Fuel Filter **124**
4. Typical Procedure for Testing Computer Command Control (3C) Performance **138**
5. Typical Procedure for Performing Ford Flash Code Diagnosis, Key On Engine Off (KOEO), and Key On Engine Running (KOER) Tests **174**
6. Typical Procedure for Scan Tester Diagnosis **208**
7. Typical Procedure for Diagnosing an EGR Vacuum Regulator Solenoid **247**
8. Typical Procedure for Five-Gas Emission Analysis **276**
9. Typical Procedure for Diagnosing Body Computer Systems **302**
10. Typical Procedure for Diagnosing Electronic Instrument Clusters **313**
11. Typical Procedure for Removing and Replacing an Inflator Module **337**
12. Typical Procedure for Performing a Scan Tester Diagnosis of a Computer-Controlled Transaxle **352**
13. Typical Procedure for Performing a Scan Tester Diagnosis of an Antilock Brake System **385**
14. Typical Procedure for Diagnosing a Programmed Ride Control System **412**
15. Typical Procedure for Performing a Scan Tester Diagnosis of an Automatic Temperature Control System **438**

PREFACE

Unlike yesterday's mechanic, the technician of today and for the future must know the underlying theory of all automotive systems and be able to service and maintain those systems. Today's technician must also know how these individual systems interact with each other. Standards and expectations have been set for today's technician, and these must be met in order to keep the world's automobiles running efficiently and safely.

The *Today's Technician* series, by Delmar Publishers, features textbooks that cover all mechanical and electrical systems of automobiles and light trucks. Principal titles correspond with the eight major areas of ASE (National Institute for Automotive Service Excellence) certification. Additional titles include remedial skills and theories common to all of the certification areas and advanced or specialized subject areas that reflect the latest technological trends.

Each title is divided into two manuals: a Classroom Manual and a Shop Manual. Dividing the material into two manuals provides the reader with the information needed to begin a successful career as an automotive technician without interrupting the learning process by mixing cognitive and performance-based learning objectives.

Each Classroom Manual contains the principles of operation for each system and subsystem. It also discusses the design variations used by different manufacturers. The Classroom Manual is organized to build upon basic facts and theories. The primary objective of this manual is to allow the reader to gain an understanding of how each system and subsystem operates. This understanding is necessary to diagnose the complex automobile systems.

The understanding acquired by using the Classroom Manual is required for competence in the skill areas covered in the Shop Manual. All of the high priority skills, as identified by ASE, are explained in the Shop Manual. The Shop Manual also includes step-by-step instructions for diagnostic and repair procedures. Photo Sequences are used to illustrate many of the common service procedures. Other common procedures are listed and are accompanied with fine-line drawings and photographs that allow the reader to visualize and conceptualize the finest details of the procedure. The Shop Manual also contains the reasons for performing the procedures, as well as when that particular service is appropriate.

The two manuals are designed to be used together and are arranged in corresponding chapters. Not only are the chapters in the manuals linked together, the contents of the chapters are also linked. Both manuals contain clear and thoughtfully selected illustrations. Many of the illustrations are original drawings or photos prepared for inclusion in this series. This means that the art is a vital part of each manual.

The page layout is designed to include information that would otherwise break up the flow of information presented to the reader. The main body of the text includes all of the "need-to-know" information and illustrations. In the side margins are many of the special features of the series. Items such as definitions of new terms, common trade jargon, tools lists, and cross-references are placed in the margin, out of the normal flow of information so as not to interrupt the thought process of the reader. Each manual in this series is organized in a like manner and contains the same features.

Jack Erjavec, Series Advisor

Classroom Manual

To stress the importance of safe work habits, the Classroom Manual dedicates one full chapter to safety. Included in this chapter are common safety practices, safety equipment, and safe handling of hazardous materials and wastes. This includes information on MSDS sheets and OSHA regulations. Other features of this manual include:

Cognitive Objectives

These objectives define the contents of the chapter and define what the student should have learned upon completion of the chapter.
Each topic is divided into small units to promote easier understanding and learning.

Marginal Notes

New terms are pulled out and defined. Common trade jargon also appears in the margin and gives some of the common terms used for components. This allows the reader to speak and understand the language of the trade, especially when conversing with an experienced technician.

Cautions and Warnings

Throughout the text, cautions are given to alert the reader to potentially hazardous materials or unsafe conditions. Warnings are also given to advise the student of things that can go wrong if instructions are not followed or if a nonacceptable part or tool is used.

References to the Shop Manual

Reference to the appropriate page in the Shop Manual is given whenever necessary. Although the chapters of the two manuals are synchronized, material covered in other chapters of the Shop Manual may be fundamental to the topic discussed in the Classroom Manual.

A Bit of History

This feature gives the student a sense of the evolution of the automobile. This feature not only contains nice-to-know information, but also should spark some interest in the subject matter.

Summaries

Each chapter concludes with summary statements that contain the important topics of the chapter. These are designed to help the reader review the contents.

Terms to Know

A list of new terms appears next to the Summary. Definitions for these terms can be found in the Glossary at the end of the manual.

Review Questions

Short answer essay, fill-in-the-blank, and multiple-choice type questions follow each chapter. These questions are designed to accurately assess the student's competence in the stated objectives at the beginning of the chapter.

Shop Manual

To stress the importance of safe work habits, the Shop Manual also dedicates one full chapter to safety. Other important features of this manual include:

Performance Objectives

These objectives define the contents of the chapter and define what the student should have learned upon completion of the chapter. These objectives also correspond with the list of required tasks for ASE certification. *Each ASE task is addressed.*

Although this textbook is not designed to simply prepare someone for the certification exams, it is organized around the ASE task list. These tasks are defined generically when the procedure is commonly followed and specifically when the procedure is unique for specific vehicle models. Imported and domestic model automobiles and light trucks are included in the procedures.

Photo Sequences

Many procedures are illustrated in detailed Photo Sequences. These detailed photographs show the students what to expect when they perform particular procedures. They also can provide a student a familiarity with a system or type of equipment, which the school may not have.

Marginal Notes

Page numbers for cross-referencing appear in the margin. Some of the common terms used for components, and other bits of information, also appear in the margin. This provides an understanding of the language of the trade and helps when conversing with an experienced technician.

Cautions and Warnings

Throughout the text, cautions are given to alert the reader to potentially hazardous materials or unsafe conditions. Warnings are also given to advise the student of things that can go wrong if instructions are not followed or if a nonacceptable part or tool is used.

References to the Classroom Manual

Reference to the appropriate page in the Classroom Manual is given whenever necessary. Although the chapters of the two manuals are synchronized, material covered in other chapters of the Classroom Manual may be fundamental to the topic discussed in the Shop Manual.

Customer Care

This feature highlights those little things a technician can do or say to enhance customer relations.

Tools Lists

Each chapter begins with a list of the Basic Tools needed to perform the tasks included in the chapter. Whenever a Special Tool is required to complete a task, it is listed in the margin next to the procedure.

Service Tips

Whenever a special procedure is appropriate, it is described in the text. These tips are generally those things commonly done by experienced technicians.

Case Studies

Case Studies concentrate on the ability to properly diagnose the systems. Each chapter ends with a case study in which a vehicle has a problem, and the logic used by a technician to solve the problem is explained.

Terms to Know

Terms in this list can be found in the Glossary at the end of the manual.

Diagnostic Chart

Chapters include detailed diagnostic charts linked with the appropriate ASE task. These charts list common problems and most probable causes. They also list a page reference in the Classroom Manual for better understanding of the system's operation and a page reference in the Shop Manual for details on the procedure necessary for correcting the problem.

ASE Style Review Questions

Each chapter contains ASE style review questions that reflect the performance objectives listed at the beginning of the chapter. These questions can be used to review the chapter as well as to prepare for the ASE certification exam.

Instructor's Guide

The Instructor's Guide is provided free of charge as part of the *Today's Technician Series* of automotive technology textbooks. It contains Lecture Outlines, Answers to Review Questions, Pretest and Test Bank including ASE style questions.

Classroom Manager

The complete ancillary package is designed to aid the instructor with classroom preparation and provide tools to measure student performance. For an affordable price, this comprehensive package contains:

Instructor's Guide	Lecture Outlines and Lecture Notes
200 Transparency Masters	Printed and Computerized Test Bank
Answers to Review Questions	Laboratory Worksheets and Practicals

Reviewers

Si Acuna
Texas State Technical College-Sweetwater
Sweetwater, TX

Rankin Barnes
Guilford Technical Community College
Jamestown, NC

Lawrence W. Breeden
Albany Technical Institute
Albany, GA

Pete Brisley
Brookhaven College
Farmers Branch, TX

Neal Clark
Erie Community College
Williamsville, NY

Earl J. Friedell, Jr.
Dekalb Technical Institute
Clarkston, GA

Daniel L. Hall
Kirkwood Community College
Cedar Rapids, IA

Tony Hernandez
Austin Community College
Austin, TX

Joseph M. Juran
Rosedale Technical Institute
Pittsburgh, Pa

Ken Kempfer
Fox Valley Technical College
Appleton, WI

Theodore Morin
Blackstone Valley Regional Vocational
  Technical High School
South Grafton, MA

Danny R. Rakes
Danville Community College
Danville, VA

Dave Spear
Niagara College of Applied Arts and
  Technology
St. Catharines, Ontario, CANADA

Jaime C. Stith
Elkhart Area Career Center
Elkhart, IN

Shop Practices

Upon completion and review of this chapter, you should be able to:

❑ Describe three requirements for shop layout and explain why these requirements are important.

❑ Observe all shop rules when working in the shop.

❑ Operate vehicles in the shop according to shop driving rules.

❑ Observe all shop housekeeping rules.

❑ Follow the necessary procedures to maintain satisfactory shop air quality.

❑ Fulfill employee obligations when working in the shop.

❑ Accept job responsibilities for each job completed in the shop.

❑ Describe the ASE technician testing and certification process, including the eight areas of certification.

Shop Layout

There are many different types of shops in the automotive service industry, including new car dealers, independent repair shops, specialty shops, service stations, and fleet shops.

CAUTION: Always know the location of all safety equipment in the shop and be familiar with the operation of this equipment.

The shop layout in any shop is important to maintaining shop efficiency and safety. Shop layout includes bays for various types of repairs, space for equipment storage, and office locations. Safety equipment such as fire extinguishers, first-aid kits, and eyewash fountains must be in easily accessible locations. The location of each piece of safety equipment must be clearly marked. Areas such as the parts department and the parts cleaning area must be located so they are easily accessible from all areas of the shop. The service manager's office should also be centrally located. All shop personnel should familiarize themselves with the shop layout, especially the location of safety equipment. If you know the exact fire extinguisher locations, you may get an extinguisher into operation a few seconds faster. These few seconds could make the difference between a fire that is quickly extinguished and one that gets out of control, causing extensive damage and personal injury! Most shops have specific bays for certain types of work, such as electrical repair, wheel alignment and tires, and machining (Figure 1-1).

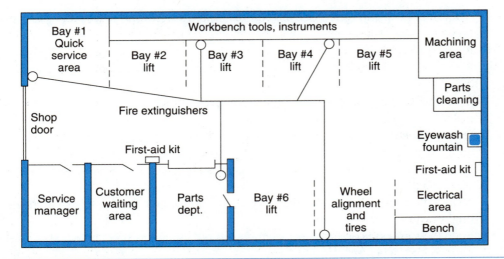

Figure 1-1 Typical shop layout

1

The tools and equipment required for a certain type of work are stored in that specific bay. For example, the equipment for electrical and electronic service work is stored in the bay allotted to that type of repair. When certain bays are allotted to specific types of repair work, unnecessary equipment movement is eliminated. Each technician has his or her own tools on a portable roll cabinet which is moved to the vehicle being repaired. Special tools are provided by the shop, and these tools may be located on tool boards attached to the wall. Other shops may have a tool room where special tools are located. Adequate workbench space must be provided in those bays where bench work is required.

Shop Rules

WARNING: Shop rules, vehicle operation in the shop, and shop housekeeping are serious business. Each year a significant number of technicians are injured and vehicles damaged by disregarding shop rules, careless vehicle operation, and sloppy housekeeping.

The application of some basic shop rules helps to prevent serious, expensive accidents. Failure to comply with shop rules may cause personal injury or expensive damage to vehicles and shop facilities. It is the responsibility of the employer and all shop employees to make sure that shop rules are understood and followed until these rules become automatic habits. The following basic shop rules should be followed:

1. Always wear safety glasses and other protective equipment that is required by a service procedure. For example, a special parts washer must be used to avoid breathing asbestos dust into the lungs. *Asbestos dust is a known cause of lung cancer.* This dust is encountered in manual transmission clutch facings and brake linings.
2. Tie long hair securely behind the head, and do not wear loose or torn clothing.
3. Do not wear rings, watches, or loose hanging jewelry. If jewelry such as a ring, metal watch band, or chain makes contact between an electrical terminal and ground, the jewelry becomes extremely hot, resulting in severe burns.
4. Set the parking brake when working on a vehicle. If the vehicle has an automatic transmission, place the gear selector in park unless a service procedure requires another selector position. When the vehicle is equipped with a manual transmission, position the gear selector in neutral with the engine running or reverse with the engine stopped.
5. Always connect a shop exhaust hose to the vehicle tailpipe and be sure the shop exhaust fan is running. If it is absolutely necessary to operate a vehicle without a shop exhaust pipe connected to the tailpipe, open the large shop door to provide adequate ventilation. Carbon monoxide in the vehicle exhaust may cause severe headaches and other medical problems. High concentrations of carbon monoxide may result in death!
6. Keep hands, clothing, and wrenches away from rotating parts such as cooling fans. Remember that electric-drive fans may start turning at any time, even with the ignition off.
7. Always leave the ignition switch off unless a service procedure requires another switch position.
8. Do not smoke in the shop. If the shop has designated smoking areas, smoke only in these areas.
9. Store oily rags and other discarded combustibles in regulation, covered metal garbage containers.
10. Always use a wrench or socket that fits properly on the bolt. Do not substitute metric for English wrenches or vice versa.
11. Keep tools in good condition. For example, do not use a punch or chisel with a mushroomed end. When struck with a hammer, a piece of the mushroomed metal could break off, resulting in severe eye or other injury.

Carbon monoxide is a poisonous gas, and when breathed into the lungs, it may cause headaches, nausea, ringing in the ears, tiredness, and heart flutter. In strong concentrations, it causes death.

12. Do not leave power tools running and unattended.
13. Serious burns may be prevented by avoiding contact with hot metal components such as exhaust manifolds, other exhaust system components, radiators, and some air conditioning hoses.
14. When lubricant such as engine oil is drained, always use caution because the oil could be hot enough to cause burns.
15. Before getting under a vehicle, be sure the vehicle is placed securely on safety stands.
16. Operate all shop equipment, including lifts, according to the equipment manufacturer's recommended procedure. Do not operate equipment unless you are familiar with the correct operating procedure.
17. Do not run or engage in horseplay in the shop.
18. Obey all state and federal fire, safety, and environmental regulations.
19. Do not stand in front of or behind vehicles.
20. Always place fender covers and a seat cover on a customer's vehicle before working on the car.
21. Inform the shop foreman of any safety dangers and suggestions for safety improvement.

Vehicle Operation

When driving a customer's vehicle, certain precautions must be observed to prevent accidents and maintain good customer relations.

1. Before driving a vehicle, always make sure the brakes are operating and fasten the safety belt.
2. Check to be sure there are no people or objects under the car before you start the engine.
3. If the vehicle is parked on a lift, be sure the lift is fully down and the lift arms or components are not contacting the vehicle chassis.
4. Check to see if there are any objects directly in front of or behind the vehicle before driving away.
5. Always drive slowly in the shop and watch carefully for personnel and other moving vehicles.
6. Make sure the shop door is up high enough so there is plenty of clearance between the top of the vehicle and the door.
7. Watch the shop door to be certain that it is not coming down as you attempt to drive under the door.
8. If a road test is necessary, obey all traffic laws, and never drive in a reckless manner.
9. Do not squeal tires when accelerating or turning corners.

If customers observe that service personnel take good care of their cars by driving carefully and installing fender and seat covers, the service department image is greatly enhanced in their eyes. These procedures impress upon customers that shop personnel respect their cars. Conversely, if grease spots are found on the upholstery or fenders after service work is completed, customers will probably think the shop is very careless, not only in car care, but also in service work quality.

Housekeeping

● **CUSTOMER CARE:** When customers see that you are concerned about their vehicles, and that you operate a shop with excellent housekeeping habits, they will be impressed and will likely keep returning for service.

Excellent housekeeping involves general shop cleanliness, proper shop safety equipment in good working condition, and the proper maintenance of all shop equipment and tools.

Careful housekeeping habits prevent accidents and increase worker efficiency. Good housekeeping also helps to impress upon customers that quality work is a priority in this shop. Follow these housekeeping rules:

1. Keep aisles and walkways clear of tools, equipment, and other items.
2. Be sure all sewer covers are securely in place.
3. Keep floor surfaces free of oil, grease, water, and loose material.
4. Place proper garbage containers in convenient locations and empty them regularly.
5. Make sure that access to fire extinguishers is unobstructed at all times, and check fire extinguishers for proper charge at regular intervals.

 SERVICE TIP: When you are finished with a tool, never set it on the customer's car. After using a tool, the best place for it is in your tool box or on the workbench. Many tools have been lost by leaving them on customers' vehicles.

6. Keep tools clean and in good condition.
7. When not in use, store tools in their proper location.
8. Place oily rags and other combustibles in properly covered garbage containers.
9. Make sure that rotating components on equipment and machinery have guards and all shop equipment has regular service and adjustment schedules.
10. Maintain benches and seats in a clean condition.
11. Keep parts and materials in their proper location.
12. When not in use, store creepers in a specific location.
13. Make sure that the shop is well-lighted, and keep all lights in working order.
14. Replace frayed electrical cords on lights or equipment.
15. Regularly clean walls and windows.
16. Keep stairs clean, well-lighted, and free of loose material.

If these housekeeping rules are followed, the shop will be a safer place to work and customers will be impressed with the appearance of the premises.

Air Quality

CAUTION: Never run the engine in a vehicle inside the shop without an exhaust hose connected to the tailpipe.

Vehicle exhaust contains small amounts of carbon monoxide, which is a poisonous gas. Strong concentrations of carbon monoxide may be fatal for human beings. All shop personnel are responsible for air quality in the shop.

Shop management is responsible for an adequate exhaust system to remove exhaust fumes from the maximum number of vehicles that may be running in the shop at the same time.

Technicians should never run a vehicle in the shop unless a shop exhaust hose is installed on the tailpipe of the vehicle. The exhaust fan must be switched on to remove exhaust fumes.

If shop heaters or furnaces have restricted chimneys, they release carbon monoxide emissions into the shop air. Therefore, chimneys should be checked periodically for restriction and proper ventilation.

Monitors are available to measure the level of carbon monoxide in the shop. Some of these monitors read the amount of carbon monoxide present in the shop air, and other monitors provide an audible alarm if the concentration of carbon monoxide exceeds the danger level.

Diesel exhaust contains some carbon monoxide, but particulates are also present in the exhaust from these engines. Particulates are basically small carbon particles, which can be harmful to the lungs.

The sulfuric acid solution in car batteries is a very corrosive, poisonous liquid. If a battery is charged with a fast charger at a high rate for a period of time, the battery becomes hot and the sulfuric

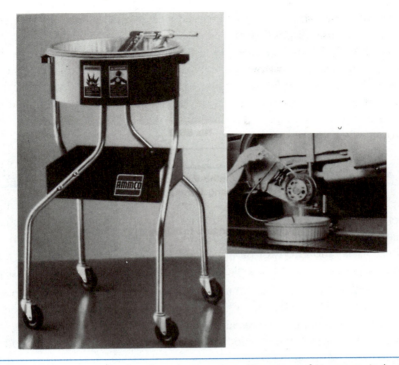

Figure 1-2 Brake assembly washer for asbestos dust (Courtesy of Hennessy Industries, Inc.)

acid solution begins to boil. Under this condition, the battery may emit a strong sulfuric acid smell, and these fumes may be harmful to the lungs. If this condition occurs in the shop, the battery charger should be turned off, or the charger rate should be reduced considerably.

WARNING: When an automotive battery is charged, hydrogen gas and oxygen gas escape from the battery. If these gases are combined, they form water, but hydrogen gas by itself is very explosive. While a battery is charged, sparks, flames, and other sources of ignition must not be allowed near the battery.

CAUTION: Breathing asbestos dust must be avoided because this dust is a known contributor to lung cancer.

Some automotive clutch facings and brake linings contain asbestos. Never use an air hose to blow dirt from these components, because this action disperses asbestos dust into the shop where it may be inhaled by technicians and other people in the shop. Parts washers approved by the Occupational Safety and Health Act (OSHA) must be used to clean the dust from these components (Figure 1-2). Brake washer concentrate is mixed with water in these parts washers (Figure 1-3). A catch basin with a removable liner is placed under the parts to be washed, and the washer sprays the cleaning solution on the parts. After the washing operation is completed, the liner containing

Figure 1-3 Brake washer concentrate (Courtesy of Hennessy Industries, Inc.)

the cleaning solution and asbestos dust is removed, sealed, and labeled for proper handling and disposal according to waste disposal laws.

Even though technicians take every precaution to maintain air quality in the shop, some undesirable gases may still get in the air. For example, exhaust manifolds may get oil on them during an engine overhaul. When the engine is started and these manifolds get hot, the oil burns off the manifolds and pollutes the shop air with oil smoke. Adequate shop ventilation must be provided to take care of this type of air contamination.

Employer and Employee Obligations

When you begin employment, you enter into a business agreement with your employer. A business agreement involves an exchange of goods or services that have value. Although the automotive technician may not have a written agreement with his or her employer, the technician exchanges time, skills, and effort for money paid by the employer. Both the employee and the employer have obligations. The automotive technician's obligations include the following:

1. *Productivity*. As an automotive technician, you have a responsibility to your employer to make the best possible use of time on the job. Each job should be done in a reasonable length of time. Employees are paid for their skills, effort, and time.
2. *Quality*. Each repair job should be a quality job! Work should never be done in a careless manner. Nothing improves customer relations like quality workmanship.
3. *Teamwork*. The shop staff are a team and everyone, including technicians and management personnel, are team members. You should cooperate with and care about other team members. Each member of the team should strive for harmonious relations with fellow workers. Cooperative teamwork helps to improve shop efficiency, productivity, and customer relations. Customers may be turned off by bickering between shop personnel.
4. *Honesty*. Employers and customers expect and deserve honesty from automotive technicians. Honesty creates a feeling of trust between technicians, employers, and customers.
5. *Loyalty*. As an employee, you are obliged to act in the best interests of your employer, both on and off the job.
6. *Attitude*. Employees should maintain a positive attitude at all times. As in other professions, automotive technicians have days when it may be difficult to maintain a positive attitude. For example, there will be days when the technical problems on a certain vehicle are difficult to solve. However, developing a negative attitude certainly will not help the situation! A positive attitude has a positive effect on the job situation as well as on the customer and employer.
7. *Responsibility*. You are responsible for your conduct on the job and your work-related obligations. These obligations include always maintaining good workmanship and customer relations. Attention to details such as always placing fender and seat covers on customer vehicles prior to driving or working on the vehicle greatly improve customer relations.
8. *Following directions*. All of us like to do things our way. Such action may not be in the best interests of the shop, and as an employee you have an obligation to follow the supervisor's directions.
9. *Punctuality and regular attendance*. Employees have an obligation to be on time for work and to be regular in attendance on the job. It is very difficult for a business to operate successfully if it cannot count on its employees to be on the job at the appointed time.
10. *Regulations*. Automotive technicians should be familiar with all state and federal regulations such as the Occupational Safety and Health Act (OSHA) and hazardous waste

disposal laws pertaining to their job situation. In Canada, employees should be familiar with workplace hazardous materials information systems (WHMIS).

Employer-to-employee obligations include:

1. *Wages.* The employer has a responsibility to inform the employee regarding the exact amount of financial remuneration they will receive and when they will be paid.
2. *Fringe benefits.* A detailed description of all fringe benefits should be provided by the employer. These benefits may include holiday pay, sickness and accident insurance, and pension plans.
3. *Working conditions.* A clean, safe workplace must be provided by the employer. The shop must have adequate safety equipment and first-aid supplies. Employers must be certain that all shop personnel maintain the shop area and equipment to provide adequate safety and a healthy workplace atmosphere.
4. *Employee instruction.* Employers must provide employees with clear job descriptions, and be sure that each worker is aware of his or her obligations.
5. *Employee supervision.* Employers should inform their workers regarding the responsibilities of their immediate supervisors and other management personnel.
6. *Employee training.* Employers must make sure that each employee is familiar with the safe operation of all the equipment that they are required to use in their job situation. Since automotive technology is changing rapidly, employers should provide regular update training for their technicians. Under the right-to-know laws, employers are required to inform all employees about hazardous materials in the shop. Employees should be familiar with workplace hazardous materials information systems (WHMIS), which detail the labeling and handling of hazardous waste, and the health problems if exposed to hazardous waste.

Job Responsibilities

An automotive technician has specific responsibilities regarding each job performed on a customer's vehicle. These job responsibilities include:

1. Do every job to the best of your ability. There is no place in the automotive service industry for careless workmanship! Automotive technicians and students must realize they have a very responsible job. During many repair jobs, you, as a student or technician working on a customer's vehicle, actually have the customer's life and the safety of the vehicle in your hands. For example, if you are doing a brake job and leave the wheel nuts loose on one wheel, that wheel may fall off the vehicle at high speed. This could result in serious personal injury for the customer and others, plus extensive vehicle damage. If this type of disaster occurs, the individual who worked on the vehicle and the shop may be involved in a very expensive legal action. As a student or technician working on customer vehicles, you are responsible for the safety of every vehicle that you work on! Even when careless work does not create a safety hazard, it leads to dissatisfied customers who often take their business to another shop, and nobody benefits when that happens.
2. Treat customers fairly and honestly on every repair job. Do not install parts that are unnecessary to complete the repair job.
3. Use published specifications; do not guess at adjustments.
4. Follow the service procedures in the service manual provided by the vehicle manufacturer or an independent manual publisher.
5. When the repair job is completed, always be sure the customer's complaint has been corrected.
6. Do not be too concerned with work speed when you begin working as an automotive technician. Speed comes with experience.

Figure 1-4 ASE certification shoulder patches worn by automotive technicians and master technicians (Courtesy of the National Institute for Automotive Service Excellence [ASE])

National Institute for Automotive Service Excellence (ASE) Certification

ASE has provided voluntary testing and certification of automotive technicians on a national basis for many years. The image of the automotive service industry has been enhanced by the ASE certification program. More than 265,000 ASE-certified automotive technicians now work in a wide variety of automotive service shops. ASE provides certification in these eight areas of automotive repair:

1. Engine repair
2. Automatic transmissions/transaxles
3. Manual drive train and axles
4. Suspension and steering
5. Brakes
6. Electrical systems
7. Heating and air conditioning
8. Engine performance

A technician may take the ASE test and become certified in any or all of the eight areas. When a technician passes an ASE test in one of the eight areas, an Automotive Technician's shoulder patch is issued by ASE. If a technician passes all eight tests, he or she receives a Master Technician's shoulder patch (Figure 1-4). Retesting at five-year intervals is required to remain certified.

The certification test in each of the eight areas contains 40 to 80 multiple choice questions. The test questions are written by a panel of automotive service experts from various areas of automotive service, such as automotive instructors, service managers, automotive manufacturer's representatives, test equipment representatives, and certified technicians. The test questions are pretested and checked for quality by a national sample of technicians. Most questions have the Technician A and Technician B format similar to the questions at the end of each chapter in this book. ASE regulations demand that each technician must have two years of working experience in the automotive service industry prior to taking a certification test or tests. However, relevant formal training may be substituted for one of the years of working experience. Contact ASE for details regarding this substitution. The contents of the Engine Performance test are listed in Table 1-1.

Shops that employ ASE-certified technicians display an official ASE blue seal of excellence. This blue seal increases the customer's awareness of the shop's commitment to quality service and the competency of certified technicians.

Classroom Manual
Chapter 1, page 1

Shop Projects to Enhance the Theories in Chapter 1 of the Classroom Manual

1. Strike the tread of an inflated tire with a rubber hammer. What reaction did you get? Why did this reaction occur?

8

Table 1-1 Engine Performance Test Summary

Content Area	Questions in Test	Percentage of Test
A. General Engine Diagnosis	19	24.0%
B. Ignition System Diagnosis and Repair	16	20.0%
C. Fuel, Air Induction, and Exhaust Systems Diagnosis and Repair	21	26.0%
D. Emissions Control Systems Diagnosis and Repair	19	24.0%
1. Positive Crankcase Ventilation (1)		
2. Spark Timing Controls (3)		
3. Idle Speed Controls (3)		
4. Exhaust Gas Recirculation (4)		
5. Exhaust Gas Treatment (2)		
6. Inlet Air Temperature Controls (2)		
7. Intake Manifold Temperature Controls (2)		
8. Fuel Vapor Controls (2)		
E. Engine Related Service	2	2.5%
F. Engine Electrical Systems Diagnosis and Repair	3	3.5%
1. Battery (1)		
2. Starting System (1)		
3. Charging System (1)		
Total	**80[1]**	**100.0%**

[Courtesy of the National Institute for Automotive Service Excellence (ASE)]

2. Place some cold water in a graduated heat-resistant container supported on a metal stand, and record the exact quantity of water. Heat the water with a propane torch until it boils and record the exact quantity of water. What happened to the water when it was heated?
3. Obtain an empty metal 1 gallon container with a tight-fitting top. Place about 1 pint of water in the container and heat the container until the water is boiling. Then install the top securely and allow the container to cool. Cooling time may be reduced by placing ice on the container. What happened to the container? Why did this happen?
4. Obtain a vacuum hand pump and a vacuum advance from a distributor. Gradually operate the pump and observe the vacuum diaphragm. How much vacuum is required to begin moving the vacuum diaphragm? How much vacuum is required to move this diaphragm to the fully advanced position? What two forces move the diaphragm?

Guidelines for Following Proper Shop Practices

1. Technicians must be familiar with shop layout, especially the location of safety equipment. This knowledge provides a safer, more efficient shop.
2. Shop rules must be observed by everyone in the shop to provide adequate shop safety, personal health protection, and vehicle protection.
3. The application of driving rules in the shop increases safety, protects customer vehicles and shop property, and improves the shop image in the eyes of the customer.
4. When good housekeeping habits are developed, shop safety is improved, worker efficiency is increased, and customers are impressed.
5. If some basic rules are followed to maintain shop air quality, the personal health of shop employees is improved.
6. When employers and employees accept and fulfill their obligations, personal relationships and general attitudes are greatly improved, shop productivity is increased, and customer relations are improved.

7. If a technician accepts certain job responsibilities, job quality improves, and customer satisfaction increases.
8. ASE technician certification improves the quality of automotive repair and improves the image of the profession.

CASE STUDY 1

A technician was removing and replacing the alternator on a General Motors car. After installing the replacement alternator and connecting the alternator battery wire, the technician proceeded to install the alternator belt. The rubber boot was still removed from the alternator battery terminal. While installing this belt, the technician's wristwatch expansion bracelet made electrical contact from the alternator battery terminal to ground on the alternator housing. Even though the alternator battery wire is protected with a fuse link, which melted, a high current flowed through the wristwatch bracelet. This heated the bracelet to a very high temperature and severely burned the technician's arm.

This technician forgot two safety rules:

1. Never wear jewelry, such as watches and rings, while working in an automotive shop.
2. Before performing electrical work on a vehicle, disconnect the negative battery cable. If the vehicle is equipped with an air bag, wait 1 minute after this cable is disconnected.

CASE STUDY 2

A technician was removing and replacing the starting motor with the solenoid mounted on top of the motor. The technician began removing the battery cable from the solenoid, and the ring on one of his fingers made contact between the end of the wrench and the engine block. The current flow through the ring was so high that the positive battery terminal melted out of the battery. The technician's finger was so badly burned that a surgeon had to cut the ring from his finger and repair the finger.

This technician forgot to disconnect the negative battery cable before working on the vehicle!

CASE STUDY 3

A technician had just replaced the engine in a Ford product, and she was performing final adjustments such as timing and air-fuel mixture. In this shop, the cars were parked in the work bays at an angle on both sides of the shop. With the engine running at fast idle, the automatic transmission suddenly slipped into reverse. The car went backwards across the shop and collided with a car in one of the electrical repair bays. Both vehicles were damaged to a considerable extent. Fortunately, no personnel were injured.

This technician forgot to apply the parking brake while working on the vehicle!

Terms to Know

Asbestos vacuum cleaner
ASE blue seal of excellence
ASE technician certification
Carbon monoxide
Diesel particulates
National Institute for Automotive Service Excellence (ASE)
Shop layout
Sulfuric acid

ASE Style Review Questions

1. While discussing shop layout and safety:
   Technician A says the location of each piece of safety equipment should be clearly marked.
   Technician B says some shops have a special tool room where special tools are located.
   Who is correct?
   A. A only
   B. B only
   C. Both A and B
   D. Neither A nor B

2. While discussing shop rules:
   Technician A says breathing carbon monoxide may cause arthritis.
   Technician B says breathing carbon monoxide may cause headaches.
   Who is correct?
   A. A only
   B. B only
   C. Both A and B
   D. Neither A nor B

3. While discussing shop rules:
   Technician A says breathing asbestos dust may cause heart defects.
   Technician B says oily rags should be stored in uncovered garbage containers.
   Who is correct?
   A. A only
   B. B only
   C. Both A and B
   D. Neither A nor B

4. While discussing shop rules:
   Technician A says English tools may be substituted for metric tools.
   Technician B says foot injuries may be caused by loose sewer covers.
   Who is correct?
   A. A only
   B. B only
   C. Both A and B
   D. Neither A nor B

5. While discussing air quality:
   Technician A says a restricted chimney on a shop furnace may cause carbon monoxide gas in the shop.
   Technician B says monitors are available to measure the level of carbon monoxide in the shop air.
   Who is correct?
   A. A only
   B. B only
   C. Both A and B
   D. Neither A nor B

6. While discussing air quality:
   Technician A says a battery gives off hydrogen gas during the charging process.
   Technician B says a battery gives off oxygen gas during the charging process.
   Who is correct?
   A. A only
   B. B only
   C. Both A and B
   D. Neither A nor B

7. While discussing air quality:
   Technician A says diesel exhaust contains particulates.
   Technician B says particulate emissions contain oxides of nitrogen.
   Who is correct?
   A. A only
   B. B only
   C. Both A and B
   D. Neither A nor B

8. While discussing employer and employee responsibilities:
   Technician A says employers are required to inform their employees about hazardous materials in the shop.
   Technician B says that employers have no obligation to inform their employees about hazardous materials in the shop.
   Who is correct?
   A. A only
   B. B only
   C. Both A and B
   D. Neither A nor B

9. While discussing ASE certification:
   Technician A says a technician must pass four of the eight ASE certification tests to receive a master technician's shoulder patch.
   Technician B says a technician must pass all eight ASE certification tests to receive a master technician's shoulder patch.
   Who is correct?
   A. A only
   B. B only
   C. Both A and B
   D. Neither A nor B

10. While discussing ASE certification:
    Technician A says retesting is required at five-year intervals to remain certified.
    Technician B says a technician must have four years of automotive repair experience before writing an ASE certification test.
    Who is correct?
    A. A only
    B. B only
    C. Both A and B
    D. Neither A nor B

Tools, Safety Practices, and Test Equipment

CHAPTER 2

Upon completion and review of this chapter, you should be able to:

- Perform automotive measurements using United States customary (USC) and international system (SI) systems of weights and measures.
- Observe all personal safety precautions while working in the automotive shop.
- Demonstrate proper lifting procedures and precautions.
- Demonstrate proper vehicle lift operating and safety procedures.
- Observe all safety precautions when hydraulic tools are used in the automotive shop.
- Follow the recommended procedure while operating hydraulic tools such as presses, floor jacks, and vehicle lifts to perform automotive service tasks.
- Follow safety precautions regarding the use of power tools.
- Demonstrate proper safety precautions during the use of compressed air equipment.
- Follow safety precautions while using cleaning equipment in the automotive shop.
- Properly connect and use fuel pressure gauges.
- Connect and operate an injector balance tester.
- Connect and use pressurized injector cleaning containers.
- Connect and use circuit testers.
- Complete proper ampere-volt tester connections.
- Complete correct multimeter connections.
- Properly connect various tach-dwellmeters and obtain meter readings.
- Operate a magnetic probe-type tachometer.
- Check ignition timing and spark advance with an advance-type timing light.
- Operate a digital timing meter with a magnetic probe.
- Connect and operate an ignition module tester.
- Connect oscilloscope leads to the ignition system.
- Properly connect a scan tester to the vehicle and program the tester for the vehicle being tested.
- Prepare an emissions analyzer to make accurate emission readings on a vehicle.
- Connect an engine analyzer and program it for a specific vehicle.

Basic Tools

Basic technician's tool set

Service manual

Circuit testers, 12-V and self-powered

Measuring Systems

Two systems of weights and measures are commonly used in the United States. One system of weights and measures is the United States customary system (USC), which is commonly referred to as the English system. Well-known measurements for length in the USC system are the inch, foot, yard, and mile. In this system, the quart and gallon are common measurements for volume, and ounce, pound, and ton are measurements for weight. A second system of weights and measures is called the international system (SI), which is often referred to as the metric system.

In the USC system, the basic linear measurement is the yard, whereas the corresponding linear measurement in the metric system is the meter. Each unit of measurement in the metric system is related to the other metric units by a factor of 10. Thus, every metric unit can be multiplied or divided by 10 to obtain larger units (multiples) or smaller units (submultiples). For example, the meter can be divided by 10 to obtain centimeters (1/100 meter) or millimeters (1/1,000 meter).

The United States customary (USC) system of weights and measures is commonly referred to as the English system.

The international system (SI) of weights and measures is called the metric system.

13

The United States government passed the Metric Conversion Act in 1975 in an attempt to move American industry and the general public to accept and adopt the metric system. The automotive industry has adopted the metric system, and in recent years most bolts, nuts, and fittings on vehicles have been changed to metric. During the early 1980s, some vehicles had a mix of USC and metric bolts. Import vehicles have used the metric system for many years. Although the automotive industry has changed to the metric system, the general public in the United States has been slow to convert from the USC system to the metric system. One of the factors involved in this change is cost. What would it cost to change every highway distance and speed sign in the United States to read kilometers? Of course the answer to that question is probably hundreds of millions or billions of dollars.

Service technicians must be able to work with both the USC and the metric system. One meter (m) in the metric system is equal to 39.37 inches (in.) in the USC system. Some common equivalents between the metric and USC system are:

1 meter (m) = 39.37 inches (in.)
1 centimeter (cm) = 0.3937 inch
1 millimeter (mm) = 0.03937 inch
1 in. = 2.54 cm
1 in. = 25.4 mm

 SERVICE TIP: A metric conversion calculator provides fast, accurate metric-to-USC conversions or vice versa.

In the USC system, phrases such as 1/8 of an inch are used for measurements. The metric system uses a set of prefixes. For example, in the word *kilometer*, the prefix *kilo* indicates 1,000, and this prefix indicates there are 1,000 meters in a kilometer. Common prefixes in the metric system are:

NAME	SYMBOL	MEANING
mega	M	one million
kilo	k	one thousand
hecto	h	one hundred
deca	da	ten
deci	d	one tenth of
centi	c	one hundreth of
milli	m	one thousandth of
micro	μ	one millionth of

Measurement of Mass

In the metric system, mass is measured in grams, kilograms, or tonnes. 1,000 grams (g) = 1 kilogram (kg). In the USC system, mass is measured in ounces, pounds, or tons. When converting pounds to kilograms, 1 pound = 0.453 kilogram.

Measurement of Length

In the metric system, length is measured in millimeters, centimeters, meters, or kilometers. 10 millimeters (mm) = 1 centimeter (cm). In the USC system, length is measured in inches, feet, yards, or miles. When distance conversions are made between the two systems, some of the conversion factors are:

1 inch = 25.4 millimeters
1 foot = 30.48 centimeters
1 yard = 0.91 meter
1 mile = 1.60 kilometers

Measurement of Volume

In the metric system, volume is measured in milliliters, cubic centimeters, and liters. 1 cubic centimeter = 1 milliliter. If a cube has a length, depth, and height of 10 centimeters (cm), the volume of the cube is 10 cm x 10 cm x 10 cm = 1,000 cm^3 = 1 liter. When volume conversions are made between the two systems, 1 cubic inch = 16.38 cubic centimeters. If an engine has a displacement of 350 cubic inches, 350 x 16.38 = 5,733 cubic centimeters, and 5,733 ÷ 1,000 = 5.7 liters.

Personal Safety

Personal safety is the responsibility of each technician in the shop. Always follow these safety practices:

1. Always use the correct tool for the job. If the wrong tool is used, it may slip and cause hand injury.
2. Follow the car manufacturer's recommended service procedures.
3. Always wear eye protection such as safety glasses or a face shield (Figure 2-1).
4. Wear protective gloves when cleaning parts in hot or cold tanks, and when handling hot parts such as exhaust manifolds.
5. Do not smoke when working on a vehicle. A spark from a cigarette or lighter may ignite flammable materials in the work area.
6. When working on a running engine, keep hands and tools away from rotating parts. Remember that electric-drive fans may start turning at any time.
7. Do not wear loose clothing, and keep long hair tied behind your head. Loose clothing or long hair is easily entangled in rotating parts.
8. Wear safety shoes or boots.
9. Do not wear watches, jewelry, or rings when working on a vehicle. Severe burns occur when jewelry makes contact between an electric terminal and ground.
10. Always place a shop exhaust hose on the vehicle tailpipe if the engine is running, and be sure the exhaust fan is running. Carbon monoxide in the vehicle exhaust is harmful or fatal to the human body.
11. Be sure that the shop has adequate ventilation.
12. Make sure the work area has adequate lighting.
13. Use trouble lights with steel or plastic cages around the bulb. If an unprotected bulb breaks, it may ignite flammable materials in the area.

Figure 2-1 Safety glasses (Courtesy of Mac Tools, Inc.)

14. When servicing a vehicle, always apply the parking brake and place the transmission in park with an automatic transmission, or neutral with a manual transmission.
15. Avoid working on a vehicle parked on an incline.
16. Never work under a vehicle unless the vehicle chassis is supported securely on safety stands.
17. When one end of a vehicle is raised, place wheel chocks on both sides of the wheels remaining on the floor.
18. Be sure that you know the location of shop first-aid kits and eyewash fountains.
19. Familiarize yourself with the location of all shop fire extinguishers.
20. Do not use any type of open flame heater to heat the work area.
21. Collect oil, fuel, brake fluid, and other liquids in the proper safety containers.
22. Use only approved cleaning fluids and equipment. Do not use gasoline to clean parts.
23. Obey all state and federal safety, fire, and hazardous material regulations.
24. Always operate equipment according to the equipment manufacturer's recommended procedure.
25. Do not operate equipment unless you are familiar with the correct operating procedure.
26. Do not leave running equipment unattended.
27. Do not use electrical equipment, including trouble lights, with frayed cords.
28. Be sure the safety shields are in place on rotating equipment.
29. Before operating electric equipment, be sure the power cord has a ground connection.
30. When working in an area where extreme noise levels are encountered, wear ear plugs or covers.

Lifting and Carrying

Many automotive service jobs require heavy lifting. You should know your maximum weight lifting ability. Do not attempt to lift more than this weight. If a heavy part exceeds your weight lifting ability, have a co-worker help with the lifting job. Follow these steps when lifting or carrying an object:

1. If the object is to be carried, be sure your path is free from loose parts or tools.
2. Position your feet close to the object and position your back reasonably straight for proper balance.
3. Your back and elbows should be kept as straight as possible. Continue to bend your knees until your hands reach the best lifting location on the object to be lifted.
4. Be certain the container is in good condition. If a container falls apart during the lifting operation, parts may drop out of the container and result in foot injury or part damage.
5. Maintain a firm grip on the object, and do not attempt to change your grip while lifting is in progress.
6. Straighten your legs to lift the object, and keep the object close to your body. Use leg muscles rather than back muscles.
7. If you have to change the direction of travel, turn your whole body instead of twisting it.
8. Do not bend forward to place an object on a workbench or table. Position the object on the front surface of the workbench and slide it back. Do not pinch your fingers under the object while setting it on the front of the bench.
9. If the object must be placed on the floor or a low surface, bend your legs to lower the object. Do not bend your back forward, because this movement strains back muscles.
10. When a heavy object must be placed on the floor, locate suitable blocks under the object to prevent jamming your fingers under the object.

Hand Tool Safety

Many shop accidents are caused by improper use and care of hand tools. These hand tool safety steps must be followed:

1. Maintain tools in good condition and keep them clean. Worn tools may slip and cause hand injury. If a hammer is used with a loose head, the head may fly off and cause personal injury or vehicle damage. Your hand may slip off a greasy tool, and this action may cause some part of your body to hit the vehicle. For example, your head may hit the vehicle hood.
2. Using the wrong tool for the job may cause damage to the tool, fastener, or your hand if the tool slips. If you use a screwdriver for a chisel or pry bar, the blade may shatter, causing serious personal injury.
3. Use sharp pointed tools with caution. Always check your pockets before sitting on the vehicle seat. A screwdriver, punch, or chisel in the back pocket may put an expensive tear in the upholstery. Do not lean over fenders with sharp tools in your pockets.
4. Tool tips that are intended to be sharp should be kept in a sharp condition. Sharp tools, such as chisels, will do the job faster with less effort.

Power Tool Safety

Power tools use electricity, shop air, or hydraulic pressure as a power source. Careless operation of power tools may cause personal injury and vehicle damage. Follow these steps for safe power tool operation:

1. Do not operate power tools with frayed power cords.
2. Be sure the power tool cord has a proper ground connection.
3. Do not stand on a wet floor while operating an electric power tool.
4. Always unplug an electric power tool before servicing the tool.
5. Do not leave a power tool running and unattended.
6. When using a power tool on small parts, do not hold the part in your hand. The part must be secured in a bench vise or with locking pliers.
7. Do not use a power tool on a job where the maximum capacity of the tool is exceeded.
8. Be sure that all power tools are in good condition, and always operate these tools according to the tool manufacturer's recommended procedure.
9. Make sure all protective shields and guards are in position.
10. Maintain proper body balance while using a power tool.
11. Always wear safety glasses or a face shield.
12. Wear ear protection.
13. Follow the equipment manufacturer's recommended maintenance schedule for all shop equipment.
14. Never operate a power tool unless you are familiar with the tool manufacturer's recommended operating procedure. Serious accidents occur from improper operating procedures.
15. Always make sure that the wheels are in good condition and are securely attached on the electric grinder.
16. Keep fingers and clothing away from grinding and buffing wheels. When grinding or buffing a small part, it should be held with a pair of locking pliers.
17. Always make sure the sanding or buffing disc is securely attached to the sander pad.

18. Special heavy-duty sockets must be used on impact wrenches. If ordinary sockets are used on an impact wrench, they may break and cause serious personal injury.
19. Never operate an air chisel unless the tool is securely connected to the chisel with the proper retaining device.
20. Never direct a blast of air from an air gun against any part of your body. If air penetrates the skin and enters the bloodstream, it may cause very serious health problems and even death.

Compressed Air Equipment Safety

The shop air supply contains high-pressure air in the shop compressor and air lines. Serious injury or property damage may result from careless operation of compressed air equipment. Follow these steps to improve safety when using compressed air equipment:

1. Wear safety glasses or a face shield for all shop tasks, including those tasks involving the use of compressed air equipment.
2. Wear ear protection when using compressed air equipment.
3. Always maintain air hoses and fittings in good condition. If an end suddenly blows off an air hose, the hose will whip around, and this may cause personal injury.
4. Do not direct compressed air against the skin. This air may penetrate the skin, particularly through small cuts or scratches. If air penetrates the skin and enters the bloodstream, it can be fatal or cause serious health complications. Use only air gun nozzles approved by Occupational Safety and Health Act (OSHA).
5. Do not use an air gun to blow off clothing or hair.
6. Do not clean the workbench or floor with compressed air. This action may blow very small parts against your skin or into your eye. Small parts blown by compressed air may cause vehicle damage. For example, if the car in the next stall has the air cleaner removed, a small part may go into the carburetor or throttle body. When the engine is started, this part will likely be pulled into the cylinder by engine vacuum, and the part will penetrate through the top of a piston.
7. Never spin bearings with compressed air, because the bearing will rotate at extremely high speed. Under this condition, the bearing may be damaged or disintegrate, causing personal injury.
8. All pneumatic tools must be operated according to the tool manufacturer's recommended operating procedure.
9. Follow the equipment manufacturer's recommended maintenance schedule for all compressed air equipment.

Hydraulic Pressing and Lifting Equipment

Hydraulic Press

CAUTION: When operating a hydraulic press, always be sure that the components being pressed are supported properly on the press bed with steel supports. Improperly supported components may suddenly move or fall, resulting in personal injury.

CAUTION: When using a hydraulic press, never operate the pump handle until the pressure gauge exceeds the maximum pressure rating of the press. If this pressure is exceeded, some part of the press may suddenly break and cause severe personal injury.

When two components have a tight precision fit between them, a hydraulic press is used to separate these components, or press them together. The hydraulic press rests on the shop floor, and an

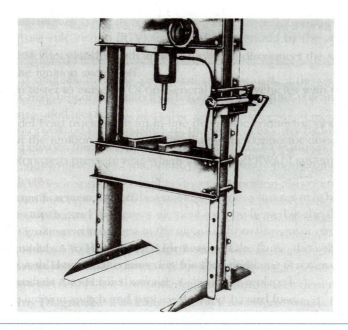

Figure 2-2 Hydraulic press (Courtesy of Mac Tools, Inc.)

adjustable steel beam bed is retained to the lower press frame with heavy steel pins. A hydraulic cylinder and ram is mounted on the top part of the press with the ram facing downward toward the press bed (Figure 2-2). The component being pressed is placed on the press bed with appropriate steel supports. A hand-operated hydraulic pump is mounted on the side of the press. When the handle is pumped, hydraulic fluid is forced into the cylinder, and the ram is extended against the component on the press bed to complete the pressing operation. A pressure gauge on the press indicates the pressure applied from the hand pump to the cylinder. The press frame is designed for a certain maximum pressure, and this pressure must not be exceeded during hand pump operation.

Floor Jack

WARNING: The maximum lifting capacity of the floor jack is usually written on a jack decal. Never lift a vehicle that exceeds the jack lifting capacity. This action may cause the jack to break or collapse, resulting in vehicle damage or personal injury.

A floor jack is a portable unit mounted on wheels. The lifting pad on the jack is placed under the chassis of the vehicle, and the jack handle is operated with a pumping action (Figure 2-3). This jack handle operation forces fluid into a hydraulic cylinder in the jack, and this cylinder extends to force the jack lift pad upward and lift the vehicle. Always be sure that the lift pad is positioned securely under one of the car manufacturer's recommended lift points. To release the hydraulic pressure and lower the vehicle, the handle or release lever must be turned slowly. Do not leave the jack handle where someone can trip over it.

Lift

CAUTION: Always be sure that the lift arms are securely positioned under the car manufacturer's recommended lift points before raising a vehicle. These lift points are shown in the service manual. Using improper lift points may cause the vehicle to slip off the lift, resulting in vehicle damage or personal injury.

WARNING: The maximum capacity of the vehicle lift is placed on an identification plate. Never lift a vehicle that is heavier than the maximum capacity of the lift.

Figure 2-3 Hydraulic floor jack (Courtesy of Mac Tools, Inc.)

A lift is used to raise a vehicle so the technician can work under the vehicle. The lift arms must be placed under the car manufacturer's recommended lift points prior to raising a vehicle. Twin posts are used on some lifts, whereas other lifts have a single post (Figure 2-4). Some lifts have an electric motor that drives a hydraulic pump to create fluid pressure and force the lift upward. Other lifts use air pressure from the shop air supply to force the lift upward. If shop air pressure is used for this purpose, the air pressure is applied to fluid in the lift cylinder. A control lever or switch is placed near the lift. The control lever supplies shop air pressure to the lift cylinder, and the switch turns on the lift pump motor. Always be sure that the safety lock is engaged after the lift is raised. When the safety lock is released, a release lever is operated slowly to lower the vehicle.

A lift may be referred to as a hoist.

An engine lift may be called a crane or a cherry picker.

Figure 2-4 Lifts are used to raise a vehicle. (Courtesy of Bear Automotive Service Equipment Company)

Lift Safety

> **CAUTION:** Do not raise a vehicle on a lift if the vehicle weight exceeds the maximum capacity of the lift. This action may result in lift damage or sudden lowering of the lift, which may result in personal injury.

> **CAUTION:** When a vehicle is raised on a lift, the vehicle must be raised high enough to allow engagement of the lift locking mechanism. If the locking mechanism is not engaged, the lift may drop suddenly, resulting in personal injury.

Special precautions and procedures must be followed when a vehicle is raised on a lift. Follow these steps for lift safety:

1. Always be sure the lift is completely lowered before driving a vehicle on or off the lift.
2. Do not hit or run over lift arms and adapters when driving a vehicle on or off the lift. Have a co-worker guide you when driving a vehicle onto the lift. Do not stand in front of a lift with the car coming towards you.
3. Be sure the lift pads on the lift are contacting the car manufacturer's recommended lift points shown in the service manual. If the proper lift points are not used, components under the vehicle such as brake lines or body parts may be damaged. Failure to use the recommended lift points may cause the vehicle to slip off the lift, resulting in severe vehicle damage and personal injury.
4. Before a vehicle is raised or lowered, close the doors, hood, and trunk lid.
5. When a vehicle is lifted a short distance off the floor, stop the lift and check the contact between the hoist lift pads and the vehicle to be sure the lift pads are still on the recommended lift points.
6. When a vehicle is raised on a lift, be sure the safety mechanism is in place to prevent the lift from dropping accidentally.
7. Before lowering a vehicle on a lift, always make sure there are no objects, tools, or people under the vehicle.
8. Do not rock a vehicle on a lift during a service job.
9. When a vehicle is raised on a lift, removal of some heavy components may cause vehicle imbalance on the lift. Since front-wheel-drive cars have the engine and transaxle located at the front of the vehicle, these cars have most of their weight on the front end. Removing a heavy rear-end component on these cars may cause the back end of the car to rise off the lift. If this action is allowed to happen, the vehicle could fall off the lift!
10. Do not raise a vehicle on a lift with people in the vehicle.
11. When raising pickup trucks and vans on a lift, remember these vehicles are higher than a passenger car. Be sure there is adequate clearance between the top of the vehicle and the shop ceiling or components under the ceiling.
12. *Do not raise a four-wheel-drive vehicle with a frame contact lift, because this may damage axle joints.*
13. Do not operate a front-wheel-drive vehicle that is raised on a frame contact lift. This action may damage the front drive axles.

Engine Lift

> **CAUTION:** An engine lift has a maximum lifting capacity that is usually indicated on a decal. Never lift anything heavier than the maximum capacity of the lift. This action may result in lift damage or personal injury.

An engine lift is used to remove and replace automotive engines. A long, pivoted arm is mounted on the top of the engine lift (Figure 2-5). When the lift handle is pumped, hydraulic fluid is forced into a cylinder under the lift arm. This action extends the cylinder ram and forces the arm upward

Figure 2-5 Engine lift (Courtesy of Mac Tools, Inc.)

to lift the engine. A lifting chain is attached to the lift arm, and this chain is bolted securely to the engine. Always be sure that these retaining bolts are strong enough to support the engine weight.

Hydraulic Jack and Jack Stand Safety

 WARNING: Always make sure the jack stand weight capacity rating exceeds the vehicle weight that is lowered onto the stands.

CAUTION: Never lift a vehicle with a floor jack if the weight of the vehicle exceeds the rated capacity of the jack. This action may cause the jack to drop suddenly, resulting in personal injury or vehicle damage.

Accidents involving the use of floor jacks and jack stands may be avoided if these safety precautions are followed:

1. Never work under a vehicle unless jack stands are placed securely under the vehicle chassis, and the vehicle is resting on these stands (Figure 2-6).
2. Before lifting a vehicle with a floor jack, be sure that the jack lift pad is positioned securely under a recommended lift point on the vehicle. Lifting the front end of a vehicle

Figure 2-6 Jack stands (Courtesy of Mac Tools, Inc.)

with the jack placed under a radiator support may cause severe damage to the radiator and support.

3. Position the jack stands under a strong chassis member such as the frame or axle housing. The jack stands must contact the vehicle manufacturer's recommended lift points.
4. Since the floor jack is on wheels, the vehicle and jack tend to move as the vehicle is lowered from a floor jack onto jack stands. Always be sure the jack stands remain under the chassis member during this operation, and be sure the jack stands do not tip. All of the jack stand legs must remain in contact with the shop floor.
5. When the vehicle is lowered from the floor jack onto jack stands, remove the floor jack from under the vehicle. Never leave a jack handle sticking out from under a vehicle. Someone may trip over the handle and injure themselves.

Cleaning Equipment Safety and Environmental Considerations

 CAUTION: Some parts cleaners contain caustic solutions. To avoid personal injury always wear protective gloves and a face shield when using this equipment.

All technicians are required to clean parts during their normal work routines. Face shields and protective gloves must be worn while operating cleaning equipment. In most states, environmental regulations require that the runoff from steam cleaning must be contained in the steam cleaning system. This runoff cannot be dumped into the sewer system. Since it is expensive to contain this runoff in the steam cleaner system, the popularity of steam cleaning has decreased. The solution in hot and cold cleaning tanks may be caustic, and contact between this solution and skin or eyes must be avoided. Parts cleaning often creates a slippery floor, and care must be taken when walking in the parts cleaning area. The floor in this area should be cleaned frequently. When the caustic cleaning solution in hot or cold cleaning tanks is replaced, environmental regulations require that the old solution be handled as hazardous waste. Use caution when placing aluminum or aluminum alloy parts in a cleaning solution. Some cleaning solutions will damage these components. Always follow the cleaning equipment manufacturer's recommendations.

Parts Washers with Electro-Mechanical Agitation

Some parts washers provide electro-mechanical agitation of the parts to provide improved cleaning action (Figure 2-7). These parts washers may be heated with gas or electricity, and various water-based hot tank cleaning solutions are available depending on the type of metals being cleaned. For example, Kleer-Flo Greasoff number 1 powdered detergent is available for cleaning iron and steel. Nonheated electro-mechanical parts washers are also available. These washers use cold cleaning solutions such as Kleer-Flo Degreasol formulas.

Many cleaning solutions, such as Kleer-Flo Degreasol 99R, contain no ingredients listed as hazardous by the Environmental Protection Agency's RCRA Act. This cleaning solution is a blend of sulphur-free hydrocarbons, wetting agents, and detergents. Degreasol 99R does not contain aromatic or chlorinated solvents, and it conforms to California's Rule 66 for clean air. Always use the cleaning solution recommended by the equipment manufacturer.

Cold Parts Washer with Agitated Immersion Tank

Some parts washers have an agitated immersion chamber under the shelves, which provides thorough parts cleaning. Folding workshelves provide a large upper cleaning area with a constant flow of solution from the dispensing hose (Figure 2-8). This cold parts washer operates on Degreasol 99R cleaning solution.

Figure 2-7 Parts washer with electro-mechanical agitator (Courtesy of Kleer-Flo Company)

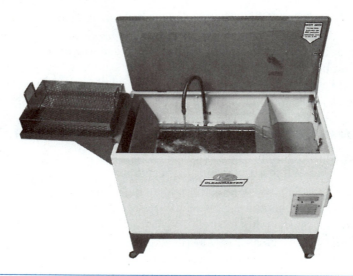

Figure 2-8 Cold parts washer with agitated immersion tank (Courtesy of Kleer-Flo Company)

Aqueous Parts Cleaning Tank

The aqueous parts cleaning tank uses a water-based environmentally friendly cleaning solution, such as Greasoff 2, rather than traditional solvents. The immersion tank is heated and agitated for effective parts cleaning (Figure 2-9). A sparger bar pumps a constant flow of cleaning solution across the surface to push floating oils away, and an integral skimmer removes these oils. This action prevents floating surface oils from redepositing on cleaned parts.

Classroom Manual
Chapter 2, page 21

Figure 2-9 Aqueous parts cleaning tank (Courtesy of Kleer-Flo Company)

Electronic and Fuel System Diagnostic Equipment

Purpose of Equipment Discussion

The purpose of this chapter is to familiarize the reader with the tune-up equipment used in the remaining chapters in the book. Our purpose in this chapter is to provide a general description of the most common types of tune-up equipment and describe briefly the basic tests that may be performed with this equipment. The use of this equipment for detailed diagnosis of various systems is provided in the appropriate following chapters.

Stethoscope

A stethoscope is used to locate the source of engine and other noises. The stethoscope pickup is placed on the suspected component, and the stethoscope receptacles are placed in the technician's ears (Figure 2-10).

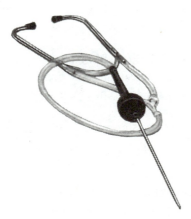

Figure 2-10 Stethoscope (Courtesy of Mac Tools, Inc.)

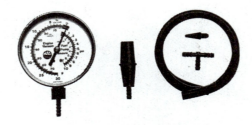

Figure 2-11 Vacuum pressure gauge (Courtesy of Mac Tools, Inc.)

Vacuum Pressure Gauge

A vacuum pressure gauge may be used to check intake manifold vacuum or low pressure such as turbocharger boost pressure or mechanical fuel pump pressure (Figure 2-11). The vacuum pressure gauge is usually supplied with a hose and various fittings and grommets.

Fuel Pressure Gauges

CAUTION: While testing fuel pressure, be careful not to spill gasoline. Gasoline spills may cause explosions and fires, resulting in serious personal injury and property damage.

CAUTION: Electronic fuel injection systems are pressurized, and these systems require depressurizing prior to fuel pressure testing and other service procedures. If the fuel system is not depressurized prior to fuel pressure testing, fuel may be spilled, causing a fire that results in personal injury and/or property damage. (Fuel system depressurizing and detailed fuel pressure testing is explained later in the appropriate chapter.)

The fuel pressure gauge is used to measure the electric fuel pump pressure in throttle body injection (TBI) and port fuel injection (PFI) systems. The mechanical fuel pump pressure on a carbureted engine is also checked with a pressure gauge.

A fuel pump pressure gauge is usually sold in a kit that contains the necessary fittings and hoses to connect the gauge to the fuel system. Since fuel system design varies depending on the manufacturer and the type of fuel system, the fuel pressure gauge requires a number of adapters. A fuel pressure gauge kit for a TBI system must have adapters to connect the gauge at the throttle body fuel inlet line (Figure 2-12).

When the fuel pump pressure is tested on a PFI system, the fuel pressure gauge must be connected to the Schrader valve on the fuel rail (Figure 2-13). Since there are several different sizes of Schrader valves on various PFI systems, a number of adapters are supplied with the PFI fuel pressure gauge kit (Figure 2-14).

If the fuel pump is defective in a PFI or TBI system, the engine usually fails to start, or stops intermittently. A plugged fuel filter may cause the same symptoms. When fuel pressure is excessive in a PFI or TBI system, the air-fuel mixture is too rich. A lean air-fuel mixture in these systems may be caused by lower-than-specified fuel pressure.

On a carbureted engine, the fuel pressure gauge is connected at the fuel inlet fitting on the carburetor to measure the mechanical fuel pump pressure. The gauge fitting is threaded into the carburetor inlet fitting, and one of the hoses on the tester is connected to the fuel inlet line. A second hose on the tester is installed in a graduated plastic container, and a special clip closes this line (Figure 2-15). With the engine idling, the clip is closed to measure the fuel pump pressure, and then the clip is released to allow fuel flow into the plastic container to measure fuel pump volume. The fuel flow into the plastic container is timed for a specific length of time, such as 30 or 45 seconds, depending on the specifications. A typical fuel pump flow specification would be 1 pint in 30 seconds.

Special Tools

Fuel pressure gauges

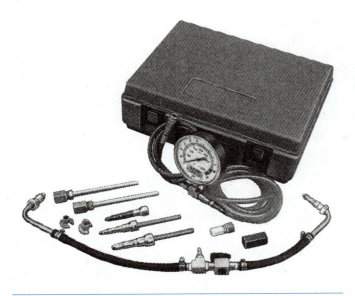

Figure 2-12 Fuel pressure gauge and adapters for throttle body injection (TBI) systems (Courtesy of OTC Division, SPX Corp.)

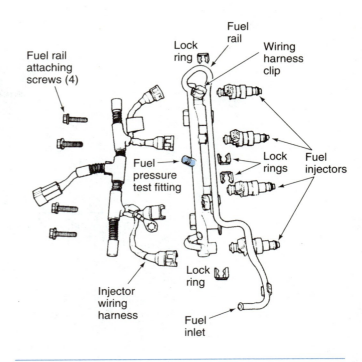

Figure 2-13 On a PFI system, the fuel pressure gauge is connected to the Schrader valve on the fuel rail to test fuel pump pressure. (Courtesy of Chrysler Corporation)

Figure 2-14 Fuel pressure gauge for port fuel injection (PFI) systems (Courtesy of OTC Division, SPX Corp.)

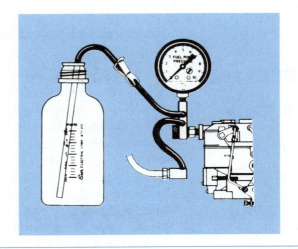

Figure 2-15 Fuel pump pressure and flow tester, carbureted engine (Courtesy of Sun Electric Corporation)

Injector Balance Tester

The injector balance tester is used to test the injectors in a port fuel injected (PFI) engine for proper operation. A fuel pressure gauge is also used during the injector balance test. The injector balance tester contains a timing circuit, and some injector balance testers have an off-on switch. A pair of leads on the tester must be connected to the battery with the correct polarity (Figure 2-16). The injector terminals are disconnected, and a second double lead on the tester is attached to the injector terminals.

Special Tools

Injector balance tester

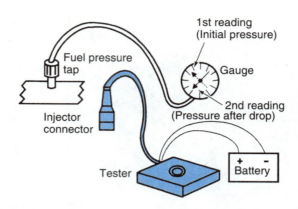

Figure 2-16 Injector balance tester (Courtesy of Oldsmobile Division, General Motors Corporation)

Before the injector test, the fuel pressure gauge is connected to the Schrader valve on the fuel rail, and the ignition switch should be cycled two or three times until the specified fuel pressure is indicated on the pressure gauge. When the tester push button is depressed, the tester energizes the injector winding for a specific length of time, and the technician records the pressure decrease on the fuel pressure gauge. This procedure is repeated on each injector.

Some vehicle manufacturers provide a specification of 3 psi (20 kPa) maximum difference between the pressure readings after each injector is energized. If the injector orifice is restricted, there is not much pressure decrease when the injector is energized. Acceleration stumbles, engine stalling, and erractic idle operation are caused by restricted injector orifices. The injector plunger is sticking open if excessive pressure drop occurs when the injector is energized. Sticking injector plungers may result in a rich air-fuel mixture.

Pressurized Injector Cleaning Container

Special Tools

Pressurized injector cleaning container

The pressurized injector cleaning container is designed for cleaning the injectors in PFI systems. The hose on this container is connected to the Schrader valve on the fuel rail, and various fittings are available for this connection. After the lid is removed from the container, a specific quantity of unleaded gasoline and injector cleaner is placed in the container. When the lid is installed, the hand pump on the container is operated until the specified fuel system pressure is indicated on the container pressure gauge (Figure 2-17). During the injector cleaning process, the electric fuel pump must be disabled and the fuel return line plugged. Under this condition, the engine runs on the fuel and injector cleaning solution in the pressurized container.

Some automotive equipment suppliers market canister-type injector cleaning containers. These containers have a valve and a hose for connection to the Schrader valve (Figure 2-18). A pre-mixed solution of unleaded gasoline and injector cleaner is placed in the container, and the specified fuel pressure is supplied from the shop air supply. With the return fuel line plugged, and the electric fuel pump disabled, the engine runs on the solution in the canister to clean the injectors.

Circuit Testers

⚠️ **WARNING:** Do not use a conventional 12-V test light to diagnose components and wires in computer systems. The current draw of these test lights may damage computers and computer system components. High-impedance test lights are available for diagnosing computer systems. Always be sure the test light you are using is recommended by the tester manufacturer for testing computer systems.

⚠️ **WARNING:** Do not use any type of test light or circuit tester to diagnose automotive air bag systems. To prevent an accidental air bag deployment, use only the vehicle manufacturer's recommended equipment on these systems.

Figure 2-17 Pressurized injector cleaning container (Courtesy of OTC Division, SPX Corp.)

Figure 2-18 Canister pressurized with shop air supply for injector cleaning (Courtesy of OTC Division, SPX Corp.)

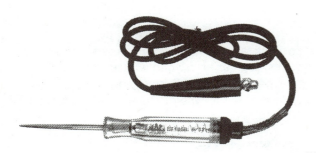

Figure 2-19 12-V test light (Courtesy of Mac Tools, Inc.)

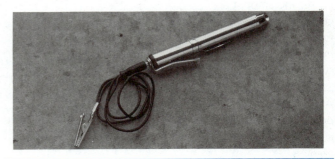

Figure 2-20 Self-powered test light (Courtesy of Mac Tools, Inc.)

Circuit testers are used to diagnose open circuits, grounds, and shorts in electric circuits. There is a large variety of circuit testers available for automotive testing, but the most common circuit tester is the 12-V test light. A sharp probe is molded into a handle on the 12-V test light, and the upper part of the handle is transparent. The 12-V bulb inside the tester is visible through the transparent handle (Figure 2-19). A ground clip extending from the handle is connected to one terminal on the bulb, and the other bulb terminal is connected to the probe. In most test situations, the ground clip is connected to a ground connection on the vehicle, and the probe is connected to a circuit to determine if voltage is available. With this type of test connection, the bulb is illuminated if voltage is available at the probe.

A self-powered test light is similar in appearance to a 12-V test light, but the self-powered test light contains an internal battery (Figure 2-20). One end of many automotive circuits is connected to ground, and battery voltage is supplied to the other end of the circuit. The ground side of the circuit may be called the negative side of the circuit, and the positive side of the circuit is connected to the battery positive terminal. The positive side of the circuit may be disconnected, and the test light lead connected to the battery positive terminal. When the self-powered test light probe is connected to this disconnected circuit, the test light should be illuminated if the other end of the circuit is connected to ground. If the circuit is open, the light remains off.

> A self-powered test light may be referred to as a continuity tester.

Volt-Ampere Tester

A volt-ampere tester is used to perform voltmeter and ammeter tests in any automotive electrical circuit. Some of the most common tests performed with a volt-ampere tester are:

1. Battery load test
2. Starter current draw test
3. Alternator maximum output test
4. Alternator normal system voltage test

Special Tools

Volt-ampere tester

The volt-ampere tester contains an ammeter, a voltmeter, and a carbon pile load. A carbon pile load is a stack of heavy carbon discs. A control knob on the tester adjusts the position of these discs. When the control knob is turned all the way counterclockwise to the off position, the carbon discs are not contacting each other. As the control knob is rotated clockwise, the discs make contact, and further clockwise rotation supplies more pressure to the discs, which reduces the resistance of the carbon pile. Two heavy leads are connected to the carbon pile load, and these leads are connected across the battery terminals with the correct polarity. The red lead is connected to the positive battery terminal, and the negative lead is attached to the negative battery terminal. Many volt-ampere testers have digital ammeter and voltmeter readings (Figure 2-21), but some of these testers are equipped with analog meters.

An inductive clamp is a type of ammeter pickup that fits over the wire in which the current is to be measured. The inductive clamp senses the amount of current flowing in the wire from the magnetic field surrounding the wire.

Some ammeters have a pair of leads that are connected in series in the circuit to be tested, but many ammeters have an inductive clamp that clips over the wire in which the current is to be measured. The inductive clamp and the ammeter read the amount of current flow from the strength of the magnetic field surrounding the wire.

Multimeter

Analog meters have a pointer and a scale to indicate a specific reading.

WARNING: A high-impedance digital multimeter must be used to test the voltage of some components and systems, such as an oxygen (O_2) sensor circuit. If a low-impedance analog meter is used in this type of circuit, the current flow through the meter is high enough to damage the sensor. Always use the type of meter specified by the vehicle manufacturer.

Digital meters have a digital reading to indicate a specific value.

Figure 2-21 Volt-ampere tester with digital ammeter and voltmeter readings (Courtesy of Snap-on Tools Corporation)

WARNING: Always be sure the proper scale is selected on the multimeter and the correct lead connections are completed for the component or system being tested. Improper multimeter lead connections or scale selections may blow the internal fuse in the meter or cause meter damage.

Multimeters are small hand-held meters that provide the following readings on several different scales:

1. DC volts
2. AC volts
3. Ohms
4. Amperes
5. Milliamperes

Since multimeters do not have heavy leads, the highest ammeter scale on this type of meter is sometimes 10 amperes. A control knob on the front of the multimeter must be rotated to the desired reading and scale (Figure 2-22). Some multimeters are autoranging, which means the meter automatically switches to a higher scale if the reading goes above the value of the scale being used. For example, if the meter is reading on the 10-V scale and the leads are connected to a 12-V battery, the meter automatically changes to the next highest scale. If the multimeter is not autoranging, the technician must select the proper scale for the component or circuit being tested.

Multimeters usually have an internal fuse that blows if the meter is connected improperly. Multimeters usually have digital readings, but analog multimeters are available (Figure 2-23). Digital multimeters have high impedance compared to analog multimeters, but digital multimeters have different impedance depending on the meter design. Always use the type of multimeter recommended by the vehicle manufacturer. The leads are plugged into terminals in the front of the multimeter, and some of these meters have several terminals. The reading provided by the terminal position is indicated beside the terminal. The technician must plug the leads into the correct terminal to obtain the desired reading. Some multimeters have a common (com) terminal, and the negative, black meter lead must be plugged into this terminal.

Special Tools
Multimeter

Meter impedance refers to the internal resistance inside the meter. Digital multimeters usually have higher impedance compared to analog meters.

Figure 2-22 The control knob is rotated to select the desired reading and scale on a digital multimeter. (Courtesy of OTC Division, SPX Corp.)

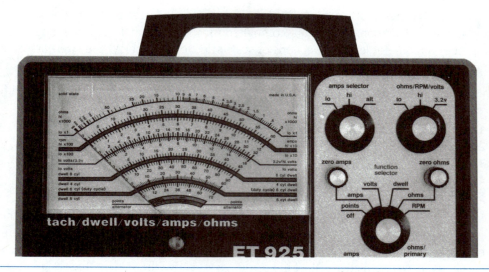

Figure 2-23 Analog multimeter (Courtesy of Mac Tools, Inc.)

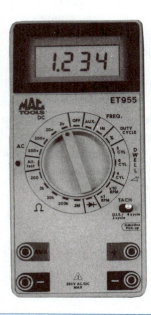

Figure 2-24 Multifunctional digital multimeter (Courtesy of Mac Tools, Inc.)

Some multimeters have additional capabilities such as testing of diode condition, frequency, temperature, engine rpm, ignition dwell, and distributor condition. This type of multimeter has more switch positions and more lead terminals (Figure 2-24).

Ignition Test Equipment

Tach-Dwellmeter

The tach-dwellmeter is one of the commonly used pieces of tune-up equipment. A small internal dry-cell battery powers most tach-dwellmeters. The red tach-dwellmeter lead is connected to the negative primary coil terminal, and the black meter lead is connected to ground. A switch on the meter must be set in the rpm or dwell position (Figure 2-25). On distributorless ignition systems, a special tachometer lead may be provided in the ignition system for the tach-dwellmeter connection.

A tachometer is designed to read engine revolutions per minute (rpm).

Special Tools

Tach-dwellmeter

A dwellmeter is designed to read dwell on the ignition system.

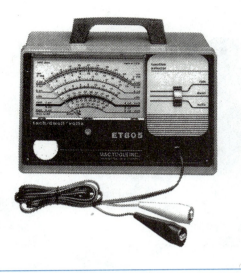

Figure 2-25 Tach-dwellmeter for checking engine rpm and dwell (Courtesy of Mac Tools, Inc.)

Figure 2-26 Digital tachometer with an inductive clamp that fits over the number 1 spark plug wire (Courtesy of Snap-on Tools Corporation)

Digital Tachometers

Since the dwell is controlled by the module in electronic ignition systems, the dwell reading is not used for diagnostic purposes. Therefore, the dwell function is eliminated on some meters, and the tachometer is available by itself. Digital tachometers are now available with an inductive pickup that is clamped over number 1 spark plug wire. These meters provide an rpm reading from the speed of spark plug firings (Figure 2-26). This type of tachometer is suitable for distributorless ignition systems.

Photoelectric tachometers are available to read engine rpm or rpm of other rotating components. An internal light source is powered by a battery in the tachometer or by the car battery. A piece of reflecting tape is applied to the crankshaft pulley and the light in the photoelectric tachometer is pointed at this tape when the engine is running. The photoelectric cell in the tachometer senses the reflected light each time the reflective tape rotates through the meter light beam. An engine rpm calculation is made by the tachometer from these reflected light pulses.

Many engines manufactured in recent years have a magnetic probe receptacle mounted above the crankshaft pulley (Figure 2-27). Digital tachometers are available with a magnetic pickup

On a point-type ignition system, dwell is the number of degrees the points remain closed on each distributor cam lobe.

Special Tools

Digital tachometer

In an electronic distributor ignition system, dwell is the degrees of distributor rotation the primary circuit is turned on prior to each cylinder firing. Since the dwell is determined electronically on these systems, it cannot be adjusted.

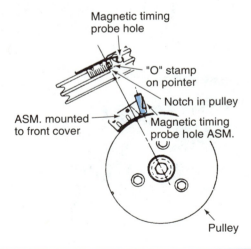

Figure 2-27 Magnetic probe receptacle mounted above the crankshaft pulley (Courtesy of Chevrolet Motor Division, General Motors Corporation)

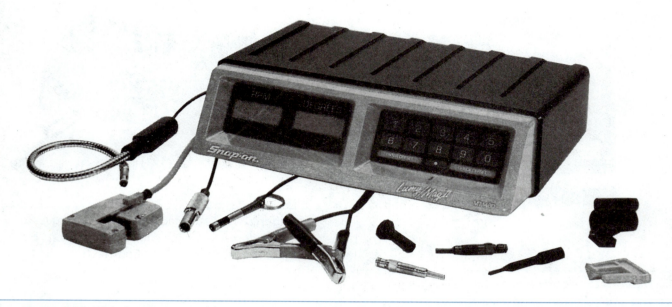

Figure 2-28 Digital tachometer with magnetic probe-type pickup (Courtesy of Snap-on Tools Corporation)

Special Tools

Magnetic probe-type tachometer

that fits in this magnetic probe receptacle, and each timing mark rotation past the pickup sends a pulse signal to the meter (Figure 2-28). The digital tachometer provides an rpm reading from these pulses. This type of digital tachometer may be used on diesel engines. The magnetic probe receptacle may also be used for engine timing purposes.

Timing Light

 CAUTION: Never pierce the number 1 spark plug wire to complete a timing light connection. This action results in high voltage leakage from the spark plug wire.

A timing light is essential for checking the ignition timing in relation to crankshaft position. Two leads on the timing light must be connected to the battery terminals with the correct polarity. Most timing lights have an inductive clamp that fits over the number 1 spark plug wire (Figure 2-29). Older timing lights have a lead that goes in series between the number 1 spark plug wire and the spark plug. A trigger on the timing light acts as an off/on switch. When the trigger is pulled with the engine running, the timing light emits a beam of light each time the spark plug fires.

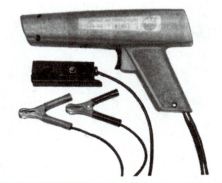

Figure 2-29 Timing light with inductive clamp that fits over the number 1 spark plug wire (Courtesy of Mac Tools, Inc.)

The timing marks are usually located on the crankshaft pulley or on the flywheel. A stationary pointer, line, or notch is positioned above the rotating timing marks. The timing marks are lines on the crankshaft pulley or flywheel that represent various degrees of crankshaft rotation when the number 1 piston is before top dead center (BTDC) on the compression stroke. The TDC crankshaft position and the degrees are usually identified in the group of timing marks. Some timing marks include degree lines representing the after top dead center (ATDC) crankshaft position (Figure 2-30).

Before checking the ignition timing, complete all of the vehicle manufacturer's recommended procedures. On fuel injected engines, special timing procedures are required, such as disconnecting a timing connector to be sure the computer does not provide any spark advance while checking basic ignition timing. These timing instructions are usually provided on the underhood emission label. (Detailed timing instructions are provided later in the appropriate chapter.)

With the engine running at the specified idle speed, the light is aimed at the timing marks. The timing mark should appear at the specified position in relation to the timing pointer. For example, if the vehicle manufacturer's timing specification is 12° BTDC, the 12° timing mark should be directly under the timing pointer. If the timing mark is not positioned properly, loosen the distributor clamp and rotate the distributor until the timing mark is in the specified location, and then retighten the clamp. The manufacturer's timing specifications are included on the underhood emission label and in the service manual. The ignition timing is not adjustable on electronic ignition (EI) systems that do not have a distributor.

If the ignition timing is advanced more than the manufacturer's specified timing, the engine may detonate on acceleration. When the ignition timing is later than specified, engine power is reduced and engine overheating may occur.

Many timing lights have a timing advance knob that can be used to check spark advance (Figure 2-31). This knob has an index line and a degree scale surrounding the knob. Before checking the spark advance, the basic timing should be checked. Accelerate the engine to 2,500 rpm or the speed recommended by the vehicle manufacturer. While maintaining this rpm, slowly rotate the advance knob toward the advanced position until the timing marks come back to the basic timing position. Under this condition, the index mark on the advance knob is pointing to the number of degrees advance provided by the computer or distributor advances. The reading on the degree scale may be compared to the vehicle manufacturer's specifications to determine if the spark advance is correct.

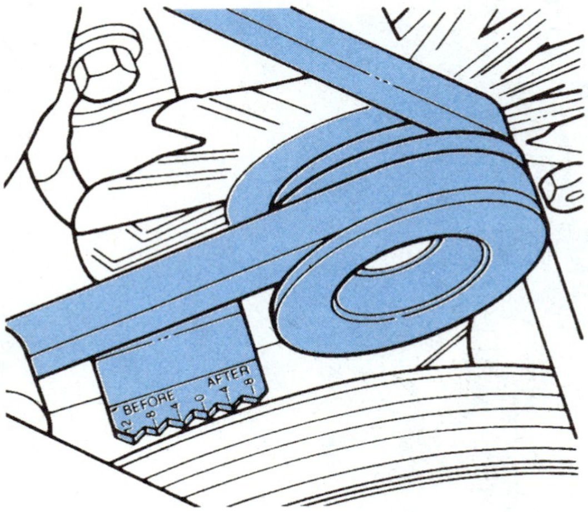

Special Tools

Timing lights, advance-type and digital advance-type

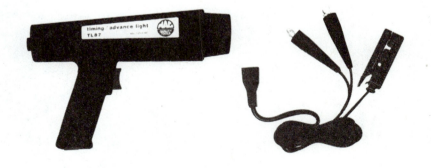

Figure 2-30 Various timing marks on the crankshaft pulley (Courtesy of Chevrolet Motor Division, General Motors Corporation)

Figure 2-31 Timing light with advance knob (Courtesy of Mac Tools, Inc.)

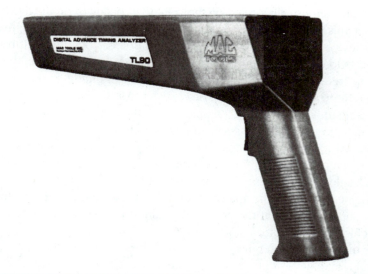

Figure 2-32 A digital advance-type timing light provides spark advance or rpm reading. (Courtesy of Mac Tools, Inc.)

Some timing lights have a digital reading in the back of the light which displays the number of degrees advance as the engine is accelerated (Figure 2-32). When the trigger on the light is squeezed, the light flashes and the digital display reads the degrees of spark advance. If the trigger is released, the digital reading indicates engine rpm.

Magnetic Timing Probe

Special Tools

Magnetic probe-type digital timing meter

Digital timing meters are available with a magnetic probe that fits into the magnetic timing receptacle above the crankshaft pulley. Some of these timing meters also have an inductive clamp for the number 1 spark plug wire (Figure 2-33). Therefore, this type of meter may be used with either pickup. The magnetic probe should be pushed into the timing receptacle until it lightly touches the

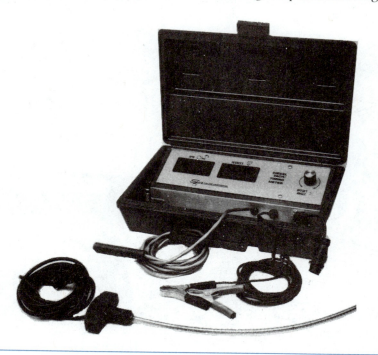

Figure 2-33 Digital timing meter with magnetic probe (Courtesy of Sun Electric Corporation)

crankshaft pulley. With the engine running, the probe senses the timing mark rotating past the probe to provide a reading on the digital display.

Since the magnetic timing receptacle is not located exactly above the number 1 TDC position on the crankshaft pulley, an offset adjustment on the meter must be completed before the timing is checked with the magnetic timing probe. The amount of offset varies depending on the engine. Always set the timing offset according to the equipment manufacturer's instructions. If the timing offset is not adjusted properly, the basic timing reading is not accurate.

Ignition Module Tester

Without an ignition module tester, the ignition module is usually tested by the process of elimination. The procedure is to test all the other components in the electronic ignition system, and if they are satisfactory, the module must be the cause of the performance problem.

The ignition module tester has a number of lead wires, and separate jumper wires are supplied with the tester (Figure 2-34). This type of tester will test a number of different ignition systems. It must always be connected according to the equipment manufacturer's recommendations. Buttons on the tester are pushed to initiate specific tests, and the tester indicates a pass or fail condition for each test.

In many cases, a defective ignition module causes a no-start problem. Sometimes, a defective ignition module causes the ignition system to intermittently stop firing. Typical tests performed by an ignition module tester are:

1. Key on/engine off test
2. Cranking current test
3. Idle current test
4. Cruise current test
5. Shorted module test
6. Cranking primary voltage test
7. Idle primary voltage test
8. Cruise primary voltage test

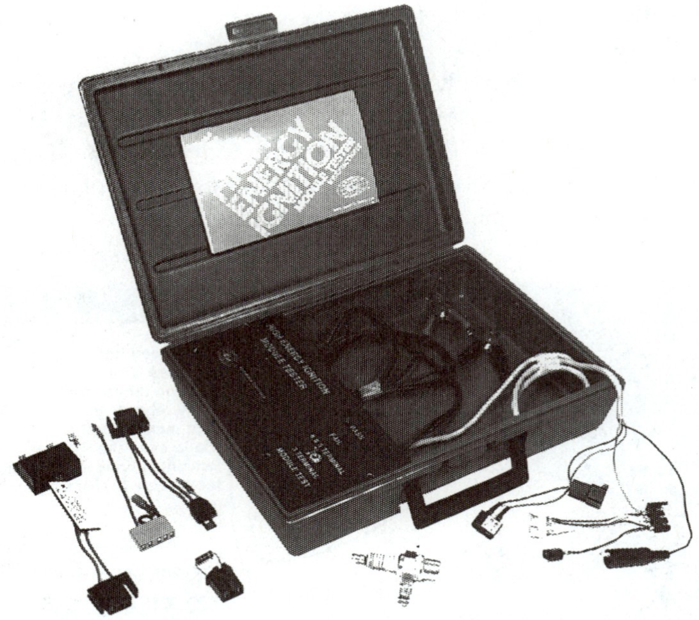

Special Tools

Ignition module tester

Figure 2-34 Ignition module tester (Courtesy of Automotive Group, Kent-Moore Division SPX Corp.)

Figure 2-35 Oscilloscope (Courtesy of Snap-on Tools Corporation)

Oscilloscope

The oscilloscope is very useful in diagnosing ignition problems quickly and accurately. Digital and analog voltmeters do not react fast enough to read secondary ignition voltages. Each time a spark plug fires, the voltage and current are only present at the spark plug electrodes for approximately 1.5 milliseconds. The oscilloscope may be considered as a very fast reacting voltmeter that reads and displays voltages in the ignition system. These voltage readings appear as a voltage trace on the oscilloscope screen (Figure 2-35). An oscilloscope screen is a cathode ray tube (CRT), which is very similar to the picture tube in a television set. High voltage from an internal source is supplied to an electron gun in the back of the CRT when the oscilloscope is turned on. This electron gun emits a continual beam of electrons against the front of the CRT. The external leads on the oscilloscope are connected to deflection plates above and below, and on each side of the electron beam (Figure 2-36). When a voltage signal is supplied from the external leads to the deflection plates, the electron beam is distorted and strikes the front of the screen in different locations to indicate the voltage signal from the external leads.

One inductive clamp on the oscilloscope fits over the secondary coil wire, and a second inductive clamp is placed on the number 1 spark plug wire. The primary scope leads are connected from the negative primary coil terminal to ground with the correct polarity. The black primary lead wire is always connected to ground. Most oscilloscopes have a pair of voltmeter lead wires that are usually connected to the battery terminals with the correct polarity.

> An oscilloscope may be called a scope.

> The negative primary coil terminal may be referred to as a "tach" terminal.

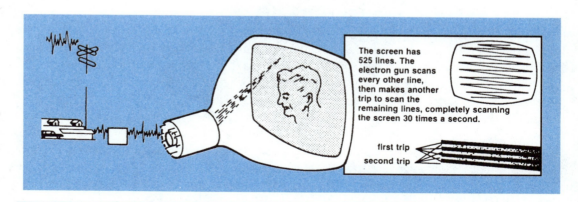

Figure 2-36 Cathode ray tube (CRT) operation (Courtesy of Ford Motor Company of Canada Limited)

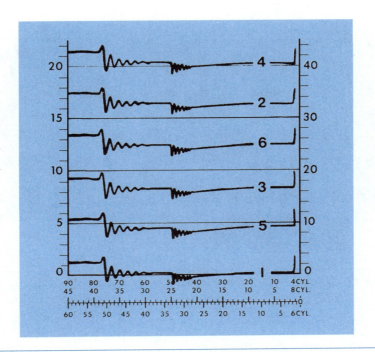

Figure 2-37 Typical scales on an oscilloscope screen (Courtesy of Sun Electric Corporation)

Oscilloscope Scales and Tests

An upward movement of the voltage trace on an oscilloscope screen indicates an increase in voltage, and a downward movement of this trace represents a decrease in voltage. The voltage scale on the left side of the screen reads secondary ignition voltages from 0 to 25 kilovolts (kV), and the scale on the right side of the screen indicates 0 to 50 kV (Figure 2-37). A control knob or push buttons on the oscilloscope allow the technician to select the desired kV scale. If the highest part of a voltage trace goes upward to 40 on the right hand scale, the secondary ignition voltage is 40 kV at that instant.

As the voltage trace moves across an oscilloscope screen, it represents a specific length of time. A horizontal scale about halfway up on the screen is graduated in milliseconds, and this scale may be used to measure the actual firing time of the spark plugs. The scale at the bottom of the screen is graduated in percentage or degrees. The horizontal degree scale may be used to measure ignition dwell. An oscilloscope may be used to perform many tests on the ignition system.

One kilovolt (kV) is equal to 1,000 volts.

Special Tools

Oscilloscope

Scan Testers

Scan testers are used to test automotive computer systems. These testers retrieve fault codes from the onboard computer memory and display these codes in the digital reading on the tester. The scan tester performs many other diagnostic functions depending on the year and make of vehicle. Most scan testers have removable modules that are updated each year. These modules are designed to test the computer systems on various makes of vehicles. For example, some scan testers have a 3-in-1 module which tests the computer systems on Chrysler, Ford, and General Motors vehicles. A 10-in-1 module is also available to diagnose computer systems on vehicles imported by 10 different manufacturers. These modules plug into the scan tester (Figure 2-38).

Scan testers have the capability to test many onboard computer systems such as engine computers, antilock brake computers, air bag computers, and suspension computers, depending on the year and make of vehicle, and the type of scan tester. In many cases, the technician must select the computer system to be tested with the scan tester after the tester is connected to the vehicle.

Special Tools

Scan tester

39

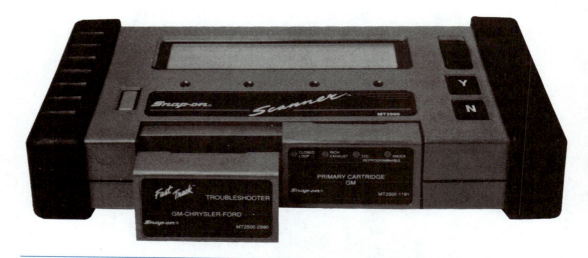

Figure 2-38 Scan tester with various modules (Courtesy of Snap-on Tools Corporation)

Figure 2-39 Scan tester connectors to fit various diagnostic connectors on different vehicles (Courtesy of OTC Division, SPX Corp.)

The scan tester is connected to specific diagnostic connectors on various vehicles. Some manufacturers have one diagnostic connector, and they connect the data wire from each onboard computer to a specific terminal in this connector. Other vehicle manufacturers have several different diagnostic connectors on each vehicle, and each of these connectors may be connected to one or more onboard computers. A set of connectors is supplied with the scan tester to allow tester connection to various diagnostic connectors on different vehicles (Figure 2-39).

The scan tester must be programmed for the model year, make of vehicle, and type of engine. With some scan testers, this selection is made by pressing the appropriate buttons on the tester, as directed by the digital tester display. On other scan testers, the appropriate memory card must be installed in the tester for the vehicle being tested (Figure 2-40). Some scan testers have a built-in printer to print test results, while other scan testers may be connected to an external printer.

As automotive computer systems become more complex, the diagnostic capabilities of scan testers continue to expand. Many scan testers now have the capability to store, or "freeze," data into the tester during a road test, and then play back this data when the vehicle is returned to the shop.

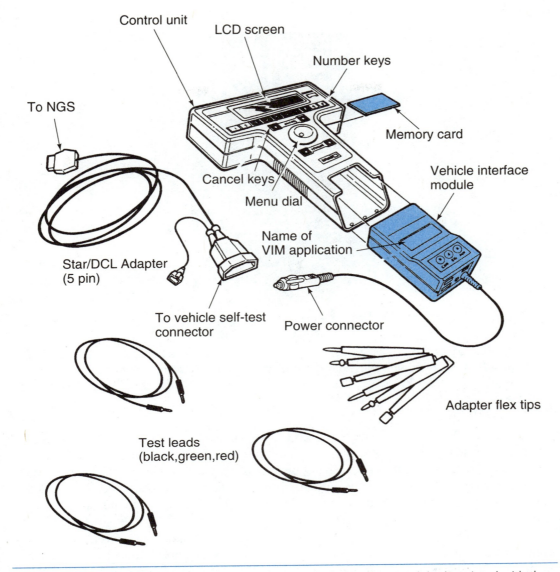

Figure 2-40 Scan tester with memory card and vehicle interface module (Reprinted with the permission of Ford Motor Company)

Some scan testers now display diagnostic information based on the fault code in the computer memory. Service bulletins published by the scan tester manufacturer may be indexed by the tester after the vehicle information is entered in the tester. Other scan testers will display sensor specifications for the vehicle being tested.

Emissions Analyzer

Vehicles produced in the United States have been required to meet specific emission standards for many years. Many states have emission inspection and maintenance (I/M) programs that require car owners to maintain their vehicles to meet emission standards. The emissions analyzer is used to measure tailpipe emissions.

Older two-gas emissions analyzers measure hydrocarbons (HC) and carbon monoxide (CO). Newer four-gas emissions analyzers measure HC and CO, plus oxygen (O_2) and carbon dioxide

Two-gas emissions analyzers measure hydrocarbon (HC) and carbon monoxide (CO).

Four-gas analyzers measure the same emissions as a two-gas analyzer plus oxygen (O_2) and carbon dioxide (CO_2).

41

Figure 2-41 Four-gas emissions analyzer (Courtesy of Bear Automotive Service Equipment Company)

Special Tools

Emissions analyzer

(CO_2) (Figure 2-41). A catalytic converter affects HC and CO readings. If a two-gas analyzer is used to measure the actual emissions levels coming out of the engine, the analyzer pickup must be installed ahead of the converter. Some exhaust pipes have a removable plug for this purpose. If the exhaust pipe does not have a plug, the pipe may be drilled, threaded, and a brass plug installed. If a four-gas analyzer pickup is installed in the tailpipe, O_2 and CO_2 readings are not affected by the catalytic converter.

Emissions analyzers require a warm-up and calibration interval of approximately 15 minutes. Some emissions analyzers perform this warm-up mode automatically, and the technician is reminded on the digital reading that the tester is in the warm-up mode. Other emissions testers do not have an automatic warm-up mode; technician must turn on the tester and wait the proper length of time. Modern four-gas emissions analyzers have an automatic calibration function, but older analog emissions analyzers had to be calibrated manually with calibration knobs.

Emissions analyzers have a filter or filters to remove water and other contaminants from the exhaust sample. These filters must be replaced periodically. A warning light on the tester is illuminated if the filter or anything else is restricting the exhaust flow through the tester. An emissions analyzer may be used to check the emission levels on a vehicle to determine if these levels meet state and federal emission standards. (Four-gas emission analysis and emission standards are explained later in the appropriate chapter.) The following items may also be checked with a four-gas analyzer:

1. Air-fuel mixture, rich or lean
2. Air pump malfunctions
3. Defective injectors or carburetor
4. Cylinder misfiring
5. Intake manifold vacuum leaks
6. Catalytic converter defects
7. Air pump defects
8. Leaking head gaskets

Figure 2-42 An engine analyzer is a combination of many automotive testers. (Courtesy of Sun Electric Corporation)

9. Defective EGR valve
10. Leaking or restricted exhaust system
11. Excessive spark advance

Engine Analyzers

SERVICE TIP: When diagnosing any engine performance or economy problem, always test the basic items first. For example, always be sure the engine has satisfactory compression, ignition, and emission component operation before diagnosing the computer system.

Engine analyzers are actually a combination of different automotive testers (Figure 2-42). Many engine analyzers have an oscilloscope screen which displays ignition voltages and many other readings. Engine analyzers have many leads that must be connected according to the analyzer manufacturer's instructions. A typical engine analyzer performs the same diagnostic functions as these individual testers:

1. Vacuum gauge
2. Pressure gauge
3. Vacuum pump
4. Compression gauge (power balance tester)
5. Voltmeter
6. Ammeter
7. Ohmmeter
8. Advance-type timing light
9. Oscilloscope
10. Scan tester
11. Emissions analyzer

Special Tools

Engine analyzer

Figure 2-43 Engine analyzer with keyboard, printer, and graphic display (Courtesy of Bear Automotive Service Equipment Company)

The engine analyzer contains a keyboard that is used to enter vehicle data, commands, and specifications (Figure 2-43). A printer in the engine analyzer provides a printout of the test results for the customer. Some engine analyzers display the test results graphically on the screen.

The normal specified sensor readings are programmed into the computer in the engine analyzer for each year and model of vehicle. During a test procedure, this computer compares the vehicle data to the specifications programmed into the analyzer computer. The engine analyzer identifies readings that are not within specifications and provides diagnostic information to find the exact cause of the defective reading.

Engine analyzers usually have manual and automatic test modes. If the technician selects the automatic mode, the analyzer automatically performs a complete series of tests. In the manual mode, the technician selects the specific tests to be performed. During the series of tests in the automatic mode, the analyzer performs tests to prove the condition of the ignition system, starting system, charging system, cylinder compression, fuel system, and emission systems.

Safety Training Exercises

After the equipment operation and safety practices are explained by your instructor, complete these safety training exercises to demonstrate your understanding of equipment operation and safety procedures:

1. Find the lift points on a specific vehicle in the vehicle manufacturer's service manual. Under the supervision of your instructor, position this vehicle properly on a lift, then use the proper lift operating procedures to raise and lower the vehicle on the lift.
2. Find the lift points on a specific vehicle in the vehicle manufacturer's service manual. Under the supervision of your instructor, raise the front and rear suspension of this vehicle with a floor jack and lower the vehicle onto jack stands. Raise the vehicle with a floor jack, and remove the jack stands, and lower the vehicle.
3. Find the lift points on a specific vehicle in the vehicle manufacturer's service manual. Under the supervision of your instructor, position this vehicle properly on a lift and

raise the vehicle. Follow proper service and safety procedures to remove and replace all four wheels. Remove two wheels with an electric impact wrench, and the other two wheels with an air impact wrench.

4. Draw a layout diagram of your automotive shop or shops, indicating the major service areas and service equipment, and clearly identify the location of all safety equipment.

● **CUSTOMER CARE:** Some automotive service centers have a policy of performing some minor service as an indication of their appreciation to the customer. This service may include cleaning all the windows and/or vacuuming the floors before the car is returned to the customer.

Although this service involves more labor costs for the shop, it may actually improve profits over a period of time. When customers find their windows cleaned and/or the floors vacuumed, it impresses them with the quality of work you do and the fact that you care about their vehicle. They will likely return for service and tell their friends about the quality of service your shop performs.

Classroom Manual
Chapter 2, page 28

Guidelines for Use of Tools, Safety Practices, and the Use of Test Equipment

1. Two systems of measurement in common use are the United States customary (USC) and international system (SI).
2. The SI system may be called the metric system.
3. In the metric system, the units can be divided or multiplied by 10.
4. Personal safety is the responsibility of everyone in the shop.
5. When lifting heavy objects, always bend your knees rather than your back.
6. Many shop accidents are caused by the improper use of hand tools.
7. Never operate any type of equipment unless you are familiar with the proper operating procedure.
8. Never exceed the rated capacity of a hydraulic press, hydraulic jack, vehicle lift, or jack stands.
9. When raising a vehicle with a lift or floor jack, always be sure the lifting equipment is contacting the vehicle on the manufacturer's recommended lift points.
10. After a vehicle is raised on a lift, be sure the lift locking mechanism is in place.
11. Never operate electric equipment with a frayed cord or without a ground wire.
12. Never direct a blast of compressed air against human flesh.
13. Always wear eye protection and protective gloves when cleaning parts in any type of cleaning solution.
14. Electronic fuel injection (EFI) systems must be depressurized before connecting a fuel gauge or disconnecting fuel system components.
15. On a port fuel injection (PFI) system, the pressure gauge is connected to the Schrader valve on the fuel rail.
16. On throttle body injection (TBI) or carburetor systems, the fuel pressure gauge must be connected in series at the fuel inlet line.
17. The injector balance tester is a timing device that opens each injector for a specific length of time.
18. During an injector balance test, the fuel system pressure drop is recorded when each injector is opened for a specific length of time.
19. Injectors are cleaned with a solution of unleaded gasoline and injector cleaner in a pressurized container connected to the Schrader valve on the fuel rail.
20. During the injector cleaning procedure, the return fuel line must be blocked, and the electric fuel pump must be disabled.

21. Conventional 12-V or self-powered test lights should not be used to test computer circuits. High-impedance test lights should be used for this purpose.
22. Test lights of any type must not be used to diagnose air bag circuits. Only the vehicle manufacturer's recommended tools should be used on these systems.
23. Volt-ampere testers are commonly used to perform battery load tests, starter current draw tests, alternator output tests, and alternator normal system voltage tests.
24. Multimeters read ac volts, dc volts, milliamperes, amperes, and ohms on various scales.
25. Digital multimeters have higher impedance than analog multimeters.
26. Only high-impedance digital multimeters should be used to test computer system components such as oxygen (O_2) sensors.
27. Tach-dwellmeters read engine rpm and ignition dwell.
28. A piece of reflective tape attached to a rotating component, such as the crankshaft pulley, provides a signal for a photoelectric tachometer.
29. The probe on a magnetic probe-type digital tachometer is installed in the magnetic timing receptacle above the crankshaft pulley.
30. The control knob on an advance-type timing light may be used to check spark advance.
31. When a magnetic probe-type digital timing meter is used, the offset on the tester must be adjusted for the engine being tested.
32. Many ignition module testers check the ignition module and primary ignition circuit voltage and current under various engine operating conditions.
33. An oscilloscope contains a cathode ray tube (CRT), which provides a voltage trace of the ignition system voltage much like a very fast reacting voltmeter.
34. Scan testers retrieve fault codes from the computer memory and perform many other diagnostic functions.
35. The scan tester must be programmed for the vehicle make, model year, and type of engine.
36. Since most scan testers have the capability to test various computer systems on the vehicle, the tester must be programmed for the computer system being tested.
37. The scan tester must be connected to the appropriate diagnostic connector on the vehicle.
38. A two-gas emissions analyzer reads carbon monoxide (CO) and hydrocarbon (HC) emission levels.
39. A four-gas emissions analyzer reads the same emission levels as a two-gas analyzer, plus oxygen (O_2) and carbon dioxide (CO_2). These two latter emissions are not affected by the catalytic converter.
40. Emissions analyzers require a 15-minute warm-up and calibration period when they are turned on.
41. An engine analyzer contains a combination of many different automotive testers.

CASE STUDY

A technician raised a vehicle on a lift to perform an oil and filter change including a chassis lubrication on a Grand Marquis. This lift was a twin post-type with separate front and rear lift posts. On this type of lift, the rear wheels must be positioned in depressions in the floor to position the rear axle above the rear lift arm. Then the front lift post and arms must be moved forward or rearward to position the front lift arm under the front suspension. The front lift arms must also be moved inward or outward so they are lifting on the vehicle manufacturer's specified lift points.

The technician carefully positioned the front lift post and arms properly, but forgot to check the position of the rear tires in the floor depressions. The car was raised on the lift, and the technician proceeded with the service work. Suddenly there was a loud thump and the rear of the car bounced up and down! The rear lift arms were positioned against the floor of the trunk rather than on the rear axle, and the lift arms punched through the floor of the trunk, narrowly missing the fuel tank. The technician was extremely fortunate the car did not fall off the lift, resulting in severe damage. If the rear lift arms had punctured the fuel tank, a disastrous fire could have occurred!. Luckily, these things did not happen.

The technician learned a very important lesson about lift operation. Always follow all the recommended procedures in the lift operator's manual! The trunk floor was repaired at no cost to the customer, and fortunately the shop and the vehicle escaped without major damage.

Terms to Know

Advance-type timing light
Analog meter
Belt tension gauge
Blowgun
Canister-type pressurized injector cleaning container
Compression gauge
Coolant hydrometer
Cooling system pressure tester
Digital meter
Engine analyzer
Engine lift
Feeler gauge
Floor jack
Four-gas emissions analyzer
Hand press
Hydraulic press
Ignition module tester
Injector balance tester
International System (SI)
Jack stand
Lift
Magnetic probe-type digital tachometer
Magnetic probe-type digital timing meter
Meter impedance
Muffler chisel
Oil pressure gauge
Oscilloscope
Photoelectric tachometer
Pipe expander
Pressurized injector cleaning container
Scan tester
Self-powered test light
Stethoscope
Tach-dwellmeter
Thermostat tester
Two-gas emissions analyzer
United States customary (USC)
Vacuum pressure gauge
Volt-ampere tester

ASE Style Review Questions

1. While discussing measurements in the metric system:
   Technician A says one decimeter is equal to 1/10 of a meter.
   Technician B says one decimeter is equal to 1/100 of a meter.
   Who is correct?
   A. A only   **C.** Both A and B
   B. B only   **D.** Neither A nor B

2. While discussing personal safety:
   Technician A says rings and jewelry may be worn in the automotive shop.
   Technician B says some electric-drive cooling fans may start turning at any time.
   Who is correct?
   A. A only   **C.** Both A and B
   B. B only   **D.** Neither A nor B

3. While discussing the proper way to lift heavy objects:
   Technician A says you should bend your back to pick up a heavy object.
   Technician B says you should bend your knees to pick up a heavy object.
   Who is correct?
   A. A only **C.** Both A and B
   B. B only **D.** Neither A nor B

4. While discussing vehicle lifts:
   Technician A says a four-wheel-drive vehicle may be lifted with a frame contact lift.
   Technician B says a vehicle should not be raised on a lift with people in the vehicle.
   Who is correct?
   A. A only **C.** Both A and B
   B. B only **D.** Neither A nor B

5. While discussing electric fuel pump pressure testing:
   Technician A says the fuel pressure gauge should be connected in series at the fuel inlet line on throttle body injection (TBI) systems.
   Technician B says the Schrader valves on the fuel rails in various fuel systems are all the same size.
   Who is correct?
   A. A only **C.** Both A and B
   B. B only **D.** Neither A nor B

6. While discussing an injector balance test:
   Technician A says the differences between the pressured drops on the injectors should not exceed 10 psi (69 kPa).
   Technician B says if there is excessive pressure drop when an injector is energized, the injector orifice is restricted.
   Who is correct?
   A. A only **C.** Both A and B
   B. B only **D.** Neither A nor B

7. While discussing injector cleaning:
   Technician A says the fuel return line must be blocked during the injector cleaning process.
   Technician B says the electric fuel pump should be disabled during the injector cleaning process.
   Who is correct?
   A. A only **C.** Both A and B
   B. B only **D.** Neither A nor B

8. While discussing circuit testers:
   Technician A says a conventional 12-V test light may be used to diagnose automotive computer circuits.
   Technician B says a self-powered test light may be used to diagnose an air bag circuit.
   Who is correct?
   A. A only **C.** Both A and B
   B. B only **D.** Neither A nor B

9. While discussing multimeters:
   Technician A says an analog multimeter should be used to test the oxygen (O_2) sensor voltage.
   Technician B says a high-impedance digital multimeter should be used to test the O_2 sensor voltage.
   Who is correct?
   A. A only **C.** Both A and B
   B. B only **D.** Neither A nor B

10. While discussing oscilloscopes:
    Technician A says the upward voltage traces on an oscilloscope screen indicate a specific length of time.
    Technician B says the cathode ray tube (CRT) in an oscilloscope is like a very fast reacting voltmeter.
    Who is correct?
    A. A only **C.** Both A and B
    B. B only **D.** Neither A nor B

Input Sensor Diagnosis and Service

CHAPTER 3

Upon completion and review of this chapter, you should be able to:

❏ Diagnose computer voltage supply wires.

❏ Diagnose computer ground wires.

❏ Test and diagnose oxygen sensors.

❏ Test and diagnose engine coolant temperature sensors.

❏ Test and diagnose air charge temperature sensors.

❏ Test, diagnose, and adjust throttle position sensors.

❏ Test and diagnose different types of manifold absolute pressure sensors.

❏ Test and diagnose various types of mass air flow sensors.

❏ Test and diagnose knock sensors.

❏ Test and diagnose exhaust gas recirculation valve position sensors.

❏ Test and diagnose vehicle speed sensors.

❏ Test and diagnose park/neutral switches.

Diagnosis of Computer Voltage Supply and Ground Wires

Computer Voltage Supply Wire Diagnosis

 SERVICE TIP: Never replace a computer unless the ground wires and voltage supply wires are proven to be in satisfactory condition.

WARNING: Always observe the correct meter polarity when connecting test meters. Since nearly all cars and light trucks have negative battery ground, the negative meter lead is connected to ground for many tests. Connecting a meter with incorrect polarity will damage the meter.

A computer will never operate properly unless it has proper ground connections and satisfactory voltage supply at the required terminals. A computer wiring diagram for the vehicle being tested must be available for these tests. Backprobe the BATT terminal on the computer and connect a pair of digital voltmeter leads from this terminal to ground (Figure 3-1). Always ground the black meter lead.

The voltage at this terminal should be 12 V with the ignition switch off. If 12 V are not available at this terminal, check the computer fuse and related circuit. Turn on the ignition switch and connect the red voltmeter lead to the +B and +B1 terminals with the black lead still grounded. The voltage measured at these terminals should be 12 V with the ignition switch on. When the specified voltage

Basic Tools

Basic technician's tool set
Service manual
Electrical probes
Propane torch
Clear, heat-resistant water container

Classroom Manual
Chapter 3, page 39

Special Tools

Digital multimeter with frequency capabilities

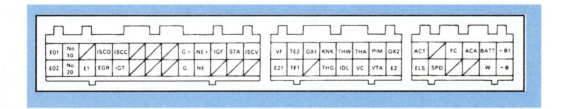

Figure 3-1 Typical computer terminals (Courtesy of Toyota Motor Corporation)

49

is not available, test the voltage supply wires to these terminals. These terminals may be connected through fuses, fuse links, or relays. Always refer to the vehicle manufacturer's wiring diagram for the vehicle being tested.

Computer Ground Wire Diagnosis

SERVICE TIP: When diagnosing computer problems, it is usually helpful to ask the customer about service work that has been performed lately on the vehicle. If service work has been performed in the engine compartment, it is possible that a computer ground wire may be loose or disconnected.

Computer ground wires usually extend from the computer to a ground connection on the engine or battery. With the ignition switch on, connect a pair of digital voltmeter leads from the E1 and E2 computer terminals to ground. The voltage drop across the ground wires should be 0.2 V or less. If the voltage reading is more than specified, repair the ground wires.

Input Sensor Diagnosis and Service

Oxygen Sensor Diagnosis

Voltage Signal Diagnosis. The engine must be at normal operating temperature before the oxygen (O_2) sensor is tested. Always follow the test procedure in the vehicle manufacturer's service manual, and use the specifications supplied by the manufacturer.

WARNING: An oxygen sensor must be tested with a digital voltmeter. If an analog meter is used for this purpose, the sensor may be damaged.

SERVICE TIP: A contaminated oxygen sensor may provide a continually high voltage reading because the oxygen in the exhaust stream does not contact the sensor.

SERVICE TIP: If the insulation on a wire must be probed to connect a meter to the wire, place a small amount of silicon sealant over the wire puncture after the test procedure.

A digital voltmeter is connected from the O_2 sensor wire to ground to test this sensor (Figure 3-2). Use an electric probe to backprobe the connector near the O_2 sensor to connect the voltmeter to the sensor signal wire. If possible, avoid probing the insulation to connect a meter to a wire. With the engine operating at normal temperature and idle speed, if the air-fuel ratio is varying slightly lean and rich from stoichiometric, the O_2 sensor voltage should be cycling from low voltage to high voltage. A typical O_2 sensor cycles from 0.3 V to 0.8 V.

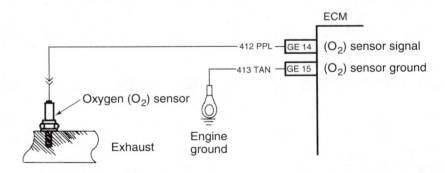

Figure 3-2 Oxygen (O_2) sensor wiring to the computer (Courtesy of Oldsmobile Division, General Motors Corporation)

WARNING: When working on an engine with an O_2 sensor, always use the room temperature vulcanizing (RTV) sealant recommended by the vehicle manufacturer. The use of other RTV sealants may contaminate the O_2 sensor.

If the voltage is continually high, the air-fuel ratio may be rich, or the sensor may be contaminated. The O_2 sensor may be contaminated with room temperature vulcanizing (RTV) sealant, antifreeze, or lead from leaded gasoline.

When the O_2 sensor voltage is continually low, the air-fuel ratio may be lean, the sensor may be defective, or the wire between the sensor and the computer may have a high-resistance problem. If the O_2 sensor voltage signal remains in a mid-range position, the computer may be in open loop or the sensor may be defective.

When the O_2 sensor is removed from the engine, a digital voltmeter may be connected from the signal wire to the sensor case, and the sensor element may be heated in the flame from a propane torch. The propane flame keeps the oxygen in the air away from the sensor element, and this causes the sensor to produce voltage. While the sensor element is in the flame, the voltage should be nearly 1 V, and the voltage should drop to zero immediately when the flame is removed from the sensor. If the sensor does not produce the specified voltage, it should be replaced.

Oxygen Sensor Wiring Diagnosis. If a defect in the O_2 sensor signal wire is suspected, backprobe the sensor signal wire at the computer and connect a digital voltmeter from the signal wire to ground with the engine idling. The difference between the voltage readings at the sensor and at the computer should not exceed the vehicle manufacturer's specifications. A typical specification for voltage drop across the average sensor wire is 0.2 V.

With the engine idling, connect a digital voltmeter from the sensor case to the sensor ground wire on the computer. A typical maximum voltage drop reading across the sensor ground circuit is 0.2 V. Always use the vehicle manufacturer's specifications. If the voltage drop across the sensor ground exceeds specifications, repair the ground wire or the sensor ground in the exhaust manifold.

Oxygen Sensor Heater Diagnosis. If the O_2 sensor heater is not working, the sensor warm-up time is extended, and the computer stays in open loop longer. In this mode, the computer supplies a richer air-fuel ratio, and the fuel economy is reduced. Disconnect the O_2 sensor connector, and connect a digital voltmeter from the heater voltage supply wire and ground. With the ignition switch on, 12 V should be supplied on this wire. If the voltage is less than 12 V, repair the fuse in this voltage supply wire or the wire itself.

With the O_2 sensor wire disconnected, connect an ohmmeter across the heater terminals in the sensor connector (Figure 3-3). If the heater does not have the specified resistance, replace the sensor.

Engine Coolant Temperature (ECT) Ohmmeter Diagnosis

WARNING: Never apply an open flame to an engine coolant temperature (ECT) sensor or air charge temperature (ACT) sensor for test purposes. This action will damage the sensor.

Open loop occurs during engine warm-up. In this mode the computer program controls air-fuel ratio while ignoring the O_2 sensor signal.

Closed loop occurs when the engine is at, or near, normal operating temperature. In this mode, the computer uses the O_2 sensor signal to control the air-fuel ratio.

Figure 3-3 Heater terminals in the (O_2) sensor (Courtesy of Honda Motor Co., Ltd.)

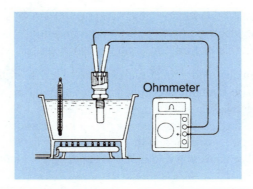

Figure 3-4 Engine coolant temperature (ECT) test connections with sensor removed and placed in hot water (Courtesy of Toyota Motor Corporation)

A defective ECT sensor may cause some of the following problems:

1. Hard engine starting
2. Rich or lean air-fuel ratio
3. Improper operation of emission devices
4. Reduced fuel economy
5. Improper converter clutch lockup
6. Hesitation on acceleration
7. Engine stalling

Special Tools

Thermometer

The ECT sensor may be removed and placed in a container of water with an ohmmeter connected across the sensor terminals (Figure 3-4). A thermometer is also placed in the water. When the water is heated, the sensor should have the specified resistance at any temperature (Figure 3-5). Always use the vehicle manufacturer's specifications. If the sensor does not have the specified resistance, replace the sensor.

Engine Coolant Temperature Sensor Wiring Diagnosis

WARNING: Before disconnecting any computer system component, be sure the ignition switch is turned off. Disconnecting components may cause high induced voltages and computer damage.

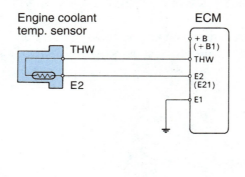
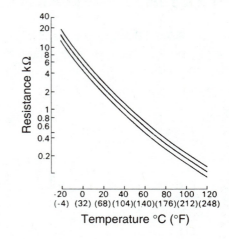

Figure 3-5 Engine coolant temperature (ECT) sensor wiring and specifications (Courtesy of Toyota Motor Corporation)

With the wiring connectors disconnected from the ECT sensor and the computer, connect an ohmmeter from each sensor terminal to the computer terminal to which the wire is connected. Both sensor wires should indicate less resistance than specified by the vehicle manufacturer. If the wires have higher resistance than specified, the wires or wiring connectors must be repaired.

Engine Coolant Temperature Sensor Voltmeter Diagnosis

With the sensor installed in the engine, the sensor terminals may be back-probed to connect a digital voltmeter to the sensor terminals. The sensor should provide the specified voltage drop at any coolant temperature (Figure 3-6).

Some computers have internal resistors connected in series with the ECT sensor. The computer switches these resistors at approximately 120°F (49°C). This resistance change inside the computer causes a significant change in voltage drop across the sensor as indicated in the specifications. This is a normal condition on any computer with this feature. This change in voltage drop is always evident in the vehicle manufacturer's specifications.

Air Charge Temperature Sensor Diagnosis

The results of a defective air charge temperature (ACT) sensor may vary depending on the vehicle make and year. A defective air charge temperature sensor may cause the following problems:

1. Rich or lean air-fuel ratio
2. Hard engine starting
3. Engine stalling or surging
4. Acceleration stumbles
5. Excessive fuel consumption

COLD CURVE 10,000-OHM RESISTOR USED		HOT CURVE CALCULATED RESISTANCE OF 909 OHMS USED	
−20°F	4.70 V	110°F	4.20 V
−10°F	4.57 V	120°F	4.00 V
0°F	4.45 V	130°F	3.77 V
10°F	4.30 V	140°F	3.60 V
20°F	4.10 V	150°F	3.40 V
30°F	3.90 V	160°F	3.20 V
40°F	3.60 V	170°F	3.02 V
50°F	3.30 V	180°F	2.80 V
60°F	3.00 V	190°F	2.60 V
70°F	2.75 V	200°F	2.40 V
80°F	2.44 V	210°F	2.20 V
90°F	2.15 V	220°F	2.00 V
100°F	1.83 V	230°F	1.80 V
110°F	1.57 V	240°F	1.62 V
120°F	1.25 V	250°F	1.45 V

Figure 3-6 Voltage drop specifications for the engine coolant temperature (ECT) sensor (Courtesy of Chrysler Corporation)

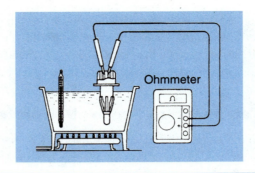

Figure 3-7 Air charge temperature (ACT) sensor placed in a container of water with an ohmmeter connected across the sensor terminals (Courtesy of Toyota Motor Corporation)

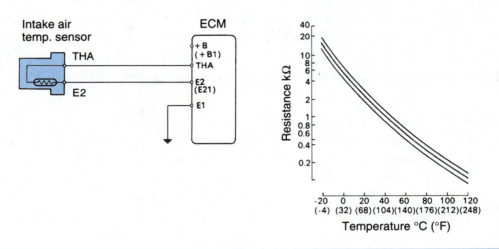

Figure 3-8 Air charge temperature (ACT) sensor wiring and resistance specifications (Courtesy of Toyota Motor Corporation)

The ACT sensor may be removed from the engine and placed in a container of water with a thermometer (Figure 3-7). When a pair of ohmmeter leads is connected to the sensor terminals and the water in the container is heated, the sensor should have the specified resistance at any temperature (Figure 3-8). If the sensor does not have the specified resistance, sensor replacement is required.

With the ACT sensor installed in the engine, the sensor terminals may be backprobed and a voltmeter connected across the sensor terminals. The sensor should have the specified voltage drop at any temperature (Figure 3-9). The wires between the air charge temperature sensor and the computer may be tested in the same way as the ECT wires.

Throttle Position Sensor Diagnosis, Three-Wire Sensor

A defective throttle position sensor (TPS) may cause acceleration stumbles, engine stalling, and improper idle speed. Backprobe the sensor terminals to complete the meter connections. With the ignition switch on, connect a voltmeter from the 5-V reference wire to ground (Figure 3-10). The voltage reading on this wire should be approximately 5 V. Always refer to the vehicle manufacturer's specifications.

If the reference wire is not supplying the specified voltage, check the voltage on this wire at the computer terminal. If the voltage is within specifications at the computer but low at the sensor, repair the 5-V reference wire. When this voltage is low at the computer, check the voltage supply wires and ground wires on the computer. If these wires are satisfactory, replace the computer.

Special Tools

Analog voltmeter

CHARGED TEMPERATURE SENSOR TEMPERATURE VS. VOLTAGE CURVE	
Temperature	Voltage
−20°F	4.81 V
0°F	4.70 V
20°F	4.47 V
40°F	4.11 V
60°F	3.67 V
80°F	3.08 V
100°F	2.51 V
120°F	1.97 V
140°F	1.52 V
160°F	1.15 V
180°F	0.86 V
200°F	0.65 V
220°F	0.48 V
240°F	0.35 V
260°F	0.28 V

Figure 3-9 Air charge temperature (ACT) sensor voltage drop specifications (Courtesy of Chrysler Corporation)

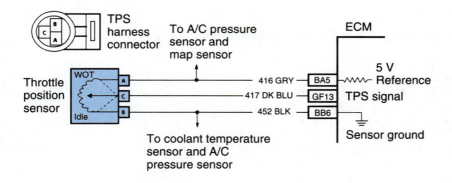

Figure 3-10 Throttle position sensor (TPS) and related wiring (Courtesy of Oldsmobile Division, General Motors Corporation)

With the ignition switch on, connect the voltmeter from the sensor ground wire to the battery ground. If the voltage drop across this circuit exceeds specifications, repair the ground wire from the sensor to the computer.

SERVICE TIP: When testing the throttle position sensor voltage signal, use an analog voltmeter, because the gradual voltage increase on this wire is quite visible on the meter pointer. If the sensor voltage increase is erratic, the voltmeter pointer fluctuates.

SERVICE TIP: When the throttle is opened gradually to check the throttle position sensor voltage signal, tap the sensor lightly and watch for fluctuations on the voltmeter pointer indicating a defective sensor.

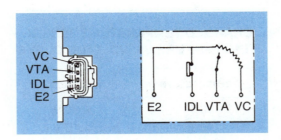

Figure 3-11 Four-wire throttle position sensor (TPS) with idle switch (Courtesy of Toyota Motor Corporation)

With the ignition switch on, connect a voltmeter from the sensor signal wire to ground. Slowly open the throttle and observe the voltmeter. The voltmeter reading should increase smoothly and gradually. Typical TPS voltage readings are 0.5 V to 1 V with the throttle in the idle position, and 4 V to 5 V at wide-open throttle. Always refer to the vehicle manufacturer's specifications. If the TPS does not have the specified voltage or if the voltage signal is erratic, replace the sensor.

Throttle Position Sensor Diagnosis, Four-Wire Sensor

Some TPSs contain an idle switch that is connected to the computer. These sensors have the same wires as a three-wire TPS, and an extra wire for the idle switch (Figure 3-11).

The four-wire TPS is tested with an ohmmeter connected from the sensor ground terminal to each of the other terminals (Figure 3-12). A specified feeler gauge must be placed between the throttle lever and the stop for some of the ohmmeter tests. When the ohmmeter is connected from the ground (E2) terminal to the VTA terminal, the throttle must be held in the wide-open position (Figure 3-13).

Throttle Position Sensor Adjustment

A TPS adjustment may be performed on some vehicles, but this adjustment is not possible on other applications. Check the vehicle manufacturer's service manual for the TPS adjustment procedure. An improper TPS adjustment may cause inaccurate idle speed, engine stalling, and acceleration stumbles. Follow these steps for a typical TPS adjustment:

1. Backprobe the TPS signal wire and connect a voltmeter from this wire to ground.
2. Turn on the ignition switch and observe the voltmeter reading with the throttle in the idle position.

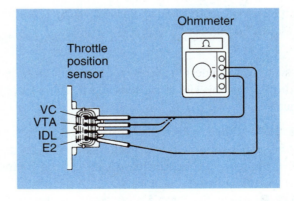

Figure 3-12 Ohmmeter test connections from the ground terminal to each of the other terminals on the throttle position sensor (TPS) (Courtesy of Toyota Motor Corporation)

CLEARANCE BETWEEN LEVER AND STOP SCREW	BETWEEN TERMINALS	RESISTANCE
0 mm (0 in.)	VTA – E2	0.28 – 6.4 kΩ
0.35 mm (0.014 in.)	IDL – E2	0.5 kΩ or less
0.70 mm (0.028 in.)	IDL – E2	Infinity
Throttle valve fully open	VTA – E2	2.0 – 11.6 kΩ
–	VC – E2	2.7 – 7.7 kΩ

Figure 3-13 Specifications for throttle position sensor (TPS) ohmmeter tests (Courtesy of Toyota Motor Corporation)

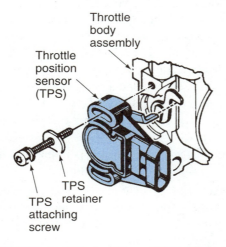

Figure 3-14 Throttle position sensor (TPS) with elongated slots for sensor adjustment (Courtesy of Chevrolet Motor Division, General Motors Corporation)

3. If the TPS does not provide the specified signal voltage, loosen the TPS mounting bolts and rotate the sensor housing until the specified voltage is indicated on the voltmeter (Figure 3-14).
4. Hold the sensor in this position and tighten the mounting bolts to the specified torque.

Manifold Absolute Pressure (MAP) Sensor Diagnosis

Barometric Pressure Voltage Signal Diagnosis. A defective manifold absolute pressure sensor may cause a rich or lean air-fuel ratio, excessive fuel consumption, and engine surging. This diagnosis applies to MAP sensors that produce an analog voltage signal. With the ignition switch on, backprobe the 5-V reference wire and connect a voltmeter from the 5-V reference wire to ground (Figure 3-15).

SERVICE TIP: Manifold absolute pressure sensors have a much different calibration on turbocharged engines compared to nonturbocharged engines. Be sure you are using the proper specifications for the sensor being tested.

If the reference wire is not supplying the specified voltage, check the voltage on this wire at the computer. If the voltage is within specifications at the computer, but low at the sensor, repair

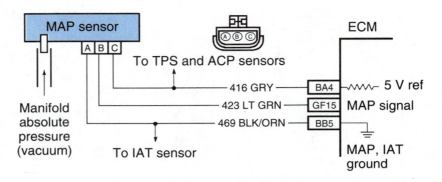

Figure 3-15 Manifold absolute pressure (MAP) sensor and connecting wires (Courtesy of Oldsmobile Division, General Motors Corporation)

the 5-V reference wire. When this voltage is low at the computer, check the voltage supply wires and ground wires on the computer. If these wires are satisfactory, replace the computer.

With the ignition switch on, connect the voltmeter from the sensor ground wire to the battery ground. If the voltage drop across this circuit exceeds specifications, repair the ground wire from the sensor to the computer.

Backprobe the MAP sensor signal wire and connect a voltmeter from this wire to ground with the ignition switch on. The voltage reading indicates the barometric pressure signal from the MAP sensor to the computer. Many MAP sensors send a barometric pressure signal to the computer each time the ignition switch is turned on and each time the throttle is in the wide-open position. If the voltage supplied by the barometric pressure signal in the MAP sensor does not equal the vehicle manufacturer's specifications, replace the MAP sensor.

The barometric pressure voltage signal varies depending on altitude and atmospheric conditions. Follow this calculation to obtain an accurate barometric pressure reading:

1. Phone your local weather or TV station and obtain the present barometric pressure reading; e.g., 29.85 inches. The pressure they quote is usually corrected to sea level.
2. Multiply your altitude by 0.001; e.g., 600 feet × 0.001 = 0.6.
3. Subtract the altitude correction from the present barometric pressure reading; e.g., 29.85 − 0.6 = 29.79.

Check the vehicle manufacturer's specifications to obtain the proper barometric pressure voltage signal in relation to the present barometric pressure (Figure 3-16).

Special Tools
Vacuum hand pump

Manifold Absolute Pressure Sensor Voltage Signal Diagnosis. Leave the ignition switch on and the voltmeter connected to the MAP sensor signal wire. Connect a vacuum hand pump to the MAP sensor vacuum connection and apply 5 inches of vacuum to the sensor. On some MAP sensors, the sensor voltage signal should change 0.7 V to 1.0 V for every 5 inches of vacuum change applied to the sensor. Always use the vehicle manufacturer's specifications. If the barometric pressure voltage signal was 4.5 V, with 5 inches of vacuum applied to the MAP sensor, the voltage should be 3.5 V to 3.8 V. When 10 inches of vacuum are applied to the sensor, the voltage signal should be 2.5 V to 3.1 V. Check the MAP sensor voltage at 5-inch intervals from 0 to 25 inches. If the MAP sensor voltage is not within specifications at any vacuum, replace the sensor.

Special Tools
MAP sensor tester

Diagnosis of Manifold Absolute Pressure Sensor with Voltage Frequency Signal. If the MAP sensor produces a digital voltage signal of varying frequency, check the 5-V reference wire and the ground wire with the same procedure used on other MAP sensors. This sensor diagnosis is based on the use of a MAP sensor tester that changes the MAP sensor varying frequency voltage to an analog voltage. Follow these steps to test the MAP sensor voltage signal:

1. Turn off the ignition switch, and disconnect the wiring connector from the MAP sensor.
2. Connect the connector on the MAP sensor tester to the MAP sensor (Figure 3-17).
3. Connect the MAP sensor tester battery leads to a 12-V battery.

Absolute Baro Reading	Lowest Allowable Voltage at −40°F	Lowest Allowable Voltage at 257°F	Lowest Allowable Voltage at 77°F	TBI MAP Sensor Designed Output Voltage	Highest Allowable Voltage at 77°F	Highest Allowable Voltage at 257°F	Highest Allowable Voltage at −40°F
31.0″	4.548 V	4.632 V	4.716 V	4.800 V	4.884 V	4.968 V	5.052 V
30.9″	4.531 V	4.615 V	4.699 V	4.783 V	4.867 V	4.951 V	5.035 V
30.8″	4.514 V	4.598 V	4.682 V	4.766 V	4.850 V	4.934 V	5.018 V
30.7″	4.497 V	4.581 V	4.665 V	4.749 V	4.833 V	4.917 V	5.001 V
30.6″	4.480 V	4.564 V	4.648 V	4.732 V	4.816 V	4.900 V	4.984 V
30.5″	4.463 V	4.547 V	4.631 V	4.715 V	4.799 V	4.883 V	4.967 V
30.4″	4.446 V	4.530 V	4.614 V	4.698 V	4.782 V	4.866 V	4.950 V
30.3″	4.430 V	4.514 V	4.598 V	4.682 V	4.766 V	4.850 V	4.934 V
30.2″	4.413 V	4.497 V	4.581 V	4.665 V	4.749 V	4.833 V	4.917 V
30.1″	4.396 V	4.480 V	4.564 V	4.648 V	4.732 V	4.816 V	4.900 V
30.0″	4.379 V	4.463 V	4.547 V	4.631 V	4.715 V	4.799 V	4.883 V

Figure 3-16 Barometric pressure voltage signal specifications at various barometric pressures (Courtesy of Chrysler Corporation)

Figure 3-17 Manifold absolute pressure (MAP) sensor tester (Courtesy of Thexton Manufacturing Company)

4. Connect a pair of digital voltmeter leads to the MAP tester signal wire and ground.
5. Turn on the ignition switch and observe the barometric pressure voltage signal on the meter. If this voltage signal does not equal the manufacturer's specifications, replace the sensor.
6. Supply the specified vacuum to the MAP sensor with a hand vacuum pump.
7. Observe the voltmeter reading at each specified vacuum. If the MAP sensor voltage signal does not equal the manufacturer's specifications at any vacuum, replace the sensor.

Photo Sequence 1 shows a typical procedure for testing a Ford MAP sensor.

Photo Sequence 1
Typical Procedure for Testing a Ford MAP Sensor

P1-1 Remove the MAP sensor wiring connector and vacuum hose.

P1-2 Connect the MAP sensor tester to the MAP sensor.

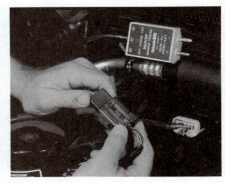

P1-3 Connect the digital voltmeter leads from the proper MAP sensor tester lead to ground.

P1-4 Connect the MAP sensor tester leads to the battery terminals with the proper polarity.

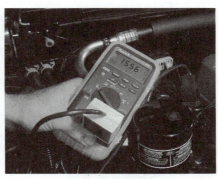

P1-5 Observe the MAP sensor barometric pressure (Baro) voltage reading on the voltmeter and compare this reading to specifications.

P1-6 Connect a vacuum hand pump hose to the MAP sensor and apply 5 in. Hg. to the MAP sensor. Observe the MAP sensor voltage signal on the voltmeter. Compare this voltmeter reading to specifications.

P1-7 Apply 10 in. Hg. to the MAP sensor with the hand pump and observe the voltmeter reading. Compare this reading to specifications.

P1-8 Apply 15 in. Hg. to the MAP sensor with the hand pump and observe the voltmeter reading. Compare this reading to specifications.

P1-9 Apply 20 in. Hg. to the MAP sensor with the hand pump and observe the voltmeter reading. Compare this reading to specifications. If any of the MAP sensor readings do not meet specifications, replace the MAP sensor.

Mass Air Flow Sensor Diagnosis

Voltmeter Diagnosis of Vane-Type Mass Air Flow Sensor. Always check the voltage supply wire and the ground wire to the MAF module before checking the sensor voltage signal. Always follow the recommended test procedure in the manufacturer's service manual and use the specifications supplied by the manufacturer. The following procedure is based on the use of a Fluke multimeter. Follow these steps to measure the MAF sensor voltage signal:

1. Set the multimeter on the Vdc scale and connect the black meter lead to the COM terminal in the meter while the red meter lead is installed in the V/rpm meter connection.
2. Connect the red meter lead to the MAF signal wire with a special piercing probe, and connect the black meter lead to ground (Figure 3-18).
3. Turn on the ignition switch and press the min/max button to activate the min/max feature (Figure 3-19).

 WARNING: While pushing the mass air flow sensor vane open and closed, be careful not to mark or damage the vane or sensor housing.

4. Slowly push the MAF vane from the closed to the wide-open position, and allow the vane to slowly return to the closed position (Figure 3-20).

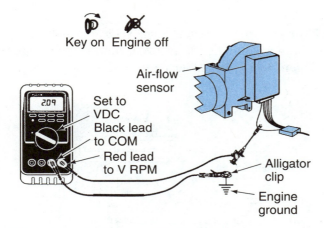

Figure 3-18 Voltmeter connected to measure mass air flow sensor (MAF) voltage signal (Courtesy of John Fluke Mfg. Co., Inc.)

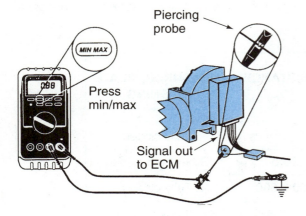

Figure 3-19 Press the min/max button to engage the min/max test mode. (Courtesy of John Fluke Mfg. Co., Inc.)

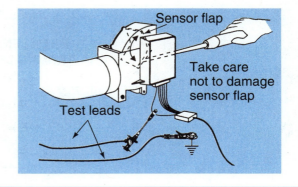

Figure 3-20 Push the mass air flow (MAF) sensor vane from the open to the closed position. (Courtesy of John Fluke Mfg. Co., Inc.)

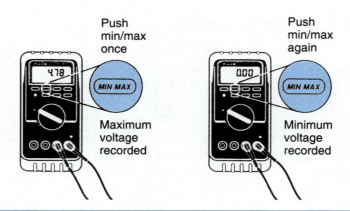

Figure 3-21 Press the min/max button to read minimum and maximum sensor voltage signals. (Courtesy of John Fluke Mfg. Co., Inc.)

5. Touch the min/max button once to read the maximum voltage signal recorded, and press this button again to read the minimum voltage signal (Figure 3-21). If the minimum voltage signal is zero, there may be an open circuit in the MAF sensor variable resistor. When the voltage signal is not within the manufacturer's specifications, replace the sensor.

Vane-Type Mass Air Flow Ohmmeter Tests. Some vehicle manufacturers specify ohmmeter tests for the MAF sensor. With the MAF sensor removed, connect the ohmmeter to the E2 and VS MAF sensor terminals (Figure 3-22). The resistance at these terminals should be 200 Ω to 600 Ω (Figure 3-23).

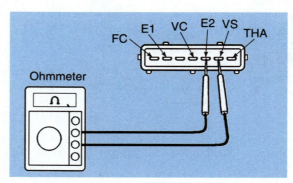

Figure 3-22 Ohmmeter connected to mass air flow (MAF) E2 and VS terminals (Courtesy of Toyota Motor Corporation)

BETWEEN TERMINALS	RESISTANCE (Ω)	MEASURING PLATE OPENING
FC – E1	Infinity	Fully closed
FC – E1	Zero	Other than closed
VS – E2	200 – 600	Fully closed
VS – E2	20 – 1,200	Fully open

Figure 3-23 Ohm specifications at mass air flow (MAF) sensor terminals (Courtesy of Toyota Motor Corporation)

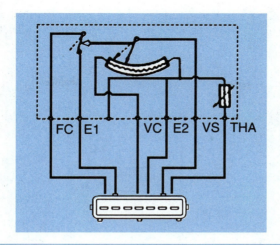

Figure 3-24 Mass air flow (MAF) sensor terminals and internal electric circuit (Courtesy of Toyota Motor Corporation)

Connect the ohmmeter leads to the other recommended MAF sensor terminals and record the resistance readings (Figure 3-24). Since the THA and E2 sensor terminals are connected internally to the thermistor, temperature affects the ohm readings at these terminals as indicated in the specifications. If the specified resistance is not available in any of the test connections, replace the MAF sensor.

Connect the ohmmeter leads to the specified MAF sensor terminals (Figure 3-25), and move the vane from the fully closed to the fully open position. With each specified meter connection and vane position, the ohmmeter should indicate the specified resistance (Figure 3-26). When the ohmmeter leads are connected to the E2 and VS sensor terminals, the ohm reading should increase smoothly as the sensor vane is opened and closed.

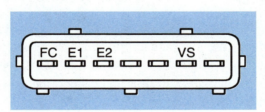

Figure 3-25 Mass air flow (MAF) sensor terminals for ohmmeter tests at specific vane positions (Courtesy of Toyota Motor Corporation)

BETWEEN TERMINALS	RESISTANCE (Ω)	TEMP. °C (°F)
VS – E2	200 – 600	—
VC – E2	200 – 400	—
THA – E2	10,000 – 20,000 4,000 – 7,000	-20 (-4) 0 (32)
FC – E1	Infinity	—

Figure 3-26 Mass air flow (MAF) sensor ohm specifications at various vane positions (Courtesy of Toyota Motor Corporation)

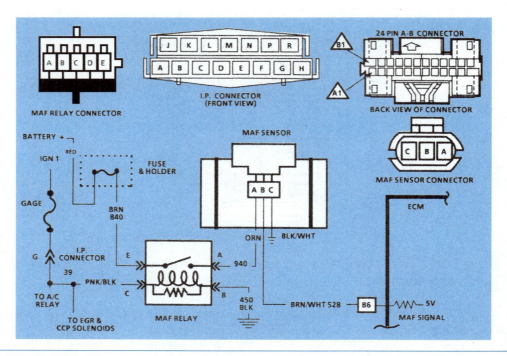

Figure 3-27 Mass air flow (MAF) sensor wiring diagram (Courtesy of Chevrolet Motor Division, General Motors Corporation)

Heated Resistor or Hot Wire Mass Air Flow Sensor Frequency Test. The test procedure for heated resistor and hot wire MAF sensors varies depending on the vehicle make and year. Always follow the test procedure in the vehicle manufacturer's service manual to determine if the MAF sensor produces a frequency signal or a dc voltage signal. Always test the MAF voltage supply and ground wires first.

A frequency test may be performed on some MAF sensors, such as the AC Delco MAF on some General Motors products. The following test procedure is based on the use of a Fluke multimeter. Follow these steps to check the MAF sensor voltage signal and frequency:

1. Place the multimeter on the V/rpm scale and connect the meter leads from the MAF voltage signal wire to the ground wire (Figure 3-27).
2. Start the engine and observe the voltmeter reading. On some MAF sensors, this reading should be 2.5 V. Always refer to the manufacturer's specifications.
3. Lightly tap the MAF sensor housing with a screwdriver handle and watch the voltmeter pointer. If the pointer fluctuates or the engine misfires, replace the MAF sensor. Some MAF sensors have experienced loose internal connections, which cause erratic voltage signals and engine misfiring and surging.
4. Be sure the meter dial is on dc volts, and press the rpm button three times so the meter displays voltage frequency. The meter should indicate about 30 hertz (Hz) with the engine idling.
5. Increase the engine speed, and record the meter reading at various speeds.
6. Graph the frequency readings. The MAF sensor frequency should increase smoothly and gradually in relation to engine speed. If the MAF sensor frequency reading is erratic, replace the sensor (Figure 3-28).

If the MAF sensor being tested produces a dc voltage signal, connect a digital voltmeter from the signal wire to ground. Test the MAF voltage signal at various engine speeds. If the MAF sensor voltage signal at any rpm does not match the specified voltage signal, replace the sensor. A scan tester displays the grams per second of air flow through MAF sensors that provide a frequency or dc voltage signal.

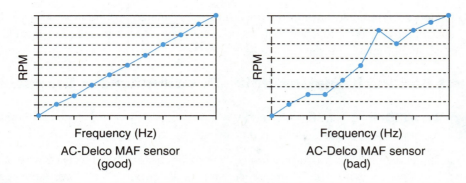

Figure 3-28 Satisfactory and unsatisfactory mass air flow (MAF) sensor frequency readings (Courtesy of John Fluke Mfg. Co., Inc.)

Knock Sensor Diagnosis

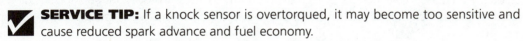

SERVICE TIP: If a knock sensor is overtorqued, it may become too sensitive and cause reduced spark advance and fuel economy.

SERVICE TIP: When the knock sensor torque is less than specified, the sensor may have reduced sensitivity, and engine detonation may occur.

SERVICE TIP: On many engines, the coolant must be drained prior to knock sensor removal.

A defective knock sensor may cause engine detonation or reduced spark advance and fuel economy. When a knock sensor is removed and replaced, the sensor torque is critical. The procedure for checking a knock sensor varies depending upon the vehicle make and year.

A quick test of the knock sensor may be performed with a timing light connected to the engine. Operate the engine at 2,000 rpm and observe the timing marks. Tap on the engine block near the knock sensor with a small hammer. If the knock sensor and PCM are operating properly, the spark advance should be reduced as indicated by the timing mark position.

Always follow the vehicle manufacturer's recommended test procedure and specifications. Follow these steps for a typical knock sensor diagnosis:

1. Disconnect the knock sensor wiring connector, and turn on the ignition switch.
2. Connect a voltmeter from the disconnected knock sensor wire to ground. The voltage should be 4 V to 6 V. If the specified voltage is not available at this wire, backprobe the knock sensor wire at the computer and read the voltage at this terminal (Figure 3-29). If the voltage is satisfactory at this terminal, repair the knock sensor wire. When the voltage is not within specifications at the computer terminal, replace the computer.

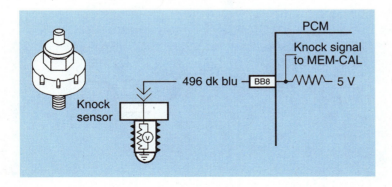

Figure 3-29 Knock sensor wiring diagram (Courtesy of Oldsmobile Division, General Motors Corporation)

65

3. Connect an ohmmeter from the knock sensor terminal to ground. Some knock sensors should have 3,300 Ω to 4,500 Ω. If the knock sensor does not have the specified ohms resistance, replace the sensor.

Exhaust Gas Recirculation Valve Position Sensor Diagnosis

CAUTION: The EGR valve and EVP sensor may be very hot if the engine has been running. Wear protective gloves if it is necessary to service these components.

Many exhaust gas recirculation valve position (EVP) sensors have a 5-V reference wire, a voltage signal wire, and a ground wire. The 5-V reference wire and the ground wire may be checked using the same procedure explained previously on TPS and MAP sensors. Connect a pair of voltmeter leads from the voltage signal wire to ground, and turn on the ignition switch. The voltage signal should be approximately 0.8 V. Connect a vacuum hand pump to the vacuum fitting on the EGR valve and slowly increase the vacuum from 0 to 20 in. Hg. The EVP sensor voltage signal should gradually increase to 4.5 V at 20 in. Hg. Always use the EVP test procedure and specifications supplied by the vehicle manufacturer. If the EVP sensor does not have the specified voltage, replace the sensor.

Vehicle Speed Sensor Diagnosis

A defective vehicle speed sensor may cause different problems depending on the computer output control functions. A defective vehicle speed sensor (VSS) may cause the following problems:

1. Improper converter clutch lockup
2. Improper cruise control operation
3. Inaccurate speedometer operation

Prior to VSS diagnosis, the vehicle should be lifted on a hoist so the drive wheels are free to rotate. Backprobe the VSS yellow wire, and connect a pair of voltmeter leads from this VSS wire to ground. Then start the engine (Figure 3-30).

Place the transaxle in drive and allow the drive wheels to rotate. If the VSS voltage signal is not 0.5 V or more, replace the sensor. When the VSS provides the specified voltage signal, backprobe the GD 14 PCM terminal and repeat the voltage signal test with the drive wheels rotating. If 0.5 V is available at this terminal, the trouble may be in the PCM.

When 0.5 V is not available at this terminal, turn off the ignition switch and disconnect the VSS terminal and the PCM terminals. Connect the ohmmeter leads from the 400 VSS terminal to the GD 14 PCM terminal. The meter should read 0 ohms. Repeat the test with the ohmmeter leads connected to the 401 VSS terminal and the GD 13 PCM terminal. This wire should also have 0 ohms resistance. If the resistance in these wires is more than specified, repair the wires.

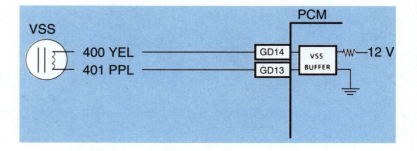

Figure 3-30 Vehicle speed sensor (VSS) wiring diagram (Courtesy of Oldsmobile Division, General Motors Corporation)

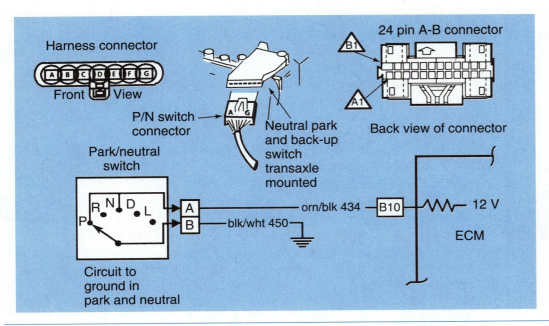

Figure 3-31 Park/neutral switch wiring diagram (Courtesy of Chevrolet Motor Division, General Motors Corporation)

Park/Neutral Switch Diagnosis

A defective park/neutral switch may cause improper idle speed and failure of the starting motor circuit. Always follow the test procedure in the vehicle manufacturer's service manual. Disconnect the park/neutral switch wiring connector and connect a pair of ohmmeter leads from the B terminal in the wiring connector to ground (Figure 3-31). If the ohmmeter does not indicate less than 0.5 Ω, repair the ground wire.

With the wiring harness connected to the switch, backprobe terminal A on the park/neutral switch and connect a pair of voltmeter leads from this terminal to ground. Turn on the ignition switch, and move the gear selector through all positions. The voltmeter should read over 5 V in all gear selector positions except neutral or park.

If the voltmeter does not indicate the specified voltage, backprobe the ECM terminal B10 and connect the meter from this terminal to ground. When the specified voltage is available at this terminal, repair the wire from the ECM to the park/neutral switch. If the specified voltage is not available, replace the ECM.

When the gear selector is placed in neutral or park, the voltmeter reading should be less than 0.5 V. If the specified voltage is not available in these gear selector positions, replace the park/neutral switch.

CUSTOMER CARE: Like everyone else, individuals involved in the automotive service industry do make mistakes. If you make a mistake that results in a customer complaint, always be willing to admit your mistake and correct it. Do not try to cover up the mistake or blame someone else. Customers are usually willing to live with an occasional mistake that is corrected quickly and efficiently.

A park/neutral switch may be referred to as a neutral drive switch.

Classroom Manual
Chapter 3, page 54

Guidelines for Input Sensor Diagnosis

1. Defective computer voltage supply wires or ground wires may cause defects in engine performance and economy. These wires must always be tested before the computer is replaced.

2. If the O_2 sensor voltage signal is higher than specified, the air-fuel ratio may be rich or the sensor may be contaminated.
3. When the O_2 sensor voltage signal is lower than specified, the air-fuel ratio may be lean or the sensor may be defective.
4. An O_2 sensor must be tested with a digital voltmeter.
5. A defective O_2 sensor heater may cause extended open loop time and reduced fuel economy.
6. An ECT sensor may be checked with an ohmmeter or a voltmeter.
7. Some computers switch an internal resistor in the ECT circuit at 120°F (49°C), and this action changes the voltage drop across the sensor.
8. A similar test procedure may be used for ECT and ACT sensors.
9. While checking the TPS voltage signal as the throttle is opened, the sensor should be tapped lightly to check for sensor defects.
10. Many four-wire TPSs contain an idle switch which informs the computer when the throttle is in the idle position.
11. On some applications, the TPS mounting bolts may be loosened and the sensor housing rotated to adjust the voltage signal with the throttle in the idle position.
12. If a MAP sensor provides a varying frequency signal, MAP sensor testers are available to change this frequency signal to an analog voltage.
13. On many MAP sensors, the barometric pressure voltage signal from the sensor should be checked with the ignition switch on.
14. The MAP sensor voltage signal should be checked at vacuum intervals of 5 in. Hg.
15. The MAF sensor voltage signal can be measured as the sensor vane is moved from the closed to the open position.
16. Some heated resistor-type or hot wire-type MAF sensors produce a frequency voltage signal that can be checked with a multimeter with frequency capabilities.
17. The EVP sensor voltage signal should change from about 0.8 V with the EGR valve closed to 4.5 V with the EGR valve wide open.
18. The VSS signal is used by the computer to control the torque converter clutch lockup, cruise control, and the electronic speedometer.
19. The voltage signal at the park/neutral switch should be low with the gear selector in park or neutral, and high in other gear selector positions.

CASE STUDY

A customer complained about hard starting on a Chevrolet Celebrity with a 2.5-L, 4-cylinder engine. The technician asked the customer when the hard starting problem occurred, and the customer replied that it happened after the car had been parked all night. The customer also revealed that the problem did not occur if the block heater was plugged in. In the cold climate where this customer lives, many vehicles have block heaters to provide easier starting in cold weather. Further questioning of the customer revealed there was evidence of black smoke from the tailpipe after a cold start. The technician informed the customer that a check of the computer input sensors would be required.

Since the computer supplies air-fuel ratio enrichment in response to the ECT signal in this system, the technician suspected a problem in this sensor. The technician checked the vehicle manufacturer's specifications for the vehicle and found an approximate ohm value for various sensor temperatures (Figure 3-32).

The technician tested the sensor and found the resistance of the sensor did vary about 200 Ω from the specifications at some temperatures. Since the specifications indicated the specified ohm values were approximate, the technician had to decide if this variation from specifications was enough to cause the problem.

COOLANT SENSOR TEMPERATURE TO RESISTANCE VALUES (APPROXIMATE)		
°F	°C	OHMS
210	100	185
160	70	450
100	38	1,800
70	20	3,400
40	4	7,500
20	−7	13,500
0	−18	25,000
−40	−40	100,700

Figure 3-32 Engine coolant temperature (ECT) sensor resistance specifications in relation to temperature (Courtesy of Chevrolet Motor Division, General Motors Corporation)

The technician obtained a new ECT sensor and compared the ohm reading on the new sensor to the reading on the old sensor. The new sensor had the specified resistance at various temperatures within 10 ohms. The new ECT sensor was installed, and a quick test was performed on the other sensors without finding any problems. Since the vehicle was needed for commuting purposes, the car was returned to the customer.

On the following day the service manager phoned the owner of the vehicle to find out the results of the sensor installation. The vehicle owner was very pleased that the car started easily and ran well when it was started with the engine cold.

Terms to Know

Air charge temperature (ACT) sensor
Barometric (Baro) pressure sensor
Closed loop
Engine coolant temperature (ECT) sensor
Exhaust gas recirculation valve position (EVP) sensor
Heated resistor-type MAF sensor

Hot wire-type MAF sensor
Knock sensor
Manifold absolute pressure (MAP) sensor
Mass air flow (MAF) sensor
Neutral/drive switch (NDS)
Open loop
Oxygen (O_2) sensor

Park/neutral switch
Reference voltage
Room temperature vulcanizing (RTV) sealant
Throttle position sensor (TPS)
Vane-type MAF sensor
Vehicle speed sensor (VSS)

ASE Style Review Questions

1. While discussing O_2 sensor diagnosis:
   Technician A says the voltage signal on a satisfactory O_2 sensor should always be cycling between 0.5 V and 1 V.
   Technician B says a contaminate O_2 sensor provides a continually low voltage signal.
   Who is correct?
   - **A.** A only
   - **B.** B only
   - **C.** Both A and B
   - **D.** Neither A nor B

2. While discussing ECT sensor diagnosis:
   Technician A says a defective ECT sensor may cause hard cold engine starting.
   Techncian B says a defective ECT sensor may cause improper operation of emission devices.
   Who is correct?
   - **A.** A only
   - **B.** B only
   - **C.** Both A and B
   - **D.** Neither A nor B

3. While discussing ECT sensor diagnosis:
   Technician A says the ECT sensor resistance should increase as the sensor temperature increases.
   Technician B says some computers have internal resistors connected in series with the ECT sensor, and the computer switches these resistors at 120°F (49°C).
   Who is correct?
   - **A.** A only
   - **B.** B only
   - **C.** Both A and B
   - **D.** Neither A nor B

4. While discussing ACT sensor diagnosis:
   Technician A says the ACT sensor resistance should decrease as the sensor temperature decreases.
   Technician B says the ACT sensor resistance should increase as the sensor temperature decreases.
   Who is correct?
   - **A.** A only
   - **B.** B only
   - **C.** Both A and B
   - **D.** Neither A nor B

5. While discussing TPS diagnosis:
   Technician A says the TPS voltage signal should increase smoothly from 1 V at idle to 6 V at wide open throttle.
   Technician B says a defective TPS may cause improper idle speed.
   Who is correct?
   - **A.** A only
   - **B.** B only
   - **C.** Both A and B
   - **D.** Neither A nor B

6. While discussing TPS diagnosis:
   Technician A says a four-wire TPS contains an idle switch.
   Technician B says in some applications the TPS mounting bolts may be loosened and the TPS housing rotated to adjust the voltage signal with the throttle in the idle position.
   Who is correct?
   - **A.** A only
   - **B.** B only
   - **C.** Both A and B
   - **D.** Neither A nor B

7. While discussing MAP sensor diagnosis:
   Technician A says with the ignition switch on, the MAP sensor produces a barometric pressure voltage signal.
   Technician B says the MAP sensor reference voltage and the ground wire should be checked with the ignition switch on.
   Who is correct?
   - **A.** A only
   - **B.** B only
   - **C.** Both A and B
   - **D.** Neither A nor B

8. While discussing MAP sensor diagnosis:
   Technician A says on some MAP sensors the voltage signal should increase 2 V for each 5 inches of vacuum applied to the sensor.
   Technician B says on some MAP sensors the voltage signal should decrease 0.7 V to 1 V for every 5 inches of vacuum increase applied to the sensor.
   Who is correct?
   - **A.** A only
   - **B.** B only
   - **C.** Both A and B
   - **D.** Neither A nor B

9. While discussing MAF sensor diagnosis:
   Technician A says on a vane-type MAF sensor, the voltage signal should be checked as the vane is moved from fully closed to fully open.
   Technician B says on a vane-type MAF sensor, the voltage signal should decrease as the vane is opened.
   Who is correct?
   A. A only
   B. B only
   C. Both A and B
   D. Neither A nor B

10. While discussing knock sensor service and diagnosis:
    Technician A says if the knock sensor torque is more than specified, the sensor has reduced sensitivity.
    Technician B says if a knock sensor torque is more than specified, the sensor may be too sensitive, resulting in reduced spark advance.
    Who is correct?
    A. A only
    B. B only
    C. Both A and B
    D. Neither A nor B

Table 3-1 ASE TASK

Inspect, test, adjust, and replace sensor and actuator components and circuits of electronic/computer-controlled systems.

Problem Area	Symptoms	Possible Causes	Classroom Manual	Shop Manual
ENGINE PERFORMANCE	Acceleration stumbles	1. Defective throttle position sensor (TPS)	58	54
		2. Improper throttle position sensor (TPS) adjustment	59	56
		3. Defective engine coolant temperature (ECT) sensor	57	51
		4. Defective air charge temperature (ACT) sensor	58	53
		5. Defective oxygen (O$_2$) sensor	54	50
	Improper idle speed, engine stalling	1. Defective engine coolant temperature (ECT) sensor	57	51
		2. Defective throttle position sensor (TPS)	58	54
		3. Improper throttle position semsor (TPS) adjustment	59	56
		4. Defective air charge temperature (ACT) sensor	58	53
	Engine surging	1. Intake manifold vacuum leak	59	57
		2. Defective manifold absolute pressure (MAP) sensor	59	57
		3. Defective mass air flow (MAF) sensor	60	61
	Detonation	1. Defective knock sensor	63	65
		2. Insufficient knock sensor torque	63	65
	Hard starting	Defective engine coolant temperature (ECT) sensor	57	51
FUEL ECONOMY	Reduced fuel mileage	1. Defective oxygen (O$_2$) sensor	54	50
		2. Defective engine coolant temperature (ECT) sensor	57	51
		3. Defective air charge temperature (ACT) sensor	58	53
		4. Defective manifold absolute pressure (MAP) sensor	59	57
		5. Defective mass air flow (MAF) sensor	60	61
		6. Defective knock sensor	63	65
		7. Excessive knock sensor torque	63	65

Ignition System Service and Diagnosis

CHAPTER 4

Upon completion and review of this chapter, you should be able to:

- ❏ Diagnose ignition system problems.
- ❏ Perform a no-start diagnosis, and determine the cause of the no-start condition.
- ❏ Diagnose primary circuit wiring.
- ❏ Remove, inspect, and service distributors.
- ❏ Inspect, service, and adjust ignition points.
- ❏ Install and time distributors to the engine.
- ❏ Inspect and test secondary ignition system wires.
- ❏ Inspect distributor caps and rotors.
- ❏ Inspect, service, and test ignition coils.
- ❏ Inspect, service, and test pickup coils.
- ❏ Inspect, service, and test ignition modules.
- ❏ Check and adjust ignition timing.
- ❏ Check ignition spark advance.
- ❏ Remove, service, and replace spark plugs.
- ❏ Perform pickup tests on distributor ignition (DI) systems.
- ❏ Perform tests on optical-type pickups.
- ❏ Perform no-start ignition tests on the cam and crankshaft sensors on electronic ignition (EI) systems.
- ❏ Perform no-start ignition tests on the coil and powertrain control module (PCM) on EI systems.
- ❏ Replace cam and crankshaft sensors on EI systems.
- ❏ Perform coil tests on EI systems.
- ❏ Adjust crankshaft sensors on EI systems.
- ❏ Perform no-start diagnoses on EI systems.
- ❏ Install and time the cam sensor on an EI system, 3.8-L turbocharged engine.
- ❏ Perform magnetic sensor tests on EI systems.
- ❏ Perform no-start tests on EI systems with magnetic sensors.
- ❏ Diagnose engine misfiring on EI-equipped engines.

Ignition System Diagnosis

No-Start Diagnosis

The following ignition defects may cause a no-start condition or hard starting:

1. Defective coil
2. Defective cap and rotor
3. Defective pickup coil
4. Open secondary coil wire
5. Low or zero primary voltage at the coil
6. Fouled spark plugs

Engine Misfiring Diagnosis

If engine misfiring occurs, check the following items:

1. Engine compression
2. Intake manifold vacuum leaks
3. High resistance in spark plug wires, coil secondary wire, or cap terminals
4. Electrical leakage in the distributor cap, rotor, plug wires, coil secondary wire, or coil tower
5. Defective coil
6. Defective spark plugs

Basic Tools

Basic technician's tool set
Service manual
12-V test light
Test spark plug
Jumper wires
Measuring tape

Ignition crossfiring occurs when the spark from one spark plug wire jumps across the distributor cap or spark plug wires and fires another spark plug.

Improperly routed spark plug wires may cause crossfiring and detonation.

73

7. Low primary voltage and current
8. Improperly routed spark plug wires
9. Worn distributor bushings

Power Loss

Check the following items to diagnosis a power loss condition:

1. Engine compression
2. Restricted exhaust or air intake
3. Late ignition timing
4. Insufficient spark advance
5. Cylinder misfiring

Engine Detonation, Spark Knock

If the engine detonates, check the following items:

1. Engine compression higher than specified
2. Ignition timing too far advanced
3. Excessive spark advance
4. Spark plug heat range too hot
5. Improperly routed spark plug wires
6. Defective knock sensor

Reduced Fuel Mileage

When the fuel consumption is excessive, check the following components:

1. Engine compression
2. Late ignition timing
3. Lack of spark advance
4. Cylinder misfiring

No-Start Ignition Diagnosis, Primary Ignition Circuit

The same no-start diagnosis may be performed on most electronic ignition systems. Follow these steps for the no-start diagnosis:

1. Connect a 12-V test lamp from the coil tachometer (tach) terminal to ground, and turn on the ignition switch (Figure 4-1). On General Motors DI systems, the test light should

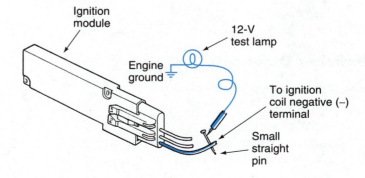

Figure 4-1 Test light connected to the negative primary coil terminal and ground (Reprinted with the permission of Ford Motor Company)

be on because the module primary circuit is open. If the test light is off, there is an open circuit in the coil primary winding or in the circuit from the ignition switch to the coil battery terminal. On some other systems, the test light should be off because the module primary circuit is closed, and since there is primary current flow, most of the voltage is dropped across the primary coil winding. This action results in very low voltage at the tach terminal, which does not illuminate the test light. On these systems, if the test light is illuminated, there is an open circuit in the module or in the wire between the coil and the module.

2. Crank the engine and observe the test light. If the test light flutters while the engine is cranked, the pickup coil signal and the module are satisfactory. When the test lamp does not flutter, one of these components is defective. The pickup coil may be tested with an ohmmeter. If the pickup coil is satisfactory, the module is defective.

No-Start Ignition Diagnosis, Secondary Ignition Circuit

1. If the test light flutters in the primary circuit no-start diagnosis, connect a test spark plug to the coil secondary wire, and ground the spark plug case (Figure 4-2). The test spark plug must have the correct voltage requirement for the ignition system being tested. For example, test spark plugs for General Motors DI systems have a 25,000-V requirement compared to a 20,000-V requirement for many other test spark plugs. A short piece of vacuum hose may be used to connect the test spark plug to the center distributor cap terminal on General Motors DI systems with an integral coil in the distributor cap.

2. Crank the engine and observe the spark plug. If the test spark plug fires, the ignition coil is satisfactory. If the test spark plug does not fire, the coil is probably defective because the primary circuit no-start test proved the primary circuit is triggering on and off.

3. Connect the test spark plug to several spark plug wires and crank the engine while observing the spark plug. If the test spark plug fired in step 2 but does not fire at some of the spark plugs, the secondary voltage and current are leaking through a defective distributor cap, rotor, or spark plug wires, or the plug wire is open. If the test spark plug fires at all the spark plugs, the ignition system is satisfactory.

Ignition Module Testing

Special Tools

Ignition module tester

A variety of ignition module testers are available from vehicle and test equipment manufacturers. These ignition module testers check the module's capability to switch the primary ignition circuit

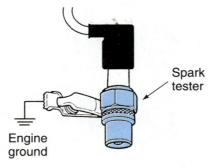

Figure 4-2 Test spark plug connected to coil secondary wire (Reprinted with the permission of Ford Motor Company)

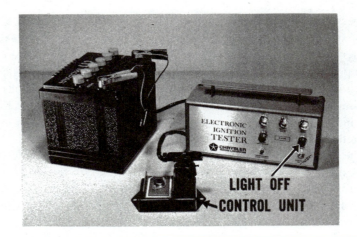

Figure 4-3 If the ignition module tester light remains off, the module is defective. (Courtesy of Chrysler Corporation)

on and off. Always follow the instructions published by the vehicle or test equipment manufacturer. The module tester leads are connected to the module, and the power supply wires are connected to the battery terminals with the correct polarity. On some testers, a green light is illuminated if the module is satisfactory, and this light remains off when the module is defective (Figure 4-3).

Pickup Coil Adjustment and Tests

Pickup Gap Adjustment

Special Tools

Nonmagnetic feeler gauge

When the pickup coil is bolted to the pickup plate, such as on Chrysler distributors, the pickup air gap may be measured with a nonmagnetic copper feeler gauge positioned between the reluctor high points and the pickup coil (Figure 4-4).

If a pickup gap adjustment is required, loosen the pickup mounting bolts and move the pickup coil until the manufacturer's specified air gap is obtained. Re-tighten the pickup coil retaining bolts to the specified torque. Some pickup coils are riveted to the pickup plate, and a pickup gap adjustment is not required.

When checking the pickup coil, always check the distributor bushing for horizontal movement, which changes the pickup gap and may cause engine misfiring.

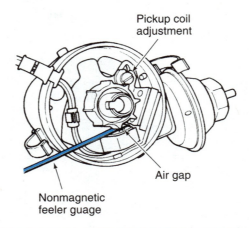

Figure 4-4 Pickup coil air gap adjustment (Courtesy of Chrysler Corporation)

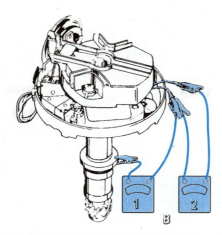

Figure 4-5 Ohmmeter-to-pickup coil test connections for grounded, open, and shorted conditions (Courtesy of Oldsmobile Division, General Motors Corporation)

Pickup Coil Ohmmeter Tests

Remove the pickup leads from the module and calibrate an ohmmeter on the X10 scale. Connect the ohmmeter to the pickup coil terminals to test the pickup coil for an open circuit or a shorted condition. While the ohmmeter leads are connected, pull on the pickup leads and watch for an erratic meter reading indicating an intermittent open in the pickup leads. Most pickup coils have 150 Ω to 900 Ω resistance, but the manufacturer's exact specifications must be used. If the pickup coil is open, the ohmmeter provides an infinite reading. A meter reading below the specified resistance indicates a shorted pickup coil.

Connect an ohmmeter from one of the pickup leads to ground to test the pickup coil for a grounded condition. If the pickup coil is not grounded, the ohmmeter provides an infinite reading. A grounded pickup coil gives a low meter reading. Ohmmeter 1 (Figure 4-5) is connected to test the pickup coil for a grounded condition. Ohmmeter 2 in the figure illustrates the connection to test the pickup coil for an open circuit or a shorted condition.

Special Tools

Ohmmeter

Ignition Coil Inspection and Tests

The ignition coil should be inspected for cracks or any evidence of leakage in the coil tower. The coil container should be checked for oil leaks. If the oil is leaking from a coil, air space is present in the coil, which allows condensation to form internally. Condensation in an ignition coil causes high voltage leaks, and engine misfiring.

Calibrate an ohmmeter on the X1 scale and connect the meter leads to the primary coil terminals to test the primary winding for an open circuit or a shorted condition (Figure 4-6). An infinite ohmmeter reading indicates an open winding. If the meter reading is below the specified resistance, the winding is shorted. Most primary windings have a resistance of 0.5 Ω to 2 Ω, but the exact manufacturer's specifications must be compared to the meter readings.

An ohmmeter must be calibrated on the X1,000 scale and then connected from the coil secondary terminal to one of the primary terminals to test the secondary winding for a shorted or open circuit (Figure 4-7). A meter reading below the specified resistance indicates a shorted secondary winding, and an infinite meter reading proves that the winding is open.

In some General Motors integral coils, the secondary winding is connected from the secondary terminal to the coil frame. When the secondary winding is tested in these coils, the ohmmeter must be connected from the secondary coil terminal to the coil frame or to the ground wire terminal extending from the coil frame to the distributor housing.

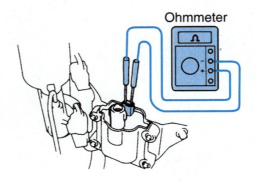

Figure 4-6 Ohmmeter connected to primary coil terminals (Courtesy of Toyota Motor Corporation)

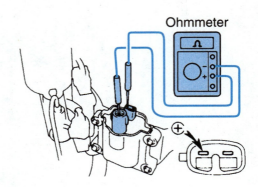

Figure 4-7 Ohmmeter connected from one primary terminal to the coil tower to test the secondary winding (Courtesy of Toyota Motor Corporation)

Many secondary windings have 8,000 Ω to 20,000 Ω resistance, but the meter readings must be compared to the manufacturer's specifications. The ohmmeter tests on the primary and secondary windings indicate satisfactory, open, or shorted windings. However, the ohmmeter tests do not indicate such defects as defective insulation around the coil windings, which causes high voltage leaks. Therefore, an accurate indication of coil condition is the coil maximum voltage output test with a test spark plug connected from the coil secondary wire to ground as explained in the no-start diagnosis.

Inspection of Distributor Cap and Rotor

 WARNING: Avoid removing excessive material from distributor cap terminals and rotor terminals with a file. This action increases the rotor gap and may increase the normal required secondary voltage.

WARNING: Cleaning distributor caps with solvent or compressed air may cause high voltage leaks in these components.

The distributor cap and rotor should be inspected for cracks, corroded terminals, and carbon tracking, indicating high voltage leaks (Figure 4-8). If the distributor cap and rotor have cracks, evidence

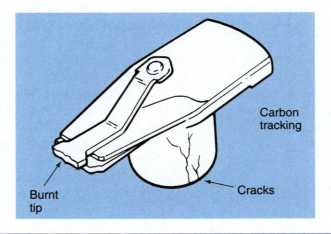

Figure 4-8 Inspecting the rotor for cracks and evidence of high voltage leaks (Courtesy of Chrysler Corporation)

78

of leakage, or worn terminals, replacement is necessary. Carefully check the distributor cap terminals for corrosion and excessive wear. Small, round wire brushes are available to clean cap terminals. Wipe the cap and rotor with a clean shop towel, but avoid cleaning these components in solvent or blowing them off with compressed air, which may contain moisture. Cleaning these components with solvent or compressed air may result in high voltage leaks.

Testing Secondary Ignition Wires

Spark Plug Wire Inspection

Inspect all the spark plug wires and the secondary coil wire for cracks and worn insulation, which cause high voltage leaks. Inspect all the boots on the ends of the plug wires and coil secondary wire for cracks and hard brittle conditions. Replace the wires and boots if they show evidence of these conditions. Some vehicle manufacturers recommend spark plug wire replacement only in complete sets.

Spark Plug Wire Testing

The spark plug wires may be left in the distributor cap for test purposes, so the cap terminal connections are tested with the spark plug wires. Calibrate an ohmmeter on the X1,000 scale, and connect the ohmmeter leads from the end of a spark plug wire to the distributor cap terminal inside the cap to which the plug wire is connected (Figure 4-9).

If the ohmmeter reading is more than specified by the vehicle manufacturer, remove the wire from the cap and check the wire alone. If the wire has more resistance than specified, replace the wire. When the spark plug wire resistance is satisfactory, check the cap terminal for corrosion. Repeat the ohmmeter tests on each spark plug wire and the coil secondary wire.

Spark Plug Wire Installation

When the spark plug wires are installed, be sure they are routed properly as indicated in the vehicle manufacturer's service manual. Two spark plug wires should not be placed side-by-side for a long span if these wires fire one after the other in the cylinder firing order. When two spark plug wires that fire one after the other are placed side-by-side for a long span, the magnetic field from the wire that is firing builds up and collapses across the other wire. This magnetic collapse may induce enough voltage to fire the other spark plug and wire when the piston in this cylinder is approaching TDC on the compression stroke. This action may cause detonation and reduced engine power.

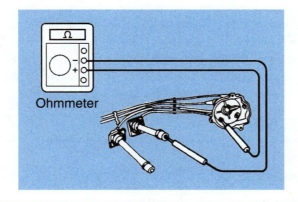

Figure 4-9 Ohmmeter connected to the spark plug wire and the distributor cap terminal to test the plug wire (Courtesy of Toyota Motor Corporation)

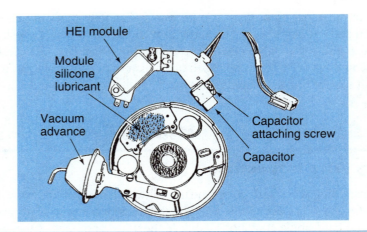

Figure 4-10 Place a light coating of silicone grease on the module mounting surface prior to module installation. (Courtesy of Oldsmobile Division, General Motors Corporation)

Ignition Module Removal and Replacement

The ignition module removal and replacement procedure varies depending on the ignition system. Always follow the procedure in the vehicle manufacturer's service manual. Follow these steps for module removal and replacement on an HEI distributor:

 WARNING: Lack of silicone grease on the module mounting surface may cause module overheating and damage.

1. Remove the battery wire from the coil battery terminal, and remove the inner wiring connector on the primary coil terminals. Remove the spark plug wires from the cap.
2. Rotate the distributor latches one-half turn and lift the cap from the distributor.
3. Remove the two rotor retaining bolts and the rotor.
4. Remove the primary leads and the pickup leads from the module.
5. Remove the two module mounting screws, and remove the module from the distributor housing.
6. Wipe the module mounting surface clean, and place a light coating of silicone heat-dissipating grease on the module mounting surface (Figure 4-10).
7. Install the module and tighten the module mounting screws to the specified torque.
8. Install the primary leads and pickup leads on the module.
9. Be sure the lug on the centrifugal advance mechanism fits into the rotor notch while installing the rotor, and tighten the rotor mounting screws to the specified torque.
10. Install the distributor cap, and be sure the projection in the cap fits in the housing notch.
11. Push down on the cap latches with a screwdriver, and rotate the latches until the lower part of the latch is hooked under the distributor housing.
12. Install the coil primary leads and battery wire on the coil terminals. Be sure the notch on the primary leads fits onto the cap projection. Install the spark plug wires.

Distributor Service

Distributor Removal

All distributor service procedures vary depending on the distributor. Always follow the recommended procedure in the vehicle manufacturer's service manual. Following is a typical distributor removal procedure:

1. Disconnect the distributor wiring connector and the vacuum advance hose.
2. Remove the distributor cap and note the position of the rotor. On some vehicles, it may be necessary to remove the spark plug wires from the cap prior to cap removal.
3. Remove the distributor hold-down bolt and clamp.
4. Note the position of the vacuum advance, and pull the distributor from the engine.
5. Install a shop towel in the distributor opening to keep foreign material out of the engine block.

Distributor Bushing Check

The distributor bushing checking procedure varies depending on the vehicle manufacturer. Always follow the procedure in the manufacturer's service manual. Some manufacturers recommend clamping the distributor housing lightly in a soft-jaw vise, and clamping a dial indicator on the top of the distributor housing. The dial indicator stem is then positioned against the top of the distributor shaft. When the shaft is pushed horizontally, observe the shaft movement on the dial indicator. If this movement exceeds the manufacturer's specifications, the distributor bushings and/or shaft are worn. Some manufacturers now recommend complete distributor replacement rather than bushing replacement.

Distributor Disassembly

Follow these steps for a typical distributor disassembly procedure:

1. Mark the gear in relation to the distributor shaft so the gear may be installed in the original position.
2. Support the distributor housing on top of a vise, and drive the roll pin from the gear and shaft with a pin punch and hammer (Figure 4-11).
3. Pull the gear from the distributor shaft, and remove any spacers between the gear and the housing. Note the position of these spacers so they may be installed in their original position.
4. Wipe the lower end of the shaft with a shop towel and inspect this area of the shaft for metal burrs. Remove any burrs with fine emery paper.
5. Pull the distributor shaft from the housing.
6. Remove the pickup coil leads from the module and the pickup retaining clip. Lift the pickup coil from the top of the distributor bushing.
7. Remove the two vacuum advance mounting screws, and remove this advance assembly from the housing (Figure 4-12).

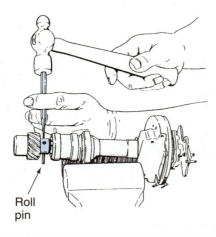

Figure 4-11 Driving the roll pin from the distributor gear (Courtesy of Oldsmobile Division, General Motors Corporation)

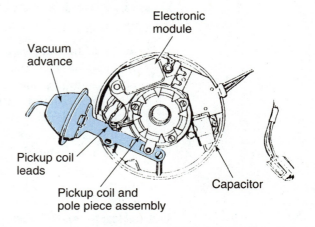

Figure 4-12 Vacuum advance and mounting screws (Courtesy of Oldsmobile Division, General Motors Corporation)

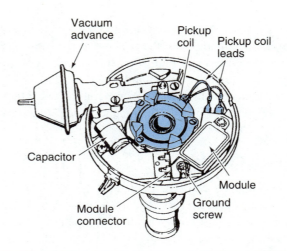

Figure 4-13 Checking pickup assembly for wear and rotation (Courtesy of Oldsmobile Division, General Motors Corporation)

Distributor Inspection

 WARNING: Distributor electrical components, and the vacuum advance may be damaged by washing them in solvent.

The housing may be washed in solvent, but do not wash electrical components or the vacuum advance. Check these items during a typical distributor inspection:

1. Inspect all lead wires for worn insulation, and loose terminals. Replace these wires as necessary.
2. Inspect the centrifugal advance mechanism for wear, particularly check the weights for wear on the pivot holes. Replace the weights, or complete shaft assembly, if necessary.
3. Inspect the pickup plate for wear and rotation (Figure 4-13). If this plate is loose or seized, replacement is required.
4. Connect a vacuum hand pump to the vacuum advance outlet and apply 20 inches of vacuum. The advance diaphragm should hold this vacuum without leaking.
5. Check the distributor gear for worn or chipped teeth.
6. Inspect the reluctor for damage. If the high points are damaged, the distributor bushing is likely worn, allowing the high points to hit the pickup coil.

Distributor Assembly

Follow these steps for a typical distributor assembly procedure:

1. Install the vacuum advance and tighten the mounting screws to the specified torque.
2. Install the pickup coil and the retaining clip. Connect the pickup leads to the module.
3. Install the module and mounting screws as discussed previously.
4. Place some bushing lubricant on the shaft and install the shaft in the distributor.
5. Install the spacers between the housing and gear in their original position.
6. Install the gear in its original position, and be sure the hole in the gear is aligned with the hole in the shaft.
7. Support the housing on top of a vise and drive the roll pin into the gear and shaft.
8. Install a new O-ring or gasket on the distributor housing.

Installing and Timing the Distributor

This procedure may be followed to install the distributor and time it to the engine:

1. Remove the spark plug from the number 1 cylinder, and place a compression gauge hose fitting in the spark plug hole.

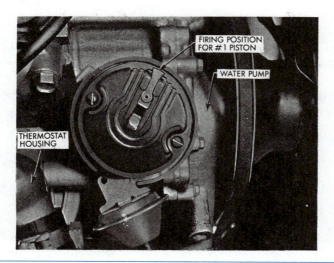

Figure 4-14 Installing the distributor with the rotor under the number 1 spark plug terminal in the distributor cap (Courtesy of Buick Motor Division, General Motors Corporation)

2. Crank the engine a small amount at a time until compression pressure appears on the gauge.
3. Crank the engine a very small amount at a time until the 0° position on the timing marks is aligned with the timing indicator.
4. Locate the number 1 spark plug wire position in the distributor cap. The wire terminals in some caps are marked, and the manufacturer's service manual provides this information.
5. Install the distributor in the block with the rotor positioned under the number 1 spark wire position in the distributor cap, and the vacuum advance in the original position (Figure 4-14). The distributor gear easily goes into mesh with the camshaft gear, but many distributors also drive the oil pump with a hex-shaped drive in the lower end of the distributor gear or shaft. It may be necessary to hold down on the distributor housing and crank the engine to get the distributor shaft into mesh with the oil pump drive. When this action is required, repeat steps 2 and 3 and be sure the rotor is under the number 1 spark plug wire terminal in the distributor cap with the timing marks aligned.
6. Rotate the distributor a small amount so the timer core teeth and pickup teeth are aligned. With a point-type distributor, rotate the distributor until the points are just beginning to open.
7. Install the distributor hold-down clamp and bolt, and leave the bolt slightly loose.

SERVICE TIP: The distributor shaft rotates in the opposite direction to which the vacuum advance pulls the pickup plate.

8. Install the spark plug wires in the direction of distributor shaft rotation and in the cylinder firing order (Figure 4-15).
9. Connect the distributor wiring connectors. The vacuum advance hose is usually left disconnected until the timing is set with the engine running.

Photo Sequence 2 shows a typical procedure for timing the distributor to the engine.

Timing Checking and Adjustment

The ignition timing procedure varies depending on the make and year of vehicle and the type of ignition system. Ignition timing specifications and instructions are included on the underhood emission label, and more detailed instructions are provided in the vehicle manufacturer's service manual. The ignition timing procedure and specifications recommended by the vehicle manufacturer must be followed. On distributors with advance mechanisms, manufacturers usually recommend

Photo Sequence 2
Typical Procedure for Timing the Distributor to the Engine

P2-1 Remove the number 1 spark plug.

P2-2 Place your thumb over the number 1 spark plug opening and crank the engine until compression is felt.

P2-3 Crank the engine a very small amount at a time until the timing marks indicate that the number 1 piston is at TDC on the compression stroke.

P2-4 Determine the number 1 spark plug wire position in the distributor cap.

P2-5 Install the distributor with the rotor under the number 1 spark plug wire terminal in the distributor cap and one of the reluctor high points aligned with the pickup coil.

P2-6 After the distributor is installed in the block, turn the distributor housing slightly so the pickup coil is aligned with the reluctor.

P2-7 Install the distributor clamp bolt, but leave it slightly loose.

P2-8 Connect the pickup leads to the wiring harness.

P2-9 Install the spark plug wires in the cylinder firing order and in the direction of distributor shaft rotation.

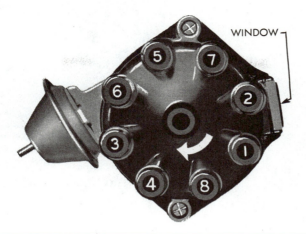

Figure 4-15 The spark plug wires are installed in the direction of distributor shaft rotation, and in the cylinder firing order. (Courtesy of Buick Motor Division, General Motors Corporation)

disconnecting and plugging the vacuum advance hose while checking ignition timing. On carbureted engines, the manufacturer usually specifies a certain engine rpm while checking the ignition timing. The timing light pickup is connected to the number 1 spark plug wire, and the power supply wires on the light are connected to the battery terminals with the proper polarity. Follow these steps for ignition timing adjustment:

1. Connect the timing light, and start the engine.
2. The engine must be idling at the manufacturer's recommended rpm and all other timing procedures must be followed.
3. Aim the timing light marks at the timing indicator, and observe the timing marks (Figure 4-16).
4. If the timing mark is not at the specified location, rotate the distributor until the mark is at the specified location.

Special Tools

Advance-type timing light

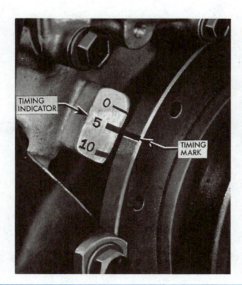

Figure 4-16 When the timing light flashes, the timing mark on the crankshaft pulley must appear at the specified location on the timing indicator above the pulley. (Courtesy of Buick Motor Division, General Motors Corporation)

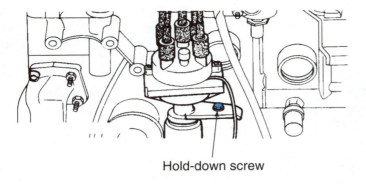

Figure 4-17 Tighten the distributor hold-down bolt after the timing is adjusted to specifications. (Courtesy of Chrysler Corporation)

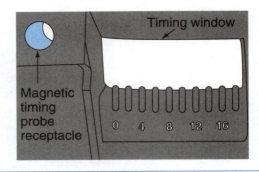

Figure 4-18 Magnetic timing probe receptacle near the timing indicator (Courtesy of Chrysler Corporation)

The magnetic timing probe receptacle is a small hole near the timing marks in the timing indicator.

Special Tools

Magnetic timing meter

5. Tighten the distributor hold-down bolt to the specified torque, and recheck the timing mark position (Figure 4-17).
6. Connect the vacuum advance hose and any other connectors, hoses, or components that were disconnected for the timing procedure.

Magnetic Timing Procedure. Many later model vehicles have a magnetic timing probe receptacle near the timing indicator (Figure 4-18). Some equipment manufacturers supply a magnetic timing meter with a pickup that fits in the probe hole. The meter pickup must be connected to the number 1 spark plug wire, and the power supply leads connected to the battery terminals (Figure 4-19). Many timing meters have two scales, timing degrees and engine rpm.

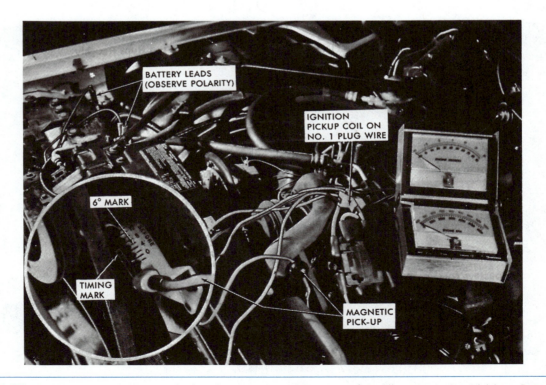

Figure 4-19 Magnetic timing meter and related connections (Courtesy of Cadillac Motor Car Division, General Motors Corporation)

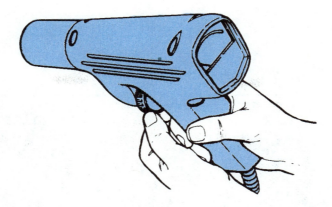

Figure 4-20 Timing light with advance control and spark advance meter (Courtesy of Chrysler Corporation)

A magnetic timing offset knob on the meter must be adjusted to the vehicle manufacturer's specifications to compensate for the position of the probe receptacle. Once the offset is adjusted, the engine may be started and the timing is indicated on the meter scale. The timing adjustment procedure is the same with the timing meter or light.

Timing Advance Check. Many timing lights have the capability to check the spark advance. An advance control on the light slows down the flashes of the light as the knob is rotated. When the light flashes are slowed with the engine running at higher speed, the timing marks move back to the basic timing setting. Follow these steps for a typical spark advance check:

1. Check the basic timing.
2. Obtain the vehicle manufacturer's spark advance specifications at a specific rpm.
3. Operate the engine at this specific rpm, and rotate the timing light advance control until the timing mark moves back to the basic timing setting (Figure 4-20).
4. Observe the spark advance on the timing light meter, and compare this reading to specifications.
5. If the spark advance is not equal to specifications, repeat steps 3 and 4 with the vacuum advance hose disconnected to determine whether the vacuum or centrifugal advance is at fault.

Spark Plug Service

Prior to spark plug removal, an air nozzle should be used to blow any foreign material from the spark plug recesses. A special spark plug socket must be used to remove and replace spark plugs. These sockets have an internal rubber bushing to prevent plug insulator breakage. When the spark plugs are removed, they should be set in order so the technician can identify the spark plug from each cylinder. Spark plug carbon conditions are an excellent indicator of cylinder conditions. If the spark plugs all have light brown or gray carbon deposits, the cylinders have been operating normally with a proper air-fuel ratio (Figure 4-21). Abnormal spark plug conditions and their causes are shown in the figure.

▲ **WARNING:** Never sandblast platinum-tipped spark plugs, and do not file the electrodes or measure the gap on these spark plugs. These service procedures will damage the platinum coating on the electrodes.

Spark plugs may be cleaned in a sandblast-type cleaner (Figure 4-22). Always consult the vehicle manufacturer's recommendations regarding spark plug cleaning. Spark plug electrodes may be filed with a small, fine-toothed file to restore the electrode surfaces. The spark plug gap

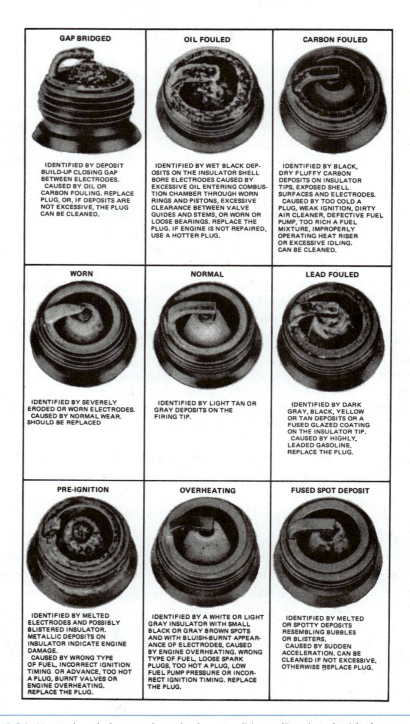

Figure 4-21 Normal and abnormal spark plug conditions (Reprinted with the permission of Ford Motor Company)

Figure 4-22 Spark plug cleaner (Courtesy of Toyota Motor Corporation)

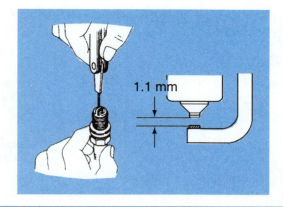

Figure 4-23 Measuring spark plug gap (Courtesy of Toyota Motor Corporation)

should be checked with a round feeler gauge and adjusted if necessary (Figure 4-23). Bend the ground electrode to adjust the spark plug gap.

Prior to spark plug installation, the spark plug threads and tapered seat must be clean. Use a wire brush to clean the threads if necessary. The spark plug seat in the cylinder head must be cleaned with a shop towel if required. If the spark plugs have sealing gaskets, these gaskets should be replaced. The spark plugs must be tightened to the specified torque.

Computer-Controlled Ignition System Service and Diagnosis

Basic Ignition Timing Tests

CAUTION: To avoid timing light damage and personal injury, keep the timing light and the timing light lead wires away from rotating parts such as fan blades and belts while adjusting basic timing.

CAUTION: To avoid personal injury, keep hands and clothing away from rotating parts such as fan blades and belts while adjusting basic timing. Remember that electric-drive cooling fans may start at any time.

Prior to any basic timing check, the engine must be at normal operating temperature. The ignition timing specifications are provided on the underhood emission control information label. On some vehicles, such as General Motors, instructions regarding the procedure for checking ignition timing are also provided on the emission control information label. On computer-controlled distributor ignition (DI) systems, the ignition timing procedure varies depending on the vehicle make and model year. On some vehicles, the manufacturer may recommend disconnecting certain components while checking the basic ignition timing. Always follow the ignition timing procedure in the vehicle manufacturer's service manual.

When the basic timing is checked on most Chrysler fuel-injected engines, the computer system must be in the limp-in mode. The coolant temperature sensor may be disconnected to place the system in the limp-in mode, and then the timing may be checked with a timing light in the normal manner. On some Chrysler 4-cylinder engines, the timing window is in the top of the flywheel housing. On many engines, the timing mark or marks are on the crankshaft pulley, and the timing indicator is mounted above the pulley. If a timing adjustment is necessary, the distributor clamp

The Society of Automotive Engineer's (SAE) J1930 terminology is an attempt to standardize terminology in automotive electronics.

In the SAE J1930 terminology, the term distributor ignition *(DI) replaces all previous terms for distributor-type ignition systems that are electronically controlled.*

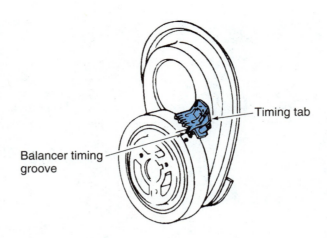

Figure 4-24 Timing mark and timing indicator (Courtesy of Chevrolet Motor Division, General Motors Corporation)

bolt must be loosened, and the distributor rotated until the timing mark appears at the specified position on the timing indicator (Figure 4-24).

After a timing adjustment, the distributor clamp bolt must be tightened to the specified torque. Removal of the coolant temperature sensor wires places a fault code in the computer memory. This code should be erased following the timing adjustment.

WARNING: When disconnecting or reconnecting wires on a computer system, always be sure the ignition switch is off. Disconnecting computer system component wires with the ignition switch on may damage system components.

On many DI systems, such as Ford and General Motors, a timing connector located in the engine compartment must be disconnected. The emission control information label usually provides the location of the timing connector on General Motor's vehicles. When the timing connector is disconnected, the PCM cannot affect spark advance, and the pickup signal goes directly to the module. The distributor clamp bolt must be loosened and the distributor rotated to adjust the basic timing. After the timing adjustment is completed, the clamp bolt must be tightened and the timing connector reconnected.

No-Start Ignition Tests

The same no-start tests may be performed on conventional DI systems with centrifugal and vacuum advances, and DI systems with computer-controlled spark advance. These tests were explained previously.

Connect a 12-V test light from the coil tachometer (TACH) terminal to ground, and crank the engine. If the 12-V test lamp does not flutter while the engine is cranked, the pickup or ignition module is likely defective. Under this condition, always check the voltage supply to the positive primary coil terminal with the ignition switch on before the diagnosis is continued.

On most Chrysler fuel-injected engines, the voltage is supplied through the automatic shutdown (ASD) relay to the coil positive primary terminal and the electric fuel pump. Therefore, a defective ASD relay may cause 0 V at the positive primary coil terminal. This relay is controlled by the PCM. On some Chrysler products, the relay closes when the ignition switch is turned on, whereas on other models, it only closes while the engine is cranking or running. If the ASD relay closes with the ignition switch on and the engine not cranking or running, it only remains closed for about 1 second. This action shuts off the fuel pump and prevents any spark from the ignition system, if the vehicle is involved in a collision with the ignition switch on and the engine stalled. A fault code should be present in the computer memory if the ASD relay is defective.

A timing connector is a single-wire connector sometimes located near the distributor which must be disconnected while checking or adjusting basic ignition timing.

The ignition coil tachometer (TACH) terminal is the negative primary coil terminal.

The automatic shutdown relay supplies voltage to the electric fuel pump, positive primary coil terminal, injectors, and oxygen sensor heater on some Chrysler vehicles.

Pickup Tests

 WARNING: Never short across or ground terminals or wires in a computer system unless instructed to do so on the vehicle manufacturer's service manual.

If a magnetic-type pickup is used, the pickup may be checked for open circuits, shorts, and grounds with an ohmmeter. These tests are performed in the same way as the pickup tests on conventional distributors described previously. If the pickup coil tests are satisfactory, and the 12-V test light connected from the coil TACH terminal to ground does not flutter while cranking the engine, the ignition module is defective.

Prior to testing a Hall Effect pickup, an ohmmeter should be connected across each of the wires between the pickup and the computer with the ignition switch off. A computer terminal and pickup coil wiring diagram are essential for these tests. Satisfactory wires have nearly 0 Ω resistance, while higher or infinite readings indicate defective wires. If the distributor has a Hall Effect pickup, the voltage supply wire and the ground wire should be checked before the pickup signal. In the following tests, the distributor connector is connected, and this connector may be backprobed to complete the necessary connections. With the ignition switch on, a voltmeter should be connected from the voltage input wire to ground, and the specified voltage must appear on the meter.

The ground wire should be tested with the ignition switch on and a voltmeter connected from the ground wire to a ground connection near the distributor. With this meter connection, the meter indicates the voltage drop across the ground wire, which should not exceed 0.2 V if the wire has a normal resistance.

Connect a digital voltmeter from the pickup signal wire to ground. If the voltmeter reading does not fluctuate while cranking the engine, the pickup is defective. A voltmeter reading that fluctuates from nearly 0 V to between 9 V and 12 V indicates a satisfactory pickup. During this test, the voltmeter reading may not be accurate, because of the short duration of the voltage signal. If the Hall Effect pickup signal is satisfactory, and the 12-V test lamp does not flutter during the no-start test, the ignition module is probably defective. The ignition module is contained in the PCM on Chrysler products. On Chrysler fuel-injected engines, the reference pickup and the SYNC pickup should be tested. If either of these pickups is defective, a fault code may be stored in the computer memory.

On Chrysler optical distributors, the pickup voltage supply and the ground wires may be tested at the four-wire connector near the distributor. With the ignition switch on, a voltmeter connected from the orange voltage supply wire to ground should indicate 9.2 V to 9.4 V (Figure 4-25). This voltage reading may vary depending on the model year. Always use the vehicle manufacturer's specifications.

A voltmeter should indicate less than 0.2 V when it is connected from the black/light blue ground wire to an engine ground connection, if the ground wire is satisfactory. When a digital voltmeter is connected from the gray/black reference pickup wire or the tan/yellow SYNC pickup wire to an engine ground connection, the voltmeter reading should cycle from nearly 0 V to approximately 5 V while the engine is cranking. This is a typical voltage figure. Always use the vehicle manufacturer's specifications for the model year being diagnosed. If the pickup signal is not within specifications, the pickup is defective. A defective SYNC pickup in this distributor should not cause a no-start problem.

Electronic Ignition (EI) System Diagnosis and Service

No-Start Ignition Diagnosis, Cam and Crank Sensors

The diagnostic procedure for EI systems varies depending on the vehicle make and model year. Always follow the procedure recommended in the vehicle manufacturer's service manual. The

Special Tools

Digital volt/ohmmeter

A Hall Effect switch contains a Hall element and a permanent magnet, and a blade representing each engine cylinder rotates between these components.

The reference pickup is a Hall Effect switch located in the distributor on some Chrysler products. This pickup is used for ignition triggering.

The synchronizer pickup is a Hall Effect switch used for injector sequencing in some Chrysler distributors.

An optical-type pickup has a slotted plate that rotates between a light emitting diode (LED) and a photo diode.

Classroom Manual
Chapter 4, page 81

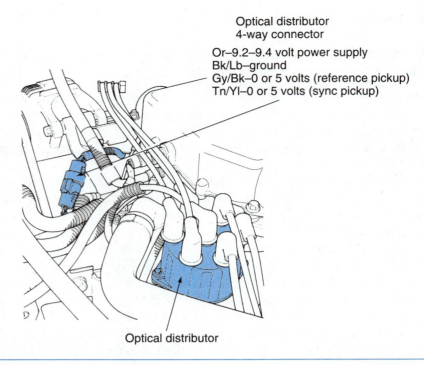

Figure 4-25 Chrysler optical distributor four-wire connector (Courtesy of Chrysler Corporation)

> In the SAE J1930 terminology, the term *electronic ignition* (EI) replaces all previous terms for distributorless ignition systems.

following procedure is based on Chrysler EI systems. The crankshaft timing sensor and camshaft reference sensor in these systems are modified Hall Effect switches.

When the engine fails to start, follow these steps:

1. Check for fault codes 11 and 43. Code 11, "Ignition Reference Signal," could be caused by a defective camshaft reference signal or crankshaft timing sensor signal. Code 43 is caused by low primary current in coil number 1, 2, or 3.
2. With the engine cranking, check the voltage from the orange wire to ground on the crankshaft timing sensor and the camshaft reference sensor (Figure 4-26). Over 7 V is

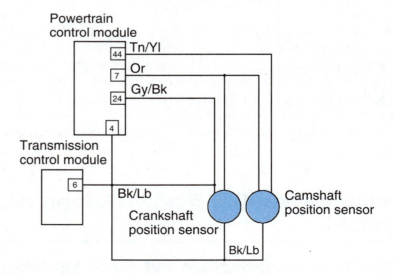

Figure 4-26 Crankshaft timing and camshaft reference sensor terminals (Courtesy of Chrysler Corporation)

satisfactory. If the voltage is less than specified, repeat the test with the voltmeter connected from PCM terminal 7 to ground. If the voltage is satisfactory at terminal 7 but low at the sensor orange wire, repair the open circuit or high resistance in the orange wire. If the voltage is low at terminal 7, replace the PCM. Be sure 12 V are supplied to PCM terminal 3 with the ignition switch off or on, and 12 V must be supplied to PCM terminal 9 with the ignition switch on. Check PCM ground connections on terminals 11 and 12 before PCM replacement.

3. With the ignition switch on, check the voltage drop across the ground circuit (black/light blue wire) on the crankshaft timing sensor and the camshaft reference sensor. A reading below 0.2 V is satisfactory.

SERVICE TIP: When using a digital voltmeter to check a crankshaft or camshaft sensor signal, crank the engine a very small amount at a time and observe the voltmeter. The voltmeter reading should cycle from almost 0 volts to a higher voltage of 9 to 12. Since digital voltmeters do not react instantly, it is difficult to see the change in voltmeter reading if the engine is cranked continually.

4. If the readings in steps 2 and 3 are satisfactory, connect a 12-V test lamp or a digital voltmeter from the gray/black wire on the crankshaft timing sensor and the tan/yellow wire on the camshaft reference sensor to ground. When the engine is cranking, a flashing 12-V lamp indicates that a sensor signal is present. If the lamp does not flash, sensor replacement is required. Each sensor voltage signal should cycle from low voltage to high voltage as the engine is cranked.

No-Start Ignition Diagnosis Coil and PCM Tests

If the sensor tests are satisfactory, proceed with these coil and PCM tests:

1. Check the spark plug wires with an ohmmeter as explained previously.
2. With the engine cranking, connect a voltmeter from the dark green/black wire on the coil to ground. If this reading is below 12 V, check the automatic shutdown (ASD) relay circuit (Figure 4-27).

SERVICE TIP: Later model Chrysler vehicles have separate fuel pump and ASD relays. Always use the proper wiring diagram for the vehicle being tested.

3. If the reading in step 2 is satisfactory, check the primary and secondary resistance in each coil with the ignition switch off. Primary resistance is 0.52–0.62 ohm, and secondary

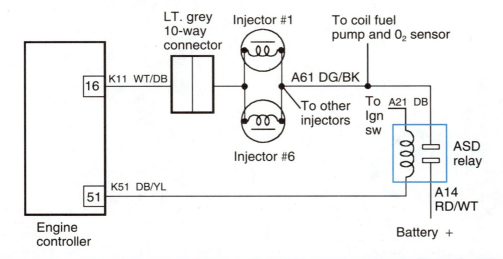

Figure 4-27 Automatic shutdown (ASD) relay circuit (Courtesy of Chrysler Corporation)

resistance is 11,000–15,000 ohms. If these ohm readings are not within specifications, replace the coil assembly.

4. With the ignition switch off, connect an ohmmeter across the three wires from the coil connector to PCM terminals 17, 18, and 19 (Figure 4-28). These terminals are connected from the coil primary terminals to the PCM. If an infinite ohmmeter reading is obtained on any of the wires, repair the open circuits.

5. Connect a 12-V test lamp from the dark blue/black wire, dark blue/gray wire, and black/gray wire on the coil assembly to ground while cranking the engine. If the test lamp does not flutter on any of the three wires, replace the PCM. Since the crankshaft and camshaft sensors, wires from the coils to the PCM, and voltage supply to the coils have been tested already, the ignition module must be defective. This module is an integral part of the PCM on Chrysler vehicles; thus, PCM replacement is necessary.

WARNING: Do not crank or run an EI-equipped engine with a spark plug wire completely removed from a spark plug. This action may cause leakage defects in the coils or spark plug wires.

CAUTION: Since EI systems have more energy in the secondary circuit, electrical shocks from these systems should be avoided. The electrical shock may not injure the human body, but such a shock may cause you to jump and hit your head on the hood or push your hand into contact with a rotating cooling fan.

6. If the tests in steps 1 through 5 are satisfactory, connect a test spark plug to each spark plug wire and ground and crank the engine. If any of the coils do not fire on the two spark plugs connected to the coil, replace the coil assembly.

Sensor Replacement

If the crankshaft timing sensor or the camshaft reference sensor is removed, follow this procedure when the sensor is replaced:

1. Thoroughly clean the sensor tip and install a new spacer (part number 5252229) on the sensor tip. New sensors should be supplied with the spacer installed (Figure 4-29).
2. Install the sensor until the spacer lightly touches the sensor ring, and tighten the sensor mounting bolt to 105 in. lb.

WARNING: Improper sensor installation may cause sensor, rotating drive plate, or timing gear damage.

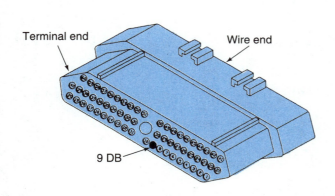

Figure 4-28 PCM terminal identification (Courtesy of Chrysler Corporation)

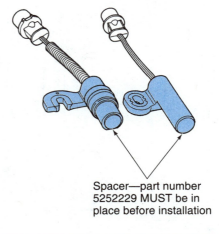

Figure 4-29 Spacer on crankshaft timing sensor and camshaft reference sensor tips (Courtesy of Chrysler Corporation)

Low Data Rate and High Data Rate EI Service and Diagnosis

No-Start Diagnosis

The diagnostic procedure for the low data rate and high data rate systems varies depending on the system being diagnosed. The technician must have the proper wiring diagram for the system being diagnosed, and the procedure in the vehicle manufacturer's service manual must be followed. Ford Motor Company provides separate diagnostic harnesses for the low data rate and high data rate EI systems. The diagnostic harness has various leads that connect in series with each component in the ignition system (Figure 4-30). A large connector on the diagnostic harness is connected to the

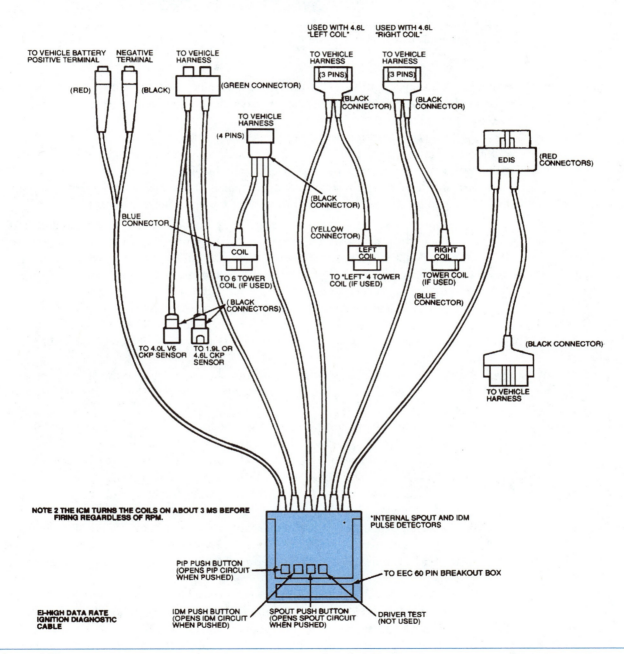

Figure 4-30 Diagnostic harness for a high data rate ignition system (Reprinted with the permission of Ford Motor Company)

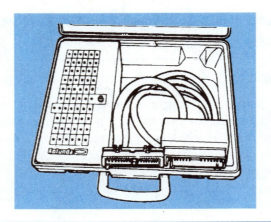

Figure 4-31 60-pin breakout box (Reprinted with the permission of Ford Motor Company)

60-pin breakout box (Figure 4-31). Overlays for the ignition system on each engine are available to fit on the breakout box terminals. These overlays identify the ignition terminals connected to the breakout box terminals.

The technician must follow the test procedures in the vehicle manufacturer's service manual, and measure the voltage or resistance at the specified breakout box terminals connected to the ignition system. If a diagnostic harness, overlay, and breakout box are not available, the technician must measure the voltage or resistance at the terminals on the individual ignition components. Following is a general no-start diagnostic procedure for a high data rate system:

1. Connect a test spark plug from each of the spark plug wires to ground and crank the engine. If the test spark plug does not fire on a pair of spark plugs connected to the same coil, test the spark plug wires connected to that coil. If these wires are satisfactory, that coil is probably defective. When the test spark plug does not fire on any spark plug, proceed to step 2.
2. Connect a digital voltmeter from terminal 2 on each coil pack to ground (Figure 4-32). With the ignition switch on, the voltmeter should indicate 12 V. If the voltage is less than specified, test the wire from the ignition switch to the coils, and the ignition switch.
3. With the ignition switch off, connect the ohmmeter leads to the primary terminals in each coil pack. If the primary resistance is not within specifications, replace the coil pack.
4. Connect the ohmmeter leads to the secondary terminals in each coil pack. Replace the coil pack if the secondary resistance in any coil is not within specifications.
5. Connect the ohmmeter leads from each primary coil terminal to the terminal on the ignition control module (ICM) to which the primary terminal is connected. Each wire should read less than 0.5 Ω. If any of these wires has more resistance than specified, repair the wire.
6. Connect a digital voltmeter from terminal 6 on the ICM to ground. With the ignition switch on, the voltmeter should indicate 12 V. If the voltage is less than specified, test the wire from the power relay to terminal 6, and the power relay. Turn off the ignition switch.
7. Connect a digital voltmeter from terminal 10 on the ICM to ground. The voltmeter should indicate 0.5 V or less with the ignition switch on. When the voltmeter reading is higher than specified, repair the ground wire.
8. Connect a digital voltmeter from terminal 7 on the ICM to ground. With the ignition switch on, the voltmeter reading should be less than specified. If the voltmeter reading is more than specified, repair the ground wire.

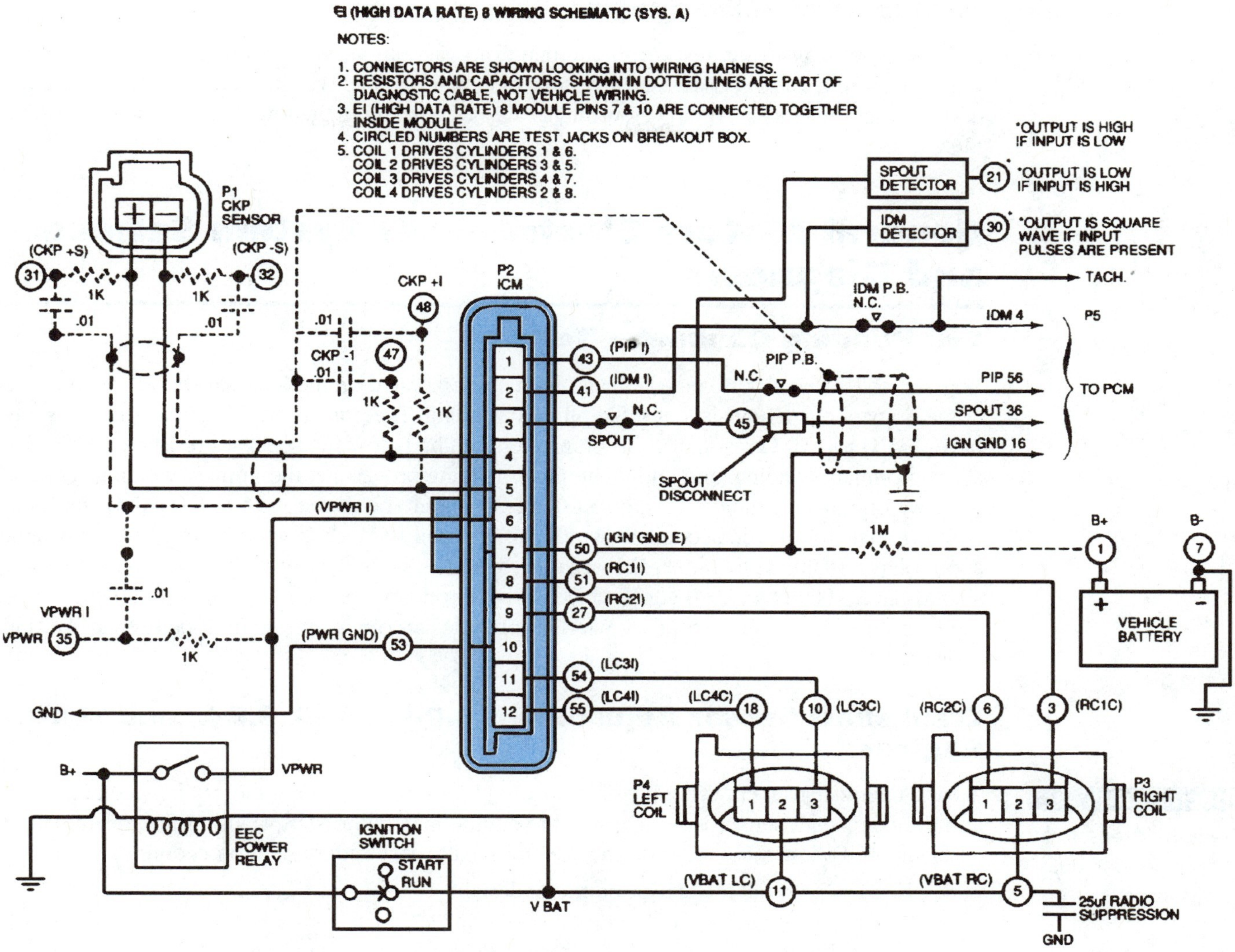

Figure 4-32 Electronic ignition (EI) system wiring diagram for a 4.6-L engine (Reprinted with the permission of Ford Motor Company)

9. Connect a digital voltmeter from terminals 8, 9, 11, and 12 on the ICM to ground and crank the engine. The voltmeter reading should fluctuate on each wire. If the voltmeter reading does not fluctuate on one of the wires, replace the ICM. When the voltmeter does not fluctuate on any of these primary wires, proceed to step 10.

10. With the ignition switch off and the ICM terminal disconnected, connect an ohmmeter to terminals 4 and 5 in the ICM wiring harness. If the ohmmeter reads 2,300 Ω to 2,500 Ω, the crankshaft position (CKP) sensor and connecting wires are satisfactory. When the ohmmeter reading is not within specifications, repeat the test with the ohmmeter leads connected to the CKP sensor terminals. If the resistance of the CKP sensor is not within specifications, replace the sensor. When the resistance reading is satisfactory at the CKP sensor terminals, but out of specifications at ICM terminals 4 and 5, repair the CKP wires.

11. Inspect the trigger wheel behind the crankshaft pulley and the CKP sensor for damage.

12. If the digital voltmeter readings do not fluctuate in step 9, but the CKP sensor test in step 10 is satisfactory, replace the ICM.

Engine Misfire Diagnosis

If the engine is misfiring, test for intake manifold vacuum leaks. Test the engine compression, spark plugs, and fuel injectors. Test the spark plug wires and coils as indicated in steps 1 through 4 of the no-start diagnosis. Timing adjustments are not possible in the low data rate or high date rate EI systems.

General Motors Electronic EI System Service and Diagnosis

Coil Winding Ohmmeter Tests

With the coil terminals disconnected, an ohmmeter calibrated on the X1 scale should be connected to the primary coil terminals to test the primary winding. The primary winding in any EI coil should have 0.35 Ω to 1.50 Ω resistance. An ohmmeter reading below the specified resistance indicates a shorted primary winding, and an infinite meter reading proves that the primary winding is open.

An ohmmeter calibrated on the X1,000 scale should be connected to each pair of secondary coil terminals to test the secondary windings. The coil secondary winding in EI type 1 systems should have 10,000 Ω to 14,000 Ω resistance, whereas secondary windings in EI type 2 systems should have 5,000 Ω to 7,000 Ω resistance. If the secondary winding is open, the ohmmeter reading is infinite. A shorted secondary winding provides an ohmmeter reading below the specified resistance.

Crankshaft Sensor Adjustment, 3.0-L, 3300, 3.8-L, and 3800 Engines

Special Tools

Crankshaft sensor adjusting tool, dual slot sensor

A basic timing adjustment is not possible on any EI system. However, if the gap between the blades and the crankshaft sensor is not correct, the engine may fail to start, stall, misfire, or hesitate on acceleration. Follow these steps during the crankshaft sensor adjustment procedure:

1. Loosely install the sensor on the pedestal.
2. Position the sensor and pedestal on the J37089 adjusting tool.
3. Position the adjusting tool on the crankshaft surface (Figure 4-33).
4. Tighten the pedestal-to-block mounting bolts to 30–35 ft. lb. (20–40 Newton meters [Nm]).
5. Tighten the pinch bolt to 30–35 in. lb. (3–4 Nm).

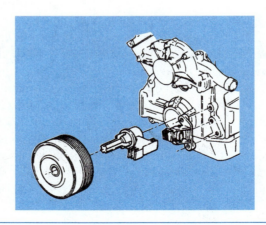

Figure 4-33 Crankshaft sensor adjustment for 3.0-L, 3300, 3.8-L, and 3800 engines (Courtesy of Oldsmobile Division, General Motors Corporation)

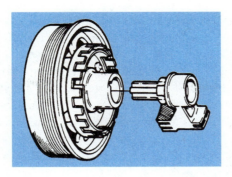

Figure 4-34 Interrupter ring checking procedure (Courtesy of Oldsmobile Division, General Motors Corporation)

The interrupter rings on the back of the crankshaft pulley should be checked for a bent condition. The same crankshaft sensor adjusting tool may be used to check these rings. Place the J37089 tool on the pulley extension surface and rotate the tool around the pulley (Figure 4-34). If any blade touches the tool, replace the pulley.

Crankshaft Sensor Adjustment, Single Slot Sensor

If a single slot crankshaft sensor requires adjustment, follow this procedure:

> **CAUTION:** Always be sure the ignition switch is off before attempting to rotate the crankshaft with a socket and breaker bar. If the ignition switch is on, the engine may start suddenly and rotate the socket and breaker bar with tremendous force. This action may result in personal injury and vehicle damage.

Special Tools

Crankshaft sensor adjusting tool, single slot sensor

1. Be sure that the ignition switch is off. Then rotate the crankshaft with a pull handle and socket installed on the crankshaft pulley nut. Continue rotating the crankshaft until one of the interrupter blades is in the sensor and the edge of the interrupter window is at the edge of the defector on the pedestal.
2. Insert adjustment tool J-36179 or its equivalent between each side of the blade and the sensor. If the tool does not fit between each side of the blade and the sensor, adjustment is required. The gap measurement should be repeated at all three blades.
3. If a sensor adjustment is necessary, loosen the pinch bolt and insert the adjusting tool between each side of the blade and the sensor. Move the sensor as required to insert the gauge.
4. Tighten the sensor pinch bolt to 30 in. lb. (3–4 Nm).
5. Rotate the crankshaft and recheck the gap at each blade.

No-Start Ignition Diagnosis for EI Type 1 and Type 2 Systems

If the engine fails to start, follow these steps for a no-start ignition diagnosis:

1. Connect a test spark plug from each spark plug wire to ground and crank the engine while observing the test spark plug.
2. If the test spark plug does not fire on any spark plug, check the 12-V supply wires to the coil module. Some coil modules have two fused 12-V supply wires. Consult the vehicle manufacturer's wiring diagrams for the car being tested to identify the proper coil module terminals.
3. If the test spark plug does not fire on a pair of spark plugs, the coil connected to that pair of spark plugs is probably defective.
4. If the test spark plug does not fire on any of the spark plugs and the 12-V supply circuits to the coil module are satisfactory, disconnect the crankshaft and camshaft sensor

connectors and connect short jumper wires between the sensor connector and the wiring harness connector. Be sure the jumper wire terminals fit securely to maintain electrical contact. Each sensor has a voltage supply wire, a ground wire, and a signal wire on 3.8-L engines. On the 3.3-L and 3300 engines, the dual crankshaft sensor has a voltage supply wire, ground wire, crank signal wire, and SYNC signal wire. Identify each of these wires on the wiring diagram for the system being tested.

5. Connect a digital voltmeter to each of the camshaft and crankshaft sensor black ground wires to an engine ground connection. With the ignition switch on, the voltmeter reading should be 0.2 V or less. If the reading is above 0.2 V, the sensor ground wires have excessive resistance.

6. With the ignition switch on, connect a digital voltmeter from the camshaft and crankshaft sensor white/red voltage supply wires to an engine ground (Figure 4-35). The voltmeter readings should be 5 V to 11 V. If the readings are below these values, check the voltage at the coil module terminals that are connected to the camshaft and crankshaft sensor voltage supply wires. When the sensor voltage supply readings are low at the coil module terminals, the coil module should be replaced. If the voltage supply readings are low at either sensor connector but satisfactory at the coil module terminal, the wire from the coil module to the sensor is defective. On the 3.3-L and 3300 engines, the crankshaft sensor ground wire and voltage supply wire are checked in the same way as explained in steps 5 and 6.

7. If the camshaft and crankshaft sensor ground and voltage supply wires are satisfactory, connect a digital voltmeter to each sensor signal wire and crank the engine. Each sensor should have a 5-V to 7-V fluctuating signal. On the 3.3-L and 3300 engines, test this voltage signal on the crank and SYNC signal wires at the crankshaft sensor. If the signal is less than specified, replace the sensor with the low signal.

8. When the camshaft and crankshaft sensor signals on 3.8-L engines or crank and SYNC signals on 3.3-L and 3300 engines are satisfactory and the test spark plug does not fire at any spark plug, the coil module is probably defective.

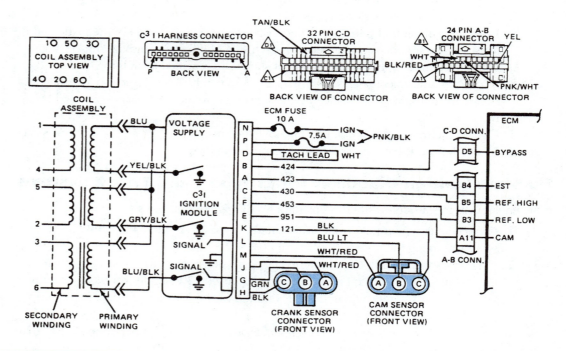

Figure 4-35 Crankshaft and camshaft sensor wiring connections for a 3.8-L engine (Courtesy of Buick Motor Division, General Motors Corporation)

9. On 3.8-L engines where the coil assembly is easily accessible, the coil assembly screws may be removed and the coil lifted up from the module with the primary coil wires still connected. Connect a 12-V test lamp across each pair of coil primary wires and crank the engine. If the test lamp does not flutter on any of the coils, the coil module is defective, assuming that the crankshaft and camshaft sensor readings are satisfactory.

No-Start Ignition Diagnosis for EI Type 1 Fast-Start Systems

Complete steps 1, 2, and 3 in the No-Start Ignition Diagnosis for EI Type 1 and Type 2 Systems, and then complete these steps:

1. If the 12-V supply circuits to the coil module are satisfactory, disconnect the crankshaft sensor connector and connect four short jumper wires between the sensor connector and the wiring harness connector.
2. Connect a digital voltmeter from the sensor ground wire to an engine ground. With the ignition switch on, the voltmeter should read 0.2 V or less. A reading above this value indicates a defective ground wire.
3. Connect the voltmeter from the sensor voltage supply wire to an engine ground. With the ignition switch on, the voltmeter reading should be 8 V to 10 V. If the reading is lower than specified, check the voltage at coil module terminal N (Figure 4-36). When the voltage at terminal N is satisfactory and the reading at the sensor voltage supply wire is low, the wire from terminal N to the sensor is defective. A low voltage reading at terminal N indicates a defective coil module.
4. If the readings in steps 3 and 4 are satisfactory, connect a voltmeter from the 3X and 18X signal wires at the sensor connector to an engine ground and crank the engine. The voltmeter reading should fluctuate from 5 V to 7 V. The exact voltage may be difficult to read, especially on the 18X signal, but the reading must fluctuate. If the voltmeter reading is steady on either sensor signal, the sensor is defective.

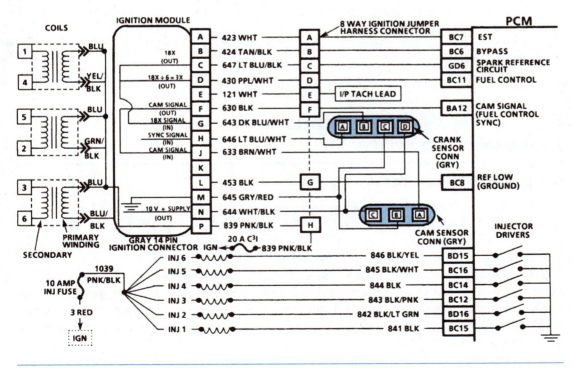

Figure 4-36 Terminal identification for an EI fast-start system 3800 engine (Courtesy of Oldsmobile Division, General Motors Corporation)

101

5. Connect a digital voltmeter from the 18X and 3X signal wires at the coil module to an engine ground and crank the engine. The voltmeter readings should be the same as in step 4. If these voltage signals are satisfactory at the sensor terminals but low at the coil module, repair the wires between the coil module and the sensor.
6. If the 18X and 3X signals are satisfactory at the coil module terminals, remove the coil assembly-to-module screws and lift the coil assembly up from the module. Connect a 12-V test lamp across each pair of coil primary terminals and crank the engine. If the test lamp does not flash on any pair of terminals, the coil module is defective.

Cam Sensor Timing for 3.8-L Turbocharged Engines

If the cam sensor is removed from the engine on 3.8-L turbocharged engines, the sensor must be timed to the engine when the sensor is installed. The cam sensor gear has a dot that must be positioned opposite the sensor disc window prior to sensor installation. As the cam sensor is installed in the engine, the gear dot must face away from the timing chain toward the passenger's side of the vehicle. When the cam sensor is installed in the engine, the sensor wiring harness must face toward the driver's side of the vehicle. Follow this procedure for cam sensor installation and timing:

1. Remove the spark plug wires from the coil assembly.
2. Remove the number 1 spark plug and crank the engine until compression is felt at the spark plug hole.
3. Slowly crank the engine until the timing mark lines up with the 0° position on the timing indicator.
4. Measure 1.47 to 1.5 in. (3.7 to 3.8 cm) from the 0° position toward the after TDC position on the crankshaft pulley, and mark the pulley at this location.
5. Slowly crank the engine until the mark placed on the pulley in step 4 is lined up with the 0° position on the timing indicator.
6. Use a weatherpack terminal removal tool to remove the center terminal B in the cam sensor connector, and connect a short jumper wire between this wire and the terminal in the connector. Terminal B is the cam sensor signal wire.
7. Connect a digital voltmeter from the cam sensor signal wire to an engine ground, and turn on the ignition switch.
8. Rotate the cam sensor until the voltmeter reading changes from high volts (5 V to 12 V) to low volts (0 V to 2 V).
9. Hold the cam sensor in this position and tighten the cam sensor-to-block retaining bolt.

Diagnosis of EI Systems with Magnetic Sensors

Magnetic Sensor Tests

With the wiring harness connector to the magnetic sensor disconnected and an ohmmeter calibrated on the X10 scale connected across the sensor terminals, the meter should read 900 Ω to 1,200 Ω on 2.0-L, 2.8-L, and 3.1-L engines. The meter should indicate 500 Ω to 900 Ω on a Quad 4 engine and 800 Ω to 900 Ω on a 2.5-L engine (Figure 4-37).

Meter readings below the specified value indicate a shorted sensor winding, whereas infinite meter readings prove that the sensor winding is open. Since these sensors are mounted in the crankcase, they are continually splashed with engine oil. In some sensor failures, the engine oil enters the sensor and causes a shorted sensor winding. If the magnetic sensor is defective, the engine fails to start.

With the magnetic sensor wiring connector disconnected, an alternating current (ac) voltmeter may be connected across the sensor terminals to check the sensor signal while the engine is cranking. On 2.0-L, 2.8-L, and 3.1-L engines, the sensor signal should be 100 millivolts (mV) ac. The

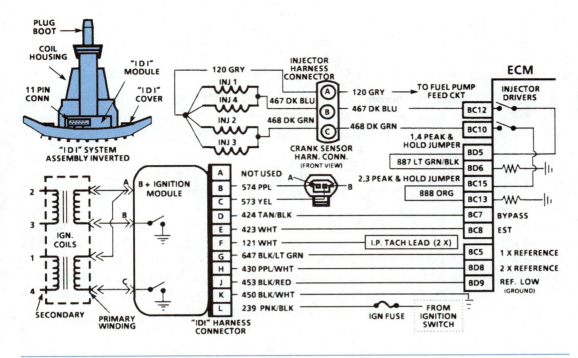

Figure 4-37 Terminal identification for an EI system 2.3-L Quad 4 engine (Courtesy of Oldsmobile Division, General Motors Corporation)

sensor voltage on a Quad 4 engine should be 200 mV ac. When the sensor is removed from the engine block, a flat steel tool placed near the sensor should be attracted to the sensor if the sensor magnet is satisfactory.

No-Start Diagnosis for an EI System with a Magnetic Sensor

When an engine with an EI system and a magnetic sensor fails to start, complete steps 1, 2, and 3 of the No-Start Diagnosis for EI Type 1 and Type 2 Systems, and then follow this procedure:

1. If the test spark plug did not fire on any of the spark plugs, check for 12 V at the coil module voltage input terminals. Consult the wiring diagram for the system being tested for terminal identification.
2. If 12 V are supplied to the appropriate coil module terminals, test the magnetic sensor as explained under Magnetic Sensor Tests.
3. When the magnetic sensor tests are satisfactory, the coil module is probably defective.

Some coil module changes were made on 1989 EI systems with magnetic sensors, and 1989 coil modules will operate satisfactorily on 1988 EI systems. However, if a 1988 module is installed on a 1989 EI system, the malfunction indicator light (MIL) will come on and code 41 will be stored in the PCM memory.

Engine Misfire Diagnosis

✓ SERVICE TIP: When diagnosing any computer system, never forget the basics. For example, always be sure the engine has satisfactory compression and ignition before attempting to diagnose the computer-controlled fuel injection.

If the engine misfires all the time or on acceleration only, test the following components:

1. Engine compression
2. Spark plugs

Classroom Manual
Chapter 4, page 95

3. Spark plug wires
4. Ignition coils—test for firing voltage with a test spark plug
5. Crankshaft sensor
6. Fuel injectors on multiport and sequential fuel injection systems

● CUSTOMER CARE: Some intermittent automotive problems are difficult to diagnose unless we can catch the car in the act. In other words, the problem is hard to diagnose if the symptoms are not present while we are performing diagnostic tests.

One solution to this problem is to have the customer leave the vehicle with the shop, and then drive the car under the conditions when the problem occurs. However, if the problem only appears once every week, this solution is not likely to work, because the problem will not occur in the short time we have to drive the car.

A second solution, especially in no-start situations, is to have the customer phone the shop immediately when the problem occurs, and send a technician to diagnose the problem. If this solution is attempted, inform the customer not to attempt starting the car until the technician arrives. This solution may be expensive, but in some cases, it may be the only way to diagnose the problem successfully. Always be willing to go the extra mile to diagnose and correct the customer's problem. By doing so, you will obtain a lot of satisfied, repeat customers.

Guidelines for Servicing Distributor and Electronic Ignition Systems

1. If a 12-V test light connected from the negative primary coil terminal to ground does not flash when the engine is cranked, the pickup coil or module is defective.
2. When a 12-V test light connected from the negative primary coil terminal to ground flashes while cranking the engine, but a test spark plug connected from the coil secondary wire to ground does not fire while cranking the engine, the coil is defective.
3. When a test spark plug connected from the coil secondary wire to ground fires while cranking the engine, but fails to fire when connected from the spark plug wires to ground, the distributor cap or rotor is defective.
4. Ignition modules may be tested with a module tester to determine if they are capable of triggering the primary ignition circuit on and off.
5. The ohmmeter leads may be connected to the pickup coil leads to test the pickup coil for open and shorted circuits.
6. The ohmmeter leads may be connected from one of the pickup leads to ground to check the pickup for a grounded condition.
7. The gap between the reluctor high points and the pickup coil may be adjusted on some distributors. This gap should be measured with a nonmagnetic feeler gauge.
8. The ohmmeter leads may be connected to the primary coil terminals to check the primary winding for open and shorted circuits.
9. The ohmmeter leads may be connected from one of the primary terminals to the coil tower to check the secondary winding for open and shorted circuits.
10. The ohmmeter tests on the ignition coil windings do not check the coil for insulation leakage.
11. The maximum secondary coil voltage test with a test spark plug is an accurate indication of coil condition.
12. Spark plug wires should be routed so two wires that fire one after the other are not positioned beside each other.
13. Ignition modules mounted on the distributor housing must have silicone grease on the module mounting surface to prevent module overheating.

14. When the distributor is timed to the engine, the number 1 piston should be at TDC on the compression stroke with the timing marks aligned, and the rotor should be under the number 1 spark plug wire terminal in the distributor cap with one of the high points aligned with the pickup coil.
15. The ignition spark advance may be checked with the advance control and the spark advance meter on the timing light.
16. A magnetic offset adjustment must be set to the manufacturer's specifications on the magnetic timing meter prior to an ignition timing check.
17. Spark plug carbon conditions are indicators of cylinder and combustion chamber operation.
18. Ignition point dwell is increased when the point gap is decreased.
19. On many vehicles, such as General Motors and Ford vehicles, a timing connector must be disconnected while checking basic timing.
20. The emission control information label provides the location of the timing connector on many General Motors vehicles.
21. On Chrysler vehicles the computer system must be in the limp-in mode while checking basic timing.
22. A defective automatic shutdown (ASD) relay will cause a no-start problem on Chrysler products.
23. The voltage supply and ground connection should be checked on Hall Effect pickups before the voltage signal from the pickup is checked with the engine cranking.
24. After the voltage supply and ground wires have been checked on an optical pickup, the voltage signal from the pickup should fluctuate between 0 V and 5 V while cranking the engine.
25. Prior to checking the voltage signal from a crankshaft or camshaft sensor on an EI system, the sensor voltage supply and ground wire should be checked.
26. The voltage signal from crankshaft or camshaft sensors may be checked with a 12-V test light or a digital voltmeter while cranking the engine.
27. On some crankshaft and camshaft sensors, such as those in Chrysler products, a paper shim must be attached to the sensor tip prior to sensor installation. The sensor must be installed until the paper shim lightly contacts the rotating ring.
28. Primary and secondary coil windings may be checked with an ohmmeter on EI coils.
29. A special tool is used to set the crankshaft sensor position in relation to the crankshaft and crankshaft pulley vanes on some EI systems.
30. The same tool used to adjust the crankshaft sensor is also used to check the crankshaft pulley vanes on some EI systems.
31. If the camshaft sensor is mounted in the previous distributor opening, this sensor must be timed to the engine when it is installed.
32. The winding in magnetic sensors on EI systems may be tested with an ohmmeter.
33. The voltage signal from a magnetic sensor in an EI system may be measured in ac millivolts.

CASE STUDY

A customer complained about a stalling problem on an Oldsmobile 88 with an EI system. When questioned about this problem, the owner said the engine stalled while driving in the city, but it only happened about once a week. Further questioning of the customer indicated the engine would restart after 5 to 10 minutes, and the owner said the engine seemed to be flooded.

The technician performed voltmeter and ohmmeter tests, and an oscilloscope diagnosis on the EI system. There were no defects in the system and the engine operation was satisfactory. The fuel pump pressure was tested and the filter checked for contamination, but no problems were discovered in these components. The customer was informed that it was difficult to diagnose this problem when the symptoms were not present. The service writer asked the customer to phone the shop immediately the next time the engine stalled, without attempting to restart the car. The service writer told the customer that a technician would be sent out to check the problem when he phoned the shop.

Approximately ten days later, the customer phoned and said the car had stalled. A technician was dispatched immediately, and she connected a test spark plug to several of the spark plug wires. The ignition system was not firing any of the spark plugs. The car was towed to the shop without any further attempts to start the engine.

The technician discovered that the voltage supply and the ground wires on the crankshaft sensor were normal, but there was no voltage signal from this sensor while cranking the engine. The crankshaft sensor was replaced and adjusted properly. When the customer returned later for other service, he reported the stalling problem had been eliminated.

Terms to Know

Automatic shutdown (ASD) relay
Distributor ignition (DI) system
Electronic ignition (EI) system
Hall Effect pickup
Ignition cross-firing
Ignition module tester
Magnetic sensor
Magnetic timing meter
Magnetic timing offset
Magnetic timing probe receptacle
Optical-type pickup
Reference pickup
Silicone grease
Synchronizer (SYNC) pickup
Tachometer (TACH) terminal
Test spark plug
Timing connector

ASE Style Review Questions

1. While discussing no-start diagnosis with a test spark plug:
   Technician A says if the test light flutters at the coil tach terminal, but the test spark plug does not fire when connected from the coil secondary wire to ground with the engine cranking, the ignition coil is defective.
   Technician B says if the test spark plug fires when connected from the coil secondary wire to ground with the engine cranking, but the test spark plug does not fire when connected from the spark plug wires to ground, the cap or rotor is defective.
   Who is correct?
   A. A only
   B. B only
   C. Both A and B
   D. Neither A nor B

2. While discussing a pickup coil test with an ohmmeter connected to the pickup leads:
   Technician A says an ohmmeter reading below the specified resistance indicates the pickup coil is grounded.
   Technician B says an ohmmeter reading below the specified resistance indicates the pickup coil is open.
   Who is correct?
   A. A only
   B. B only
   C. Both A and B
   D. Neither A nor B

3. While discussing ignition coil ohmmeter tests:
   Technician A says the ohmmeter should be placed on the X1,000 scale to test the secondary winding.
   Technician B says the ohmmeter tests on the coil check the condition of the winding insulation.
   Who is correct?
   A. A only
   B. B only
   C. Both A and B
   D. Neither A nor B

4. While discussing timing of the distributor to the engine with the number 1 piston at TDC compression and the timing marks aligned:
   Technician A says the distributor must be installed with the rotor under the number 1 spark plug terminal in the distributor cap and one of the reluctor high points aligned with the pickup coil.
   Technician B says the distributor must be installed with the rotor under the number 1 spark plug terminal in the distributor cap and the reluctor high points out of alignment with the pickup coil.
   Who is correct?
   A. A only
   B. B only
   C. Both A and B
   D. Neither A nor B

5. While discussing EI system coil test results in which one primary winding has 0.5 Ω resistance and the specified resistance is 1 Ω:
   Technician A says the primary winding in this coil is grounded.
   Technician B says the primary winding in this coil is shorted.
   Who is correct?
   A. A only
   B. B only
   C. Both A and B
   D. Neither A nor B

6. While discussing basic ignition timing adjustment on vehicles with computer-controlled distributor ignition (DI):
   Technician A says on some DI systems, a timing connector must be disconnected.
   Technician B says the distributor must be rotated until the timing mark appears at the specified location on the timing indicator.
   Who is correct?
   A. A only
   B. B only
   C. Both A and B
   D. Neither A nor B

7. While discussing the diagnosis of an electronic ignition (EI) system in which the crankshaft and camshaft sensor tests are satisfactory, but a test spark plug connected from the spark plug wires to ground does not fire:
   Technician A says the coil assembly may be defective.
   Technician B says the voltage supply wire to the coil assembly may be open.
   Who is correct?
   A. A only
   B. B only
   C. Both A and B
   D. Neither A nor B

8. While discussing EI service and diagnosis:
   Technician A says the crankshaft sensor may be rotated to adjust the basic ignition timing.
   Technician B says the crankshaft sensor may be moved to adjust the clearance between the sensor and the rotating blades on some EI systems.
   Who is correct?
   A. A only
   B. B only
   C. Both A and B
   D. Neither A nor B

9. While discussing the EI system crankshaft and camshaft sensors that require a paper spacer on the sensor tip prior to installation:
   Technician A says the sensor should be installed so the paper spacer lightly touches the rotating sensor ring.
   Technician B says the sensor should be installed so the paper spacer lightly touches the rotating sensor ring and then pulled outward 0.125 in.
   Who is correct?
   A. A only
   B. B only
   C. Both A and B
   D. Neither A nor B

10. While discussing engine misfire diagnosis:
    Technician A says a defective EI coil may cause cylinder misfiring.
    Technician B says the engine compression should be verified first if the engine is misfiring continually.
    Who is correct?
    A. A only
    B. B only
    C. Both A and B
    D. Neither A nor B

Table 4-1 ASE Task

Diagnose no-starting, hard starting, engine misfire, poor driveability, spark knock, power loss, and poor mileage problems on vehicles with electronic ignition systems and determine needed repairs.

Problem Area	Symptoms	Possible Causes	Classroom Manual	Shop Manual
ENGINE PERFORMANCE	No-starting or hard starting	1. Defective coil	75	77
		2. Defective cap and rotor	75	78
		3. Open secondary coil wire	75	79
		4. Low or zero primary voltage at the coil	75	75
		5. Fouled spark plugs	76	88
		6. Defective pickup	73	76
	Engine misfire	1. Low cylinder compression	76	73
		2. Intake manifold vacuum leaks	76	73
		3. High resistance in spark plug wires, coil secondary wire, or cap terminals	75	79
		4. Electrical leakage in the cap, rotor, plug wires, coil wire, or coil tower	75	78
		5. Defective coil	75	77
		6. Defective spark plugs	75	87
		7. Low primary voltage and current	75	74
		8. Improperly routed spark plug wires	75	79
		9. Worn distributor bushings	75	81
	Poor driveability, power loss	1. Low engine compression	76	73
		2. Restricted exhaust or air intake	71	74
		3. Late ignition timing	71	83
		4. Insufficient centrifugal or vacuum advance	76	87
		5. Cylinder misfiring	76	73
	Spark knock	1. Higher than normal compression	71	74
		2. Ignition timing too far advanced	71	85
		3. Excessive centrifugal or vacuum advance	71	85
		4. Spark plug heat range too hot	76	87
		5. Improperly routed spark plug wires	75	79
FUEL ECONOMY	Low fuel mileage	1. Low engine compression	71	74
		2. Late ignition timing	71	74
		3. Lack of spark advance	71	83
		4. Cylinder misfiring	71	73

Table 4-2 ASE Task

Inspect, test, repair, or replace ignition primary circuit wiring and components.

Problem Area	Symptoms	Possible Causes	Classroom Manual	Shop Manual
ENGINE PERFORMANCE	No-start	1. Defective pickup coil	73	76
		2. Defective ignition coil	75	77
		3. Defective module	74	75
		4. Defective primary circuit wiring	72	74
	Engine misfiring	1. Defective coil	76	77
		2. Defective pickup coil or improper gap	73	76
		3. High resistance in primary circuit wiring	72	74

Table 4-3 ASE Task

Inspect, test, service, repair, or replace ignition system secondary circuit wiring and components.

Problem Area	Symptoms	Possible Causes	Classroom Manual	Shop Manual
ENGINE PERFORMANCE	No-start	1. Defective coil	76	77
		2. Defective cap and rotor	75	78
		3. Fouled spark plugs	75	88
		4. Defective secondary coil wire	73	79
	Engine misfiring	1. Defective coil	75	77
		2. Defective cap and rotor	75	78
		3. Defective spark plug wires or secondary coil wire	75	79

Table 4-4 ASE Task

Inspect, test, and replace ignition coils.

Problem Area	Symptoms	Possible Causes	Classroom Manual	Shop Manual
ENGINE PERFORMANCE	No-start	Defective coil	75	77
	Engine misfiring	Defective coil	75	77

Table 4-5 ASE Task

Check and adjust ignition system timing and timing advance/retard.

Problem Area	Symptoms	Possible Causes	Classroom Manual	Shop Manual
NOISE	Spark knock	1. Ignition timing too far advanced	71	83
		2. Excessive centrifugal or vacuum advance	71	74
ENGINE PERFORMANCE	Power loss, poor driveability	1. Late ignition timing	71	74
		2. Insufficient centrifugal or vacuum advance	71	83
FUEL ECONOMY	Low fuel mileage	1. Late ignition timing		
		2. Insufficient centrifugal or vacuum advance	71	74

Table 4-6 ASE Task

Inspect, test, and replace electronic ignition wiring harness and connectors.

Problem Area	Symptoms	Possible Causes	Classroom Manual	Shop Manual
ENGINE PERFORMANCE	No-start	Open circuit in primary ignition wiring	72	74

Table 4-7 ASE Task

Inspect, test, and replace electronic ignition system pickup sensor or triggering devices.

Problem Area	Symptoms	Possible Causes	Classroom Manual	Shop Manual
ENGINE PERFORMANCE	No-start	Defective or improperly adjusted pickup coil	73	76

Table 4-8 ASE Task

Inspect, test, and replace electronic ignition system control unit (module).

Problem Area	Symptoms	Possible Causes	Classroom Manual	Shop Manual
ENGINE PERFORMANCE	No-start	Defective ignition module	74	75

Fuel Tank, Line, Filter, and Pump Diagnosis and Service

CHAPTER 5

Upon completion and review of this chapter, you should be able to:

- ❏ Test alcohol content in fuel.
- ❏ Relieve fuel system pressure.
- ❏ Inspect fuel tanks.
- ❏ Drain fuel tanks.
- ❏ Remove and replace fuel tanks.
- ❏ Remove, inspect, service, and replace electric fuel pumps and gauge sending units.
- ❏ Flush fuel tanks.
- ❏ Purge fuel tanks.
- ❏ Inspect and service nylon fuel lines.
- ❏ Inspect and service steel fuel tubing.
- ❏ Remove and replace fuel filters.
- ❏ Service and test mechanical fuel pumps.
- ❏ Remove and replace mechanical fuel pumps.

Alcohol in Fuel Test

Some gasoline may contain a small quantity of alcohol. The percentage of alcohol mixed with the fuel usually does not exceed 10%. An excessive quantity of alcohol mixed with gasoline may result in fuel system corrosion, fuel filter plugging, deterioration of rubber fuel system components, and a lean air-fuel ratio. These fuel system problems caused by excessive alcohol in the fuel may cause driveability complaints such as lack of power, acceleration stumbles, engine stalling, and no-start.

Some vehicle manufacturers supply test equipment to check the level of alcohol in the gasoline. The following alcohol in fuel test procedure requires only the use of a calibrated cylinder:

1. Obtain a 100-mL cylinder graduated in 1-mL divisions.
2. Fill the cylinder to the 90-mL mark with gasoline.
3. Add 10-mL of water to the cylinder so it is filled to the 100-mL mark.
4. Install a stopper in the cylinder, and shake it vigorously for 10 to 15 seconds.
5. Carefully loosen the stopper to relieve any pressure.
6. Install the stopper and shake vigorously for another 10 to 15 seconds.
7. Carefully loosen the stopper to relieve any pressure.
8. Place the cylinder on a level surface for 5 minutes to allow liquid separation.
9. Any alcohol in the fuel is absorbed by the water and settles to the bottom. If the water content in the bottom of the cylinder exceeds 10-mL, there is alcohol in the fuel. For example, if the water content is now 15-mL, there was 5% alcohol in the fuel.

Since this procedure does not extract 100% of the alcohol from the fuel, the percentage of alcohol in the fuel may be higher than indicated.

Basic Tools

Basic technician's tool set

Service manual

100-mL graduated cylinder

Approved gasoline containers

Fuel System Pressure Relief

CAUTION: Failure to relieve the fuel pressure on electronic fuel injection (EFI) systems prior to fuel system service may result in gasoline spills, serious personal injury, and expensive property damage.

CAUTION: Make sure no one smokes near a vehicle while servicing gasoline fuel system components. Such action may result in serious personal injury and property damage.

113

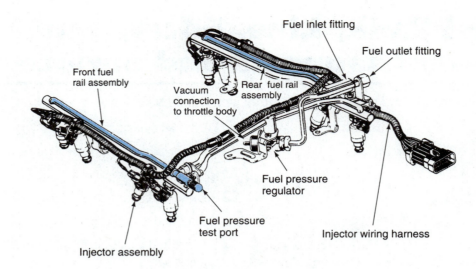

Figure 5-1 Fuel rail with pressure test port on a port fuel injected (PFI) engine (Courtesy of Cadillac Motor Car Division, General Motors Corporation)

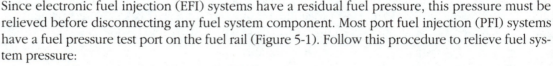

CAUTION: When servicing gasoline fuel system components, flames, sparks, or sources of ignition in the area may result in an explosion causing serious personal injury and property damage.

CAUTION: Always wear eye protection and observe all other safety rules to avoid personal injury when servicing fuel system components.

Since electronic fuel injection (EFI) systems have a residual fuel pressure, this pressure must be relieved before disconnecting any fuel system component. Most port fuel injection (PFI) systems have a fuel pressure test port on the fuel rail (Figure 5-1). Follow this procedure to relieve fuel system pressure:

1. Disconnect the negative battery cable to avoid fuel discharge if an accidental attempt is made to start the engine.
2. Loosen the fuel tank filler cap to relieve any fuel tank vapor pressure.
3. Wrap a shop towel around the fuel pressure test port on the fuel rail and remove the dust cap from this valve.
4. Connect the fuel pressure gauge to the fuel pressure test port on the fuel rail (Figure 5-2).

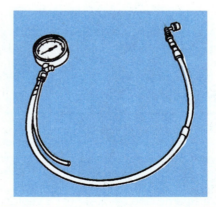

Figure 5-2 Fuel pressure gauge for engines with port fuel injection (PFI) (Courtesy of Cadillac Motor Car Division, General Motors Corporation)

Electronic fuel injection (EFI) is a generic term that may be applied to any computer-controlled fuel injection system.

Port fuel injection (PFI) systems have a fuel injector located in each intake port, and these systems usually have a fuel pressure test port on the fuel rail to which a pressure gauge may be connected for test purposes.

The fuel pressure test port may be called a Schrader valve.

Special Tools

Fuel pressure gauge, electronic fuel injection (EFI)

5. Install the bleed hose on the gauge in an approved gasoline container, and open the gauge bleed valve to relieve fuel pressure from the system into the gasoline container. Be sure all the fuel in the bleed hose is drained into the gasoline container.

On EFI systems that do not have a fuel pressure test port, such as most throttle body injection (TBI) systems, follow these steps to relieve fuel system pressure:

1. Loosen the fuel tank filler cap to relieve any tank vapor pressure.
2. Remove the fuel pump fuse.
3. Start and run the engine until the fuel is used up in the fuel system and the engine stops.
4. Engage the starter for 3 seconds to relieve any remaining fuel pressure.
5. Disconnect the negative battery terminal to avoid possible fuel discharge if an accidental attempt is made to start the engine.

Throttle body injection (TBI) systems have one or two fuel injectors positioned above the throttle, or throttles, in the throttle body assembly.

Fuel Tank Service

Fuel Tank Inspection

The fuel tank should be inspected for leaks, road damage, corrosion and rust on metal tanks, loose, damaged, or defective seams, loose mounting bolts, and damaged mounting straps. Leaks in the fuel tank, lines, or filter may cause gasoline odor in and around the vehicle, especially during low-speed driving and idling. In most cases, the fuel tank must be removed for servicing.

Fuel Tank Draining

 SERVICE TIP: When a fuel tank must be removed, if possible, inform the customer to bring the vehicle to the shop with a minimum amount of fuel in the tank.

 WARNING: Always drain gasoline into an approved container and use a funnel to avoid gasoline spills.

 WARNING: When servicing fuel system components, always place a Class B fire extinguisher near the work area.

The fuel tank must be drained prior to tank removal. If the tank has a drain bolt, this bolt may be removed and the fuel drained into an approved container (Figure 5-3).

If the fuel tank does not have a drain bolt, follow these steps to drain the fuel tank:

1. Remove the negative battery cable.
2. Raise the vehicle on a hoist.
3. Locate the fuel tank drain pipe and remove the drain pipe plug.
4. Install the appropriate adapter in the fuel tank drain pipe, and connect the intake hose from a hand-operated or air-operated pump to this adapter (Figure 5-4). If the fuel tank does not have a drain pipe, install the pump hose through the filler pipe into the fuel tank.
5. Install the discharge hose from the hand-operated or air-operated pump into an approved gasoline container, and operate the pump until all the fuel is removed from the tank.

Special Tools

Hand- or air-operated pump

Fuel Tank Removal

The fuel tank removal procedure varies depending on the vehicle make and year. Always follow the procedure in the vehicle manufacturer's service manual. Following is a typical fuel tank removal procedure:

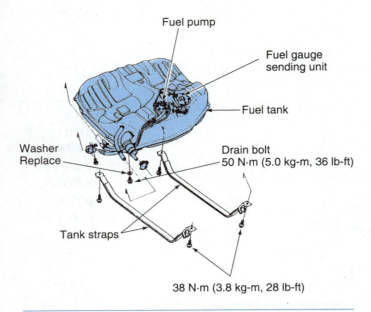

Figure 5-3 Fuel tank with drain bolt (Courtesy of Honda Motor Co., Ltd.)

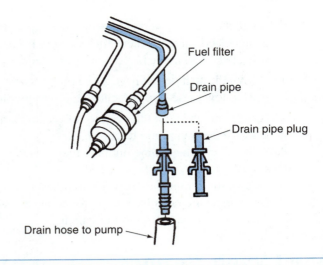

Figure 5-4 Fuel tank drain pipe with adapter (Courtesy of Cadillac Motor Car Division, General Motors Corporation)

Special Tools

Quick-disconnect fuel line fitting removal tools

Classroom Manual
Chapter 5, page 113

1. Relieve the fuel system pressure, and drain the fuel tank.
2. Raise the vehicle on a hoist, or lift the vehicle with a floor jack and lower the chassis onto jack stands.
3. Use compressed air to blow dirt from the fuel line fittings and wiring connectors.
4. Remove the fuel tank wiring harness connector from the body harness connector.
5. Remove the ground wire retaining screw from the chassis if used.
6. Disconnect the fuel lines from the fuel tank. If these lines have quick-disconnect fittings, follow the manufacturer's recommended removal procedure in the service manual. Some quick-disconnect fittings are hand-releasable, and others require the use of a special tool (Figure 5-5).
7. Wipe the filler pipe and vent pipe hose connections with a shop towel, and then disconnect the hoses from the filler pipe and vent pipe to the fuel tank.
8. Support the fuel tank with a transmission jack, and remove the front and rear tank strap attaching bolts (Figure 5-6).
9. Remove the tank straps, and then lower the transmission jack to remove the fuel tank.

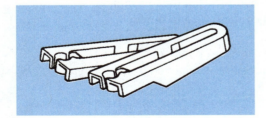

Figure 5-5 Tools required for quick-disconnect fuel line fittings (Courtesy of Cadillac Motor Car Division, General Motors Corporation)

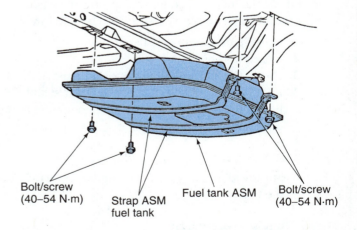

Figure 5-6 Front and rear tank strap retaining bolts (Courtesy of Oldsmobile Division, General Motors Corporation)

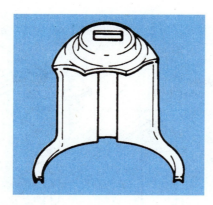

Figure 5-7 Special tool for removing fuel pump and gauge sending unit from the fuel tank (Courtesy of Oldsmobile Division, General Motors Corporation)

Electric Fuel Pump Removal and Replacement and Fuel Tank Cleaning

1. Remove the fuel tank from the vehicle.
2. Follow the vehicle manufacturer's recommended procedure to remove the fuel pump and gauge sending unit from the fuel tank. In many cases, a special tool must be used to remove this assembly (Figure 5-7).
3. Check the filter on the fuel pump inlet. If the filter is contaminated or damaged, replace the filter.
4. Inspect the fuel pump inlet for dirt and debris. Replace the fuel pump if these foreign particles are found in the pump inlet.
5. If the pump inlet filter is contaminated, flush the tank with hot water for at least 5 minutes.
6. Dump all the water from the tank through the pump opening in the tank. Shake the tank to be sure all the water is removed.
7. Check all fuel hoses and tubing on the fuel pump assembly. Replace fuel hoses that are cracked, deteriorated, or kinked. When fuel tubing on the pump assembly is damaged, replace the tubing or the pump.
8. Be sure the sound insulator sleeve is in place on the electric fuel pump, and check the position of the sound insulator on the bottom of the pump (Figure 5-8).
9. Clean the pump and sending unit mounting area in the fuel tank with a shop towel, and install a new gasket or O-ring on the pump and sending unit (Figure 5-9). On some tanks, the gauge sending unit and fuel pump are mounted separately (Figure 5-10).
10. Install the fuel pump and gauge sending unit assembly in the fuel tank and secure this assembly in the tank using the vehicle manufacturer's recommended procedure. On some vehicles, this procedure involves the use of a special tool. On some vehicles with a separate fuel pump and gauge sending unit, a lock ring must be rotated into place with a brass drift and a hammer to secure each of these units (Figure 5-11).

Fuel Tank Purging

CAUTION: Always wear eye protection and protective gloves when purging a fuel tank.

CAUTION: When handling emulsifying agents, always follow the precautions recommended by the agent manufacturer. Failure to follow these precautions may result in personal injury.

Classroom Manual Chapter 5, page 121

Special Tools

Electric fuel pump removal and replacement tools

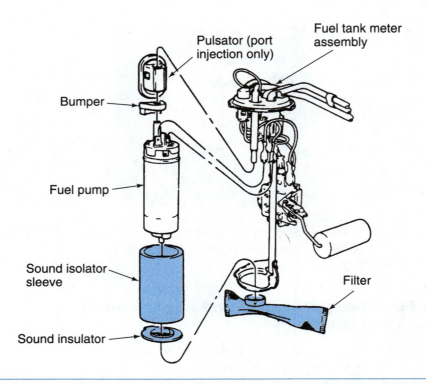

Figure 5-8 Electric fuel pump and gauge sending unit with filter, sound insulator sleeve, and sound insulator (Courtesy of Oldsmobile Division, General Motors Corporation)

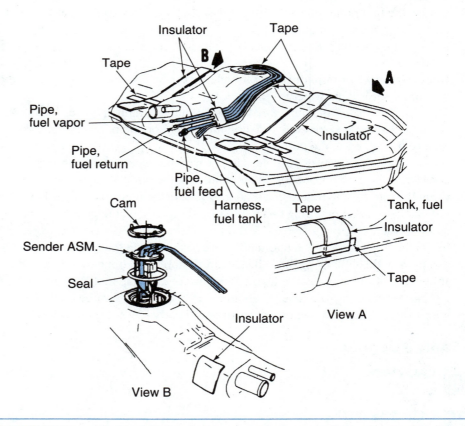

Figure 5-9 New gasket installed on fuel pump and gauge sending unit assembly (Courtesy of Oldsmobile Division, General Motors Corporation)

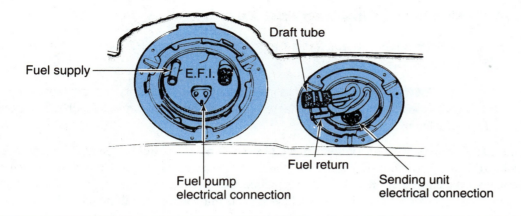

Figure 5-10 Electric fuel pump and gauge sending unit mounted separately in the fuel tank (Courtesy of Chrysler Corporation)

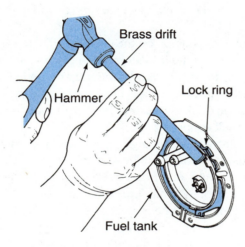

Figure 5-11 Installing the fuel pump lock ring with a brass drift and a hammer (Courtesy of Chrysler Corporation)

 WARNING: When disposing of contaminated fuel or emulsifying agents, obey all local environmental regulations.

The fuel tank purging procedure provides additional cleaning and removal of gasoline vapors. Following is a typical fuel tank purging procedure:

1. Remove the fuel tank from the vehicle.
2. Follow the vehicle manufacturer's recommended procedure to remove the fuel pump and gauge sending unit from the fuel tank.
3. Be sure all the gasoline is removed from the tank, and fill the tank with tap water.
4. Vigorously agitate the tank, and drain the tank.
5. Mix an emulsifying agent such as Product-Sol No. 913 or its equivalent with the amount of water recommended by the emulsifying agent manufacturer, and pour this mixture into the fuel tank. Then fill the tank with water.
6. Agitate the tank for 10 minutes, and then drain the tank.
7. Refill the tank completely with water, and then completely empty the tank.

After the purging procedure, an explosion meter should be used to determine if gasoline vapors remain in the fuel tank.

Fuel Tank Steam Cleaning and Repairing

WARNING: Do not steam clean plastic fuel tanks. This procedure may damage these tanks.

WARNING: Empty fuel tanks may contain gasoline vapors, making them extremely dangerous! Do not allow any flames, sparks, or other sources of ignition near an empty gasoline fuel tank.

Some manufacturers of plastic or metal fuel tanks recommend tank replacement if the tank is leaking. Repair kits are available for plastic fuel tanks, and if these kits are used, the kit manufacturer's instructions must be carefully followed. Metal tanks may be steam cleaned prior to tank repairs to remove all gasoline residue and vapors. The steam cleaning and repair of metal fuel tanks should be done by a radiator and fuel tank specialty shop.

Fuel Tank Installation

1. Be sure the electric fuel pump and gauge sending unit are securely and properly installed in the fuel tank.
2. Raise the fuel tank into position on the chassis with a transmission jack if the vehicle is raised on a hoist. Be sure the insulators are in place on top of the fuel tank.
3. Install the fuel tank straps, and tighten the strap mounting bolts to the specified torque.
4. If the fuel lines have quick-disconnect fittings, be sure the large collar on these fittings is rotated back to the original position. Be sure the springs are visible on the inside diameter of the quick connector.
5. Place one or two drops of clean engine oil on the male tube ends where the quick connectors will be installed. Install all the fuel lines and electrical connections to the fuel pump and gauge sending unit. Be sure these lines and wires are properly secured, and check to be sure they do not interfere with other components.
6. Install and tighten the filler pipe hose and vent hose connections to the fuel tank.
7. Check the filler cap for damage. If the cap has pressure and vacuum valves, be sure these valves are working freely and are not damaged.
8. Install some fuel in the tank and cycle the ignition switch several times on an EFI system to pressurize the fuel system.
9. Start the engine and check the fuel tank and all line connections for leaks.

Fuel Line Service

Nylon Fuel Pipe Inspection and Service

CAUTION: Always cover a nylon fuel pipe with a wet shop towel before using a torch or other source of heat near the line. Failure to observe this precaution may result in fuel leaks, personal injury, and property damage.

WARNING: If a vehicle has nylon fuel pipes, do not expose the vehicle to temperatures above 239°F (115°C) for more than 1 hour or to temperatures above 194°F (90°C) for any extended period to avoid damage to the fuel pipes.

WARNING: Do not nick or scratch nylon fuel pipes. If damaged, these fuel pipes must be replaced.

Nylon fuel pipes should be inspected for leaks, nicks, scratches, cuts, kinks, melting, and loose fittings. If these fuel pipes are kinked or damaged in any way, they must be replaced. Nylon fuel pipes must be secured to the chassis at regular intervals to prevent fuel pipe wear and vibration.

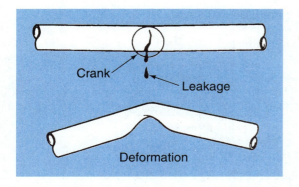

Figure 5-12 Steel tubing should be inspected for leaks, kinks, and deformation. (Courtesy of Toyota Motor Corporation)

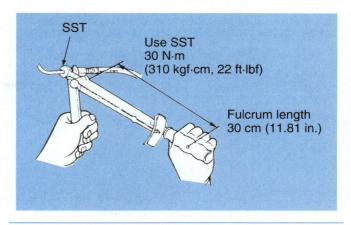

Figure 5-13 If fittings on fuel tubing are loose, they must be tightened to the specified torque. (Courtesy of Toyota Motor Corporation)

Nylon fuel pipes provide a certain amount of flexibility and can be formed around gradual curves under the vehicle. Do not force a nylon fuel pipe into a sharp bend, because this action may kink the pipe and restrict the flow of fuel. When nylon fuel pipes are exposed to gasoline, they may become stiffer, making them more subject to kinking. Be careful not to nick or scratch nylon fuel pipes.

Steel Fuel Tubing Inspection and Service

Steel fuel tubing should be inspected for leaks, kinks, and deformation (Figure 5-12). This tubing should also be checked for loose connections and proper clamping to the chassis. If the fuel tubing threaded connections are loose, they must be tightened to the specified torque (Figure 5-13). Damaged fuel tubing should be replaced.

 WARNING: O-rings in fuel line fittings are usually made from fuel-resistant Viton. Other types of O-rings must not be substituted for fuel fitting O-rings.

Some threaded fuel line fittings contain an O-ring. If the fitting is removed, the O-ring should be replaced. Flared ends are used on some steel fuel tubing. If a new flare is required, the old flare can be cut from the tubing with a pipe cutter, and a new flare can be made on the end of the tubing with a double flaring tool (Figure 5-14).

Special Tools

Double flaring tool

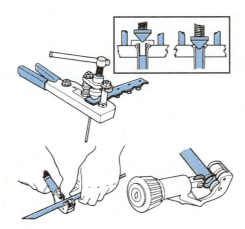

Figure 5-14 Cutting old flare from steel fuel tubing and making a new double flare (Courtesy of Chrysler Corporation)

121

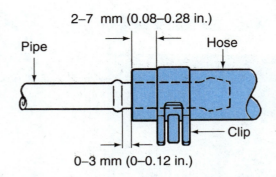

Figure 5-15 Rubber fuel hose installation on steel fitting or line (Courtesy of Toyota Motor Corporation)

Rubber Fuel Hose Inspection and Service

Rubber fuel hose should be inspected for leaks, cracks, cuts, kinks, oil soaking, and soft spots or deterioration. If any of these conditions are found, the fuel hose should be replaced. When rubber fuel hose is installed, the hose should be installed to the proper depth on the metal fitting or line (Figure 5-15).

The rubber fuel hose clamp must be properly positioned on the hose in relation to the steel fitting or line as illustrated in the figure. Fuel hose clamps may be spring-type or screw-type. Screw-type fuel hose clamps must be tightened to the specified torque.

Classroom Manual
Chapter 5, page 118

Fuel Filter Service

Fuel Filter Removal

Some vehicle manufacturers recommend fuel filter replacement at 30,000 miles (48,000 km). Always replace the fuel filter at the vehicle manufacturer's recommended mileage. If dirty or contaminated fuel is placed in the fuel tank, the filter may require replacing before the recommended mileage. A plugged fuel filter may cause the engine to surge and cut out at high speed, or to hesitate on acceleration. If a plastic fuel filter is used on a carbureted engine, contaminants may be seen in the filter. A restricted fuel filter causes low fuel pump pressure.

The fuel filter replacement procedure varies depending on the make and year of vehicle, and the type of fuel system. Always follow the filter replacement procedure in the vehicle manufacturer's service manual. Following is a typical filter replacement procedure on a vehicle with EFI:

CAUTION: On an engine with electronic fuel injection, never turn on the ignition switch or crank the engine with a fuel line disconnected. This action will result in gasoline discharge from the disconnected line, which may result in a fire causing personal injury and/or property damage.

1. Relieve fuel system pressure as mentioned previously.
2. Raise the vehicle on a hoist.
3. Flush the quick connectors on the filter with water and use compressed air to blow debris from the connectors.
4. Disconnect the inlet connector first. Grasp the large connector collar, twist in both directions, and pull the connector off the filter.
5. Disconnect the outlet connector using the same procedure used on the inlet connector (Figure 5-16).
6. Loosen and remove the filter mounting bolts, and remove the filter from the vehicle.

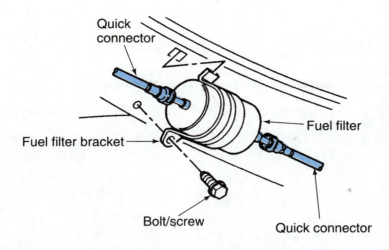

Figure 5-16 Removing fuel filter quick connectors (Courtesy of Oldsmobile Division, General Motors Corporation)

Fuel Filter Installation

Following is a typical filter installation procedure on a vehicle with EFI:

1. Use a clean shop towel to wipe the male tube ends of the new filter.
2. Apply a few drops of clean engine oil to the male tube ends on the filter.
3. Check the quick connectors to be sure the large collar on each connector has rotated back to the original position. The springs must be visible on the inside diameter of each quick connector.
4. Install the filter on the vehicle in the proper direction, and leave the mounting bolt slightly loose.
5. Install the outlet connector onto the filter outlet tube and press the connector firmly in place until the spring snaps into position.
6. Grasp the fuel line and try to pull this line from the filter to be sure the quick connector is locked in place.
7. Repeat steps 5 and 6 on the inlet connector.
8. Tighten the filter retaining bolt to the specified torque.
9. Lower the vehicle, start the engine, and check for leaks at the filter.

Photo Sequence 3 shows a typical procedure for relieving fuel pressure and removing a fuel filter.

Classroom Manual
Chapter 5, page 119

Mechanical Fuel Pump Service and Diagnosis

Mechanical Fuel Pump Inspection

A mechanical fuel pump should be inspected for fuel leaks. If gasoline is leaking at the line fittings, these fittings should be tightened to the specified torque. If the fuel leak is still present, replace the fittings and/or fuel line. When gasoline is leaking from the vent opening in the pump housing, the pump diaphragm is leaking, and fuel pump replacement is necessary.

If engine oil is leaking from the vent opening in the pump housing, the pull rod seal is worn, and pump replacement is required. A loose pivot pin may cause oil leaks between the pin and the pump housing. When oil is leaking between the pump housing and the engine block, the pump mounting bolts should be tightened to the specified torque. If the oil leak continues, the gasket between the pump housing and the engine block must be replaced.

Photo Sequence 3
Typical Procedure for Relieving Fuel Pressure and Removing a Fuel Filter

P3-1 Disconnect the negative battery cable.

P3-2 Loosen the fuel tank filler cap to relieve any fuel tank vapor pressure.

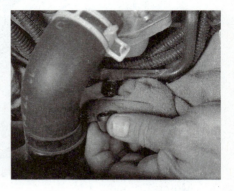

P3-3 Wrap a shop towel around the Schrader valve on the fuel rail, and remove the dust cap from this valve.

P3-4 Connect a fuel pressure gauge to the Schrader valve.

P3-5 Install the guage bleed hose into an approved gasoline container, and open the guage bleed valve to relieve the fuel pressure.

P3-6 Place the lift arms under the manufacturer's specified lift points on the vehicle, and lift the vehicle.

P3-7 Flush the fuel filter line connectors with water, and use compressed air to blow debris from the connectors.

P3-8 Follow the vehicle manufacturer's recommended procedure to remove the inlet connector.

P3-9 Follow the vehicle manufacturer's recommended procedure to remove the outlet connector, and remove the fuel filter.

The mechanical fuel pump should be checked for excessive noise. If the pump makes a clicking noise when the engine is operating at a fast idle, the small spring between the rocker arm and the pump housing is broken or weak.

Mechanical Fuel Pump Testing

The mechanical fuel pump should be tested for pressure and volume. Mechanical fuel pump testers vary depending on the tester manufacturer. Always use the fuel pump tester according to the tester manufacturer's recommended procedure. Following is a mechanical fuel pump test procedure with a typical fuel pump tester:

1. Connect the fuel pump tester between the carburetor inlet fuel line and the inlet nut (Figure 5-17). Since the tester fittings contain heavy rubber washers, the tester fittings only require hand tightening.
2. Place the clipped fuel hose from the tester into the calibrated container supplied with the tester. Be sure the clip is closing the fuel hose. Have a co-worker hold the calibrated container in an upright position away from rotating or hot components.
3. Start the engine and immediately check for fuel leaks at the gauge connections. If fuel leaks exist, shut off the engine immediately and repair the leaks.
4. If no fuel leaks exist, record the fuel pressure on the tester gauge with the engine idling.
5. Release the clip on the fuel discharge hose, and allow fuel to discharge into the calibrated container for the time specified in the vehicle manufacturer's fuel pump volume specifications. Usually this time is 30 to 45 seconds. At the end of the specified time, immediately close the clip on the fuel discharge hose, and check the amount of fuel in the calibrated container.
6. Pour the fuel in the calibrated container back into the fuel tank. Use a funnel to avoid gasoline spills.
7. Compare the fuel pump pressure recorded in step 4 and the volume obtained in step 5 to the vehicle manufacturer's specifications.

If the pressure or volume are less than specified, check the fuel filter for restrictions, and check the fuel lines for restrictions and leaks before replacing the fuel pump. When air leaks into the fuel line between the fuel pump and the fuel tank, the fuel pump discharges some air with the fuel, which reduces pump volume. When the fuel pump is removed, always check the camshaft lobe or eccentric for wear.

Special Tools
Fuel pressure and volume tester, carbureted engines

Fuel pump volume is the amount of fuel the pump will deliver in a specific length of time.

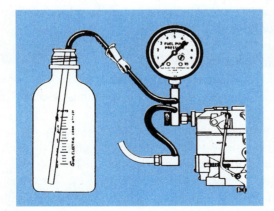

Figure 5-17 Tester connected to test mechanical fuel pump pressure and volume (Courtesy of Sun Electric Corporation)

Remove and Replace the Mechanical Fuel Pump

Follow these steps for a typical mechanical fuel pump removal and replacement procedure:

1. Place shop towels under the inlet and outlet fittings, and loosen these fittings. Remove the inlet and outlet lines from the pump.
2. Remove the fuel pump mounting bolts, and remove the fuel pump from the engine.
3. Use a scraper to clean the pump mounting surface on the engine block. Remove all the old gasket material.
4. Check the fuel pump cam lobe for wear.
5. If the fuel pump is to be reused, clean the pump mounting surface with a scraper.
6. Install a new gasket on the fuel pump mounting surface.
7. Install the fuel pump in the engine block, and alternately tighten the mounting bolts to the specified torque.
8. Connect the inlet and outlet fuel lines to the fuel pump, and tighten these fittings to the specified torque.

Classroom Manual
Chapter 5, page 120

Electric Fuel Pump Testing

The electric fuel pump test procedure varies depending on the type of fuel injection. Always follow the electric fuel pump test procedure in the vehicle manufacturer's service manual. Since electric fuel pumps in EFI systems are computer-controlled, the testing of these pumps is discussed later in the appropriate chapter.

● **CUSTOMER CARE:** When performing any undercar service, always perform a quick visual inspection of the fuel tank, lines, filter, and pump for fuel leaks and damaged components. If there is any evidence of a fuel leak or damaged components, advise the customer regarding the potential danger and the necessary repairs. In most cases, the customer will approve the necessary repairs. The customer will usually be impressed with your thorough inspection, and will likely return to the shop for other service.

Guidelines for Fuel Tank, Line, Filter, and Pump Service and Diagnosis

1. Excessive amounts of alcohol mixed with gasoline may cause fuel system corrosion, fuel filter plugging, deterioration of rubber fuel system components, and a lean air-fuel ratio.
2. Excessive amounts of alcohol in the fuel may cause lack of power, acceleration stumbles, engine stalling, or no-start.
3. On electronic fuel injection (EFI) systems, fuel system pressure relief is necessary before disconnecting fuel system components.
4. Fuel tanks should be inspected for leaks, road damage, corrosion or rust on metal tanks, loose, damaged, or defective seams, loose mounting bolts, and damaged mounting straps.
5. Fuel tanks may be drained through the drain bolt opening, or the hose on a hand-operated pump many be connected to the tank drain pipe. If the tank does not have a drain pipe, the pump hose may be installed through the filler pipe.
6. Electric in-tank fuel pumps should be inspected for a contaminated filter, deteriorated rubber hoses, dirt in the pump inlet, and damaged steel fuel tubing.
7. Fuel tanks containing dirt or contaminants may be flushed with hot water.
8. Fuel tanks may be purged with an emulsifying agent mixed with water.
9. Some vehicle manufacturers recommend fuel tank replacement if the tank is leaking.

10. Fuel tank repairing should be done by a specialty radiator and fuel tank repair shop.
11. Nylon fuel pipes must not be subjected to excessive heat above 239°F (115°C) for 1 hour, or above 194°F (90°C) for an extended time period.
12. When a torch or heat source must be used near a nylon fuel line, cover the fuel line with a wet shop towel.
13. Steel fuel tubing should be inspected for leaks, kinks, and deformation.
14. Rubber fuel hose should be inspected for leaks, cracks, cuts, kinks, oil soaking, soft spots, and deterioration.
15. Fuel filters must be installed in the proper direction.
16. Mechanical fuel pumps should be inspected for gasoline leaks, oil leaks, and excessive clicking noise.
17. Mechanical fuel pumps should be tested for pressure and volume.

CASE STUDY

A customer complained about severe engine surging on a Dodge station wagon with a 5.2-L carbureted engine. When the technician lifted the hood, he noticed many fuel system and ignition system components had been replaced recently. The customer was asked about previous work done on the vehicle. She indicated that the problem had existed for some time, and several shops had worked on the vehicle, but the problem still persisted. The carburetor had been overhauled, and the fuel pump replaced. Many ignition components, such as the coil, distributor cap and rotor, spark plugs, and spark plug wires, had also been replaced.

The technician road tested the vehicle and found it did have a severe surging problem at freeway cruising speeds. From past experience, the technician thought this severe surging problem was caused by a lack of fuel supply. The technician decided to connect a fuel pressure gauge at the carburetor inlet nut. The gauge was securely taped to one of the windshield wiper blades so the gauge could be observed from the passenger compartment. The technician drove the vehicle on a second road test, and found when the surging problem occurred, the fuel pump pressure dropped well below the vehicle manufacturer's specifications. Since the fuel pump and filter had been replaced, the technician concluded the problem must be in the fuel line or tank.

The technician returned to the shop and raised the vehicle on a hoist. The steel fuel tubing appeared to be in satisfactory condition. However, a short piece of rubber fuel hose between the steel fuel tubing and the fuel line entering the fuel tank was flattened and soft in the center. This fuel hose was replaced and routed to avoid kinking. Another road test proved the surging problem was eliminated.

The flattened fuel hose restricted fuel flow, and at higher speeds, the increased vacuum from the fuel pump made the flattened condition worse, which restricted the fuel flow and caused the severe surging problem.

Terms to Know

Electronic fuel injection (EFI)
Fuel pressure test port
Fuel pump volume
Fuel tank purging
Port fuel injection (PFI)
Quick-disconnect fuel line fittings
Schrader valve
Throttle body injection (TBI)

ASE Style Review Questions

1. While discussing alcohol content in gasoline:
   Technician A says excessive quantities of alcohol in gasoline may cause fuel filter plugging.
   Technician B says excessive quantities of alcohol in gasoline may cause lack of engine power.
   Who is correct?
   - **A.** A only
   - **B.** B only
   - **C.** Both A and B
   - **D.** Neither A nor B

2. While discussing an alcohol-in-fuel test with a 100-mL cylinder filled with 90 mL of gasoline and 10 mL of water:
   Technician A says water and alcohol in the gasoline remain separate during the test.
   Technician B says if the water content is 20 mL at the end of the test, the gasoline contains 10% alcohol.
   Who is correct?
   - **A.** A only
   - **B.** B only
   - **C.** Both A and B
   - **D.** Neither A nor B

3. While discussing fuel system service on EFI systems:
   Technician A says the fuel system pressure must be relieved before fuel system components are removed.
   Technician B says the fuel system pressure may be relieved by connecting a pressure gauge to the fuel pressure test port and opening the bleed valve on the gauge with the bleed hose installed in an approved container.
   Who is correct?
   - **A.** A only
   - **B.** B only
   - **C.** Both A and B
   - **D.** Neither A nor B

4. While discussing fuel tank draining:
   Technician A says the fuel tank may be drained with a hand-operated pump.
   Technician B says some fuel tanks have a drain pipe for draining the tank.
   Who is correct?
   - **A.** A only
   - **B.** B only
   - **C.** Both A and B
   - **D.** Neither A nor B

5. While discussing quick-disconnect fuel line fittings:
   Technician A says some quick-disconnect fittings may be disconnected with a pair of snap ring pliers.
   Technician B says some quick-disconnect fittings are hand-releasable.
   Who is correct?
   - **A.** A only
   - **B.** B only
   - **C.** Both A and B
   - **D.** Neither A nor B

6. While discussing fuel tank and electric pump service:
   Technician A says if the filter on the pump inlet is contaminated, the fuel tank should be flushed with hot water.
   Technician B says if there is dirt in the pump inlet, the inlet may be cleaned and the pump reused.
   Who is correct?
   - **A.** A only
   - **B.** B only
   - **C.** Both A and B
   - **D.** Neither A nor B

7. While discussing nylon fuel pipes:
   Technician A says nylon fuel pipes may be subjected to temperatures up to 300°F (149°C).
   Technician B says nylon fuel pipes may be bent at a 90° angle.
   Who is correct?
   - **A.** A only
   - **B.** B only
   - **C.** Both A and B
   - **D.** Neither A nor B

8. While discussing quick-disconnect fuel line fittings:
   Technician A says on some hand-releasable quick-disconnect fittings, the fitting may be removed by pulling on the fuel line.
   Technician B says some hand-releasable quick-disconnect fittings may be disconnected by twisting the large connector collar in both directions and pulling on the connector.
   Who is correct?
   - **A.** A only
   - **B.** B only
   - **C.** Both A and B
   - **D.** Neither A nor B

9. While discussing mechanical fuel pump diagnosis:
Technician A says if gasoline is leaking from the vent opening in the pump housing, the pump diaphragm is leaking.
Technician B says if oil is leaking from the vent opening in the pump housing, the pump diaphragm is leaking.
Who is correct?
A. A only
B. B only
C. Both A and B
D. Neither A nor B

10. While discussing mechanical fuel pump testing:
Technician A says if the fuel pump pressure is lower than specified, the fuel pump camshaft lobe may be worn.
Technician B says if the fuel pump volume is lower than specified, there may be an air leak in the fuel line between the tank and the pump.
Who is correct?
A. A only
B. B only
C. Both A and B
D. Neither A nor B

Table 5-1 ASE Task

Inspect fuel tank, tank filter, and gas cap; inspect and replace fuel lines, fittings, and hoses; check fuel for contaminants and quality.

Problem Area	Symptoms	Possible Causes	Classroom Manual	Shop Manual
ENGINE PERFORMANCE	No-start, stalling	Excessive alcohol or contaminants in the fuel	113	113
	Lack of power	Excessive alcohol or contaminants in the fuel	113	113
	Acceleration stumbles	Excessive alcohol or contaminants in the fuel	113	113
ODOR	Gasoline odor, low-speed driving, idling	Fuel leaks in tank, filter, lines	113	115
ENGINE PERFORMANCE	Surging, cutting out at high speed	1. Restricted fuel lines 2. Contaminated fuel pump inlet filter	117 122	120 119

Table 5-2 ASE Task

Inspect, test, and replace mechanical and electrical fuel pumps and pump control systems; inspect, service, and replace fuel filters.

Problem Area	Symptoms	Possible Causes	Classroom Manual	Shop Manual
ENGINE PERFORMANCE	Surging, cutting out at high speed	1. Restricted fuel filter 2. Defective fuel pump	118 120	122 123

Computer-Controlled Carburetor Diagnosis and Service

Upon completion and review of this chapter, you should be able to:

❏ Adjust idle speed.
❏ Adjust fast idle speed.
❏ Adjust idle mixture.
❏ Adjust A/C idle speed.
❏ Adjust engine idle speed.
❏ Adjust idle air-fuel mixture.
❏ Perform an antidieseling adjustment.
❏ Perform a computer-controlled carburetor system performance test.

❏ Diagnose the results of a computer-controlled carburetor system performance test.
❏ Perform a flash code diagnosis.
❏ Obtain fault codes with an analog voltmeter.
❏ Perform a computed timing test.
❏ Perform an output state test.
❏ Erase fault codes.
❏ Perform a continuous self-test.
❏ Diagnose a computer-controlled carburetor system with a scan tester.

Computer-Controlled Carburetor Diagnosis and Service

Preliminary Diagnostic Procedure

When a problem occurs in an automotive computer system, the tendency is to blame the computer system. However, before any computer system diagnosis is performed, the following must be checked first, and proven to be in satisfactory condition:

1. Engine compression
2. Intake manifold vacuum leaks
3. Emission devices such as the EGR valve
4. Ignition system components

Prior to a computer system diagnosis, all wiring harness connections to computer system components should be checked for damage or loose connections. The engine must be warmed up to normal operating temperature before fault codes are obtained on any computer system.

Idle Speed Adjustment

A/C Idle Speed Check. The idle speed adjustment procedure varies with each vehicle, engine, or model year. Always follow the vehicle manufacturer's recommended procedure in the service manual. The following A/C idle rpm check applies to a carburetor with a combined vacuum and electric throttle kicker. The vacuum throttle kicker is activated when the driver selects the A/C mode, and the electric kicker solenoid is energized when the ignition switch is turned on. Prior to any idle speed check, the engine must be at normal operating temperature. Follow these steps for the A/C idle speed check:

1. Select the A/C mode and set the temperature control to the coldest position.
2. Each time the A/C compressor cycles off and on, the kicker solenoid should be energized and the vacuum kicker plunger should move in and out. If the plunger reacts properly, the system is satisfactory. There is no adjustment on the vacuum kicker stem. When the vacuum kicker plunger does not react properly, check all the vacuum hoses and the solenoid, and test the vacuum kicker diaphragm for leaks.

Basic Tools

Basic technician's tool set
Service manual
Jumper wires

Engine Idle Speed Check. The ignition timing should be checked and adjusted as necessary prior to the idle speed check. Follow these steps for the idle rpm check:

1. With the transaxle in neutral and the parking brake applied, turn off all the accessories and lights. Be sure the engine is at normal operating temperature, and connect a tachometer from the coil negative primary terminal to ground.
2. Disconnect the cooling fan motor connector and connect 12 V to the motor terminal so the fan runs continually.
3. Remove the PCV valve from the crankcase vent module and allow this valve to draw in underhood air.
4. Disconnect the O_2 feedback test connector on the left fender shield.
5. Disconnect the wiring connector from the kicker solenoid on the left fender shield.
6. If the idle rpm is not as specified on the underhood emission label, adjust the idle rpm with the screw on the kicker solenoid (Figure 6-1).
7. Reconnect the O_2 connector, PCV valve, and kicker solenoid connector on the left fender shield.
8. Increase the engine rpm to 2,500 for 15 seconds and then allow the engine to idle. If the idle speed changes slightly, this is normal and a readjustment is not required.
9. Disconnect the jumper wire and reconnect the fan motor connector.

Idle Mixture Adjustment, Propane-Assisted Method

Concealment Plug Removal. The idle mixture adjustment procedure varies depending on the vehicle model and year. Always follow the procedure in the vehicle manufacturer's service manual. The carburetor concealment plug must be removed prior to an idle mixture adjustment. Following is a typical concealment plug removal procedure:

1. Center punch the carburetor casting 1/4 in. (6.35 mm) from the end of the mixture screw housing (Figure 6-2).
2. Drill the outer housing at the center punched location with a 3/16-in. (4.762-mm) drill bit.
3. Pry out the concealment plug and save the plug for reinstallation.

CAUTION: Tampering with emission devices is a serious federal offense in the United States. Altering any component or system that affects emission systems on a vehicle may result in a fine and/or imprisonment.

Propane-Assisted Idle Mixture Adjustment. The propane-assisted idle mixture adjustment may vary depending on the vehicle make and year. Always follow the idle mixture adjustment proce-

> Tampering with a carburetor may be defined as any carburetor adjustment that is not recommended by the vehicle manufacturer and causes higher-than-specified exhaust emissions levels for that year of vehicle.

> The concealment plug is a metal plug placed in the outer end of the idle mixture screw bore to try to prevent improper idle air-fuel mixture adjustments.

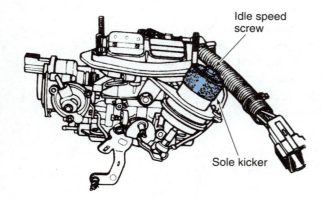

Figure 6-1 Idle speed screw on kicker solenoid (Courtesy of Chrysler Corporation)

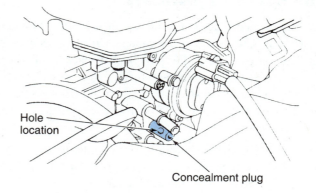

Figure 6-2 Concealment plug removal (Courtesy of Chrysler Corporation)

dure in the vehicle manufacturer's service manual. After the concealment plug is removed, follow these steps for a typical propane-assisted idle mixture adjustment:

> **WARNING:** During the propane-assisted idle mixture adjustment, the propane bottle must be kept in an upright position and located in a safe place where it will not contact rotating parts.

1. With the transaxle in neutral and the parking brake on, turn off all the lights and accessories. Be sure the engine is at normal operating temperature, and connect a tachometer from the negative primary coil terminal to ground.
2. Disconnect the cooling fan motor connector, and connect a jumper wire from a 12-V source to the fan motor terminal so the fan motor runs all the time. Remove the PCV valve from the crankcase vent module and allow this valve to draw in underhood air. Disconnect the O_2 feedback system test connector on the left fender shield.
3. Disconnect the vacuum hoses from the coolant vacuum switch cold closed (CVSCC) valve on the driver's side of the thermostat housing and plug both hoses. This valve controls the vacuum to the EGR valve on some engines. On 2.2-L engines, disconnect the wiring harness connector from the throttle kicker solenoid on the left fender shield.
4. Locate the vacuum hose from the air cleaner heated air sensor to the intake manifold and disconnect this hose from the intake manifold. Connect the supply hose from the propane bottle to the intake manifold connection where the air cleaner hose was removed. Be sure the propane bottle valves are closed and the bottle is in an upright position. Position the propane bottle in a safe place where it will not come in contact with rotating parts.
5. With the air cleaner in place, slowly open the main valve on the propane bottle (Figure 6-3). Slowly open the propane metering valve until the highest engine rpm is reached. When excessive propane is added, the engine starts to slow down. Slowly rotate the propane metering valve until the highest possible rpm is reached.
6. With the propane still flowing, adjust the idle speed screw on the carburetor electric kicker solenoid to obtain the specified propane idle speed. Increase the engine speed to 2,500 rpm for 15 seconds, and allow the engine speed to return to idle. Slowly rotate the propane metering valve in either direction, recheck the propane rpm, and adjust as necessary.

Special Tools

Propane bottle with metering valve

Special Tools

Tachometer

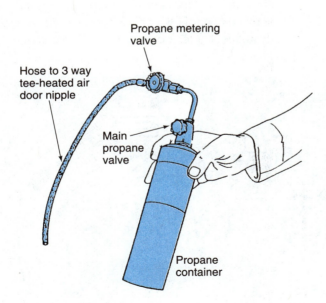

Figure 6-3 Propane bottle with hose and metering valve (Courtesy of Chrysler Corporation)

7. Close the main valve on the propane bottle and allow the engine to idle. Adjust the mixture screw to obtain the specified idle set rpm. Increase the engine rpm to 2,500 for 15 seconds and allow the engine to return to idle speed. Recheck the idle set rpm and adjust the idle mixture screw to correct the idle set rpm if necessary.
8. Turn on the main propane bottle valve and slowly rotate the metering valve to obtain the highest rpm. If the rpm is more than 25 rpm above or below the propane set rpm, repeat steps 5 through 8.
9. Turn off both propane bottle valves, remove the propane supply hose, and install the vacuum hose from the air cleaner to the intake manifold. Install the concealment plug.
10. Place the fast idle screw on the lowest step of the fast idle cam and rotate this screw until the specified fast idle speed is obtained.
11. Allow the engine to idle and reconnect the O_2 feedback system test connector, PCV valve, throttle kicker solenoid wiring connector, and CVSCC valve hoses.

After an idle air-fuel mixture adjustment is completed, the vehicle must conform to all federal and state emission standards.

Antidieseling Adjustment

Follow these steps for a typical antidieseling adjustment:

1. Be sure the engine is at normal operating temperature and place the transaxle in neutral with the parking brake applied. Turn off all the lights and accessories.
2. Remove the red wire from the six-way carburetor connector on the carburetor side of the connector, and disconnect the O_2 feedback test connector on the left fender shield.
3. Adjust the throttle stop screw to 700 rpm (Figure 6-4).
4. Reconnect the red wire in the six-way carburetor connector and the O_2 feedback test connector.

Computer-Controlled Carburetor System Performance Test

During a computer-controlled carburetor performance test, the engine rpm is checked under specific conditions with the mixture control solenoid connected and disconnected to determine if this solenoid is functioning properly. The following performance test procedure applies to General Motors vehicles. However, a similar performance test may be recommended on other vehicles.

Prior to any computer command control (3C) diagnosis, complete the preliminary diagnostic procedure mentioned previously in this chapter. Connect a dwell meter to the mixture control (MC)

Special Tools
Dwellmeter

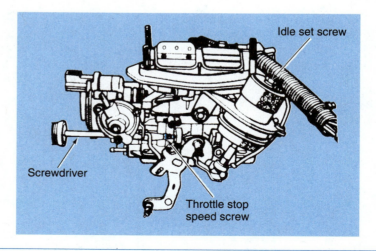

Figure 6-4 Throttle stop screw (Courtesy of Chrysler Corporation)

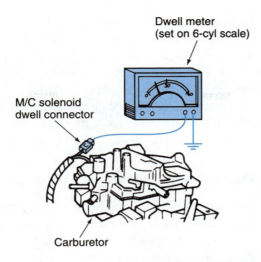

Figure 6-5 Dwell meter connection to the mixture control (MC) solenoid dwell connector (Courtesy of Chevrolet Motor Division, General Motors Corporation)

solenoid dwell connector and ground to begin the performance test (Figure 6-5). The dwell connector is a green plastic connector on a blue wire located near the carburetor on many models. Since the MC solenoid operates on a duty cycle of ten times per second regardless of the engine type, the dwell meter must be used on the six-cylinder scale on all engines.

> **WARNING:** Never ground or short across any terminals in a computer system unless instructed to do so in the vehicle manufacturer's service manual. This action may damage computer system components.

> **WARNING:** Never disconnect any computer system electrical connectors with the ignition switch on, because this action may damage system components.

Be sure that the engine is at normal operating temperature, and then complete the following in the performance test:

1. Start the engine.
2. Connect a jumper wire between terminals A and B in the data link connector (DLC) under the dash (Figure 6-6).
3. Connect a tachometer from the ignition coil tachometer terminal to ground.
4. Disconnect the MC solenoid wiring connector and ground the MC solenoid dwell lead. This action moves the MC solenoid plunger upward and provides a rich air-fuel ratio.
5. Run the engine at a constant 3,000 rpm.
6. Reconnect the MC solenoid connector and note the engine rpm. This action moves the MC solenoid plunger downward and creates a lean air-fuel ratio because the MC solenoid

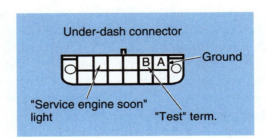

Figure 6-6 Data link connector (DLC) (Courtesy of Chevrolet Motor Division, General Motors Corporation)

The SAE J1930 terminology is an attempt to standardize terminology in automotive electronics.

In the SAE J1930 terminology, the term *data link connector* (DLC) replaces the terms *self-test connector* and *assembly line diagnostic link* (ALDL).

dwell connector is grounded. If the rpm decreases 300 or more, the MC solenoid plunger is reacting properly. When the rpm drop is less than 300, check the MC solenoid wires, and inspect the evaporative canister for excessive fuel loading. If these items are satisfactory, the MC solenoid and carburetor require servicing. Remove the ground wire from the MC solenoid dwell connector.

7. Be sure that the engine is at normal operating temperature, and observe the dwell meter at idle speed and 3,000 rpm. If the 3C system is operating correctly, the dwell should be varying between 10° and 54°.

Diagnostic Procedure if the Dwell is Fixed Below 10°

1. If the dwell reading is fixed below 10° in the performance test, the MC solenoid is upward most of the time, and the system is trying to provide a richer air-fuel ratio. Therefore, the PCM may be receiving a continually lean signal from the O_2 sensor. Run the engine at 2,000 rpm and momentarily choke the engine. Wait 1 minute while observing the dwell meter. If the dwell meter reading increases above 50°, the PCM and O_2 sensor are responding to the rich mixture provided by the choking action. When the dwell meter responds properly, check for vacuum leaks, an air management system that is continually pumping air upstream to the exhaust ports, vacuum hose routing, EGR operation, and deceleration valve, if so equipped. When these items and components are satisfactory, the carburetor is providing a continually lean mixture.

2. If the dwell reading did not reach 50° degrees when the engine was choked, the PCM and O_2 sensor are not responding properly. Select the 20-V scale on a digital voltmeter. Disconnect the O_2 sensor wire and connect the digital voltmeter from the battery positive terminal to the purple O_2 sensor wire from the PCM. Operate the engine at part throttle and observe the dwell meter. If the meter reading increases, the PCM is responding to a rich signal. Check the O_2 sensor ground wire for an open circuit. If the ground wire is satisfactory, the O_2 sensor is defective. When the dwell does not increase with the digital voltmeter connected to the purple wire from the PCM to the O_2 sensor, check the coolant temperature sensor and sensor wires, and the TPS and TPS wires. If these items are satisfactory, replace the PCM.

In the SAE J1930 terminology, the term *powertrain control module* (PCM) replaces all other terms for engine computers.

Special Tools

Digital voltmeter

Diagnostic Procedure if the Dwell is Fixed Above 50°

When the dwell is fixed above 50° in the performance test, the MC solenoid plunger is downward most of the time, and the system is trying to provide a leaner air-fuel ratio. Therefore, the O_2 sensor signal must be continually rich. If the dwell meter reading is fixed above 50° in the performance test, follow these steps to locate the problem:

1. Start the engine and connect a jumper wire between terminals A and B in the DLC. Connect a dwell meter to the MC solenoid dwell connector and operate the engine at part throttle for 2 minutes. Return the throttle to idle speed.
2. Remove the PCV valve from the PCV hose and cover the hose with your thumb. Slowly uncover the PCV hose and create a large intake manifold vacuum leak. Run the engine for 2 minutes and observe the dwell meter. If the dwell reading drops 20° or more, the O_2 sensor and the PCM are responding to the lean air-fuel ratio. Check the evaporative fuel canister for fuel loading, and check the carburetor bowl venting system. If these components and systems are satisfactory, the carburetor is providing a continually rich air-fuel ratio, and carburetor service is required.
3. If the dwell meter reading does not drop 20° in step 2, the PCM and the O_2 sensor are not responding to a lean air-fuel ratio. Disconnect the O_2 sensor connector and ground the O_2 sensor signal wire from the PCM to the sensor. Operate the engine at part throttle and observe the dwell meter. If the dwell reading does not change, the PCM is not

responding to a simulated lean O₂ sensor signal, and the PCM is faulty. When the dwell drops to 10° or less, the PCM is responding to a simulated lean O₂ sensor signal, and this sensor or the sensor signal wire is defective. Remove the O₂ sensor wire from ground and connect the digital voltmeter to this wire and ground. With the ignition switch on and the engine stopped, the voltmeter reading should be under 0.55 V. If this voltage reading is satisfactory, replace the O₂ sensor. If the voltage reading is above 0.55 V, the signal wire to the O₂ sensor may be shorted to battery voltage. If this wire is satisfactory, the PCM is faulty.

Diagnostic Procedure if the Dwell is Fixed Between 10° and 50°

1. Start the engine and connect a jumper wire between terminals A and B in the DLC. Run the engine at 2,000 rpm for 2 minutes and then allow the engine to idle. Disconnect the O₂ sensor wire and ground the purple wire from the PCM to the sensor. Increase the engine rpm to 2,000 for 2 minutes, then allow the engine to idle, and observe the MC solenoid dwell reading. If the dwell reading decreases, the PCM is responding to a simulated lean O₂ sensor signal. This sensor or the sensor ground wire must be defective. If the sensor ground wire is not open, leave the purple wire grounded, and connect a digital voltmeter to the O₂ sensor wire and ground. Run the engine at part throttle. The sensor voltage should be over 0.8 V. If the voltmeter reading is satisfactory, check for a loose sensor connection. If the voltmeter reading is below 0.8 V, replace the O₂ sensor.
2. If the dwell meter reading does not change in step 1, check the coolant temperature sensor and sensor wires. A defective coolant temperature sensor or sensor wires causes the system to remain in open loop with a fixed dwell. Check for an open circuit in the purple wire from the PCM to the O₂ sensor, and be sure that the coolant sensor and O₂ sensor wiring connections at the PCM are satisfactory. If these items are satisfactory, the PCM is faulty.

Photo Sequence 4 shows a typical procedure for testing computer command control (3C) performance.

Flash Code Diagnosis

General Motors Computer Command Control System Fault Code Diagnosis

The preliminary diagnostic procedure mentioned previously in this chapter must be completed before the diagnostic trouble code (DTC) diagnosis. If a defect occurs in the 3C system and a DTC is set in the PCM memory, the malfunction indicator lamp (MIL) on the instrument panel is illuminated. Should the fault disappear, the MIL lamp will go out, but a DTC is likely set in the computer memory. If a 3C system defect occurs and the MIL lamp is illuminated, the system is usually in a limp-in mode. In this mode, the PCM provides a rich air-fuel ratio and a fixed spark advance. Therefore, engine performance and economy decrease, and emission levels increase, but the vehicle may be driven to a service center.

Connect a jumper wire from terminals A to B in the DLC and turn on the ignition switch to obtain the DTCs. The DTCs are flashed out by the MIL lamp. Two lamp flashes followed by a brief pause and two more flashes indicate code 22. Each code is repeated three times, and when more than one code is present, the codes are given in numerical order. Code 12 is given first, indicating that the PCM is capable of diagnosis. The DTC sequence continues to repeat until the ignition switch is turned off. A DTC indicates a problem in a specific area. Some voltmeter or ohmmeter tests may be necessary to locate the exact defect.

In the SAE J1930 terminology, the term *malfunction indicator light* (MIL) replaces all other terms for check engine light.

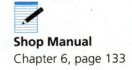

Shop Manual
Chapter 6, page 133

Photo Sequence 4
Typical Procedure for Testing Computer Command Control (3C) Performance

P4-1 Start the engine.

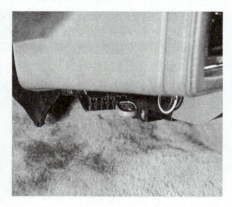

P4-2 Connect terminals A and B in the data link connector (DLC).

P4-3 Connect the tachometer leads from the coil tachometer terminal to ground.

P4-4 Disconnect the mixture control (MC) solenoid and ground the MC solenoid dwell lead.

P4-5 Operate the engine at 3,000 rpm.

P4-6 Reconnect the MC solenoid connector and note the engine rpm on the tachometer.

P4-7 Compare the rpm in step 6 to the vehicle manufacturer's specifications in the service manual.

P4-8 Return the engine to idle speed and disconnect the ground wire from the MC solenoid dwell lead.

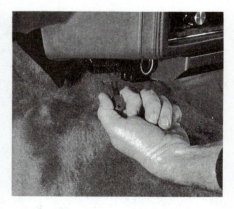

P4-9 Remove the connection between the A and B terminals in the DLC.

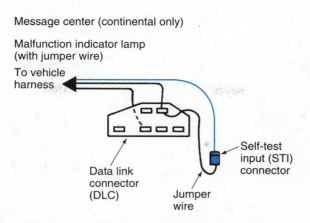

Figure 6-7 DLC with separate self-test input wire (Reprinted with the permission of Ford Motor Company)

Erasing Fault Codes

A quick-disconnect wire connected to the battery positive cable may be disconnected for 10 seconds with the ignition switch off to erase DTCs. If the vehicle does not have a quick-disconnect wire, remove the PCM battery (B) fuse to erase codes. DTCs vary depending on the vehicle model and year. The technician must be sure that the DTC list is correct for the vehicle being tested.

Voltmeter Diagnosis

Key On Engine Off (KOEO) Test

A DLC is located under the hood on Ford vehicles with electronic engine control III (EEC III) or EEC IV systems with computer-controlled carburetors. A separate self-test input wire is located near the DLC on many of these systems (Figure 6-7). However, on some Ford systems, the self-test input wire is integral with the DLC (Figure 6-8).

Many Ford products with computer-controlled carburetors do not have an MIL light in the instrument panel. On these systems, an analog voltmeter may be connected to the DLC to obtain fault codes. The positive voltmeter lead must be connected to the positive battery terminal, and the negative voltmeter lead must be connected to the proper DLC terminal. With the ignition switch off, connect a jumper wire from the separate self-test input wire to the proper DLC terminal (Figure 6-9). If

Special Tools

Analog voltmeter

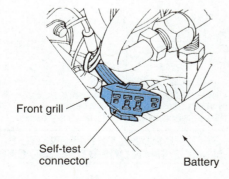

Figure 6-8 DLC with integral self-test input wire (Reprinted with the permission of Ford Motor Company)

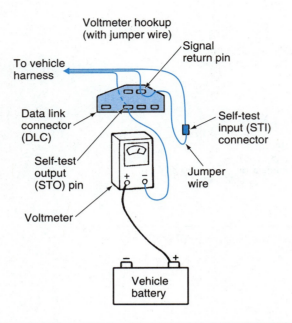

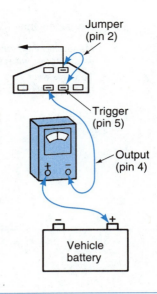

Figure 6-9 Voltmeter and jumper wire connections to the DLC and self-test input wire (Reprinted with the permission of Ford Motor Company)

Figure 6-10 Voltmeter and jumper wire connections to the DLC (Reprinted with the permission of Ford Motor Company)

the vehicle does not have a separate self-test input wire, connect the jumper wire to the proper self-test connector terminals (Figure 6-10).

During the Ford diagnosis, the voltmeter pointer sweeps out the DTCs. For example, if the voltmeter pointer sweeps upward twice, pauses, and then sweeps upward three times, code 23 is displayed (Figure 6-11).

Perform the preliminary diagnosis mentioned previously in this chapter, and then follow these steps for the key on engine off (KOEO) test:

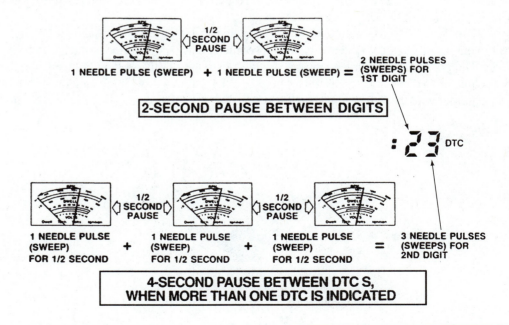

Figure 6-11 Analog voltmeter fault code reading (Reprinted with the permission of Ford Motor Company)

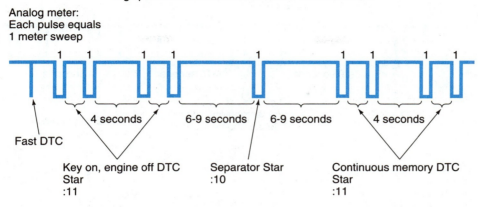

Figure 6-12 Key on engine off (KOEO) test sequence (Reprinted with the permission of Ford Motor Company)

1. Be sure the ignition switch is off and then connect the jumper wire and voltmeter to the DLC and the self-test input wire.
2. Turn on the ignition switch, and observe the voltmeter. Hard fault DTCs are displayed, followed by a separator code 10, and then memory fault codes. Each DTC is displayed twice. If no DTCs are present, code 11, system pass code, is shown (Figure 6-12).

Memory codes represent intermittent faults. These codes may also be called history codes.

Key On Engine Running (KOER) Test

Before any test is repeated or another test is initiated, the ignition switch must be turned off for 10 seconds on these computer-controlled carburetor systems. When the key on engine running (KOER) test is performed, leave the voltmeter and jumper wire connected to the DLC. Turn off the ignition switch, wait 10 seconds, and then start the engine.

An engine identification (ID) code is displayed first, and this code is equal to half the engine cylinders. For example, on a V6 engine, the engine ID code is 30. This code is displayed on a voltmeter as 3 upward pointer sweeps. After the engine ID code, a separator code 10 is provided on some Ford products. When this code is received, the technician has 20 seconds to momentarily move the throttle to the wide-open position. During this time, the PCM monitors the input sensors for defects. After the separator code 10, engine running DTCs are displayed representing faults that are present at the time of testing (Figure 6-13).

DTCs vary depending on the model year and vehicle. The technician must use the correct fault code list for the vehicle being tested.

Computed Timing Test

The computed timing test checks the spark advance provided by the PCM. Complete these steps for the computed timing check:

1. Disconnect the in-line timing connector near the distributor and check the basic timing with a timing light. Leave the timing light connected, and reconnect the in-line timing connector (Figure 6-14).
2. Leave the analog voltmeter connected, and disconnect the self-test input wire. Be sure the ignition switch has been off for 10 seconds.
3. Start the engine and connect the jumper wire to the same terminals used for the KOEO and KOER tests.

Hard fault DTCs represent faults that are present at the time of testing.

Memory DTCs represent intermittent faults that occurred sometime in the past and were placed in the computer memory.

Memory DTCs may be referred to as history codes.

Special Tools

Advance-type timing light

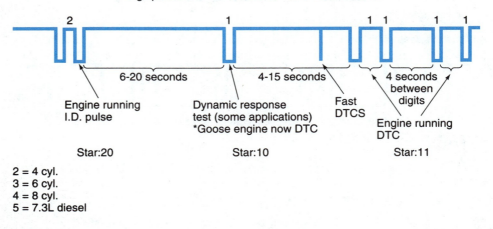

Figure 6-13 Key on engine running (KOER) test sequence (Reprinted with the permission of Ford Motor Company)

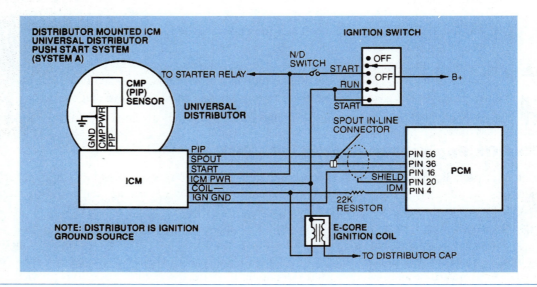

Figure 6-14 In-line timing connector (Reprinted with the permission of Ford Motor Company)

4. The spark advance should be 20° more than the basic timing setting if the PCM is providing the correct spark advance. The timing marks may be observed with the timing light to check the advance. If an advance-type timing light is used, the timing knob on the light may be rotated until the timing marks return to the basic timing setting. The scale on the light then indicates the degrees of advance.

Output State Test

The output state test may be referred to as an output cycling test. Complete the KOEO test, and then push the throttle wide open and release the throttle to enter the output state test. When this action is taken, the PCM energizes all the actuators in the EEC IV system. Normally closed actuators go to the open position, and normally open actuators move to the closed position. If the throttle is pushed wide open and released a second time, the PCM de-energizes all the actuators. During the output state test, the technician may listen to suspected actuators for a clicking noise. Voltmeter tests may be performed at suspected actuators to locate the cause of the problem.

Erase Code Procedure

Do not erase DTCs until the defects represented by the codes have been corrected. To erase DTCs, perform a KOEO test and disconnect the jumper wire at the self-test connector during the code display. Turn off the ignition switch for 10 seconds and repeat the KOEO and KOER tests to be sure the codes are erased. If DTCs are left in the PCM memory, they will be erased after the problem is corrected and the engine is stopped and started 50 times. This action applies to many computer systems.

Continuous Self-Test

Test procedures may vary depending on the vehicle year and model. Always use the procedure recommended in the vehicle manufacturer's service manual. This test allows the technician to wiggle suspected wiring harness connectors while the EEC IV system is monitored. The continuous monitor test may be performed at the end of the KOER test. Leave the jumper wire connected to the self-test connector. Approximately 2 minutes after the last code is displayed, the continuous monitor test is started. If suspected wiring connectors are wiggled, the voltmeter pointer deflects when a loose connection is present.

The continuous self-test may be referred to as a wiggle test.

Scan Tester Diagnosis

Chrysler Oxygen (O₂) Feedback System Fault Code Diagnosis

A scan tester may be used to diagnose different types of computer-controlled carburetor systems. Scan testers are supplied by the vehicle manufacturer and several independent suppliers. Many scan testers have a removable module plugged into the tester. This module is designed for a specific vehicle make and model year. Always be sure the proper module is installed in the tester for the vehicle being diagnosed.

Some Chrysler O₂ feedback systems have self-diagnostic capabilities, whereas older models of these systems had to be diagnosed with voltmeter and ohmmeter tests. Always use the diagnostic procedure recommended in the vehicle manufacturer's appropriate service manual. Some Chrysler products with O₂ feedback systems and self-diagnostic capabilities do not have a check engine light. These systems have to be diagnosed with a scan tester (Figure 6-15). The scan tester must be connected to the diagnostic connector in the engine compartment (Figure 6-16).

Special Tools

Scan tester

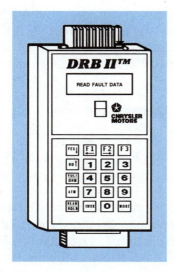

Figure 6-15 Chrysler digital readout box II (DRB II) (Courtesy of Chrysler Corporation)

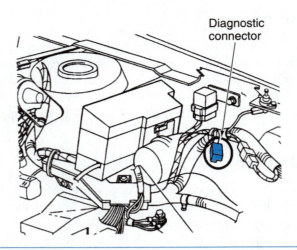

Figure 6-16 DLC in the engine compartment (Courtesy of Chrysler Corporation)

After the scan tester is connected to the diagnostic connector, follow the scan tester manufacturer's recommended procedure to obtain the DTCs. These fault codes are displayed once in numerical order. If the technician wants to repeat the DTC display, the ignition switch must be turned off and the fault code procedure repeated. DTCs in any computer system indicate a fault in a specific area, not necessarily in a specific component. For example, if a code representing a throttle position sensor (TPS) is present, the wires from the sensor to the computer may be defective, the sensor may require replacement, or the computer may not be able to receive this sensor signal. The technician may have to perform some voltmeter or ohmmeter tests on the sensor wiring to locate the exact cause of the code.

Switch Test Mode

If a defective input switch is suspected, the switch test mode may be used to check the switch inputs. Switch inputs include the brake switch, neutral park switch, and A/C switch. The switch inputs vary depending on the vehicle and model year.

When the fault code display is completed, be sure all the input switches are off, and then follow the scan tester manufacturer's recommended procedure to enter the switch test mode. During the switch test mode, each switch is turned on and off, and the reading on the scan tester should change, indicating this switch input signal was received by the computer. For example, the brake switch is turned on and off by depressing and releasing the brake pedal. If the scan tester display does not change when any input switch is activated, the switch is defective or connecting wires are defective.

Actuation Test Mode (ATM)

If a fault code is obtained representing a solenoid or relay, the ATM may be used to cycle the component on and off to locate the exact cause of the problem. When code 55 is displayed at the end of the fault code diagnosis, follow the scan tester manufacturer's recommended procedure to enter the ATM mode.

Typical components that may be cycled during the ATM are the canister purge solenoid, cooling fan relay, and throttle kicker solenoid. The actual components that may be cycled in the ATM vary depending on the vehicle and model year. Each relay or solenoid may be cycled for 5 minutes, or until the ignition switch is turned off. When the appropriate scan tester button is pressed, the tester begins cycling the next relay or solenoid. The following chapter provides more information on scan tester diagnosis of electronic fuel injection systems.

Classroom Manual
Chapter 6, page 141

● **CUSTOMER CARE:** Always concentrate on quality workmanship and customer satisfaction. Most customers do not mind paying for vehicle repairs if the work is done properly and their vehicle problem is corrected. A follow-up phone call to determine customer satisfaction a few days after servicing a vehicle indicates that you consider quality work and satisfied customers a priority.

Guidelines for Conventional and Computer-Controlled Carburetor System Diagnosis and Service

1. The engine must be at normal operating temperature before any idle speed, fast idle speed, and idle mixture adjustment.
2. After any carburetor idle speed and mixture adjustment, the vehicle must meet emission level standards for that vehicle year.
3. An A/C idle speed check or adjustment may be required on vehicles equipped with A/C.
4. The ignition timing must be checked and adjusted as necessary prior to the engine idle speed check.
5. The concealment plug must be removed to gain access to the idle mixture screw.
6. After the idle mixture adjustment is completed, the vehicle must conform to federal and state emission standards.
7. During the propane-assisted idle mixture adjustment, the propane set rpm is adjusted with the idle speed screw, and the idle set rpm is adjusted with the mixture screw.
8. The throttle stop screw is adjusted to complete the antidieseling adjustment.
9. During a computer-controlled carburetor system performance test, the engine rpm is checked with the mixture control solenoid disconnected and connected to determine the operation of the mixture control solenoid and carburetor.
10. If the computer-controlled carburetor system is working properly with the engine at normal operating temperature, the mixture control solenoid dwell should be fluctuating between 10° and 54°.
11. If the dwell on the mixture control solenoid is fixed below 10°, the system is trying to provide a rich air-fuel ratio, and the O_2 sensor signal must be continually lean.
12. When the mixture control solenoid dwell is fixed above 50°, the system is attempting to provide a lean air-fuel ratio, and the O_2 sensor signal must be continually rich.
13. If the mixture control solenoid dwell is fixed between 10° and 54°, the system may be in open loop, or there may be a defect in one of the input sensors or the computer.
14. On General Motors vehicles, terminals A and B may be connected in the DLC to read DTCs on the MIL light.
15. Some vehicles with computer-controlled carburetors do not have an MIL light. On some of these vehicles, an analog voltmeter may be connected to the self-test connector to read the DTCs.
16. A scan tester may be connected to the DLC on many vehicles to obtain DTCs and perform other system tests.
17. A computed timing test allows the technician to check the spark advance provided by the computer.
18. An output state test allows the technician to cycle the solenoids and relays on and off in a computer-controlled carburetor system.
19. A continuous self-test, or wiggle test, allows the technician to check for loose wiring connections on system components.
20. A switch test checks the switch inputs to the computer such as the brake switch and park/neutral switch.

21. An actuation test mode (ATM) allows the technician to cycle the solenoids and relays in the system individually.

CASE STUDY

A customer complained about the check engine light coming on on a Buick Riviera with a computer-controlled carburetor system. When questioned further about the driving conditions when this light came on, the customer indicated the light came on when the car was driven at lower speeds. However, when driving on the freeway, the light did not come on. The technician asked the customer if the car had any other performance and economy problems. The customer replied that the fuel consumption had increased recently, and the engine performance seemed erratic at low speed.

The technician checked the computer for fault codes and found a code 45, indicating a rich exhaust condition. A performance test was completed on the computer-controlled carburetor system. During this test, the system did not provide the specified rpm change when the mixture control solenoid was disconnected and connected. This test result indicated a possible internal carburetor problem.

The technician remembered the diagnostic procedure learned during Automotive Technology training. This procedure said something about always performing the easiest, quickest tests first when eliminating the causes of a problem. The technician thought about other causes of this problem, and it occurred to him that one cause might be the vapor control system. This car had a vacuum-operated purge control valve. When the technician checked this valve with the engine idling, he found the valve was wide open and pulling large quantities of fuel vapors from the carburetor float bowl once the engine was at normal operating temperature.

The purge control valve was replaced and the car was driven at lower speeds during a road test. The check engine light remained off and the engine performance was normal. From this experience, the technician proved the diagnostic procedure he had learned was very helpful. Part of this procedure stated that a technician should always think about the causes of a problem and test to locate the defect beginning with the easiest, quickest tests first. If the technician had overhauled the carburetor, the customer's money would have been wasted, and the problem would have remained unsolved.

Terms to Know

Actuation test mode
Antidieseling adjustment
Computed timing test
Computer-controlled carburetor performance test
Concealment plug

Continuous self-test
Data link connector (DLC)
Diagnostic trouble code (DTC)
Flash code diagnosis
Key on engine off (KOEO) test
Key on engine running (KOER) test

Output state test
Propane-assisted idle mixture adjustment
Self-test input wire
Switch test

ASE Style Review Questions

1. While discussing idle mixture adjustment:
   Technician A says the vehicle must meet emission standards for that year of vehicle after the idle mixture adjustment.
   Technician B says during the propane enrichment method of idle mixture adjustment, the idle set rpm is adjusted with the idle mixture screws.
   Who is correct?
   - **A.** A only
   - **B.** B only
   - **C.** Both A and B
   - **D.** Neither A nor B

2. While discussing propane-assisted idle mixture adjustment:
   Technician A says the proper propane idle speed is adjusted with the idle speed screw.
   Technician B says the propane metering valve must be set to obtain the highest idle speed prior to adjusting the idle speed screw.
   Who is correct?
   - **A.** A only
   - **B.** B only
   - **C.** Both A and B
   - **D.** Neither A nor B

3. While discussing a computer-controlled carburetor system performance test:
   Technician A says if the system is operating normally, the mixture control solenoid dwell should be varying between 10° and 20°.
   Technician B says if the system is operating normally, the mixture control solenoid dwell should be varying between 10° and 54°.
   Who is correct?
   - **A.** A only
   - **B.** B only
   - **C.** Both A and B
   - **D.** Neither A nor B

4. While discussing a computer-controlled carburetor system performance test:
   Technician A says if the mixture control solenoid dwell is fixed below 10°, the system is trying to provide a rich air-fuel ratio.
   Technician B says if the mixture control solenoid dwell is fixed below 10°, the O_2 sensor voltage signal is continually high.
   Who is correct?
   - **A.** A only
   - **B.** B only
   - **C.** Both A and B
   - **D.** Neither A nor B

5. While discussing flash code diagnosis:
   Technician A says the check engine light flashes fault codes in numerical order.
   Technician B says if a fault code is set in the computer memory, the system may be in the limp-in mode.
   Who is correct?
   - **A.** A only
   - **B.** B only
   - **C.** Both A and B
   - **D.** Neither A nor B

6. While discussing erase code procedures:
   Technician A says on many computer-controlled carburetor systems, fault codes are erased by disconnecting battery power from the computer for 10 seconds.
   Technician B says if fault codes are left in a computer after the fault is corrected, the code is erased after the vehicle is started 50 times.
   Who is correct?
   - **A.** A only
   - **B.** B only
   - **C.** Both A and B
   - **D.** Neither A nor B

7. While discussing diagnostic trouble codes (DTCs):
   Technician A says memory codes represent intermittent faults.
   Technician B says intermittent faults may be called history codes.
   Who is correct?
   - **A.** A only
   - **B.** B only
   - **C.** Both A and B
   - **D.** Neither A nor B

8. While discussing a key on engine off (KOEO) test:
   Technician A says the codes representing intermittent faults are provided before the separator code.
   Technician B says the codes representing intermittent faults are provided after the separator code.
   Who is correct?
   - **A.** A only
   - **B.** B only
   - **C.** Both A and B
   - **D.** Neither A nor B

9. While discussing a key on engine running (KOER) test:
   Technician A says an engine ID code is provided first in this test.
   Technician B says if a separator code is provided, the throttle must be momentarily pushed wide open.
   Who is correct?
   A. A only
   B. B only
   C. Both A and B
   D. Neither A nor B

10. While discussing scan tester diagnosis:
    Technician A says some computer-controlled carburetor systems do not have a check engine light in the instrument panel.
    Technician B says an actuation test mode (ATM) allows the technician to obtain fault codes.
    Who is correct?
    A. A only
    B. B only
    C. Both A and B
    D. Neither A nor B

Table 6-1 ASE TASK

Perform analytic/diagnostic procedures on vehicles with on-board or self-diagnostic computer systems; determine needed repairs.

Problem Area	Symptoms	Possible Causes	Classroom Manual	Shop Manual
ENGINE PERFORMANCE	Low fuel economy, performance problems, MIL light on	Engine compression, computer system faults, ignition system defects	129, 133	131, 137

Electronic Fuel Injection Diagnosis and Service

CHAPTER 7

Upon completion and review of this chapter, you should be able to:

- ❏ Inspect and test for missing, modified, or tampered computerized engine control components.
- ❏ Locate and utilize relevant service information.
- ❏ Determine appropriate diagnostic procedures based on available vehicle data and service information; determine if available information is adequate to proceed with effective diagnosis.
- ❏ Perform a preliminary diagnostic procedure on a throttle body injection (TBI), multiport fuel injection (MFI), or sequential fuel injection (SFI) system.
- ❏ Locate and utilize relevant service information.
- ❏ Test and analyze fuel pump pressure and volume.
- ❏ Check for leakage in the fuel pump check valve and the pressure regulator valve.
- ❏ Determine the need for fuel injector performance testing.
- ❏ Diagnose, test, and clean fuel injectors.
- ❏ Remove and replace the fuel rail, injectors, and pressure regulator.
- ❏ Remove, test, and replace cold-start injectors.
- ❏ Perform a minimum idle speed adjustment.
- ❏ Inspect and clean throttle body assemblies.
- ❏ Perform a flash code diagnosis on various vehicles.
- ❏ Differentiate between computerized engine control electronic problems and mechanical problems.
- ❏ Differentiate between fuel system and air induction system mechanical and electrical/electronic problems.
- ❏ Evaluate the integrity of the air induction system.
- ❏ Perform voltage drop tests on power and ground circuits.
- ❏ Diagnose specific driveability problems on electronic fuel injection systems and determine the needed action.

Throttle Body, Multiport, and Sequential Fuel Injection Service and Diagnosis

Preliminary Inspection

Computer-controlled fuel, ignition, and emission system components should be inspected for tampering or modification. Follow these steps during a preliminary computerized engine control inspection:

1. Be sure all wiring connections are securely connected to all system components.
2. Refer to the underhood vacuum hose diagram and check all vacuum hoses for proper connection.
3. Check vacuum hoses for restrictions, such as ball bearings, which may have been installed by someone tampering with the system.
4. Check the catalytic converter or converters to be sure they have not been removed.
5. Check all the hoses on the air pump system, including the hose connected to the catalytic converter.
6. Check the air pump belt condition and tension.

Basic Tools

Basic technician's tool set

Service manual

Special jumper wires for connection to various DLCs and other components

7. Check all other emission system components, such as EGR valves, air pumps, vapor purge canisters, and PCV valves, to be sure they have not been removed, disconnected, or modified.

Preliminary Diagnostic Procedure

When engine performance or economy complaints occur on fuel injected vehicles, the tendency of many technicians is to think that the problem is in the fuel injection and computer system. However, many other defects can affect the engine and fuel injection system operation. For example, an intake manifold vacuum leak causes a rough idle condition and engine surging at low speed. All vacuum hose connections to the intake manifold must be securely connected. Intake air hoses between the air cleaner and the throttle body must be securely clamped and leak-free. Intake manifold bolts must be torqued to the specified torque. An intake manifold vacuum leak may result in faster-than-specified idle speed.

If an intake manifold vacuum leak is present, a vacuum gauge connected to the intake manifold indicates a low, steady reading. Squirt a small amount of oil near suspected vacuum leak locations. When a vacuum leak is present, the oil is pulled into the intake.

If the engine has a MAF sensor, an intake manifold vacuum leak allows additional air into the intake. This air does not flow through the MAF sensor; therefore, the MAF sensor signal indicates to the PCM that less air flow is entering the engine in relation to the throttle opening. Under this condition, the PCM supplies less fuel through the injectors, and the air-fuel ratio is lean, which results in engine surging and acceleration stumbles. Expensive hours of diagnostic time may be saved if the following items are proven to be satisfactory before the fuel injection system is diagnosed:

1. Intake manifold vacuum leaks
2. Emission devices such as the EGR valve and related controls
3. Ignition system condition
4. Engine compression
5. Battery fully charged
6. Engine at normal operating temperature
7. All accessories turned off

Service Precautions

The following precautions must be observed when TBI, MFI, and SFI systems are diagnosed and serviced:

1. Always relieve the fuel pressure before disconnecting any component in the fuel system.
2. Never turn on the ignition switch when any fuel system component is disconnected.
3. Use only the test equipment recommended by the vehicle manufacturer.
4. Always turn off the ignition switch before connecting or disconnecting any system component or test equipment.
5. When arc welding is necessary on a computer-equipped vehicle, disconnect both battery cables before welding is started. Always disconnect the negative cable first.
6. Never allow electrical system voltage to exceed 16 V. This could be done by disconnecting the circuit between the alternator and the battery with the engine running.
7. Avoid static electric discharges when handling computers, modules, and computer chips.

Disconnecting Battery Cables

In a sequential fuel injection (SFI) system, each injector has an individual ground wire connected into the computer.

During the diagnosis of TBI, MFI, and SFI systems, many procedures indicate the removal of the negative battery cable or both battery cables. The negative battery cable may be disconnected during diagnostic and service procedures, but disconnecting the battery has the following effects:

1. Deprograms the radio
2. Deprograms other convenience items such as memory seats or mirrors

3. Erases the trip odometer if the vehicle has digital instrumentation
4. Erases the adaptive strategy in the computer

Disconnecting the battery has the same effect on any vehicle with TBI, MFI, or SFI and adaptive strategy in the PCM. If the adaptive strategy in the computer is erased, the engine operation may be rough at low speeds when the engine is restarted, because the computer must relearn about computer system defects. Under this condition, the vehicle should be driven for 5 minutes on the road with the engine at normal operating temperature. Some manufacturers recommend that a 12-V dry cell battery be connected from the positive battery cable to ground if the battery is disconnected. The 12 V supplied by the dry cell prevents deprogramming and memory erasing. Some 12-V sources for this purpose are designed to plug into the cigarette lighter socket.

Locating Service Information

The first step in accurate computer system diagnosis is to locate the proper service information for the vehicle being diagnosed. The technician must have specifications for each specific vehicle. Do not guess at specifications! The technician must have service bulletin information for the vehicle that is being serviced. If a service bulletin recommends a change-up component to correct the problem being diagnosed, the technician must have this information or much time may be wasted during the diagnosis.

The technician must have a wiring diagram or diagrams for the vehicle. It is extremely difficult or impossible to diagnose computer systems without the proper wiring diagrams. Service procedure information must be available in the vehicle manufacturer's service manual, generic service manual, or electronic data system. The diagnosis may indicate a problem in a specific input sensor, but the diagnosis may not indicate whether the defect is in the sensor or connecting wires. The technician must have the proper service procedures available to locate the exact cause of the problem. A parts locator book often saves time by helping the technician to find a component on the vehicle. Sometimes the diagnostic procedures inform the technician that a certain component is the problem, but the technician may spend time locating the component on today's complex electronic systems. A parts locator book allows the technician to find components quickly.

If the computer has a removable PROM in the vehicle being diagnosed, the technician must have the latest PROM change-up information. When the vehicle manufacturer recommends a replacement PROM to correct a specific driveability problem, the technician may waste diagnostic time if he or she does not have this information.

When the specifications, service bulletins, wiring diagrams, service procedure information, and PROM change-up information are not available, it is probably advisable not to attempt a diagnosis of the vehicle.

Fuel Pressure Testing

✓ SERVICE TIP: Remember that Ford products have an inertia switch in the fuel pump circuit. If there is no fuel pump pressure, always push the inertia switch reset button first and determine if this is a problem.

✓ SERVICE TIP: Many fuel pump circuits are connected through a fuse in the fuse panel or a fusible link. If there is no fuel pump pressure, always check the fuel pump fuse or fuse link first.

When tests are performed to diagnose any automotive problem, always start with the tests that are completed quickly and easily. The fuel pressure test is usually one of the first tests to consider when TBI, MFI, and SFI systems are diagnosed. Remember that low fuel pressure may cause lack of power, acceleration stumbles, engine surging, and limited top speed, whereas high fuel pressure results in excessive fuel consumption, rough idle, engine stalling, and excessive sulphur smell from the catalytic converter. *It is important to remember that in most computer systems, the computer diagnostics do not have the capability to diagnose fuel pressure.*

In a multiport fuel injection (MFI) system, the injector ground wires are connected to the computer in pairs, or in groups of three or four, depending on the engine.

The SAE J1930 terminology is an attempt to provide a universal terminology for automotive electronics.

In the SAE J1930 terminology, the terms *MFI* and *SFI* replace the previous EFI and PFI terminology.

In the SAE J1930 terminology, the term *powertrain control module* (PCM) replaces all previous terms for engine computers.

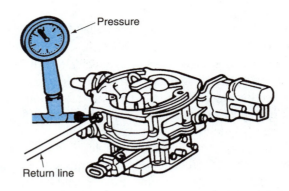

Figure 7-1 Connecting a fuel pressure gauge in series in the fuel inlet line at the throttle body assembly (Courtesy of Chrysler Corporation)

In some cases, in-tank fuel pumps have the specified pressure when the ignition switch is turned on or when the engine is idling, but the fuel pump cannot meet the engine demand for fuel at or near wide-open throttle. Therefore, if the customer complains about the engine quitting momentarily or completely at higher speeds, the fuel pump pressure should be tested at higher speeds during a road test. The hose is long enough on some fuel pressure gauges to allow the gauge hose to be connected under the hood and the gauge placed in the passenger compartment.

Relieving Fuel Pressure. Before a pressure gauge is connected on TBI, MFI, or SFI systems, the fuel pressure should be relieved. This is accomplished by momentarily supplying 12 V to one injector terminal and grounding the other injector terminal. This action lifts the injector plunger and the fuel discharges from the injector to relieve the fuel pressure. Do not supply 12 V and a ground to an injector for more than 5 seconds unless a vehicle manufacturer's recommended procedure specifies a longer time period.

⚠ **WARNING:** Never energize an injector with a 12-V source for more than 5 seconds unless a vehicle manufacturer's recommended procedure specifies a longer time period. This action may damage the injector winding.

Special Tools

Fuel pressure gauge

Connecting a Fuel Pressure Gauge. In a TBI system, the inlet fuel line at the throttle body assembly must be removed and the pressure gauge hose installed in series between the inlet line and the inlet fitting (Figure 7-1). On other TBI systems, the vehicle manufacturer recommends connecting the fuel pressure gauge at the fuel filter inlet (Figure 7-2). Use new gaskets on the union bolt when the pressure gauge is connected at this location. Some equipment manufacturers market an adapter that is mounted in place of the fuel filter to check fuel pressure on TBI systems. This adapter contains a Schrader valve. A port fuel injection pressure tester can be easily connected to this valve (Figure 7-3).

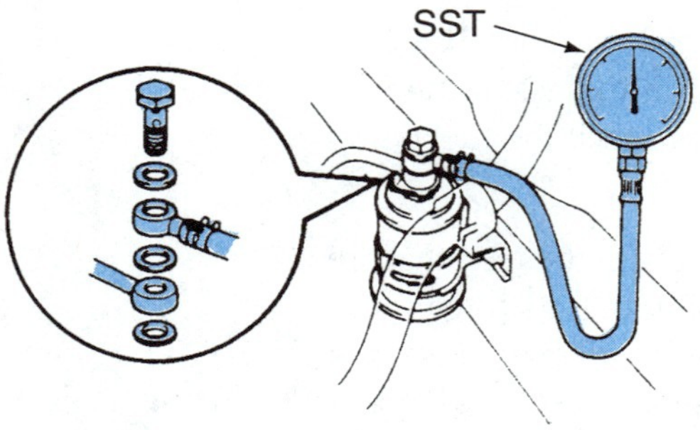

Figure 7-2 Connecting fuel pressure gauge at the fuel filter (Courtesy of Toyota Motor Corporation)

Figure 7-3 Adapter with Schrader valve that replaces fuel filter for port pressure gauge connection to a TBI system (Courtesy of OTC Division, SPX Corporation)

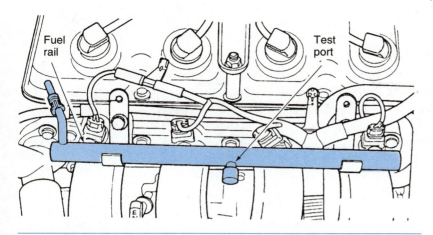

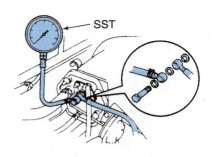

Figure 7-5 Connecting a fuel pressure gauge to the cold-start injector fuel line (Courtesy of Toyota Motor Corporation)

Figure 7-4 Connecting a fuel pressure gauge to the Schrader valve on the fuel rail (Courtesy of Chrysler Corporation)

CAUTION: Never turn on the ignition switch or crank the engine with a fuel line disconnected. This action causes the fuel pump to discharge fuel from the disconnected line, which may result in a fire causing personal injury and/or property damage.

In an MFI or SFI system, the pressure gauge must be connected to the Schrader valve on the fuel rail (Figure 7-4). On some SFI systems, the vehicle manufacturer recommends connecting the fuel pressure gauge to the cold-start injector fuel line (Figure 7-5). Install new gaskets on the union bolt when the pressure gauge is installed at this location.

Operating the Fuel Pump to Test Pressure. The technician must have pressure specifications for the make and model year of vehicle being tested. Once the pressure gauge is connected, the ignition switch may be cycled several times to read the fuel pressure, or the pressure may be read with the engine idling. In cases where the engine will not start or when further diagnosis of the fuel pump circuit is required, it may be helpful to operate the fuel pump continually. Many fuel pump circuits have a provision for operating the fuel pump continually to test fuel pump pressure, if the engine will not run. Some manufacturers of import vehicles, such as Toyota, recommend operating the fuel pump with a jumper wire connected across the appropriate terminals in the DLC to operate the fuel pump continually and check fuel pump pressure. On many Toyota products, the jumper wire must be connected across the B+ and FP terminals in the DLC with the ignition switch on (Figure 7-6).

In the SAE J1930 universal terminology, the term for self-test connector is *data link connector* (DLC).

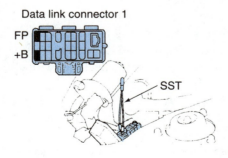

Figure 7-6 A jumper wire may be connected across terminals B+ and FP in the DLC with the ignition switch on to operate the fuel pump while testing fuel pressure. (Courtesy of Toyota Motor Corporation)

Fuel pump test procedures vary depending on the year and make of vehicle. Always follow the recommended procedure in the vehicle manufacturer's service manual. Following is a typical fuel pump test procedure on a Toyota vehicle:

1. Connect a 12-V power supply to the cigarette lighter socket and disconnect the negative battery cable. If the vehicle is equipped with an air bag, wait one minute.
2. Bleed pressure from the fuel system as mentioned previously.
3. Connect the fuel pressure gauge as outlined previously. Use a shop towel to wipe up any spilled gasoline.
4. Connect the jumper wire across the B+ and FP terminals in the DLC.
5. Reconnect the battery negative cable, and turn on the ignition switch.
6. Observe the fuel pressure on the gauge.
7. Disconnect the jumper wire from the DLC terminals.
8. Disconnect and plug the vacuum hose from the pressure regulator, and start the engine.
9. Observe the fuel pressure on the gauge with the engine idling.
10. Reconnect the vacuum hose to the pressure regulator and observe the fuel pressure on the gauge.

The fuel pump pressure must equal the manufacturer's specifications under all conditions. This pressure is usually about 10 psi (70 kPa) higher with the vacuum hose removed from the pressure regulator compared to when this vacuum hose is connected. If the pressure is higher than specified, check the return fuel line and pressure regulator.

When there is no fuel pump pressure, check the fusible link, fuses, SFI main relay, fuel pump, PCM, and wiring connections. If the fuel pump pressure is lower than specified, check the fuel lines and hoses, fuel pump, fuel filter, pressure regulator, and cold-start injector.

On many Ford products, a self-test connector is located in the engine compartment. This connector is tapered on both ends, but one tapered end is longer compared to the other end. A wire is connected from the fuel pump relay to the outer terminal in the short tapered end of the DLC (Figure 7-7).

WARNING: When instructed to ground a wire for diagnostic purposes, always be sure you are grounding the proper wire under the specified conditions. Improper grounding of computer system terminals may damage computer system components.

The PCM normally grounds this wire to close the fuel pump relay points. If this wire is grounded with a jumper wire when the ignition switch is on, the fuel pump runs continually for diagnostic purposes.

On many Chrysler products, a square DLC is located in the engine compartment. This connector has a notch in one corner (Figure 7-8). The terminal in the corner of the diagnostic connector directly opposite the notch may be grounded with a 12-V test lamp to operate the fuel pump continually. This terminal could be grounded with a jumper wire, but there is a 12-V power wire in one of the other diagnostic connector terminals. If this power wire is accidentally grounded with a jumper wire, severe computer and wiring harness damage may result. On some Chrysler products, the fuel pump test wire is discontinued. Always check the wiring diagram for the vehicle being diagnosed.

On many General Motors products, a 12-terminal DLC is located under the instrument panel. In most of these connectors, the terminals are lettered A to F across the top row, and G to M across the bottom row (Figure 7-9).

On some General Motors vehicles, a fuel pump test connector is located in terminal G on the DLC. On other General Motors vehicles, this fuel pump test wire is located in the engine compartment. If 12 V are supplied to the fuel pump test wire with the ignition switch off, voltage is supplied through a pair of fuel pump relay points to the fuel pump. The technician may observe the fuel pump pressure under this condition, or listen at the fuel tank filler neck for the fuel pump run-

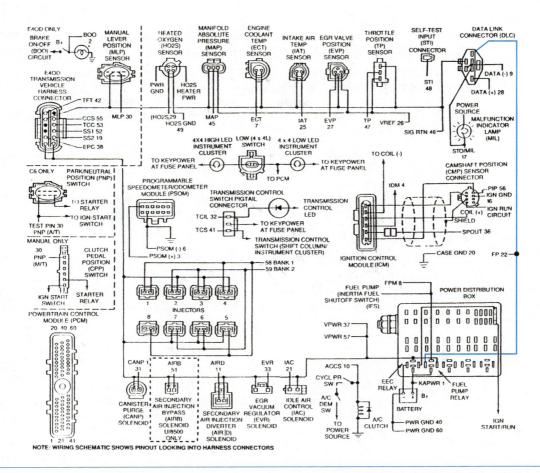

Figure 7-7 The wire from the fuel pump relay to the outer terminal in the short tapered end of the DLC may be grounded to operate the fuel pump continually. (Reprinted with the permission of Ford Motor Company)

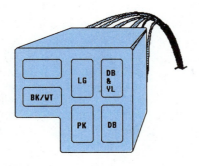

Figure 7-8 On many Chrysler products, the terminal directly opposite the notch in the corner of the DLC may be grounded with a 12-V test lamp to operate the fuel pump continually. (Courtesy of Chrysler Corporation)

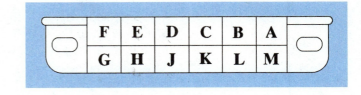

Figure 7-9 Data link connector (DLC)

ning. If the fuel pump operates satisfactorily under this test condition, the fuel pump and the wire from the relay to the pump are satisfactory. When the fuel pump does not run when the ignition switch is turned on, the fuel pump relay or PCM is defective, or the connecting wires are defective.

Causes of Low Fuel Pump Pressure. If the fuel pressure is low, always check the filter and fuel lines for restrictions before the fuel pump is diagnosed as the cause of the problem. In some cases, water or dirt in the fuel tank covers and plugs the pickup sock on the in-tank fuel pump. This action shuts off the fuel supply to the pump and the engine stops. This problem usually occurs at highway speeds. Technicians must keep this problem in mind when fuel pump pressure is tested.

Fuel Pump Volume Testing

Always follow the vehicle manufacturer's recommended fuel pump volume test procedure. A typical fuel pump volume test procedure follows:

1. Relieve fuel pressure from the system as described previously.
2. Disconnect the return hose from the fuel rail.
3. Connect a hose from the fuel rail return fuel connection to a graduated container with a 2-quart capacity. Secure the container and the fuel line to avoid a gasoline spill.
4. Energize the fuel pump for 30 seconds as explained previously.
5. Observe the fuel level in the graduated container.

Classroom Manual
Chapter 7, page 149

If the fuel pump volume is less than specified, check the fuel lines and filter for restrictions, and inspect the fuel lines for air leaks. When the lines and filter are satisfactory, the voltage at the fuel pump should be checked with the pump energized. A low voltage supply at the fuel pump results in low pump volume. If the lines, filter, and pump voltage supply are satisfactory, replace the fuel pump.

Injector Testing

SERVICE TIP: It is important to remember that most computer diagnostics do not have the capability to diagnose restricted or sticking injectors. Some computer diagnostic trouble codes (DTCs) indicate open or shorted injector windings or connecting wires.

Since injectors on MFI and SFI systems are subject to more heat than TBI injectors, port injectors have more problems with tip deposits. The symptoms of restricted injectors are:

1. Lean surge at low speeds
2. Acceleration stumbles
3. Hard starting
4. Acceleration sag, cold engine
5. Engine misfiring
6. Rough engine idle
7. Lack of engine power
8. Slow starting when cold

An injector balance test may be performed to diagnose restricted injectors on MFI and SFI systems. A fuel pressure gauge and an injector balance tester are required for this test. The fuel pressure should be checked before the injector balance test is performed. The injector balance tester contains a timer circuit, which energizes each injector for an exact time period when the timer button is pressed. When the injector balance test is performed, follow these steps:

1. Connect the fuel pressure gauge to the Schrader valve on the fuel rail.
2. Connect the injector tester leads to the battery terminals with the correct polarity. Remove one of the injector wiring connectors and install the tester lead to the injector terminals (Figure 7-10).
3. Cycle the ignition switch on and off until the specified fuel pressure appears on the fuel gauge. Many fuel pressure gauges have an air bleed button that must be pressed to bleed air from the gauge. Cycle the ignition switch or start the engine to obtain the specified pressure on the fuel gauge, and then leave the ignition switch off.
4. Push the timer button on the tester and record the gauge reading. When the timer energizes the injector, fuel is discharged from the injector into the intake port, and the fuel pressure drops in the fuel rail.

Special Tools

Injector balance tester

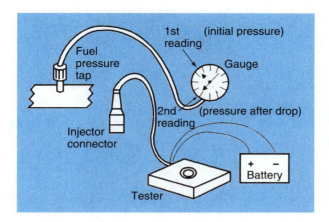

Figure 7-10 Injector balance tester and pressure gauge connections for injector balance test (Courtesy of Oldsmobile Division, General Motors Corporation)

5. Repeat steps 2, 3, and 4 on each injector, and record the fuel pressure after each injector is energized by the timer.
6. Compare the gauge readings on each injector. When the injectors are in satisfactory condition, the fuel pressure is the same after each injector is energized by the timer. If an injector orifice or tip is restricted, the fuel pressure does not drop as much when the injector is energized by the timer. When an injector plunger is sticking in the open position, the fuel pressure drop is excessive. If the fuel pressure on an injector is 1.4 psi (10 kPa) below or above the average pressure when the injectors are energized by the timer, the injector is defective (Figure 7-11).

Injector Service and Diagnosis

Injector Cleaning

If the injector balance test indicates that some of the injectors are restricted, the injectors may be cleaned. Tool manufacturers market a variety of injector cleaning equipment. The injector cleaning solution is poured into a canister on some injector cleaners, and the shop air supply is used to pressurize the canister to the specified pressure. The injector cleaning solution contains unleaded fuel mixed with injector cleaner. The container hose is connected to the Schrader valve on the fuel rail (Figure 7-12).

Automotive parts stores usually sell a sealed pressurized container of injector cleaner with a hose for Schrader valve attachment. During the cleaning process, the engine is operated on the pressurized container of unleaded fuel and injector cleaner. The fuel pump operation must be stopped to prevent the pump from forcing fuel up to the fuel rail, and the fuel return line must be

Special Tools
Injector cleaner

CYLINDER	1	2	3	4	5	6
HIGH READING	225	225	225	225	225	225
LOW READING	100	100	100	90	100	115
AMOUNT OF DROP	125	125	125	135	125	110
	OK	OK	OK	FAULTY, RICH (TOO MUCH) (FUEL DROP)	OK	FAULTY, LEAN (TOO LITTLE) (FUEL DROP)

Figure 7-11 Pressure readings from injector balance test indicating defective injectors (Courtesy of Oldsmobile Division, General Motors Corporation)

Figure 7-12 Injector cleaner connected to Schrader valve on fuel rail and pressurized by the shop air supply (Courtesy of OTC Division, SPX Corporation)

plugged to prevent the solution in the cleaning container from flowing through the return line into the fuel tank. Follow these steps for the injector cleaning procedure:

1. Disconnect the wires from the in-tank fuel pump or the fuel pump relay to disable the fuel pump. If you disconnect the fuel pump relay on General Motors products, the oil pressure switch in the fuel pump circuit must also be disconnected to prevent current flow through this switch to the fuel pump.
2. Plug the fuel return line from the fuel rail to the tank.
3. Connect a can of injector cleaner to the Schrader valve on the fuel rail, and run the engine for about 20 minutes on the injector solution.

After the injectors are cleaned or replaced, rough engine idle may still be present. This problem occurs because the adaptive memory in the computer has learned previously about the restricted injectors. If the injectors were supplying a lean air-fuel ratio, the computer increased the pulse width to try bring the air-fuel ratio back to stoichiometric. With the cleaned or replaced injectors, the adaptive computer memory is still supplying the increased pulse width, which makes the air-fuel ratio too rich now that the restricted injector problem does not exist. With the engine at normal operating temperature, drive the vehicle for at least 5 minutes to allow the adaptive computer memory to learn about the cleaned or replaced injectors. After this time, the computer should supply the correct injector pulse width, and the engine should run smoothly. This same problem may occur when any defective computer system component is replaced.

Injector Sound Test

Special Tools

Stethoscope

A port injector that is not functioning may cause a cylinder misfire at low engine speeds. With the engine idling, a stethoscope pickup may be placed on the side of the injector body (Figure 7-13). Each injector should produce the same clicking noise. If an injector does not produce any clicking noise, the injector, connecting wires, or PCM may be defective. When the injector clicking noise is erratic, the injector plunger may be sticking. If there is no injector clicking noise, proceed with the injector ohms test and noid light test to locate the cause of the problem.

Injector Ohmmeter Test

Special Tools

Digital volt-ohmmeter

An ohmmeter may be connected across the injector terminals to check the injector winding (Figure 7-14) after the injector wires are disconnected. If the ohmmeter reading is infinite, the injector winding is open. An ohmmeter reading below the specified value indicates the injector winding is shorted. A satisfactory injector winding has the amount of resistance specified by the manufacturer. Injector replacement is necessary if the injector winding does not have the specified resistance.

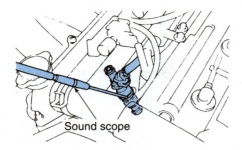

Figure 7-13 Checking for a clicking noise at each injector with a stethoscope (Courtesy of Toyota Motor Corporation)

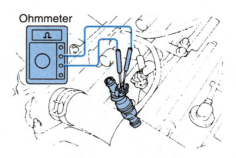

Figure 7-14 An ohmmeter may be connected across the injector terminals to test the injector winding. (Courtesy of Toyota Motor Corporation)

Figure 7-15 Noid light for testing on-off switching capabilities of the PCM (Courtesy of OTC Division, SPX Corporation)

Noid Light Test

Some manufacturers of automotive test equipment market noid lights, which have terminals designed to plug into most injector wiring connectors after these connectors are disconnected from the injector (Figure 7-15). When the engine is cranked, the noid light flashes if the computer is cycling the injector on and off. If the light is not flashing, the computer or connecting wires are defective.

Special Tools

Noid light

Injector Flow Testing

Some vehicle manufacturers recommend an injector flow test rather than the balance test. Follow these steps to perform an injector flow test:

1. Connect a 12-V power supply to the cigarette lighter socket and disconnect the negative battery cable. If the vehicle is equipped with an air bag, wait one minute.
2. Remove the injectors and fuel rail, and place the tip of the injector to be tested in a calibrated container. Leave the injectors in the fuel rail.
3. Connect a jumper wire between the B+ and FP terminals in the DLC as in the fuel pump pressure test.
4. Turn on the ignition switch.
5. Connect a special jumper wire from the terminals of the injector being tested to the battery terminals (Figure 7-16).
6. Disconnect the jumper wire from the negative battery cable after 15 seconds.
7. Record the amount of fuel in the calibrated container.

Special Tools

Graduated plastic container

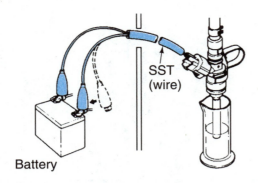

Figure 7-16 Special jumper wire connected from the injector terminals to the battery terminals (Courtesy of Toyota Motor Corporation)

Figure 7-17 Checking injector leakage with the fuel system pressurized and the injectors in the fuel rail (Courtesy of Toyota Motor Corporation)

8. Repeat the procedure on each injector. If the volume of fuel discharged from any injector varies more than 0.3 cu. in. (5 cc) from the specifications, the injector should be replaced.
9. Connect the negative battery cable and disconnect the 12-V power supply.

Injector, Fuel Pump, and Pressure Regulator Leakage Test

Connect the fuel pressure gauge to the fuel system as explained previously in this chapter. While the fuel system is pressurized with the injectors and fuel rail removed after the flow test, observe each injector for leakage from the injector tip (Figure 7-17). Injector leakage must not exceed the manufacturer's specifications. Injector leakage may cause slow starting when the engine is hot or cold.

If the injectors leak into the intake ports on a hot engine, the air-fuel ratio may be too rich when a restart is attempted a short time after the engine is shut off. When the injectors leak, they drain all the fuel out of the rail after the engine is shut off for several hours. This may result in slow starting after the engine has been shut off for a longer period of time.

While checking leakage at the injector tips, observe the fuel pressure in the pressure gauge. If the fuel pressure drops off and the injectors are not leaking, the fuel may be leaking back through the check valve in the fuel pump. Repeat the test with the fuel line plugged. If the fuel pressure no longer drops, the fuel pump check valve is leaking. When the fuel pressure drops off and the injectors are not leaking, the fuel pressure may be leaking through the pressure regulator and the return fuel line. Repeat the test with the return line plugged. If the fuel pressure no longer drops off, the pressure regulator valve is leaking.

Removing and Replacing Fuel Rail, Injectors, and Pressure Regulator

Fuel Rail, Injector, and Pressure Regulator Removal

 WARNING: Cap injector openings in the intake manifold to prevent the entry of dirt and other particles.

 WARNING: After the injectors and pressure regulator are removed from the fuel rail, cap all fuel rail openings to keep dirt out of the fuel rail.

 WARNING: Do not use compressed air to flush or clean the fuel rail. Compressed air contains water, which may contaminate the fuel rail.

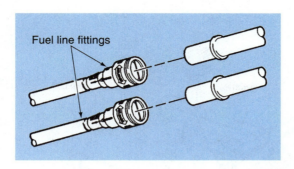

Figure 7-18 Quick-disconnect fuel line fittings (Courtesy of Oldsmobile Division, General Motors Corporation)

⚠️ **WARNING:** Do not immerse the fuel rail, injectors, or pressure regulator in any type of cleaning solvent. This action may damage and contaminate these components.

The procedure for removing and replacing the fuel rail, injectors, and pressure regulator varies depending on the vehicle. On some applications, certain components must be removed to gain access to these components. Always follow the procedure recommended in the vehicle manufacturer's service manual. Following is a typical removal and replacement procedure for the fuel rail, injectors, and pressure regulator on a General Motors 3800 engine:

1. Connect a 12-V power supply to the cigarette lighter and disconnect the battery negative cable. If the vehicle is equipped with an air bag, wait one minute.
2. Bleed the pressure from the fuel system.
3. Wipe excess dirt from the fuel rail with a shop towel.
4. Loosen fuel line clamps on the fuel rail if clamps are present on these lines. If these lines have quick-disconnect fittings, grasp the larger collar on the connector and twist in either direction while pulling on the line to remove the fuel supply and return lines (Figures 7-18 and 7-19).
5. Remove the vacuum line from the pressure regulator.
6. Disconnect the electrical connectors from the injectors.
7. Remove the fuel rail hold-down bolts (Figure 7-20).
8. Pull with equal force on each side of the fuel rail to remove the rail and injectors.

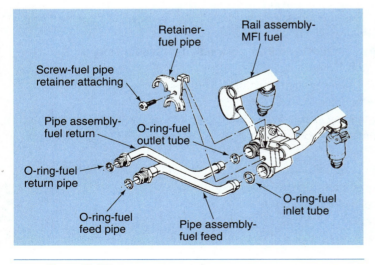

Figure 7-19 Fuel supply and return lines on fuel rail (Courtesy of Chevrolet Division, General Motors Corporation)

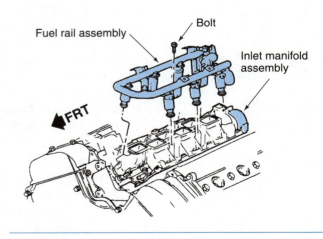

Figure 7-20 Fuel rail hold-down bolt locations (Courtesy of Chevrolet Division, General Motors Corporation)

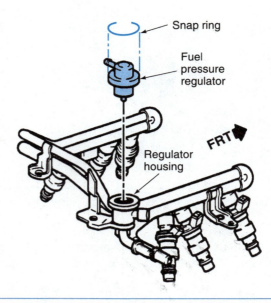

Figure 7-21 Removing the snap ring and pressure regulator from the fuel rail (Courtesy of Oldsmobile Division, General Motors Corporation)

Fuel Rail, Injector, and Pressure Regulator Cleaning and Inspection

1. Before removing the injectors and pressure regulator, clean the fuel rail with a spray-type engine cleaner such as AC Delco X-30A or its equivalent.
2. Pull the injectors from the fuel rail.
3. Use snap ring pliers to remove the snap ring from the pressure regulator cavity. Note the original direction of the vacuum fitting on the pressure regulator, and pull the pressure regulator from the fuel rail (Figure 7-21).
4. Clean all components with a clean shop towel. Be careful not to damage fuel rail openings and injector tips.
5. Check all injector and pressure regulator openings in the fuel rail for metal burrs and damage.

Installation of Fuel Rail, Injectors, and Pressure Regulator

1. If the same injectors and pressure regulator are reinstalled, replace all O-rings, and coat each O-ring lightly with engine oil.
2. Install the pressure regulator in the fuel rail, and position the vacuum fitting on the regulator in the original direction.
3. Install the snap ring above the pressure regulator.
4. Install the injectors in the fuel rail.
5. Install the fuel rail while guiding each injector into the proper intake manifold opening. Be sure the injector terminals are positioned so they are accessible to the electrical connectors.
6. Alternately tighten the fuel rail hold-down bolts, and torque them to specifications.
7. Reconnect the vacuum hose on the pressure regulator.
8. Install the fuel supply and fuel return lines on the fuel rail.
9. Install the injector electrical connectors.
10. Connect the negative battery terminal, and disconnect the 12-V power supply from the cigarette lighter.
11. Start the engine and check for fuel leaks at the rail, and be sure the engine operation is normal.

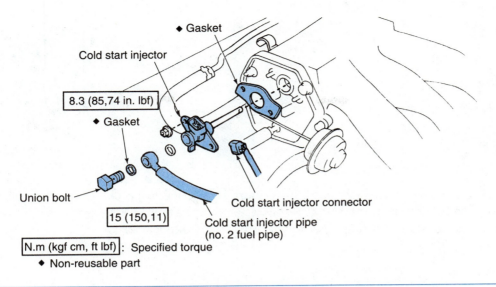

Figure 7-22 Removing the cold-start injector union bolt and fuel line (Courtesy of Toyota Motor Corporation)

Cold-Start Injector Diagnosis and Service

Cold-Start Injector Removal and Testing

WARNING: Energizing the cold-start injector for more than 5 seconds may damage the injector winding.

The cold-start injector service and diagnosis procedure varies depending on the vehicle. A typical cold-start injector removal and testing procedure follows:

1. Bleed the pressure from the fuel system.
2. Connect a 12-V power supply to the cigarette lighter socket and disconnect the negative battery cable. If the vehicle is equipped with an air bag, wait one minute.
3. Wipe excess dirt from the cold-start injector with a shop towel.
4. Remove the electrical connector from the cold-start injector.
5. Remove the union bolt and the cold-start injector fuel line (Figure 7-22).
6. Remove the cold-start injector retaining bolts, and remove the cold-start injector.
7. Connect an ohmmeter across the cold-start injector terminals (Figure 7-23). If the resistance is more or less than specified, replace the injector.
8. Connect the fuel line and union bolt to the cold-start injector and place the injector tip in a container.

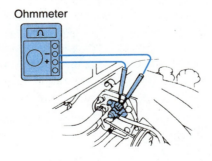

Figure 7-23 Testing the cold-start injector resistance with an ohmmeter (Courtesy of Toyota Motor Corporation)

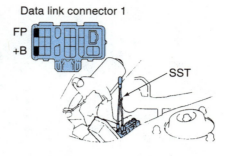

Figure 7-24 A jumper wire is connected from the B+ terminal to the FP terminal in the DLC to operate the fuel pump and check the cold-start injector. (Courtesy of Toyota Motor Corporation)

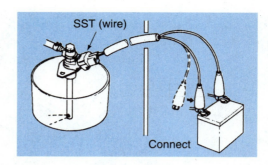

Figure 7-25 Connecting a jumper wire from cold start injector terminals to the battery terminals (Courtesy of Toyota Motor Corporation)

9. Connect a jumper wire to the B+ and FP terminals in the data link connector (DLC) and turn on the ignition switch (Figure 7-24).
10. Connect a special jumper wire from the cold-start injector terminals to the battery terminals (Figure 7-25).
11. Check the fuel spray pattern from the injector. This pattern should be as illustrated in (Figure 7-25). If the pattern is not as shown in the figure, replace the injector. Do not energize the cold-start injector for more than 5 seconds.

Cold-Start Injector Installation

1. Replace all cold-start injector gaskets, and check all mounting surfaces for metal burrs, scratches, and warping.
2. Install the cold-start injector gasket and injector, and tighten the injector mounting bolts to the specified torque.
3. Install the cold-start injector fuel line, gaskets, and union bolt, and tighten this bolt to the specified torque.
4. Connect the cold-start injector electrical connector.
5. Connect the negative battery cable and disconnect the 12-V power supply.
6. Start the engine and check for fuel leaks at the cold-start injector.

Minimum Idle Speed Adjustment and Throttle Position Sensor Adjustment

The minimum idle speed adjustment may be referred to as a minimum air rate or air flow adjustment.

The minimum idle speed adjustment may be performed on some MFI and SFI systems with a minimum idle speed screw in the throttle body. This screw is factory adjusted and the head of the screw is covered with a plug. This adjustment should only be required if throttle body parts are replaced. If the minimum idle speed adjustment is not adjusted properly, engine stalling may result. The procedure for performing a minimum idle speed adjustment varies considerably depending on the vehicle. Always follow the vehicle manufacturer's recommended procedure in the service manual. Following is a typical minimum idle speed adjustment procedure for a General Motors vehicle:

1. Be sure the engine is at normal operating temperature, and turn off the ignition switch.
2. Connect terminals A and B in the DLC, and connect a tachometer from the ignition tachometer terminal to ground.
3. Turn on the ignition switch and wait 30 seconds. Under this condition, the idle air control (IAC) motor is driven completely inward by the PCM.
4. Disconnect the IAC motor connector.

Special Tools

Tachometer

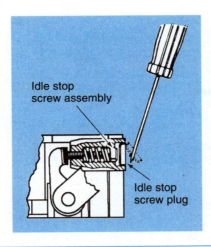

Figure 7-26 Adjusting the idle stop screw for minimum air adjustment (Courtesy of Chevrolet Motor Division, General Motors Corporation)

5. Remove the connection between terminals A and B in the DLC and start the engine.
6. Place the transmission selector in drive with an automatic transmission or neutral with a manual transmission.
7. Adjust the idle stop screw if necessary to obtain 500 to 600 rpm with an automatic transmission or 550 to 650 rpm with a manual transmission. A plug must be removed to access the idle stop screw (Figure 7-26).
8. Turn off the ignition switch and reconnect the IAC motor connector.
9. Turn on the ignition switch and connect a digital voltmeter from the TPS signal wire to ground. If the voltmeter does not indicate the specified voltage of 0.55 V, loosen the TPS mounting screws and rotate the sensor until this voltage reading is obtained. Hold the TPS in this position and tighten the mounting screws.

Before an IAC motor is installed in a General Motors throttle body on TBI, MFI, and SFI systems, the distance from the end of the valve to the shoulder on the motor body must not exceed 1.125 in. (28 mm). If the IAC motor is installed with the plunger extended beyond this measurement, the motor may be damaged.

Minimum Idle Speed Adjustment Throttle Body Injection

The minimum idle speed adjustment procedure varies depending on the year and make of the vehicle. Always follow the adjustment procedure in the vehicle manufacturer's service manual. The minimum idle speed adjustment is only required if the TBI assembly or TBI assembly components are replaced. If the minimum idle speed adjustment is not adjusted properly, engine stalling may result. Proceed as follows for a typical minimum idle speed adjustment on a General Motors TBI system:

1. Be sure that the engine is at normal operating temperature, and remove the air cleaner and TBI-to-air cleaner gasket. Plug the air cleaner vacuum hose inlet to the intake manifold.
2. Disconnect the throttle valve (TV) cable to gain access to the minimum air adjustment screw. A tamper-resistant plug in the TBI assembly must be removed to access this screw.
3. Connect a tachometer from the ignition tachometer terminal to ground, and disconnect the IAC motor connector.

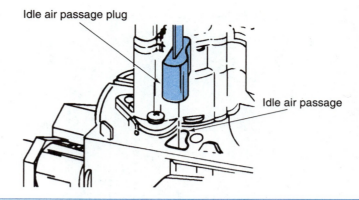

Figure 7-27 Special tool to plug the air passage to the IAC motor while checking minimum idle speed adjustment on a TBI system (Courtesy of Chevrolet Motor Division, General Motors Corporation)

4. Start the engine and place the transmission in park with an automatic transmission or neutral with a manual transmission.
5. Plug the air intake passage to the IAC motor. Tool J-33047 is available for this purpose (Figure 7-27).
6. On 2.5-L 4-cylinder engines, use the appropriate torx bit to rotate the minimum air adjustment screw until the idle speed on the tachometer is 475 to 525 rpm with an automatic transaxle or 750 to 800 rpm with a manual transaxle.
7. Stop the engine and remove the plug from the idle air passage. Cover the minimum air adjustment screw opening with silicone sealant, reconnect the TV cable, and install the TBI gasket and air cleaner.

Throttle Body Service

Throttle Body On-Vehicle Cleaning

 WARNING: Use only approved throttle body cleaners. Other cleaners may damage throttle body components.

WARNING: When cleaning a throttle body on the vehicle, be careful not to get the cleaner into the TPS or idle air control (IAC) valve. Throttle body cleaner will damage these components.

Special Tools

Throttle body cleaner

After many miles or kilometers of operation, an accumulation of gum and carbon deposits may occur around the throttle area in TBI, MFI, and SFI systems. This condition may cause rough idle operation. A pressurized can of throttle body cleaner may be used to spray around the throttle area without removing and disassembling the throttle body. If this cleaning method does not remove the deposits, the throttle body will have to be removed, disassembled, and placed in an approved cleaning solution. Never place the IAC motor or the TPS in cleaning solution, or damage to these components will result! Always remove the TPS, MAF, IAC motor, injectors, seals, gaskets, and pressure regulator before the throttle body is placed in a cleaning solution. Since MFI and SFI systems do not have injectors and a pressure regulator in the throttle body, removal of these components is not required.

Throttle Body On-Vehicle Inspection

Throttle body inspection and service procedures vary widely depending on the year and make of vehicle. However, some components such as the TPS are found on nearly all throttle bodies. Since throttle bodies have some common components, inspection procedures often involve checking

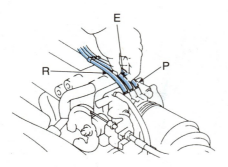

Port name	At idle	Other than idle
P	No vacuum	Vacuum
E	No vacuum	Vacuum
R	No vacuum	No vacuum

Figure 7-28 Throttle body vacuum ports and appropriate vacuum in relation to throttle position (Courtesy of Toyota Motor Corporation)

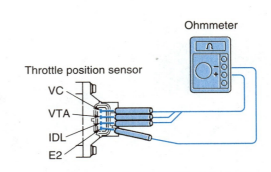

Figure 7-29 Ohmmeter connected to various TPS terminals to test TPS condition (Courtesy of Toyota Motor Corporation)

common components with the procedure recommended in the vehicle manufacturer's service manual. The following throttle body service procedures are based on a Toyota MFI system:

1. Check for smooth movement of the throttle linkage from the idle position to the wide-open position. Check the throttle linkage and cable for wear and looseness.
2. With the engine idling and operating at higher speed, check for vacuum with your finger at each vacuum port in the throttle body (Figure 7-28).
3. Apply vacuum from a hand vacuum pump to the throttle opener, and disconnect the TPS connector. Test the TPS with an ohmmeter connected across the appropriate terminals (Figure 7-29), and the specified thickness gauge inserted between the throttle stop screw and the stop lever (Figure 7-30).
4. Check the ohmmeter reading when the ohmmeter is connected to each of the specified terminals on the TPS (Figure 7-31).

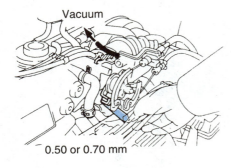

Figure 7-30 Thickness gauge inserted between the throttle stop screw and the stop lever while testing the TPS (Courtesy of Toyota Motor Corporation)

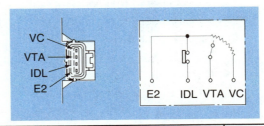

Clearance between lever and stop screw	Between terminals	Resistance
0 mm (0 in.)	VTA–E2	0.2–5.7 kΩ
0.50 mm (0.20 in.)	IDL–E2	2.3 kΩ or less
0.70 mm (0.028 in.)	IDL–E2	Infinity
Throttle valve fully open	VTA–E2	2.0–10.2 kΩ
—	VC–E2	2.5–5.9 kΩ

Figure 7-31 Specified ohmmeter reading at the TPS terminals (Courtesy of Toyota Motor Corporation)

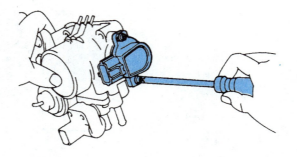

Figure 7-32 Loosening the TPS mounting screws to adjust the TPS until the specified ohmmeter readings are obtained (Courtesy of Toyota Motor Corporation)

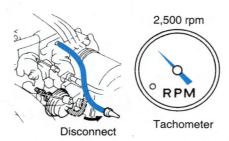

Figure 7-33 Throttle opener hose disconnected and plugged, and engine running at 2,500 rpm prior to throttle opener test (Courtesy of Toyota Motor Corporation)

5. Loosen the two TPS mounting screws and rotate the TPS as required to obtain the specified ohmmeter readings (Figure 7-32). Then retighten the mounting screws. If the TPS cannot be adjusted to obtain the proper ohmmeter readings, replace the TPS.
6. Operate the engine until it reaches normal operating temperature, and check the idle speed on a tachometer. The idle speed should be 700 to 800 rpm.
7. Disconnect and plug the vacuum hose from the throttle opener, and maintain 2,500 engine rpm (Figure 7-33).
8. Be sure the cooling fan is off. Release the throttle valve, and observe the tachometer reading. When the throttle linkage strikes the throttle opener stem, the engine rpm should be 1,300 to 1,500.
9. Adjust the throttle opener as necessary (Figure 7-34), and reconnect the throttle opener vacuum hose.

Fuel Cut RPM Check

If the fuel cut mode is not operating properly, emission levels are high during deceleration, and an increase in fuel consumption is experienced. The checking procedure for the fuel cut operation varies depending on the vehicle year and model. Following is a typical procedure for checking fuel cut operation:

1. Operate the engine until it is at normal operating temperature.
2. Connect a tachometer pickup lead to the IG terminal in the DLC, and connect the other tachometer leads as recommended by the tachometer manufacturer (Figure 7-35). Consult the vehicle manufacturer's information to be sure the tachometer is compatible with the vehicle electrical system.

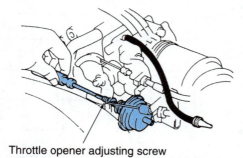

Figure 7-34 Throttle opener adjustment (Courtesy of Toyota Motor Corporation)

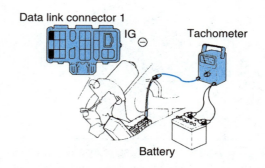

Figure 7-35 Tachometer connection to the IG terminal in the DLC (Courtesy of Toyota Motor Corporation)

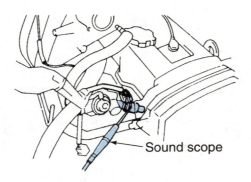

Figure 7-36 Stethoscope pickup placed on the injector body (Courtesy of Toyota Motor Corporation)

3. Increase the engine rpm to 2,500, and place a stethoscope pickup against the body of a fuel injector.
4. Allow the engine speed to return to idle, and listen to the injector operation with the stethoscope (Figure 7-36). The injector should stop clicking momentarily, and then resume clicking as the engine speed decreases. The injector should begin clicking again at 1,500 rpm as the engine decelerates.
5. Shut off the engine and disconnect the tachometer.

Flash Code Diagnosis of TBI, MFI, and SFI Systems

Chrysler Flash Code Diagnosis

If a TBI, MFI, or SFI system is working normally, the malfunction indicator light (MIL) is illuminated when the ignition switch is turned on, and it goes out a few seconds after the engine is started. The MIL light should remain off while the engine is running.

If a defect occurs in a sensor and a diagnostic trouble code (DTC) is set in the computer memory, the computer may enter a limp-in mode. In this mode, the MIL, or check engine light, is on, the air-fuel ratio is rich, and the spark advance is fixed, but the vehicle can be driven to an automotive service center. When a vehicle is operating in the limp-in mode, fuel consumption, and emission levels increase, and engine performance may decrease.

Prior to any DTC diagnosis, the Preliminary Diagnostic Procedure mentioned previously in this chapter must be completed, and the engine must be at normal operating temperature. If the engine is not at normal operating temperature, the computer may provide erroneous DTCs. The battery in the vehicle must be fully charged prior to DTC diagnosis.

Follow these steps to read the DTCs from the flashes of the MIL light on most Chrysler products:

1. Cycle the ignition switch on and off, on and off, and on in a 5-second interval.
2. Observe the MIL lamp flashes to read the DTCs. Two quick flashes followed by a brief pause and three quick flashes indicates code 23. The DTCs are flashed once in numerical order.
3. When code 55 is flashed, the DTC sequence is completed. The ignition switch must be turned off, and steps 1 and 2 repeated to read the DTCs a second time.

On any TBI or PFI system, a DTC indicates a defect in a specific area. For example, a TPS code indicates a defective TPS, defective wires between the TPS and the computer, or the computer may be unable to receive the TPS signal. Specific ohmmeter or voltmeter tests may be necessary to locate the exact cause of the fault code. On logic module and power module systems, disconnect

In the SAE J1930 terminology, the term malfunction indicator light *(MIL) replaces other terms such as check engine light and service engine soon light.*

Hard fault diagnostic trouble codes (DTCs) are present in the computer memory at the time of testing.

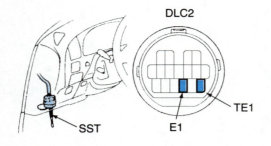

Figure 7-37 E1 and TE1 terminals in round DLC located under the instrument panel (Courtesy of Toyota Motor Corporation)

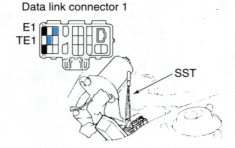

Figure 7-38 E1 and TE1 terminals in rectangular DLC positioned in the engine compartment (Courtesy of Toyota Motor Corporation)

the quick-disconnect connector at the positive battery cable for 10 seconds with the ignition switch off to erase DTCs. On later module PCMs, this connector must be disconnected for 30 minutes to erase fault codes.

Toyota Flash Code Diagnosis

Flash Code Output. Prior to the flash code output, the Preliminary Diagnostic Procedure must be performed as mentioned at the beginning of this chapter. Follow these steps for DTC diagnosis:

1. Turn on the ignition switch and connect a jumper wire between terminals E1 and TE1 in the DLC. Some round DLCs are located under the instrument panel (Figure 7-37), while other rectangular-shaped DLCs are positioned in the engine compartment (Figure 7-38).
2. Observe the MIL light flashes. If the light flashes on and off at 0.26-second intervals, there are no DTCs in the computer memory (Figure 7-39).
3. If there are DTCs in the computer memory, the MIL light flashes out the DTCs in numerical order. For example, one flash followed by a pause and three flashes is code 13, and three flashes followed by a pause and one flash represents code 31 (Figure 7-40). The codes will be repeated as long as terminals E1 and TE1 are connected and the ignition switch is on.
4. Remove the jumper wire from the DLC.

Driving Test Mode. Follow this procedure to obtain fault codes during a driving test mode:

1. Turn on the ignition switch and then connect terminals E1 and TE2 in the DLC (Figure 7-41).
2. Start the engine and drive the vehicle at speeds above 6 mph (10 km/h). Simulate the conditions when the problem occurs.

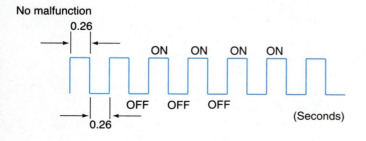

Figure 7-39 If the MIL light flashes at 0.26-second intervals, there are no DTCs in the computer memory. (Courtesy of Toyota Motor Corporation)

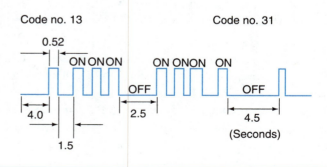

Figure 7-40 DTCs 13 and 31 (Courtesy of Toyota Motor Corporation)

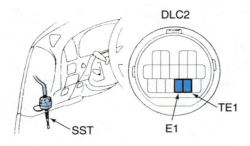

Figure 7-41 Terminals E1 and TE2 in the DLC (Courtesy of Toyota Motor Corporation)

3. Connect a jumper wire between terminals E1 and TE1 on the DLC.
4. Observe the flashes of the MIL light to read the DTCs, and remove the jumper wire from the DLC.

Ford Flash Code Diagnosis

Most Ford vehicles have an MIL light on the instrument panel, and this light flashes the DTCs in the diagnostic mode. When a defect occurs in a major sensor, the PCM illuminates the MIL light and enters the limp-in mode in which the air-fuel ratio is rich and the spark advance is fixed. In this mode, engine performance decreases, and fuel consumption and emission levels increase.

Jumper Wire Connection. Prior to any fault code diagnosis, the engine must be at normal operating temperature and the Preliminary Diagnostic Procedure mentioned previously in the chapter must be completed. A jumper wire must be connected from the self-test input wire to the appropriate DLC terminal to enter the self-test mode. When the ignition switch is turned on after this jumper wire connection, the MIL light begins to flash any DTCs in the PCM memory.

Optional Voltmeter Connection. If the vehicle does not have a check engine light, a voltmeter can be connected from the positive battery terminal to the proper DLC terminal (Figure 7-42). The voltmeter must be connected with the correct polarity as indicated in the figure.

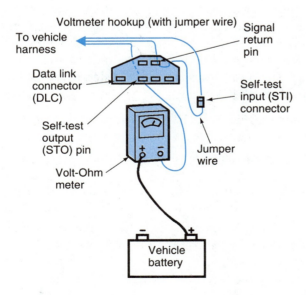

Figure 7-42 Jumper wire and voltmeter connection to Ford DLC (Reprinted with the permission of Ford Motor Company)

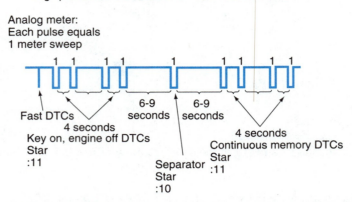

Figure 7-43 Key on engine off (KOEO) test procedure (Reprinted with the permission of Ford Motor Company)

When the ignition switch is turned on after the jumper wire and voltmeter connections are completed, the DTCs may be read from the sweeps of the voltmeter pointer or the flashes of the check engine light. For example, if three upward sweeps of the voltmeter pointer are followed by a pause and then four upward sweeps, code 34 is displayed.

Key On Engine Off (KOEO) Test. Follow these steps for the key on engine off (KOEO) fault code diagnostic procedure:

1. With the ignition switch off, connect the jumper wire to the self-test input wire and the appropriate terminal in the DLC.
2. If the vehicle does not have an MIL light, connect the voltmeter to the positive battery terminal and the appropriate DLC terminal.
3. Turn on the ignition switch and observe the MIL light or voltmeter. Hard fault DTCs are displayed, followed by a separator code 10, and continuous memory DTCs (Figure 7-43).

Hard fault DTCs are present at the time of testing, whereas memory DTCs represent intermittent faults that occurred some time ago and are set in the computer memory. Separator code 10 is displayed as one flash of the MIL light or one sweep of the voltmeter pointer. Each fault DTC is displayed twice, and the DTCs are provided in numerical order. If there are no DTCs, system pass code 11 is displayed. If the technician wants to repeat the test or proceed to another test, the ignition switch must be turned off for 10 seconds.

Key On Engine Running (KOER) Test. Follow these steps to obtain the fault codes in the Key On Engine Running (KOER) test sequence:

1. Connect the jumper wire and the voltmeter as explained in steps 1 and 2 of the KOEO test.
2. Start the engine and observe the MIL lamp or voltmeter. The engine identification code is followed by the separator code 10 and the hard fault codes (Figure 7-44).

The engine identification (ID) code represents half of the engine cylinders. On a V8 engine, the MIL light flashes four times, or the voltmeter pointer sweeps upward four times during the engine ID display.

On some Ford products, the brake on/off (BOO) switch and the power steering pressure switch (PSPS) must be activated after the engine ID code, or DTCs 52 and 74, representing these switches, are present. Step on the brake pedal and turn the steering wheel to activate these switches immediately after the engine ID display.

Hard fault diagnostic trouble codes (DTCs) are present in the computer memory at the time of testing.

Memory DTCs represent intermittent faults that occurred previously and were set in the computer memory at that time.

Memory DTCs may be called continuous memory codes or history codes.

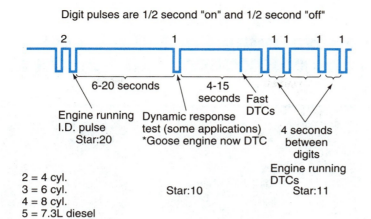

Figure 7-44 Key on engine running (KOER) test (Reprinted with the permission of Ford Motor Company)

Separator code 10 is presented during the KOER test on many Ford products. When this code is displayed, the throttle must be pushed momentarily to the wide-open position. The best way to provide a wide-open throttle is to momentarily push the gas pedal to the floor. On some Ford products, the separator code 10 is not displayed during the KOER test, and this throttle action is not required.

Hard fault DTCs are displayed twice in numerical order. If no faults are present, system pass code 11 is given.

Fault Code Erasing Procedure. DTCs may be erased by entering the KOEO test procedure and disconnecting the jumper wire between the self-test input wire and the DLC during the code display.

Photo Seqence 5 shows a typical procedure for performing Ford flash code diagnosis, key on engine off (KOEO), and key on engine running (KOER) tests.

General Motors Flash Code Testing

When a fault occurs in a major sensor, the PCM illuminates the MIL light, and a fault code is set in the PCM memory. Once this action takes place, the PCM is usually operating in a limp-in mode, and the air-fuel ratio is rich with a fixed spark advance. In this mode, driveability is adversely affected and fuel consumption and emission levels increase. Prior to any fault code diagnosis, the engine must be at normal operating temperature and the Preliminary Diagnostic Procedure mentioned previously must be completed. Follow these steps to obtain the DTCs with the MIL light flashes:

1. With the ignition switch off, connect a jumper wire between terminals A and B in the DLC under the instrument panel (Figure 7-45). A special tool that has two lugs that fit between these terminals is available. Terminals A and B are usually located at the top right corner of the DLC, but some DLCs are mounted upside down or vertically. Always consult the vehicle manufacturer's service manual for exact terminal location.
2. Turn on the ignition switch, and observe the MIL lamp.
3. One lamp flash followed by a brief pause and two more flashes indicates code 12, and this code indicates that the PCM is capable of diagnosis. Each code is flashed three

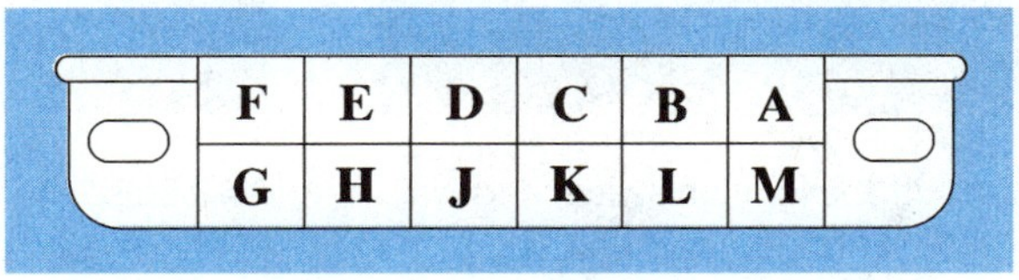

Figure 7-45 Terminals in the DLC

Photo Sequence 5
Typical Procedure for Performing Ford Flash Code Diagnosis, Key On Engine Off (KOEO), and Key On Engine Running (KOER) Tests

P5-1 Be sure the engine is at normal temperature and the ignition switch is off.

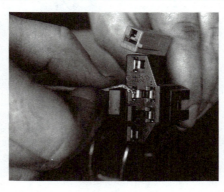

P5-2 Connect a jumper wire to the proper terminals in the DLC.

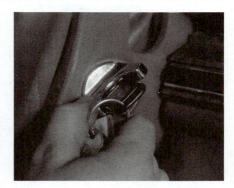

P5-3 Turn on the ignition switch.

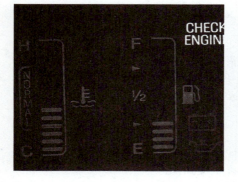

P5-4 Observe the MIL light flashes in the instrument panel. The sequence of codes in the KOEO test is hard fault codes, separator code 10, and intermittent fault codes.

P5-5 Turn off the ignition switch and wait 10 seconds.

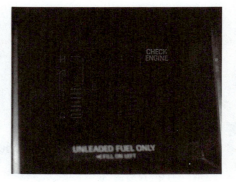

P5-6 Start the engine and observe the MIL light flashes in the instrument panel. The sequence of codes in the KOER test is engine ID code, separator code 10 (some systems), and hard faults.

P5-7 Record all the fault codes and determine the fault code interpretation from the fault code list in the vehicle manufacturer's appropriate service manual.

P5-8 Turn off the ignition switch and wait 10 seconds.

P5-9 Erase the fault codes by turning the ignition switch on and removing the jumper wire when the MIL light is flashing the KOEO sequence.

times, and codes are given in numerical order. If there are no DTCs in the PCM, only code 12 is provided. The code sequence keeps repeating until the ignition switch is turned off.

Complete DLC Terminal Explanation. The actual number of wires in the DLC varies depending on the vehicle make and year. The purpose of each DLC terminal may be explained as follows:

- **A** Ground terminal
- **B** Diagnostic request
- **C** Air injection reactor—When this terminal is grounded, the air injection reactor (AIR) pump air is continually directed upstream to the exhaust ports because this connection grounds the AIR system port solenoid. This applies to AIR systems with a converter solenoid and a port solenoid; however, this action does not apply to newer AIR systems with an electric diverter valve (EDV) solenoid.
- **D** MIL light —When this terminal is grounded on some systems, the check engine light is illuminated continually.
- **E** Serial data slow-speed 160-baud PCM—This terminal supplies input sensor data to the scan tester on 160-baud PCM systems.
- **F** Torque converter clutch (TCC)—If the vehicle is lifted and the engine accelerates until the transmission shifts through all the gears, a 12-V test light may be connected from this terminal to ground to diagnose the TCC system. The light is on when the TCC is not locked up, and the light goes out when TCC lockup occurs.
- **G** Fuel pump test—On some models, when 12 V are supplied to this terminal with the ignition off, current flows through the top fuel pump relay contacts to the fuel pump and the pump should run. Other models have a separate fuel pump test lead located under the hood.
- **H** Antilock brake system (ABS), cars and trucks—When a jumper wire is connected from this terminal to terminal A, the ABS computer flashes the ABS warning light to provide fault codes.
- **J** Not used.
- **K** Air bag, supplemental inflatable restraint (SIR) system—When this terminal is connected to terminal A, the SIR computer flashes fault codes on the SIR warning light.
- **L** Not used.
- **M** High-speed serial data P4 PCM—This terminal supplies sensor data to the scan tester on P4 PCM systems.

DTC Erasing Procedure. The fault codes may be erased by disconnecting the quick-disconnect connector at the positive battery terminal for 10 seconds with the ignition switch off. If the vehicle does not have a quick-disconnect connector, the PCM B fuse may be disconnected to erase fault codes. On later model General Motors vehicles with P4 PCMs, the quick-disconnect, or PCM B, fuse may have to be disconnected for a longer time to erase codes.

If DTCs are left in a computer after the defect is corrected, the codes are erased automatically when the engine is stopped and started 30 to 50 times. This also applies to most computer-equipped vehicles.

Field Service Mode. If the A and B terminals are connected in the DLC and the engine is started, the PCM enters the field service mode. In this mode, the speed of the MIL lamp flashes indicate whether the system is in open loop or closed loop. If the system is in open loop, the MIL lamp flashes quickly. When the system enters closed loop, the MIL lamp flashes at half the speed of the open loop flashes.

Nissan Flash Code Testing

In some Nissan electronic concentrated engine control systems (ECCS), the PCM has two light emitting diodes (LEDs), which flash a fault code if a defect occurs in the system. One of these LEDs is

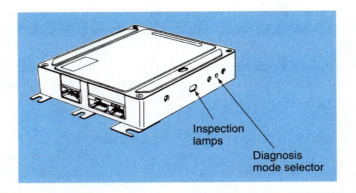

Figure 7-46 Diagnosis mode selector on PCM (Courtesy of Nissan Motor Co., Ltd.)

red and the second LED is green. The technician observes the flashing pattern of the two LEDs to determine the DTC. If there are no DTCs in the ECCS, the LEDs flash a system pass code. The flash code procedure varies depending on the year and model, and the procedure in the manufacturer's appropriate service manual must be followed. Later model Nissan engine computers have a five-mode diagnostic procedure. Be sure the engine is at normal operating temperature, and complete the Preliminary Diagnostic Procedure explained previously in this chapter. Turn the diagnosis mode selector in the PCM to obtain the diagnostic modes (Figure 7-46).

The following diagnostic modes are available on some Nissan products:

Mode 1—This mode checks the oxygen sensor signal. With the system in closed loop and the engine idling, the green light should flash on each time the oxygen sensor detects a lean condition. This light goes out when the oxygen sensor detects a rich condition. After 5–10 seconds, the PCM "clamps" on the ideal air-fuel ratio and pulse width. The green light may be on or off. This PCM clamping of the pulse width only occurs at idle speed.

Mode 2—In this mode, the green light comes on each time the oxygen sensor detects a lean mixture, and the red light comes on when the PCM receives this signal and makes the necessary correction in pulse width.

Mode 3—This mode provides DTCs representing various defects in the system.

Mode 4—Switch inputs to the PCM are tested in this mode. Mode 4 cancels codes available in mode 3.

Mode 5—This mode increases the diagnostic sensitivity of the PCM for diagnosing intermittent faults while the vehicle is driven on the road.

After the defect has been corrected, turn the ignition switch off, rotate the diagnosis mode selector counterclockwise, and install the PCM securely in the original position.

General TBI, MFI, and SFI Diagnosis

Differentiating Between Mechanical Problems and Electronic Problems

The technician must remember that engine mechanical problems may cause a DTC to be displayed representing a component in the engine computer system. For example, if a burned exhaust valve results in low cylinder compression, a DTC indicating an always lean O_2 sensor condition may be provided. Under this condition, much of the air-fuel mixture from this cylinder is unburned, and the O_2 sensor senses the additional oxygen in the exhaust stream. The technician must determine if the always lean O_2 sensor code is caused by a defective sensor or some mechanical condition.

The burned exhaust valve and low cylinder compression will result in a rough idle condition. The technician must verify the low compression with a compression test. Once the low compression and burned valve condition is verified, the technician should inform the customer that a valve job is required on the engine rather than engine computer system repairs.

If the compression is satisfactory, the technician must prove the ignition system is not causing the rough idle problem. An oscilloscope diagnosis will check the various ignition components.

As mentioned previously, an intake manifold vacuum leak results in rough idle operation. An intake manifold vacuum leak test must be performed to locate vacuum leaks as was explained at the beginning of this chapter.

The technician must obtain the DTCs to determine if a code indicates the cause of the rough idle problem. Since injectors may result in a rough idle condition, the injectors may be diagnosed, tested, and serviced as explained previously in this chapter.

 WARNING: Never run combustion chamber cleaner through a MAF sensor. This action will ruin the sensor.

Intake valve deposits and throttle body deposits may result in erratic idle operation and stalling. If these deposits are causing the rough idle problem, there are usually no DTCs in the computer memory. A can of combustion chamber cleaner may be sprayed into the intake with the engine running to remove intake valve deposits. Always follow the combustion chamber cleaner manufacturer's recommended service procedure. On engines with a MAF sensor, remove the PCV valve and spray the combustion chamber cleaner through the PCV hose with the engine running. Do not allow combustion chamber cleaner to flow through the MAF sensor. In many cases, throttle body deposits may be removed with a throttle body cleaner and the throttle body installed on the engine.

The technician must always verify engine compression, ignition, vacuum leaks, and any other mechanical causes of the problem before proceeding with the electrical/electronic diagnosis.

Service Bulletin Information

When diagnosing problems in TBI, MFI, and SFI systems, service bulletin information is absolutely essential. If a technician does not have service bulletin information, many hours of diagnostic time may be wasted. Of course, we cannot include service bulletin information on all domestic and imported vehicles in this publication. This information is available from different suppliers on CD. We will discuss the solution to three problems found in service bulletin information to emphasize the importance of this information.

Many General Motors engines are equipped with Multec injectors. Some of these injectors have experienced shorting problems in the windings, especially if the fuel used contained some alcohol content. If the injectors become shorted, they draw excessive current. General Motors P4 PCMs have a sense line connected to the quad driver that operates the injectors. When this sense line experiences excessive current flow from the shorted injectors, the quad driver shuts off and stops operating the injectors. This action protects the quad driver, but also causes the engine to stall. After a few minutes, the engine will usually restart. If a technician does not have this information available in a service bulletin, a great deal of time may be wasted locating the problem.

On 160-baud General Motors computers, the pins on the internal components extend through the circuit board tracks, and soldering is done on the opposite side of the board from where the components are located. On P4 PCMs, a surface mount technology (SMT) was developed in which the component pins are bent at a 90° angle and then soldered on top of the tracks on the circuit boards. In some cases, loose connections have developed in the computers with the SMT. These loose connections usually cause the engine to quit. If a technician suspects this problem, the PCM may be removed, with the wiring harness connected. Start the engine and give the PCM a slap with the palm of your hand. If the engine stalls or the engine operation changes, a loose connection is present on the circuit board. When a technician does not have this information available in service bulletins, much diagnostic time may be wasted.

In 1991, Chrysler experienced some low-speed surging during engine warm-up on 3.3-L and 3.8-L engines. On these engines, the port fuel injectors sprayed against a hump in the intake port. As a result, fuel puddled behind this hump, especially while the engine was cold. When the engine temperature increased, this fuel evaporated and caused a rich air-fuel ratio and engine surging. Chrysler corrected this problem by introducing angled injectors with the orifices positioned at an angle so the fuel sprayed over the hump in the intake. When angled injectors are installed, the wiring connector must be positioned vertically. Angled injectors have beige exterior bodies. Technicians must have service bulletin information regarding problems like this.

Diagnosis of Computer Power and Ground Wires

SERVICE TIP: Never replace a computer until the ground wires and voltage supply wires to the computer are checked and proven to be in satisfactory condition. High resistance in computer ground wires may cause unusual problems.

The condition of the PCM power and ground wires must be verified prior to PCM replacement. Low voltage supply to the PCM or high resistance in the PCM ground wires may result in unusual problems. Follow this procedure to check the PCM ground and power wires:

1. Obtain the appropriate PCM wiring diagram for the vehicle being diagnosed, and identify the PCM power and ground wire terminals on the diagram.
2. With the ignition switch on, backprobe the PCM connector and individually connect the digital voltmeter leads to PCM terminals 5, 11, and 12 and the battery ground (Figure 7-47). In any of the three tests, a voltmeter reading above the vehicle manufacturer's specifications indicates excessive resistance in that ground wire.
3. Turn on the ignition switch and connect the digital voltmeter leads from PCM terminal 9 to the battery ground. If the vehicle manufacturer's specified voltmeter reading appears on the voltmeter, the PCM power wire is satisfactory. A lower-than-specified voltmeter reading indicates high resistance in the PCM voltage supply wire or low battery voltage.

Diagnosis of Specific Problems and Necessary Corrections

No-Start

1. Low compression—perform compression test, repair engine if necessary.
2. Improper valve timing—check valve timing, correct if necessary.
3. Defective ignition—test and repair as required.
4. Defective fuel system, fuel pump, filter, injectors—test fuel pressure and injectors.

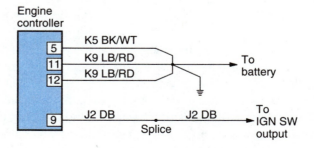

Figure 7-47 Testing PCM ground and power wires (Courtesy of Chrysler Corporation)

Hard Starting

1. Low compression—perform compression test, repair engine as required.
2. Lean air-fuel ratio, vacuum leak, injectors—test intake vacuum leaks, injectors.
3. Rich air-fuel ratio, injectors, cold-start injector, input sensors—test injectors, and obtain DTCs, perform voltmeter or ohmmeter tests on input sensors.
4. Leaking pressure regulator, fuel pump check valve, injectors—test fuel pressure and injectors.
5. Defective ignition system—perform oscilloscope test of ignition system.

Rough Idle

1. Low compression—perform compression test, repair engine as required.
2. EGR valve (stuck open)—perform EGR valve test.
3. Vacuum leak—test intake vacuum leaks, repair as necessary.
4. Dirty injectors—clean injectors.
5. Dirty throttle body—clean throttle body as required.
6. Intake valve deposits—remove valve deposits with combustion chamber cleaner.
7. Defective cold-start injector—test cold-start injector and related components, repair or replace as necessary.

High Idle Speed

1. Engine coolant temperature (ECT) sensor—test ECT sensor and connecting wires, repair or replace as necessary.
2. Inlet air temperature (IAT) sensor—test IAT sensor and connecting wires, repair or replace as necessary.
3. Thermostat stuck open—test thermostat, replace as necessary.
4. Low coolant level—check coolant level, correct as required.
5. P/N switch—test P/N switch and connecting wires, repair or replace as necessary.
6. Low battery and charging system voltage—test battery, charging system, and connecting wires, repair or replace as required.
7. Low voltage to computer (resistance in battery 12 V, or ignition on 12 V, wires)—test power supply wires, repair as necessary.
8. Vacuum leak—test and repair intake vacuum leaks as required.
9. Sticking or defective idle air control (IAC) motor—test, clean, or replace IAC motor.
10. Improper TPS adjustment or faulty TPS—test TPS and connecting wires, repair or replace as necessary.

Low Idle Speed

1. ECT sensor—test ECT sensor and connecting wires, repair or replace as required.
2. IAT sensor—test IAT sensor and connecting wires, repair or replace as necessary.
3. Sticking or defective idle air control motor—test, clean, or replace IAC motor.
4. Improper TPS adjustment or faulty TPS—test TPS sensor and connecting wires, adjust or replace as required.
5. P/N switch—test P/N switch and connecting wires, repair or replace as necessary.

Rich Air-Fuel Mixture, Low Fuel Economy, Excessive Catalytic Converter Odor

1. Low compression—test engine compression, perform engine repairs as required.
2. Defective ignition—perform oscilloscope diagnosis, repair or replace components as necessary.
3. High fuel pump pressure—test pump pressure, repair as necessary.

4. Running in limp-in mode (defective sensor) (MAP sensor)—obtain DTCs, test and replace sensors as required.
5. ECT sensor—test ECT sensor and connecting wires, repair or replace as necessary.
6. Low coolant level—check coolant level, correct as required.
7. IAT sensor—test IAT sensor and connecting wires, repair or replace as required.
8. Insufficient spark advance—test spark advance, determine if PCM is in limp-in, repair or replace as necessary.
9. Air pump air always upstream to exhaust ports with engine hot—test air pump system, repair or replace necessary components.
10. Defective injectors—diagnose, test, clean, or replace injectors.

Lean Air-Fuel Mixture

1. Low fuel pump pressure, pump, filter, regulator—test fuel pump pressure, replace components as necessary.
2. Vacuum leak, especially on MAF applications (PCV valve)—test intake manifold vacuum leaks, repair as required.
3. Dirty injectors—diagnose, test, clean, or replace injectors.

Surging at Idle

1. Vacuum leak—test intake vacuum leaks, repair as necessary.
2. Defective MAP sensor—test MAP sensor and connecting wires, repair or replace as necessary.
3. Defective MAF sensor (also surges on acceleration)—test MAF sensor and connecting wires, repair or replace as required.
4. Dirty injectors—diagnose, test, clean, or replace injectors.

Detonation

1. Lean air-fuel mixture—test intake vacuum leaks, oxygen sensor, and connecting wires, repair or replace as necessary.
2. Excessive spark advance—test spark advance knock sensor and connecting wires, repair or replace as required.
3. Defective knock sensor or ESC module—test knock sensor module and connecting wires, repair or replace as necessary.
4. Spark plug heat range too hot—check plug heat range, replace as required.
5. Plug wires routed incorrectly—check plug wire routing, correct as necessary.
6. PROM change required GM—check PROM update information, replace as required.
7. Remove octane adjust connector—reduces spark advance on some Ford products.

Engine Stalling

1. Defective injectors—diagnose, test, clean, or replace injectors as necessary.
2. Defective cold-start injector—test cold-start injector and related components, repair or replace as required.
3. Defective IAC motor—test IAC motor and connecting wires, repair or replace as required.
4. Deposits in IAC motor passages—remove carbon and clean as necessary with throttle body cleaner.
5. Improper idle speed—test TPS voltage and other causes.
6. Improper minimum idle speed adjustment—perform minimum idle speed adjustment and correct as necessary.
7. Improper TPS adjustment or faulty TPS—test TPS voltage, adjust or replace as required.
8. Carbon and gum deposits in throttle body—clean with throttle body cleaner.

Engine Surging After Torque Converter Clutch Lockup

1. Spark plugs—test, replace as necessary.
2. Spark plug wires—test with an ohmmeter and replace as required.
3. Distributor cap and rotor (distributor-type ignition)—test and replace as necessary.
4. Ignition coil—test with an ohmmeter or oscilloscope and replace as required.
5. Fuel injectors—diagnose, test, clean, and replace injectors as necessary.
6. Vacuum leaks—test and correct intake vacuum leaks as required.
7. EGR valve—test EGR valve and clean or replace.
8. MAF or MAP sensor—test sensor and connecting wires, repair or replace as necessary.
9. Worn camshaft lobes—visually inspect cam lobes, replace camshaft as necessary.
10. Oxygen sensor—test sensor and connecting wires, repair or replace as required.
11. Low fuel pump pressure—test fuel pump pressure, replace necessary components.
12. Worn engine mounts—check engine mounts and replace as necessary.
13. Front drive axle joints—check front axle joints and replace as required.
14. TPS sensor—test sensor and connecting wires, repair or replace as necessary.
15. Low cylinder compression—test engine compression, perform engine repairs as required.
16. Contaminated fuel—test fuel for water and alcohol content, replace as necessary.

Engine Dieseling

1. Leaking injectors—diagnose, test, clean, or replace injectors as necessary.
2. Leaking cold-start injector—test cold start injector and related components, repair or replace as required.

Cylinder Misfiring

1. Low compression—test engine compression, perform engine repairs as necessary.
2. Defective ignition system, spark plugs, plug wires, coil—perform oscilloscope diagnosis, replace components as required.
3. Defective injectors—diagnose, test, clean, or replace injectors as necessary.
4. Vacuum leak—test intake vacuum leaks, and correct as required.

Engine Power Loss

1. Low compression—test engine compression, perform engine repairs as necessary.
2. Improper EGR valve operation—test EGR valve and clean or replace as required.
3. Ignition defects—perform oscilloscope diagnosis and replace necessary components.
4. Reduced spark advance—test spark advance and knock sensor to determine if the PCM is in limp-in mode.
5. Computer operating in limp-in mode—evidenced by fixed, low spark advance, rich air-fuel ratio, and MIL light illuminated.
6. Low fuel pump pressure—test fuel pump pressure and replace components as required.
7. Injectors—diagnose, test, clean, or replace injectors as necessary.
8. Restricted exhaust—evidenced by very low intake vacuum as engine is loaded.

Hesitation on Acceleration

1. Lean air-fuel ratio, low fuel pump pressure, injectors, filter, vacuum leak—test fuel pump pressure and intake vacuum leaks.
2. Improper TPS adjustment or faulty TPS—test TPS and connecting wires, repair or replace components as necessary.
3. Reduced spark advance—test spark advance, knock sensor, and connecting wires, determine if PCM is in limp-in mode, perform necessary repairs.

Classroom Manual
Chapter 7, page 178

4. Computer operating in limp-in mode—obtain DTCs, test input sensor and connecting wires, repair or replace necessary components.
5. Defective negative backpressure EGR valve GM—replace EGR valve.

● **CUSTOMER CARE:** Diagnosing is an extremely important part of a technician's job on today's high-tech vehicles. Always take time to diagnose a customer's vehicle accurately. Fast, inaccurate diagnosis of automotive problems leads to unnecessary, expensive repairs, and unhappy customers who may take their business to another shop. Accurate diagnosis may take more time, but in the long term, it will improve customer relations, and bring customers back to the shop.

Guidelines for Servicing TBI, MFI, and SFI Systems

1. Prior to any diagnosis of TBI, MFI, and SFI systems, a preliminary diagnostic procedure must be performed, which includes checking such items as vacuum leaks and emission devices.
2. If there is no fuel pump pressure, always check the inertia switch, fuse, or fuse link first.
3. Fuel pressure should be relieved prior to disconnecting fuel system components.
4. Many vehicles have a test connector, which allows the technician to operate the fuel pump continually while testing the fuel pump.
5. Low fuel pump pressure may be caused by a restricted line or filter, or a defective fuel pump.
6. Low fuel pump pressure causes a lean air-fuel ratio.
7. High fuel pump pressure is caused by a sticking pressure regulator or a restricted return fuel line.
8. High fuel pump pressure results in a rich air-fuel ratio.
9. Port injectors may be tested with a balance test or a flow test.
10. Injectors should be tested for leakage and ohms resistance in the winding.
11. When injectors are cleaned with a pressurized container connected to the fuel rail, the fuel pump must be disabled and the fuel return line must be plugged.
12. A minimum idle speed adjustment is possible on some TBI, MFI, and SFI systems.
13. Throttle body components such as the IAC motor, TPS, and O-rings must not come in contact with throttle body cleaner.
14. On many vehicles, two terminals in the DLC must be connected to obtain flash codes from the check engine light.
15. On Chrysler products, the flash codes are obtained by cycling the ignition switch three times in a 5-second interval.

CASE STUDY

A customer complained about erratic idle speed on a Cadillac Brougham with port fuel injection. The customer said that on one occasion the engine speed increased significantly while he was stopped at a stop light. Although the customer was able to apply the brakes and keep the car from moving ahead, this experience made him very concerned about the problem. When the customer was questioned about other problems with the car, he replied the engine sometimes had a detonation problem. Questioning the customer about previous work done on the vehicle indicated the only work done recently on the vehicle was replacement of the rocker arm cover gaskets.

During a preliminary inspection, the technician noticed that a ground wire was attached to one of the rocker arm cover bolts and this bolt was loose. The technician obtained the wiring diagram for the engine computer system and discovered the ground wire was connected to the PCM. The technician measured the voltage drop from the ground wire terminal on the PCM to the battery ground and found it to be 1.2 V. After the rocker arm cover bolt was tightened to the specified torque, this voltage drop was 0.2 V.

The technician completed the preliminary inspection and diagnosis without finding any other problems. A check of the basic timing indicated it was set to the vehicle manufacturer's specifications. The technician checked the computer for DTCs and found there were no codes in the PCM memory. During an extensive road test, the idle speed remained stable with no evidence of engine detonation.

Terms to Know

Data link connector (DLC)
Diagnostic trouble code (DTC) diagnosis
Field service mode
Fuel cut rpm
Idle air control (IAC) motor
Malfunction indicator light (MIL)
Multiport fuel injection (MFI)
Powertrain control module (PCM)
Schrader valve
Sequential fuel injection (SFI)
Throttle body injection (TBI)

ASE Style Review Questions

1. While discussing fuel pump pressure diagnosis:
   Technician A says higher-than-specified fuel pump pressure may be caused by a sticking pressure regulator.
   Technician B says the water in the fuel tank may prevent the fuel pump from pumping fuel.
   Who is correct?
   A. A only   **C.** Both A and B
   B. B only   **D.** Neither A nor B

2. While discussing injector testing:
   Technician A says a defective injector may cause cylinder misfiring at idle speed.
   Technician B says restricted injector tips may result in acceleration stumbles.
   Who is correct?
   A. A only   **C.** Both A and B
   B. B only   **D.** Neither A nor B

3. While discussing flash code diagnosis:
   Technician A says the ignition switch must be turned off for 10 seconds between test sequences on a Ford EEC IV system.
   Technician B says after one test sequence is completed on a Ford EEC IV system, another test may be started immediately.
   Who is correct?
   A. A only   **C.** Both A and B
   B. B only   **D.** Neither A nor B

4. While discussing flash code diagnosis:
   Technician A says in a General Motors MFI system, the MIL light flashes each fault code four times.
   Technician B says in a General Motors MFI system, terminals A and D must be connected in the DLC to obtain the fault codes.
   Who is correct?
   A. A only   **C.** Both A and B
   B. B only   **D.** Neither A nor B

5. While discussing the effects of disconnecting battery cables:
   Technician A says in most later model SFI systems, the battery cables may be disconnected without any adverse effects on the vehicle electronic system.
   Technician B says disconnecting the battery cables on these systems erases the adaptive memory in the computer.
   Who is correct?
   A. A only   **C.** Both A and B
   B. B only   **D.** Neither A nor B

6. While discussing a high idle speed problem:
   Technician A says higher-than-normal idle speed may be caused by low electrical system voltage.
   Technician B says higher-than-normal idle speed may be caused by a defective coolant temperature sensor.
   Who is correct?
   A. A only   **C.** Both A and B
   B. B only   **D.** Neither A nor B

7. While discussing the causes of a rich air-fuel ratio:
   Technician A says a rich air-fuel ratio may be caused by low fuel pump pressure.
   Technician B says a rich air-fuel ratio may be caused by a defective coolant temperature sensor.
   Who is correct?
   A. A only
   B. B only
   C. Both A and B
   D. Neither A nor B

8. While discussing flash code diagnosis:
   Technician A says in a Chrysler MFI system, the MIL light flashes each fault code twice.
   Technician B says in a Chrysler MFI system, terminals B and C must be connected in the DLC to obtain the fault codes.
   Who is correct?
   A. A only
   B. B only
   C. Both A and B
   D. Neither A nor B

9. While discussing EFI diagnosis:
   Technician A says engine surging at idle may be caused by a vacuum leak.
   Technician B says engine surging at idle may be caused by dirty injectors.
   Who is correct?
   A. A only
   B. B only
   C. Both A and B
   D. Neither A nor B

10. While discussing the effects of low voltage at the fuel pump:
    Technician A says low voltage at the fuel pump may result in a rich air-fuel ratio.
    Technician B says low voltage at the fuel pump may result in low fuel pump volume.
    Who is correct?
    A. A only
    B. B only
    C. Both A and B
    D. Neither A nor B

Table 7-1 ASE Task

Inspect and test for missing, modified, or tampered computerized engine control components.

Problem Area	Symptoms	Possible Causes	Classroom Manual	Shop Manual
FAILURE TO MEET STATE EMISSION TEST	Possible driveability problems such as engine hesitation or reduced fuel economy	Disconnected, modified, or tampered emission and fuel system components	148	149

Table 7-2 ASE Task

Differentiate between fuel system and air induction system mechanical and electrical, electronic problems.

Problem Area	Symptoms	Possible Causes	Classroom Manual	Shop Manual
DRIVEABILITY	Rough idle, surging hesitation, fast idle speed	Intake manifold vacuum leaks	165	176

Table 7-3 ASE Task

Locate and utilize relevant service information.

Problem Area	Symptoms	Possible Causes	Classroom Manual	Shop Manual
DIAGNOSTIC ACCURACY AND TIME	Inaccurate, lengthy diagnosis	Service information not available or not utilized	154	151

Table 7-4 ASE Task

Determine appropriate diagnostic procedures based on available vehicle data and service information; determine if available information is adequate to proceed with effective diagnosis.

Problem Area	Symptoms	Possible Causes	Classroom Manual	Shop Manual
DIAGNOSTIC PROCEDURE	Improper procedure, inaccurate diagnosis, wasted time	Service information, data not available	165	151

Table 7-5 ASE Task

Test and analyze fuel system pressure and delivery rate.

Problem Area	Symptoms	Possible Causes	Classroom Manual	Shop Manual
ENGINE PERFORMANCE	Hesitation, engine stopping, lean air-fuel ratio	Low fuel pump pressure, flow	157	151
	Rich air-fuel ratio, reduced fuel economy	High fuel pump pressure	157	151

Table 7-6 ASE Task

Determine the need for fuel injector performance testing (fuel flow and pattern).

Problem Area	Symptoms	Possible Causes	Classroom Manual	Shop Manual
ENGINE PERFORMANCE	Engine surge, hesitation, hard starting, misfiring, rough idle, reduced power, stalling	Internal injector deposits	166	156

Table 7-7 ASE Task

Determine the integrity of the air induction system.

Problem Area	Symptoms	Possible Causes	Classroom Manual	Shop Manual
ENGINE PERFORMANCE	Rough idle, surging, hesitation, fast idle speed	1. Intake manifold vacuum leaks 2. Intake valve deposits	172 167	150 179

Table 7-8 ASE Task

Perform voltage drop tests on power circuits and ground circuits.

Problem Area	Symptoms	Possible Causes	Classroom Manual	Shop Manual
ENGINE PERFORMANCE	Erratic idle, stalling, intermittent detonation, no-start	High resistance or open circuit in computer ground	179	178
	Fast idle speed with engine warm	High resistance in computer power supply	179	178

Table 7-9 ASE Task

Differentiate between computerized engine controls electrical/electronic and mechanical problems.

Problem Area	Symptoms	Possible Causes	Classroom Manual	Shop Manual
ENGINE	Rough idle	Low compression, vacuum leak, injector deposits, intake valve or throttle body deposits	181	176

Table 7-10 ASE Task

Diagnose hot or cold no-starting, hard starting, poor driveability, incorrect idle speed, poor idle, flooding, hesitation, surging, engine misfire, power loss, stalling, poor mileage, dieseling, and emission problems due to malfunctions with fuel injection fuel systems; determine needed action.

Problem Area	Symptoms	Possible Causes	Classroom Manual	Shop Manual
ENGINE PERFORMANCE	No-start	1. Low compression, improper valve timing	165	178
		2. Defective ignition or fuel system	165	178
	Hard starting	1. Low compression, defective ignition	167	179
		2. Vacuum leak, injectors, cold-start injector, pressure regulator, fuel pump check valve	167	179
	Incorrect idle speed	1. ECT, IAT, or TPS sensor, low coolant temperature	148	179
		2. Low power supply to computer, vacuum leak	179	179
		3. Sticking, defective IAC motor or P/N switch	165	179
	Rough idle	1. Low compression, EGR valve stuck open, vacuum leak	178	179

Table 7-10 ASE Task (continued)

Problem Area	Symptoms	Possible Causes	Classroom Manual	Shop Manual
ENGINE PERFORMANCE	Rough idle	2. Dirty injectors or throttle body, defective cold-start injector	165	179
		3. Intake valve deposits	165	179
	Flooding, rich air-fuel ratio, poor fuel economy	1. Low compression, defective ignition, high fuel pump pressure, defective injectors	182	180
		2. PCM in limp-in mode, defective input sensor	182	180
		3. Insufficient spark advance	182	180
		4. Air flow from pump always upstream	182	180
	Surging, lean air-fuel ratio	Vacuum leak, defective MAP or MAF sensor, dirty injectors, low fuel pump pressure	182	181
	Hesitation	1. Lean air-fuel ratio, low fuel pump pressure, defective TPS or adjustment	184	181
		2. Reduced spark advance, PCM in limp-in	184	181
	Misfire	Low compression, defective ignition, vacuum leak, defective injectors	187	181
	Power loss	1. Low compression, defective ignition, reduced spark advance, PCM in limp-in	165	181
		2. Low fuel pump pressure, defective injectors, improper EGR operation, restricted exhaust	165	181
	Dieseling and stalling	Leaking injectors or cold-start injector, defective IAC motor or deposits in IAC motor passages, improper minimum idle speed adjustment, defective TPS or adjustment, dirty throttle body	165	181

Scan Tester and Digital Storage Oscilloscope Diagnosis of Electronic Fuel Injection and On-Board Diagnostics II

Upon completion and review of this chapter, you should be able to:

❏ Perform a scan tester diagnosis on various vehicles.

❏ Interpret scan tester data to determine system condition.

❏ Determine the relative importance of displayed serial data.

❏ Evaluate serial data and confirm accuracy.

❏ Research system operation using technical information to determine diagnostic procedure.

❏ Test input sensors and related circuits using serial data.

❏ Test output actuators and related circuits using serial data.

❏ Test PCM control circuits using serial data.

❏ Determine the root cause of computer system failures.

❏ Determine the root cause of repeated component failures.

❏ Determine the root cause of multiple component failures.

❏ Remove and replace computer chips.

❏ Diagnose OBD II systems

Scan Tester Diagnosis

Scan Tester Precautions

Several makes of scan testers are available to read the fault codes and perform other diagnostic functions. The exact tester buttons and test procedures vary on these testers, but many of the same basic diagnostic functions are completed regardless of the tester make. When test procedures are performed with a scan tester, the following precautions must be observed:

1. Always follow the directions in the manual supplied by the scan tester manufacturer.
2. Do not connect or disconnect any connectors or components with the ignition switch on. This includes the scan tester power wires and the connection from the tester to the vehicle diagnostic connector.
3. Never short across or ground any terminals in the electronic system except those recommended by the vehicle manufacturer.
4. If the computer terminals must be removed, disconnect the scan tester diagnostic connector first.

Scan Tester Initial Entries

CAUTION: While using a scan tester, keep the leads and the tester away from rotating components. Failure to do this may result in personal injury, and tester or vehicle damage.

Scan tester operation varies depending on the make of the tester, but the following is a typical example of initial entries:

1. Be sure the engine is at normal operating temperature and the ignition switch is off. With the correct module in the scan tester (Figure 8-1), connect the power cord to the vehicle battery.

Basic Tools

Basic technician's tool set

Service manual

Ground wire for static charges

Special Tools

Scan tester

Figure 8-1 Installing the proper module in the scan tester for the vehicle being tested (Courtesy of OTC Division, SPX Corporation)

Special Tools

Scan tester modules

2. Enter the vehicle year. The tester displays enter 199X, and the technician presses the correct single digit and the ENTER key.
3. Enter the VIN code. This is usually a two-digit code based on the model year and engine type. These codes are listed in the scan tester operator's manual. The technician enters the appropriate two-digit code and presses the ENTER key.
4. Connect the scan tester adapter cord to the diagnostic connector on the vehicle being tested.

Scan Tester Initial Selections

When the technician has programmed the scan tester by performing the initial entries, some entry options appear on the screen. These entry options vary depending on the scan tester and the vehicle being tested. Following is a typical list of initial entry options:

1. Engine
2. Antilock brake system (ABS)
3. Suspension
4. Transmission
5. Data line
6. Deluxe CDR
7. Test mode

The technician presses the number beside the desired selection to proceed with the test procedure. In the first four selections, the tester is asking the technician to select the computer system to be tested. If data line is selected, the scan tester provides a voltage reading from each input sensor in the system. If the technician selects Chrysler digital readout (CDR) on a Chrysler product, some of the next group of selections appear on the screen. When the technician selects the test mode number on a General Motors product, each test mode provides voltage or status readings from specific sensors, or components.

Scan Tester Test Selections

When the technician makes a selection from the initial test selection, the scan tester moves on to the actual test selections. These selections vary depending on the scan tester and the vehicle being tested. The following list includes many of the possible test selections and a brief explanation:

1. Faults codes—displays fault codes on the scan tester display.
2. Switch tests—allows the technician to operate switch inputs, such as the brake switch to the PCM. Each switch input should change the reading on the tester.
3. ATM tests—forces the PCM to cycle all the solenoids and relays in the system for 5 minutes, or until the ignition switch is turned off.
4. Sensor tests—provides a voltage reading from each sensor.
5. Automatic idle speed (AIS) motor—forces the PCM to operate the AIS motor when the up and down arrows are pressed. The engine speed should increase 100 rpm each time the up arrow is touched. RPM is limited to 1,500 or 2,000 rpm.
6. Solenoid state tests or output state tests—displays the on or off status of each solenoid in the system.
7. Emission maintenance reminder (EMR) tests—allows the technician to reset the EMR module. The EMR light reminds the driver when emission maintenance is required.
8. Wiggle test—allows the technician to wiggle solenoid and relay connections, and provides an audible beep from the scan tester if a loose connection is present.
9. Key on engine off (KOEO) test—allows the technician to perform this test with the scan tester on Ford products.
10. Computed timing check—forces the PCM to move the spark advance 20° ahead of the initial timing setting on Ford products.
11. Key on engine running (KOER) test—allows the technician to perform this test with the scan tester on Ford products.
12. Clear memory or erase code—quickly erases fault codes in the PCM memory.
13. Code library—reviews fault codes.
14. Basic test—allows the technician to perform a faster test procedure without prompts.
15. Cruise control test—allows the technician to test the cruise control switch inputs if the cruise module is in the PCM.

Interpreting Scan Tester Serial Data

The scan tester data varies depending on the vehicle being tested. Always refer to the vehicle manufacturer's data specifications. Following is an example of typical scan tester serial data (Figure 8-2):

1. Engine speed is the rpm at which the engine is rotating.
2. Desired rpm is the engine speed commanded by the PCM.
3. Coolant temperature is displayed in degrees Celsius or Fahrenheit. With the engine at normal operating temperature, the coolant temperature should be 185° to 223°F.
4. IAT temperature is the intake air temperature sensed by the IAT sensor. When the engine is at normal operating temperature, this temperature should be 50° to 176°F.
5. MAP volts is the voltage signal from the MAP sensor. With the engine idling, this voltage signal should be 1 to 2 volts.
6. BARO is the voltage signal from the BARO sensor in the MAP sensor. With the ignition switch on, this voltage signal should be 2.5 V to 5.5 V.
7. Throttle position is the voltage signal from the TPS. This voltage signal should be 0.29 V to 0.98 V with the engine idling.
8. Throttle angle is the amount of throttle opening displayed in degrees.
9. Oxygen sensor is the voltage signal from the oxygen sensor. With the computer system in closed loop, this signal should be varying from 100 to 1,000 mV.

> Serial data is information regarding computer inputs and outputs that is supplied through a wire to the DLC. A scan tester connected to the DLC displays this serial data.

"SCAN" DATA
Idle / Upper Radiator Hose Hot / Closed Throttle / Park or Neutral / Closed Loop / Acc. off

"SCAN" Position	Units Displayed	Typical Data Value
Engine Speed	RPM	± 100 RPM from desired RPM (± 50 RPM in drive)
Desired RPM	RPM	ECM idle command (varies with temperature)
Coolant Temperature	C°/F°	85° - 109° (185°F - 223°F)
IAT Temperature	C°/F°	10° - 80° (50°F - 176°F) depends on underhood temperature.
MAP	kPa/Volts	29 - 48kPa (1 - 2 volts) depends on Vac. & Baro pressure.
BARO	kPa/Volts	58 - 114kPa (2.5 - 5.5) depends on altitude & Baro pressure.
Throttle Position	Volts	.29 - .98
Throttle Angle	0 - 100%	0
Oxygen Sensor	M/Volts	100-1000 and varying
Inj. Pulse Width	M/Sec	1-4 and varying
Spark Advance	# of Degrees	Varies
Engine Speed	RPM	± 100 RPM from desired RPM (± 50 RPM in drive)
Fuel Integrator	Counts	Varies
Block Learn	Counts	110 - 156
Open/Closed Loop	Open/Closed	Closed Loop (may go open with extended idle)
Block Learn Cell	Cell Number	0 or 1 (depends on Air Flow & RPM)
Knock Retard	Degrees of Retard	0*
Knock Signal	Yes/No	No
EGR 1/EGR 2	Off/On	Off (On when commanded by ECM)
EGR 3	Off/On	Off (On when commanded by ECM)
Idle Air Control	Counts (steps)	5 - 50
Park/Neutral	P/N and RDL	Park/Neutral (P/N)
MPH/KPH	MPH/KPH	0
Torque Converter Clutch	On/Off	Off/ (on with TCC commanded)
Battery Voltage	Volts	13.5 - 14.5 volts
Fuel Pump volts	Volts	13.5 - 14.5 volts
Crank RPM	RPM	Varies
Battery Voltage	Volts	13.5-14.5 volts
A/C Request	Yes/No	No (Yes, with A/C requested)
A/C Clutch	On/Off	Off (On, with A/C commanded on)
A/C Clutch	On/Off	Off (On, with A/C commanded on)
A/C Pressure	psi/Volt	Varies (depends on temperature)
Fan (Fan if applicable)	On/Off	109°C, 228°F) with A/C Off/106°C (223°F) with A/C On.
Coolant Temperature	C°/F°	85° - 109° (185°F - 223°F)
Power Steering	Normal/Hi Press.	Normal
Purge Duty Cycle	0-100	0%
Park Neutral	P/N and RDL	Park/Neutral (P/N)
2nd Gear	Yes/No	No (yes when in 2nd, 3rd or 4th gear)
3rd Gear	Yes/No	No (yes, when in 3rd or 4th gear)
4th Gear	Yes/No	No (yes, when in 4th gear)
Prom ID	0-999	Varies
Time from Start	Hrs/Min	Varies

*NOTE: If maximum retard is indicated, go to CHART C-5

Figure 8-2 Typical scan tester serial data (Courtesy of Oldsmobile Division, General Motors Corporation)

10. Injector pulse width is the length of time in milliseconds when the injector is energized. In Figure 8-2, the pulse width should be 1–4 ms. The injector pulse width is lower at idle speed and increases as engine speed increases.
11. Spark advance is the spark advance provided by the PCM. This spark advance should increase as engine speed increases.
12. Fuel integrator is a short-term fuel control chip. The fuel integrator is expressed as a range of 0 to 255 with an ideal value of 128.
13. Block learn is a long-term fuel control chip. Block learn is expressed as a range of 0 to 255 with preferred value of 128.
14. Open/closed loop indicates the loop status. If the engine coolant temperature is above a specific value, the O_2 sensor signal is valid, and the throttle is not in the wide-open position, the system should be in closed loop. Open loop occurs when any of these conditions are not present. The system may enter open loop during periods of extended idle.
15. Block learn cell indicates the cell being used in the block learn chip.
16. Knock retard is the amount of spark retard supplied by the PCM in response to the knock sensor signal.

17. Knock signal indicates whether a knock sensor signal is present or not.
18. EGR1/EGR2 indicates whether solenoids 1 and 2 in the electronically operated EGR valve are on or off.
19. EGR3 indicates whether EGR solenoid 3 in the electronically operated EGR valve is on or off.
20. Idle air control indicates the position of the IAC motor pintle expressed in a range of 5 to 50.
21. Park/neutral indicates the position of the gear selector lever.
22. MPH/KMH indicates the vehicle speed from the VSS signal.
23. Torque converter clutch indicates the status of this clutch as on or off.
24. Battery voltage is the electrical system voltage sensed at the PCM.
25. Fuel pump voltage is the voltage between the fuel pump relay and the fuel pump.
26. Crank rpm is the cranking speed of the engine.
27. A/C request is the status of the A/C as requested by the driver expressed as yes or no.
28. A/C clutch is the status of this clutch indicated by on or off.
29. A/C pressure is the pressure between the condenser and the evaporator.
30. Fan is the status of the cooling fan indicated by on or off.
31. Power steering is the pressure between the power steering pump and the steering gear expressed as normal or high pressure.
32. Purge duty cycle is the duty cycle of the purge solenoid expressed in percentage.
33. Second gear indicates the closed or open status of this switch in the transaxle as yes or no.
34. Third gear indicates the closed or open status of this switch in the transaxle as yes or no.
35. Fourth gear indicates the closed or open status of this switch in the transaxle as yes or no.
36. PROM ID provides the identification number for the PROM, which may be compared to the latest PROM change-up information.
37. Time from start indicates the length of time the ignition switch has been on since the last engine start.

Mass Air Flow Sensor Testing, General Motors

SERVICE TIP: During the MAF sensor test, tap the sensor lightly and observe the grams per second reading. If this reading becomes erratic when the sensor is tapped, the sensor is defective or the wiring connections are defective.

While diagnosing a General Motors vehicle, one test mode displays grams per second from the MAF sensor. This mode provides an accurate test of the MAF sensor. The grams per second reading should be 4 to 7 with the engine idling, and this reading should gradually increase as the engine speed increases. When the engine speed is constant, the grams per second reading should remain constant. If the grams per second reading is erratic at a constant engine speed or if this reading varies when the sensor is tapped lightly, the sensor is defective. A MAF sensor fault code may not be present with an erratic grams per second reading, but the erratic reading indicates a defective sensor.

Block Learn and Integrator Diagnosis, General Motors

While using a scan tester on General Motors vehicles, one test mode displays block learn and integrator. A scale of 0 to 255 is used for both of these displays and a mid-range reading of 128 is preferred. The oxygen (O_2) sensor signal is sent to the integrator chip and then to the pulse width calculation chip in the PCM. The block learn chip is connected parallel to the integrator chip. If the O_2 sensor voltage changes once, the integrator chip and the pulse width calculation chip change the injector pulse width. If the O_2 sensor provides four continually high or low voltage signals, the block learn chip makes a further injector pulse width change. When the integrator or block learn numbers are considerably above 128, the PCM is continually attempting to increase fuel; therefore, the O_2 sensor voltage signal must be continually low, or lean. If the integrator or block learn numbers are considerably below 128, the PCM is continually decreasing fuel, which indicates that the O_2 sensor voltage must be always high, or rich.

Snapshot Testing

Many scan testers have snapshot capabilities on some vehicles, which allow the technician to operate the vehicle under the exact condition when a certain problem occurs and freeze the sensor voltage readings into the tester memory. The vehicle may be driven back to the shop, and the technician can play back the recorded sensor readings. During the playback, the technician watches closely for a momentary change in any sensor reading, which indicates a defective sensor or wire. This action is similar to taking a series of sensor reading snapshots, and then reviewing the pictures later. The snapshot test procedure may be performed on most vehicles with a data line from the computer to the DLC.

Cylinder Output Test, Ford

A cylinder output test may be performed on electronic engine control IV (EEC IV) SFI engines at the end of the KOER test. The cylinder output test is available on many SFI Ford products regardless of whether the KOER test was completed with a scan tester or with the flash code method.

When the KOER test is completed, momentarily push the throttle wide open to start the cylinder output test. After this throttle action, the PCM may require up to 2 minutes to enter the cylinder output test. In the cylinder output test, the PCM stops grounding each injector for about 20 seconds. This action causes each cylinder to misfire.

While the cylinder is misfiring, the PCM looks at the rpm that the engine slows down. If the engine does not slow down, there is a problem in the injector, ignition system, engine compression, or a vacuum leak. If the engine does not slow down as much on one cylinder, a fault code is set in the PCM memory. For example, code 50 indicates a problem in the number 5 cylinder. The correct DTC list must be used for each model year.

Ford Breakout Box Testing

On Chrysler and General Motors products, a data line is connected from the computer to the DLC, and the sensor voltage signals are transmitted on this data line and read on the scan tester. If these sensor voltage signals are normal on the scan tester, the technician knows that these signals are reaching the computer.

In 1989, Ford introduced data links on some Lincoln Continental models. Since that time, Ford has gradually installed data links on their engine computers. If a Ford vehicle has data links, there are two extra pairs of wires in the DLC. Ford has introduced a new generation star (NGS) tester which has the capability to read data on some Ford engine computers. Other scan testers now have the capability to read data on Ford engine computers (Figure 8-3).

Since most Ford products previous to 1990 do not have a data line from the computer to the DLC, the sensor voltage signals cannot be displayed on the scan tester. Therefore, some other method must be used to prove that the sensor voltage signals are received by the computer. A breakout box is available from Ford Motor Company and some other suppliers (Figure 8-4).

Two large wiring connectors on the breakout box allow the box to be connected in series with the PCM wiring. Once the breakout box is connected, each PCM terminal is connected to a corresponding numbered breakout box terminal. These terminals match the terminals on Ford PCM wiring diagrams. For example, on many Ford EEC IV systems, PCM terminals 37 and 57 are 12-V supply terminals from the power relay to the PCM. Therefore, with the ignition switch on and the power relay closed, a digital voltmeter may be connected from breakout box terminals 37 and

Special Tools
Breakout box

Special Tools
Digital multimeter

```
Function Menu
1-QUICK TESTS
   1-KOEO TEST
   2-KOER TEST
   3-TIMING CHECK
   4-REVIEW CODES
      1-KOEO
      2-KOER
   5-CLEAR CODES
   6-Options
      1-WIGGLE TEST
      2-OUTPUT STATE
      3-FUEL PUMP TEST
      4-IDLE AIR ADJUST
      5-BASIC/STAR MODE
→ 2-DCL DATA STREAM
3-PATHFINDER
   1-GUIDED PATH
   2-QUICK PATH
      1-READ FAULT CODES
         1-KOEO
         2-KOER
      2-CODE INFORMATION
         1-KOEO On Demand
            1-CKT DESCRIPTION
            2-POSSIBLE CAUSES
            3-DIAGNOSTIC TEST
               1-FAULT ISOLATION
               2-FUNCTIONAL
            4-RELATED TSBs
            5-T&T REFERENCE
            6-WIRING INFO
            7-WIGGLE TEST
         2-KOEO Memory
            1-CKT DESCRIPTION
            2-POSSIBLE CAUSES
            3-DIAGNOSTIC TEST
               1-FAULT ISOLATION
               2-FUNCTIONAL
            4-RELATED TSBs
            5-T&T REFERENCE
            6-WIRING INFO
            7-WIGGLE TEST
         3-KOER
            1-CKT DESCRIPTION
            2-POSSIBLE CAUSES
            3-DIAGNOSTIC TEST
               1-FAULT ISOLATION
               2-FUNCTIONAL
            4-RELATED TSBs
            5-T&T REFERENCE
            6-WIRING INFO
            7-WIGGLE TEST
      3-SYMPTOMS
         1-POSSIBLE CAUSES
         2-RELATED TSBs
         3-T&T REFERENCE
      4-DATA/SENSOR INFO
         1-DESCRIPTION
         2-FUNCTIONAL TEST
         3-B.O.B. PIN VALUE
         4-RELATED TSBs
         5-T&T REFERENCE
         6-WIRING INFO
      5-TSB REFERENCES
      6-B.O.B. INFO
         1-DESCRIPTION
         2-FUNCTIONAL TEST
         3-B.O.B. PIN VALUE
         4-TSB REFERENCE
         5-T&T REFERENCE
         6-WIRING INFO
         7-CLEAR CODES
      3-SUMMARY REVIEW
         1-VIEW SUMMARY
            1-RECOMMENDED TEST
            2-FAULT CODE REVIEW
            3-SENSOR RANGE CHK
            4-SYMPTOMS
               1-POSSIBLE CAUSES
               2-RELATED TSBs
               3-T&T REFERENCE
            5-RERUN LAST TEST
            6-PRINT SUMMARY
         2-DELETE SUMMARY
         3-NAME SUMMARY
         4-PRINT SUMMARY
      4-SPECIFICATIONS
         Select Emissions (this is for
         certain vehicles, most vehicles
         will go to tune-up specs)
         1-FEDERAL
         2-CALIFORNIA
         3-CANADA
```

Figure 8-3 Many scan testers now display Ford PCM data. (Courtesy of OTC Division, SPX Corporation)

Figure 8-4 A breakout box may be connected in series with the PCM terminals on Ford vehicles to allow meter test connections. (Courtesy of OTC Division, SPX Corporation)

57 to ground, and 12 V should be available at these terminals. The wiring diagram for the model year and system being tested must be used to identify the breakout box terminals.

Testing Input Sensors Using Serial Data

☑ **SERVICE TIP:** Never forget the basics when diagnosing computer systems. Always be sure the engine has proper compression, ignition, and no vacuum leaks before proceeding with the diagnosis.

Prior to any computer system diagnosis, the technician must have vehicle specifications, computer system wiring diagrams, service manual diagnostic procedures, service bulletin information, PROM ID information, and parts location information. When this information is not available for a certain vehicle, it is advisable to turn down the opportunity to diagnose the vehicle. If this essential information is available, never forget the basics when diagnosing electronic systems. Always be sure the engine has proper compression, ignition, and no vacuum leaks before proceeding with an electronic diagnosis. Perform a tampering inspection before the electronic diagnosis.

Connect the scan tester to the DLC and read the serial data. Diagnose the input sensors by comparing the data reading on each sensor to the vehicle manufacturer's specifications. If one or more of the sensors is not within these specifications, the technician must determine the diagnostic procedure to locate the exact cause of the problem. It is not our purpose to include the diagnosis for every possible input sensor problem. This information is in the vehicle manufacturer's service manuals. We will discuss an example of using the serial data to diagnose a specific problem.

The same basic procedure may be followed when diagnosing input sensor data:

1. Perform tampering inspection and diagnosis of compression, ignition, and vacuum leaks.
2. Retrieve DTCs.
3. Obtain sensor data and determine if any sensor reading is not within specifications.
4. Perform voltmeter or ohmmeter tests on the suspected sensor and connecting wires to locate the exact cause of the problem.

Vehicle manufacturers usually provide diagnostic charts or trees for each DTC. These charts provide a diagnostic procedure to pinpoint the cause of the DTC (Figures 8-5 and 8-6). In the diagnostic chart in Figure 8-6, DTC 15 is obtained, indicating the engine coolant temperature sensor signal is excessively low. In step 2 of the chart, the technician is instructed to remove the ECT sensor wires and connect the terminals together with a jumper wire. This action should simulate a high-temperature condition if the ECT wires and PCM are satisfactory. When the ECT wires are jumped together and the scan tester displays high coolant temperature, the chart informs the technician that the ECT sensor is defective or the wires have an intermittent faulty connection. If the scan tester does not indicate high coolant temperature, the technician is instructed to proceed with ECT wiring tests.

Example of Input Sensor Diagnosis Using Serial Data. The customer complains of hard starting when the engine is cold, and after the engine starts, black smoke is emitted from the tailpipe. The technician verifies engine compression, ignition, and the absence of vacuum leaks. A tampering inspection is performed. The technician reads the DTCs and finds there are no DTCs in the computer memory.

When the serial data is displayed on the scan tester, the engine coolant temperature is always below 165°F, and the system never enters closed loop. The technician tapes a thermometer to the upper radiator hose and discovers the coolant temperature is actually 195°F with the engine

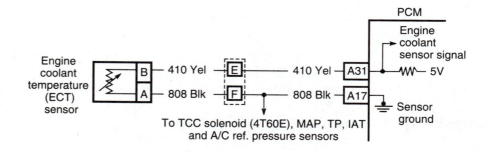

Figure 8-5 Engine coolant temperature (ECT) sensor wiring connections to the PCM (Courtesy of Chevrolet Motor Division, General Motors Corporation)

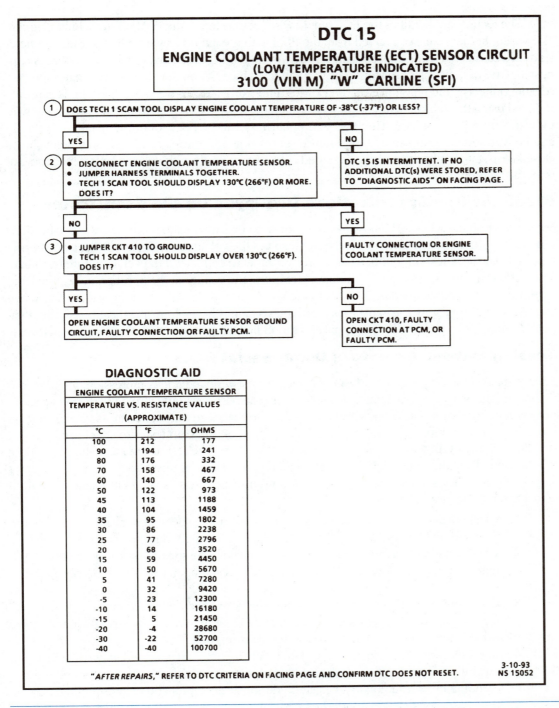

Figure 8-6 Diagnostic chart for locating the cause of an ECT sensor DTC (Courtesy of Chevrolet Motor Division, General Motors Corporation)

warmed up. The technician concludes that the engine coolant temperature (ECT) sensor or connecting wires are probably defective.

The technician shuts off the engine and drains some of the engine coolant from the radiator. The ECT sensor is removed and cooled to room temperature. With an ohmmeter connected across the ECT sensor terminals, the sensor has the specified resistance. The ECT sensor is immersed in a container of water with the ohmmeter connected to the sensor terminals. When the water in the container is heated with a propane torch, the ECT sensor has the specified resistance at various temperatures. The technician concludes the ECT sensor is in satisfactory condition.

The technician identifies the ECT sensor wires connected to the PCM on the wiring diagram. The wiring harness is disconnected from the PCM with the ignition switch off, and the ohmmeter leads are connected from each coolant sensor wire to the PCM terminal to which the wire is connected. One of the coolant sensor wires has 5-Ω resistance. When the technician wiggles the ECT sensor connector, the resistance reading varies, indicating a defective wire. The wire is carefully repaired using the vehicle manufacturer's recommended procedure. An ohmmeter test on the wire indicates normal resistance. The PCM wiring harness and ECT sensor are reinstalled. With the engine warmed up, the coolant temperature sensor now indicates 195°F, and the system shows closed loop status. When the engine is started cold, it starts easily with no evidence of black smoke from the tailpipe.

Breakout Box Specifications Displayed on the Scan Tester

Some scan testers such as OTC now have the ability to display breakout box (BOB) specifications (Figure 8-7). In this test mode, the scan tester displays the voltage specification for each computer terminal. The scan tester identifies the source of the wire connected to each computer terminal. The technician must use a digital volt-ohmmeter to test the voltage at the appropriate terminal and determine if the voltage reading is equal to the specifications displayed on the scan tester. When the test on an input sensor is satisfactory and the voltage is not within specifications at the computer terminals to which the sensor signal and ground wires are connected, the wires are defective.

Testing Output Actuators Using Serial Data

Prior to diagnosing output actuators, the technician must verify engine compression, ignition condition, and the absence of vacuum leaks. A tampering inspection should also be performed prior to output actuator diagnosis. When serial data is used to diagnose output actuators, the technician must compare the output actuator serial data displayed on the scan tester to specifications, and determine if any output actuator serial data is unsatisfactory. The technician must check the input sensor serial data, and determine if a sensor reading is causing the unsatisfactory output actuator serial data. Some examples of out-of-specification sensor readings causing unsatisfactory output actuator serial data are:

1. A TPS voltage signal that is lower than specified at idle speed may cause the IAC motor counts to be higher than normal at idle, resulting in higher than specified idle speed.
2. A TPS voltage signal that is higher than specified at idle speed may cause the IAC motor counts to be lower than specified at idle, resulting in a slower than specified idle speed.
3. An O_2 sensor voltage signal that is continually lower than specified may result in a higher than normal injector pulse width and higher than normal interrogator and block learn numbers, resulting in excessive fuel consumption.
4. An O_2 sensor voltage signal that is continually higher than specified may result in a lower than normal injector pulse width and lower than normal interrogator and block learn numbers, resulting in a hesitation during acceleration and engine surging.
5. An ECT sensor signal that indicates a coolant temperature below the actual coolant temperature may increase injector pulse width and spark advance. This type of ECT sensor signal may result in an inoperative EGR valve, cooling fan, torque converter clutch, and vapor purge system.
6. A defective VSS may cause improper operation of the torque converter clutch and cruise control.

It is impossible for us to discuss the diagnosis of every possible output actuator. This information is in the vehicle manufacturer's service manuals, and must be followed by the technician. A general diagnostic procedure when using serial data to diagnose output actuators is:

1. Perform a tampering inspection and verify engine compression, ignition, and the absence of vacuum leaks.

```
1-DATA STREAM
    1-STANDARD MODES
    2-CUSTOM DISPLAY
    3-DELUXE DISPLAY
    4-ALDL MENU
2-FAULT CODES
    1-CURRENT CODES
    2-HISTORY CODES
    3-CLEAR CODES
→ 3-PATHFINDER
    1-GUIDED PATH
 →  2-QUICK PATH
        1-READ FAULT CODES
            1-CURRENT CODES
            2-HISTORY CODES
            3-CLEAR CODES
        2-CODE INFORMATION
            1-CKT DESCRIPTION
            2-POSSIBLE CAUSES
            3-DIAGNOSTIC TEST
                1-FAULT ISOLATION
                2-FUNCTIONAL
            4-RELATED TSBs
            5-T&T REFERENCE
            6-B.O.B. INFO MENU
                1-B.O.B. VALUES
                2-WIRING INFO
        3-SYMPTOMS
            1-POSSIBLE CAUSES
            2-RELATED TSBs
            3-T&T REFERENCE
        4-DATA/SENSOR INFO
            1-DATA ITEM MENU
                1-DESCRIPTION
                2-FUNCTIONAL TEST
                3-SCAN TOOL VALUE
                4-RELATED TSBs
                5-T&T REFERENCE
            2-B.O.B. INFO MENU
                1-B.O.B. VALUES
                2-WIRING INFO
        5-TSB REFERENCES
     →  6-B.O.B. INFO
         →  1-SCROLL ALL PINS
         →  2-KEY PAD SELECTION
                1-DATA ITEM MENU
                    1-DESCRIPTION
                    2-FUNCTIONAL TEST
                    3-SCAN TOOL VALUE
                    4-RELATED TSBs
                    5-T&T REFERENCE
             →  2-B.O.B. INFO MENU
                 →  1-B.O.B. VALUES
                 →  2-WIRING INFO
     →  7-CLEAR CODES
        3-SUMMARY REVIEW
            1-VIEW SUMMARY
                1-RECOMMENDED TEST
                2-FAULT CODE REVIEW
                3-SENSOR RANGE CHK
                4-SYMPTOMS
                    1-POSSIBLE CAUSES
                    2-RELATED TSBs
                    3-T&T REFERENCE
                5-RERUN LAST TEST
                6-PRINT SUMMARY
            2-DELETE SUMMARY
            3-NAME SUMMARY
            4-PRINT SUMMARY
4-RECORD/PLAYBACK
    1-RECORD
    2-PLAYBACK
    3-OPTIONS
5-PROM ID
6-SPECIFICATIONS
    1-AUTOMATIC
    2-MANUAL
        1-FEDERAL
        2-CALIFORNIA
        3-CANADA
7-SNAPSHOT
```

Figure 8-7 Breakout box (BOB) values displayed on the scan tester (Courtesy of OTC Division, SPX Corporation)

2. Retrieve DTCs from the computer memory.
3. Use the scan tester to obtain input and output serial data.
4. Check the input sensor data for defective sensor readings that could result in unsatisfactory output actuator data, and repair inputs as required.
5. If the input sensor readings are normal, perform the vehicle manufacturer's recommended voltmeter or ohmmeter tests on the suspected output and connecting wires to locate the cause of the problem.

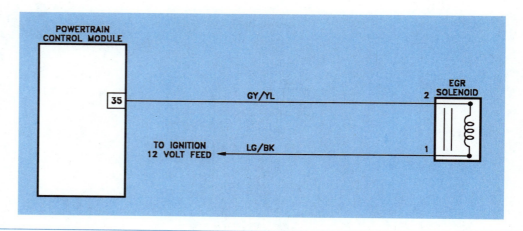

Figure 8-8 EGR solenoid wiring (Courtesy of Chrysler Corporation)

Example of Output Actuator Diagnosis. The MIL is illuminated with the engine running on a 1995 Chrysler Concorde with a 3.5-L engine, but there are no driveability complaints. The technician performs a tampering inspection, and verifies engine compression, ignition, and the absence of vacuum leaks. When the DTCs are retrieved, the technician finds the diagnostic message "EGR solenoid circuit" on the scan tester. The technician obtains the serial data on the scan tester, and all the input and output data appears to be normal. The actuation test mode is entered on the scan tester and the EGR solenoid is selected. In this mode, there is no evidence of cycling, or clicking, from the EGR valve, indicating the PCM is not cycling the EGR solenoid on and off.

The technician locates the EGR solenoid wiring diagram in the vehicle manufacturer's service information (Figure 8-8). When the digital voltmeter leads are connected from the 12-V feed wire on the EGR solenoid to ground with the ignition switch on, a 12.5-V reading is obtained. This reading indicates a normal 12-V feed wire and fuse.

The technician turns off the ignition switch and disconnects the EGR solenoid connector. With a digital ohmmeter connected across the EGR solenoid terminals, an infinite reading is obtained, indicating an open EGR solenoid winding (Figure 8-9). The technician connects the digital ohmmeter leads from terminal 1 on the EGR solenoid connector to terminal 35 on the disconnected PCM connector to make sure this wire is in satisfactory condition (Figure 8-10). The technician replaces the EGR solenoid and reconnects the PCM and EGR solenoid connectors. With

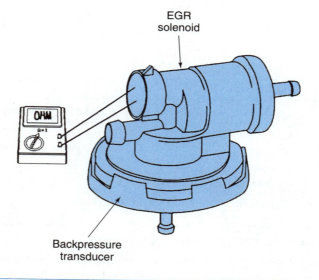

Figure 8-9 Testing the EGR solenoid winding (Courtesy of Chrysler Corporation)

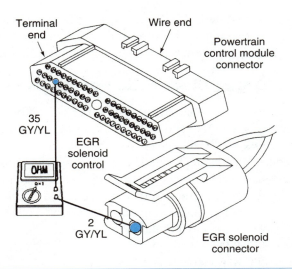

Figure 8-10 Testing the wire from the EGR solenoid to the PCM (Courtesy of Chrysler Corporation)

the scan tester in the ATM mode, the EGR solenoid cycles normally with an audible clicking noise. The MIL remains off with the engine running.

Testing PCM Control Circuit Using Serial Data

SERVICE TIP: Never replace a computer unless the specified voltage is verified at the computer voltage supply terminals and the specified voltage drop is verified at the computer ground terminals.

The first step in computer diagnosis is to check the service bulletin information for the vehicle. In some cases, a service bulletin directs the technician to install a revised PCM to correct a specific problem. In some computer systems, a faulty computer is indicated by a DTC. The next step in computer diagnosis is to verify the specified voltage on the computer voltage supply terminals and the specified voltage drop on the computer ground terminals. Computer diagnosis is a continuation of input sensor and output actuator diagnosis. We will continue from the diagnosis of the 1995 Chrysler inoperative EGR solenoid described previously to computer diagnosis related to the same problem. The vehicle manufacturer's service information directs the technician to perform the following tests:

1. Measure the voltage on the EGR solenoid 12-V feed wire.
2. Measure the resistance in the wire from EGR solenoid connector terminal 1 to PCM terminal 35 to check this wire for an open circuit.
3. Check the resistance from EGR solenoid connector terminal 1 to ground with this solenoid connector and the PCM connector disconnected. This checks the wire for a grounded condition.
4. Connect the ohmmeter leads across the EGR solenoid terminals to check the solenoid winding.

If the tests in steps 1 through 4 are satisfactory, the technician is instructed to backprobe PCM terminal 35 and measure the voltage from this terminal to ground with the ignition switch on and the EGR solenoid and PCM connectors in place. If the voltage at PCM terminal 35 is above 10 V, the EGR solenoid circuit is satisfactory and the PCM should be replaced.

The same basic procedure is usually followed when diagnosing the computer in relation to input sensor problems. For example, when diagnosing the computer in relation to an ECT sensor problem, the technician is usually instructed to check the ECT sensor wires from the PCM to the sensor. The ECT sensor may be tested with an ohmmeter at various temperatures. When the ECT sensor and the connecting wires are tested and proven to be satisfactory, the technician is usually instructed to replace the PCM when there is a problem in relation to this sensor.

Additional Scan Tester Diagnostic Functions

Some scan testers display code information for each DTC. This information includes a circuit description, possible causes of the DTC, and diagnostic tests to isolate the problem (Figure 8-11). Many scan testers display service bulletin numbers that apply to the vehicle year and VIN entered in the scan tester. These service bulletins are located in a service bulletin book available from the scan tester manufacturer. Some scan testers provide a listing of pages in other manuals that supply information related to specific symptoms. These books may be called Tools and Techniques (T & T) manuals. The possible causes of driveability symptoms are displayed on some scan testers.

Most scan testers display the number of the PROM currently installed in the PCM. This number can be identified in a PROM ID book supplied by the scan tester manufacturer. This book will indicate if this is the latest PROM available from the manufacturer, or if an updated PROM is available to correct a specific driveability problem. Some scan testers display basic tune-up specifications for the vehicle year and VIN entered in the tester.

```
1-DATA STREAM
    1-STANDARD MODES
    2-CUSTOM DISPLAY
    3-DELUXE DISPLAY
    4-ALDL MENU
2-FAULT CODES
    1-CURRENT CODES
    2-HISTORY CODES
    3-CLEAR CODES
→ 3-PATHFINDER
    1-GUIDED PATH
→ 2-QUICK PATH
    1-READ FAULT CODES
        1-CURRENT CODES
        2-HISTORY CODES
        3-CLEAR CODES
    2-CODE INFORMATION
        1-CKT DESCRIPTION
        2-POSSIBLE CAUSES
        3-DIAGNOSTIC TEST
            1-FAULT ISOLATION
            2-FUNCTIONAL
        4-RELATED TSBs
        5-T&T REFERENCE
        6-B.O.B. INFO MENU
            1-B.O.B. VALUES
            2-WIRING INFO
    3-SYMPTOMS
        1-POSSIBLE CAUSES
        2-RELATED TSBs
        3-T&T REFERENCE
→   4-DATA/SENSOR INFO
→       1-DATA ITEM MENU
            1-DESCRIPTION
            2-FUNCTIONAL TEST
→           3-SCAN TOOL VALUE
→           4-RELATED TSBs
→           5-T&T REFERENCE
        2-B.O.B. INFO MENU
            1-B.O.B. VALUES
            2-WIRING INFO
    5-TSB REFERENCES
    6-B.O.B. INFO
        1-SCROLL ALL PINS
        2-KEY PAD SELECTION
            1-DATA ITEM MENU
                1-DESCRIPTION
                2-FUNCTIONAL TEST
                3-SCAN TOOL VALUE
                4-RELATED TSBs
                5-T&T REFERENCE
            2-B.O.B. INFO MENU
                1-B.O.B. VALUES
                2-WIRING INFO
    7-CLEAR CODES
    3-SUMMARY REVIEW
        1-VIEW SUMMARY
            1-RECOMMENDED TEST
            2-FAULT CODE REVIEW
            3-SENSOR RANGE CHK
            4-SYMPTOMS
                1-POSSIBLE CAUSES
                2-RELATED TSBs
                3-T&T REFERENCE
            5-RERUN LAST TEST
            6-PRINT SUMMARY
        2-DELETE SUMMARY
        3-NAME SUMMARY
        4-PRINT SUMMARY
    4-RECORD/PLAYBACK
        1-RECORD
        2-PLAYBACK
        3-OPTIONS
    5-PROM ID
    6-SPECIFICATIONS
        1-AUTOMATIC
        2-MANUAL
            1-FEDERAL
            2-CALIFORNIA
            3-CANADA
    7-SNAPSHOT
```

Figure 8-11 Additional scan tester functions (Courtesy of OTC Division, SPX Corporation)

HESITATION, SAG, STUMBLE

Definition: Momentary lack of response as the accelerator is pushed down. Can occur at all vehicle speeds. Usually most severe when first trying to make the vehicle move, as from a stop sign. May cause engine to stall if severe enough.

PRELIMINARY CHECKS
- Perform the careful visual/physical checks as described at start of "Symptoms," Section "6E3-B".

SENSORS
- CHECK: Throttle Position (TP) sensor - Check TP sensor for binding or sticking. Voltage should increase at a steady rate as throttle is moved toward Wide Open Throttle (WOT).
- CHECK: MAP sensor - use CHART C-1D.
- CHECK: For sensors that share the 5 volt reference line for being skewed causing reference voltage to be pulled low.
- CHECK: 24x signal for being erratic, caused by a damaged or wobbing reluctor wheel.

IGNITION SYSTEM
- CHECK: Spark plugs for being fouled or for faulty spark plug wire routing.
- CHECK: Electronic ignition system ground, CKT 51.

FUEL SYSTEM
- CHECK: Fuel pressure, use CHART A-7.
- CHECK: Contaminated fuel.
- CHECK: EVAP canister purge system for proper operation. Use CHART C-3.
- CHECK: Fuel injectors. Perform fuel injector coil/balance test, use CHART C-2A.

ADDITIONAL CHECKS
- CHECK: Service Bulletins for updates.
- CHECK: EGR operation, use CHART C-7.
- CHECK: Engine thermostat functioning correctly and proper heat range.
- CHECK: Generator output voltage. Repair if less than 9 volts or more than 16 volts.

Figure 8-12 Diagnosis-by-symptom chart for hesitation, sag, and stumble (Courtesy of Chevrolet Motor Division, General Motors Corporation)

Diagnosis By Symptom

In some cases, there are no DTCs and the serial data displayed on the scan tester is normal. However, the vehicle still has a driveability problem. Some vehicle manufacturers supply diagnosis-by-symptom charts in their service manuals to solve these problems. A diagnosis-by-symptom chart for hesitation, sag, and stumble is provided (Figure 8-12). Figure 8-13 illustrates a diagnosis-by-symptom chart for poor fuel economy.

Idle Air Control Motor Service and Diagnosis

Improper Idle Speed General Diagnosis

Since the PCM operates the IAC or IAC BPA motor in response to the input signals, idle speed is controlled automatically. If the idle speed is not correct, the problem is in one of these areas:
1. An intake manifold vacuum leak
2. A defective input sensor or switch

POOR FUEL ECONOMY

Definition: Fuel economy, as measured by an actual road test, is noticeably lower than expected. Also, economy is noticeably lower than it was on this vehicle at one time, as previously shown by an actual road test.

PRELIMINARY CHECKS

- Perform the careful visual checks as described at start of "Symptoms," Section "6E3-B".
- Visually (physically) check: Vacuum hoses for splits, kinks, and proper connections as shown on "Vehicle Emission Control Information" label.
- Check owner's driving habits.
  - Is A/C "ON" full time (Defroster mode "ON")?
  - Are tires at correct pressure?
  - Are excessively heavy loads being carried?
  - Is acceleration too much, too often?
- Check for a dirty or plugged air cleaner.
  - Fuel leaks.

ADDITIONAL CHECKS

- **CHECK:** TCC operation. Use CHART C-8. A Tech 1 scan tool should indicate an RPM drop, when the TCC is commanded "ON."
- **CHECK:** For exhaust system restriction. Use CHART B-1.
- **CHECK:** For proper calibration of speedometer.
- **CHECK:** Induction system and crankcase for air leaks.

IGNITION SYSTEM

- **CHECK:** Spark plugs. Remove spark plugs, check for wet plugs, cracks, wear, improper gap, burned electrodes, or heavy deposits. Repair or replace as necessary.
- **CHECK:** For open ignition control CKT 423. Refer to "Engine Components/Wiring Diagrams/Diagnostic Charts," Section "6E3-A", DTC 41.

COOLING SYSTEM

- **CHECK:** Engine coolant level.
- **CHECK:** Engine thermostat for faulty part (always open) or for wrong heat range. Refer to SECTION 6B.

ENGINE MECHANICAL

- **CHECK:** Compression. Refer to SECTION 6A.

Figure 8-13 Diagnosis-by-symptom chart for poor fuel economy (Courtesy of Chevrolet Motor Division, General Motors Corporation)

3. The IAC or IAC BPA motor
4. The connecting wires from the IAC or IAC BPA motor to the PCM
5. The PCM

If an intake manifold vacuum leak occurs, the PCM senses the increase in manifold pressure caused by the vacuum leak. Under this condition, the PCM supplies more fuel and the idle speed increases. When the idle speed is higher than specified, always check for intake manifold vacuum leaks before proceeding with any further diagnosis.

If the IAC or IAC BPA motor is defective or seized, the idle speed is fixed at all temperatures. All other possible causes of the problem must be eliminated and the PCM diagnosed as the problem before PCM replacement.

Idle Air Control Motor Adjustment

Some IAC motors, such as those on Chrysler TBIs, have a hex bolt on the end of the motor plunger. This hex bolt is not for idle speed adjustment. If this hex bolt is turned, it will not affect idle speed, because the PCM will correct the idle speed. If the hex bolt is turned and the length of the IAC motor plunger is changed in an attempt to adjust idle speed, the throttle may not be in the proper position for starting, and hard starting at certain temperatures may occur. On some applications, an idle rpm specification is provided with the plunger fully extended. The plunger could be adjusted under this condition. On Chrysler products, the plunger may be fully extended by turning off the ignition switch, and then disconnecting the IAC motor connector. The IAC motor hex bolt should only require adjustment if it has been improperly adjusted or if a new motor is installed.

Idle Contact Switch Test

WARNING: Never connect a 12-V source across the idle contact switch terminals on the IAC motor. If these contacts are closed, the contacts will be ruined.

CAUTION:: Connecting pair of jumper wires from the terminals of a 12-V battery to the IAC motor idle contact switch terminals may result in very high current flow and jumper wire heating, which could result in burns to your hands.

The procedure for diagnosing the idle contact switch varies depending on the vehicle make and model year. Always follow the procedure recommended in the vehicle manufacturer's service manual. Following is a typical idle contact switch test:

1. Backprobe terminal B on the IAC motor, and turn on the ignition switch. Connect a digital voltmeter from terminal B to ground, and hold the throttle approximately half open (Figure 8-14). The voltage supplied from the PCM to the idle contact switch should be 4.5 V to 8 V depending on the system. Always refer to the vehicle manufacturer's specifications.
2. If the voltage at terminal B on the IAC motor is not within specifications, turn off the ignition switch and disconnect the IAC motor connector and the PCM connector. Connect the ohmmeter leads from terminal B in the IAC motor connector to terminal 2D8

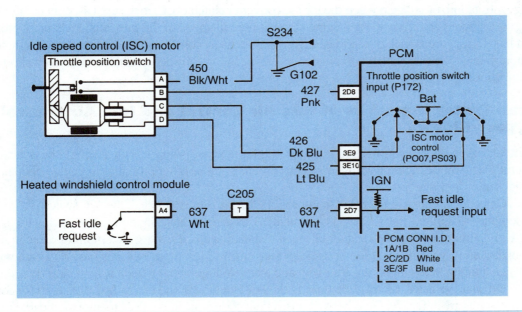

Figure 8-14 Idle air control motor circuit (Courtesy of Cadillac Motor Car Division, General Motors Corporation)

in the PCM connector. The ohm reading should be less than 0.5 Ω. If the reading is more than this value, repair the resistance problem or open circuit in the wire from the PCM to the idle contact switch.

3. Connect the ohmmeter leads from terminal B on the IAC motor connector to ground. The ohmmeter reading should be infinite. If the reading is not infinite, repair the wire from the PCM to the idle contact switch.
4. If the voltage at terminal B on the IAC motor is not within specifications and the ohmmeter readings in steps 2 and 3 are satisfactory, reconnect the PCM and IAC motor connectors. Backprobe terminal 2D8 at the PCM, and turn on the ignition switch. Connect a digital voltmeter from terminal 2D8 to ground, and observe the voltage. If the voltage is not within specifications, check all the power supply and ground wires to the PCM. When the power supply and ground wires are satisfactory, replace the PCM.
5. If the voltage in step 1 is satisfactory, return the throttle to the idle position, and observe the voltmeter. The reading should be less than 1 V with the throttle in the idle position and the idle contact switch closed. If the voltage reading is within specifications, the idle contact switch is satisfactory.
6. When the voltage in step 5 is not within specifications, connect the voltmeter from terminal A on the IAC motor to ground. If the voltage at this point is above 0.2 V, repair the resistance problem or open circuit in the ground wire connected to terminal A. When the voltage at terminal A is 0.2 V or less, and the voltage in step 5 is above specifications, replace the IAC motor.

Basic Idle Air Control Bypass Motor Diagnosis

With the engine at normal operating temperature, observe the idle air control bypass motor counts on the scan tester. Quickly accelerate the engine to 2,500 rpm and allow the throttle to return to idle. Repeat this procedure four times. Each time the throttle is returned to idle, the idle air control motor counts should be within five counts of the original reading. When the idle air control bypass motor counts are not within this range, replace the motor.

When the idle air control bypass motor is removed, you should be able to push the motor pintle in and out with some hand pressure. If the pintle does not move by hand, replace the motor.

Intermittent shorting of an idle air control bypass motor winding may cause an engine stall, surge, hesitation, or chuggle when the torque converter clutch (TCC) locks up. Disconnect the idle air control bypass motor with the engine idling at normal speed. If the previous driveability problem has disappeared, replace the idle air control bypass motor. The idle air control bypass motor windings may be tested with an ohmmeter. A reading below the specified value indicates a shorted motor winding.

Scan Tester Diagnosis of Idle Air Control and Idle Air Control Bypass Motors

If the idle speed is not within specifications, the input sensors and switches should be checked carefully with the scan tester. Many input sensor defects cause other problems in engine operation besides improper idle rpm.

Defective input switches result in improper idle rpm. For example, if the A/C switch is always closed, the PCM thinks the A/C is on continually. This action results in the PCM operating the IAC or IAC BPA motor to provide a higher idle rpm. On many vehicles, the scan tester indicates the status of the input switches as closed or open, or high or low. Most input switches provide a high voltage signal to the PCM when they are open, and a low voltage signal when they are closed.

On some vehicles, a fault code is set in the PCM memory if the IAC or IAC BPA motor or connecting wires are defective. On other systems, a fault code is set in the PCM memory if the idle rpm is out of range. On Chrysler products, the actuation test mode (ATM) or actuate outputs mode may be entered with the ignition switch on. The IAC or IAC BPA motor may be selected in the actuate outputs mode, and the PCM is forced to extend and retract the IAC or IAC BPA motor plunger every 2.8 seconds. When this plunger extends and retracts properly, the motor, connecting wires, and PCM are in normal condition. If the plunger does not extend and retract, further diagnosis is necessary to locate the cause of the problem.

WARNING: When performing a set engine rpm mode test on an IAC motor, always be sure the transmission selector is in the park position and the parking brake is applied.

On some IAC and IAC BPA motors, a set engine rpm mode may be entered on the scan tester. In this mode, each time a specified scan tester button is touched, the rpm should increase 100 rpm to a maximum of 2,000 rpm. Another specified scan tester button may be touched to decrease the speed in 100-rpm steps. On some scan testers, the up and down arrows are used to increase and decrease the engine rpm during this test. If the IAC or IAC BPA motor responds properly during this diagnosis, the PCM, motor, and connecting wires are in satisfactory condition and further diagnosis of the inputs is required.

SERVICE TIP: If the IAC or IAC BPA motor counts are zero on the scan tester, the circuit between the PCM and the motor is likely open. Wiggle the wires on the IAC or IAC BPA motor and observe the scan tester reading. If the count reading changes while wiggling the wires, you have found the problem.

On some systems, the scan tester reads the IAC or IAC BPA motor counts, and the count range is provided in the scan tester instruction manual. Some of the input switches, such as the A/C, may be operated and the scan tester counts should change. If the scan tester counts change when the A/C is turned on and off, the motor, connecting wires, and PCM are operational. When the scan tester counts do not change under this condition, further diagnosis is required.

A typical procedure for scan tester diagnosis is shown in Photo Sequence 6.

Classroom Manual
Chapter 8, page 193

Digital Storage Oscilloscope Diagnosis of Electronic Fuel Injection Systems

Diagnosis of Input Sensor Signals

Oxygen Sensor. Since the controls are different on various DSOs, it is not our objective to provide a step-by-step procedure for using the DSO. This information is in the DSO operator's manual. Our objective is to discuss normal and defective waveforms displayed by any DSO while diagnosing EFI systems. The vertical voltage scale must be adjusted in relation to the voltage expected in the signal being displayed. The horizontal time base or milliseconds per division must be adjusted so the waveform appears properly on the screen. Many waveforms are clearly displayed when the horizontal time base is adjusted so three waveforms are displayed on the screen.

WARNING: When using a digital storage oscilloscope (DSO) always follow the DSO manufacturer's recommended operating procedures. Improper DSO use may result in tester damage.

The trigger is the signal that tells a DSO to start drawing a waveform. When measuring signals from 0 V to 5 V, a 50% trigger may be used. A marker indicates the trigger line on the screen, and minor adjustments of the trigger line may be necessary to position the waveform in the desired vertical position. Trigger slope indicates the direction in which the voltage signal is moving when it crosses the trigger line. A positive trigger slope means the voltage signal is moving upward as it crosses the trigger line, whereas a negative trigger slope indicates the voltage signal is moving downward when it crosses the trigger line. The DSO provides more accurate diagnosis of automotive computer systems because it has a much faster signal sampling speed than other types of test equipment.

Special Tools

Digital storage oscilloscope

The trigger is the signal that tells a DSO to start drawing a waveform.

A positive trigger slope means the voltage signal is moving upward as it crosses the trigger line.

A negative trigger slope indicates the voltage signal is moving downward when it crosses the trigger line.

Photo Sequence 6
Typical Procedure for Scan Tester Diagnosis

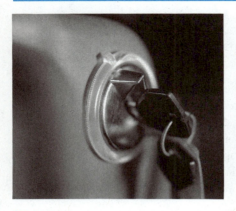

P6-1 Be sure the engine is at normal operating temperature and the ignition switch is off.

P6-2 Install the proper module in the scan tester for the vehicle and system being diagnosed.

P6-3 Connect the scan tester power cables to the battery terminals with the proper polarity.

P6-4 Enter the model year and the VIN code in the scan tester for the vehicle being tested.

P6-5 Select the proper scan tester data cable end for the vehicle being tested.

P6-6 Connect the scan tester data cable to the DLC on the vehicle.

P6-7 Retrieve the DTCs with the scan tester.

P6-8 Start the engine and then obtain and print out the input sensor and output actuator data with the scan tester.

P6-9 Compare the input sensor and output actuator data to specifications in the appropriate service manual. Determine data that is not within specifications.

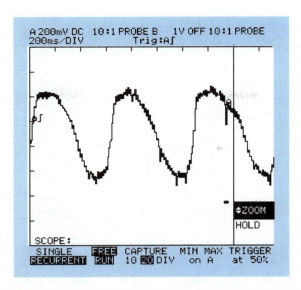

Figure 8-15 Normal DSO waveform, O₂ sensor (Courtesy of Fluke Corporation)

Before diagnosing an oxygen (O₂) sensor, the engine must be at normal operating temperature and the computer must be in closed loop. When diagnosing input sensor waveforms, adjust the horizontal time base on the DSO so three waveforms appear on the screen (Figure 8-15). When testing any input sensor, the DSO leads must be connected from the sensor signal wire to ground. The O₂ sensor voltage should be cycling continually in the 0–1-V range (Figure 8-16). O₂ voltage signal cross counts are the number of times the O₂ voltage signal changes above or below 0.45 V per second. With the engine running at 2,000 rpm and no load, 2 to 3 cross counts on the O₂ sensor signal are normal. If the O₂ cross counts are reduced, the sensor may be defective or partially contaminated. When the voltage remains about 0.5 V with very little fluctuation, the signal wire or sensor may have an open circuit. If the voltage remains below 0.5 V with very little fluctuation, the intake manifold may have a vacuum leak.

Propane enrichment may be used to force the fuel system into a very rich condition, and verify the O₂ sensor operation. A propane cylinder with a precision metering valve must be used for

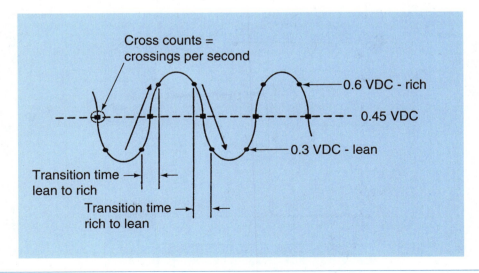

Figure 8-16 O₂ sensor voltage signal cross counts are the number of times the O₂ sensor voltage signal changes above or below 0.45 V per second. (Courtesy of OTC Division, SPX Corporation)

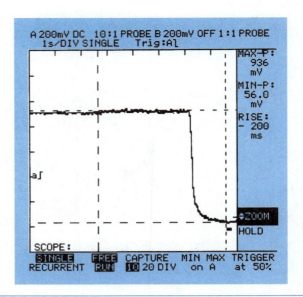

Figure 8-17 Rich to lean O₂ sensor switching (Courtesy of Fluke Corporation)

this test. Always maintain the propane cylinder in an upright position, and position the cylinder and hose away from rotating components. To verify rich-to-lean O₂ sensor switching, connect the propane hose to a large intake manifold port such as the brake booster hose. Be sure there are no leaks between the propane hose and the brake booster hose. Open the propane cylinder valve and introduce propane into the intake manifold. If necessary, hold the throttle open to maintain engine speed. The O₂ sensor signal should become a flat line above 600 mV. Quickly disconnect the propane hose and expose the brake booster hose to atmosphere. The O₂ sensor signal should drop from 600 mV to 300 mV in less than 125 ms (Figure 8-17).

Hold the throttle open to maintain a reasonable engine rpm with the brake booster hose exposed to atmosphere. Quickly install the propane hose into the brake booster hose and introduce propane into the intake manifold. Under this condition the O₂ sensor signal should increase from 300 mV to 600 mV in 100 ms (Figure 8-18). If the O₂ sensor voltage signal does not switch from rich to lean or lean to rich in the specified time, the sensor is defective. (The specifications on O₂ sensor switching are supplied courtesy of General Motors and Fluke Corporation.)

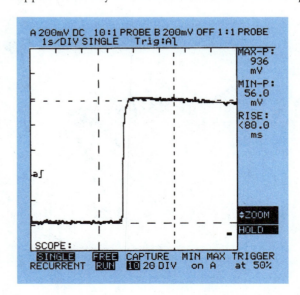

Figure 8-18 Lean to rich O₂ sensor switching (Courtesy of Fluke Corporation)

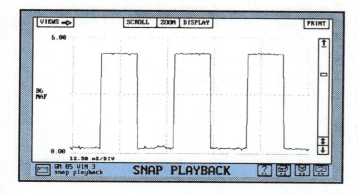

Figure 8-19 Normal DSO waveform, frequency-type MAF sensor (Courtesy of EDGE Diagnostic Systems)

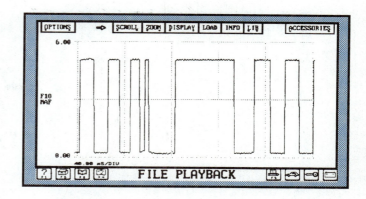

Figure 8-20 Defective frequency-type MAF sensor waveform (Courtesy of EDGE Diagnostic Systems)

Mass Air Flow Sensor. The DSO leads must be connected from the MAF sensor signal wire to ground. On frequency-type MAF sensors such as General Motors and Ford, the waveform should appear as a series of digital signals (Figure 8-19). When engine speed and intake air flow are increased, the frequency of the MAF sensor signals increases proportionally. If the MAF or connecting wires are defective, there is an intermittent change in signal frequency (Figure 8-20).

 SERVICE TIP: Tap the MAF sensor while observing the waveform. If this action causes an erratic waveform, the sensor is defective or the connecting wires are defective.

A vane-type MAF sensor should provide an analog voltage signal when the engine is accelerated (Figure 8-21). A defective vane-type MAF sensor waveform displays sudden and erratic voltage changes (Figure 8-22).

Throttle Position Sensor. Each time the throttle is opened, the TPS sensor should provide a smooth analog voltage signal (Figure 8-23). If the TPS sensor is defective, glitches will appear in the sensor signal as the throttle is opened (Figure 8-24).

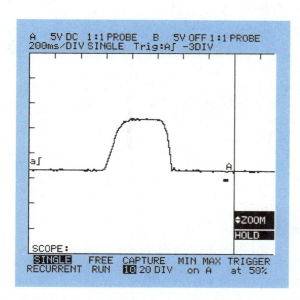

Figure 8-21 Normal DSO waveform, vane-type MAF sensor (Courtesy of Fluke Corporation)

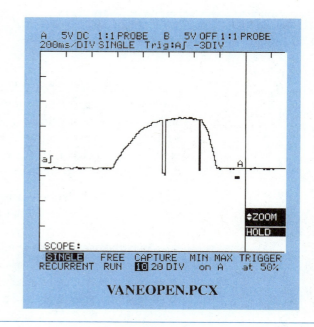

Figure 8-22 Defective vane-type MAF sensor waveform (Courtesy of Fluke Corporation)

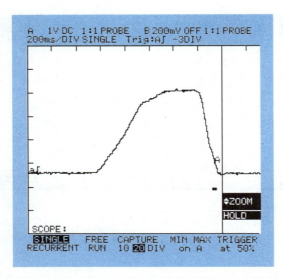

Figure 8-23 Normal DSO waveform, TPS sensor (Courtesy of Fluke Corporation)

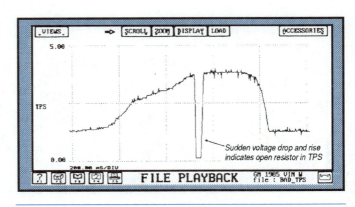

Figure 8-24 Defective TPS waveform (Courtesy of EDGE Diagnostic Systems)

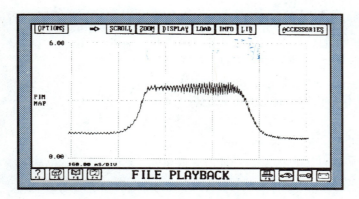

Figure 8-25 Normal MAP sensor DSO waveform (Courtesy of EDGE Diagnostic Systems)

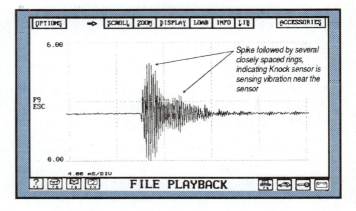

Figure 8-26 Normal knock sensor DSO waveform (Courtesy of EDGE Diagnostic Systems)

Manifold Absolute Pressure Sensor. When the engine is accelerated and returned to idle, the MAP sensor voltage signal should increase and decrease (Figure 8-25). If the engine is accelerated and the MAP sensor voltage signal does not rise and fall, or the signal is erratic, the sensor or connecting wires are defective. A Ford MAP sensor produces a frequency-type signal, and the frequency should change as the engine speed increases. If the frequency is erratic or the frequency does not change in relation to engine speed, the sensor is defective.

Knock Sensor. A knock sensor signal may be obtained by tapping the engine near the knock sensor with a heavy metal object. A normal knock sensor waveform has a group of closely spaced oscillations with a gradually decreasing voltage (Figure 8-26). If the sensor or connecting wire is defective, the oscillations are higher, lower, or erratic.

On some General Motors vehicles, the knock sensor signal is conditioned by a knock sensor module before this signal is sent to the PCM. If the DSO leads are connected from the knock sensor module signal wire to ground and the engine is tapped near the sensor, the signal from this module should be pulled low (Figure 8-27). The low voltage should continue for the duration of the tapping. If this voltage signal is defective and the knock sensor signal is satisfactory, the knock sensor module may be defective, or the power supply to this module may be open.

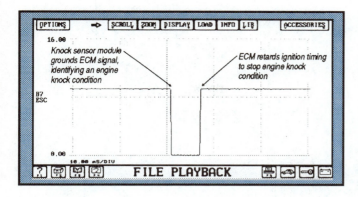

Figure 8-27 Normal knock sensor module DSO waveform (Courtesy of EDGE Diagnostic Systems.

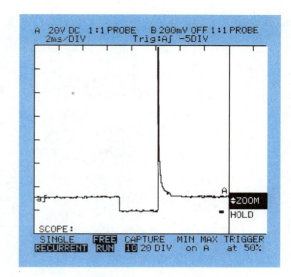

Figure 8-28 Normal DSO injector waveform (Courtesy of Fluke Corporation)

Diagnosis of Output Actuators

Injectors. The DSO leads must be connected from the switched side of the injector to ground to obtain an injector waveform. When the PCM grounds the injector, the voltage should decrease (Figure 8-28). The voltage should remain low while the PCM keeps the injector grounded. When the PCM opens the injector ground and the injector shuts off, a voltage spike should occur from the induced voltage in the injector winding. When the voltage spike is lower on one injector compared to the other injectors, the injector with the low spike probably has a shorted winding. If the PCM operates the injectors with a peak-and-hold current, the waveform has a voltage spike when the PCM reduces the injector current and a second spike when the injector shuts off (Figure 8-29). When the waveform is aligned properly with the horizontal grids on the screen, the injector pulse width can be measured (Figure 8-30). When the injector waveform is ragged, the injector driver in

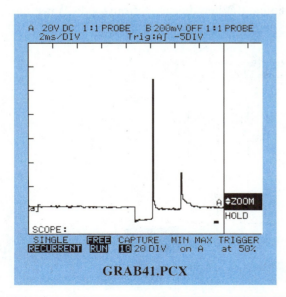

Figure 8-29 Normal DSO peak-and-hold injector waveform (Courtesy of Fluke Corporation)

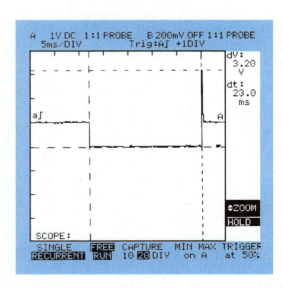

Figure 8-30 Measuring injector pulse width with a DSO injector waveform (Courtesy of Fluke Corporation)

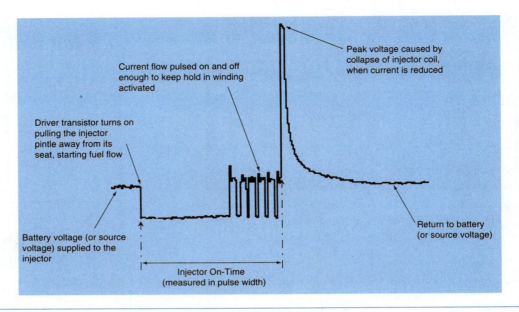

Figure 8-31 Pulse width modulated (PWM) signal used to reduce injector current (Courtesy of Fluke Corporation)

the PCM may be degraded. In some fuel systems, the PCM grounds the injector winding and then uses a pulse width modulated (PWM) signal to reduce injector current (Figure 8-31).

Idle Air Control Motor

Some IAC motors contain two windings and four terminals. These windings are pulsed alternately by the PCM to move the motor pintle in and out, and this action controls the amount of air flow around the throttle to control idle speed. When the PCM activates one IAC motor winding, the other winding is deactivated. Connect the DSO test probes to the signal wires for each coil in the IAC motor. When the engine is idling, turn on the A/C or place the transmission in drive with the parking brake applied. This action causes the PCM to send command signals to the IAC motor (Figure 8-32). If the signals from the PCM are satisfactory, but there is no change in rpm, the IAC motor is defective or the motor air passages are plugged. When the signal from the PCM is erratic, the PCM is defective or the connecting wires are defective.

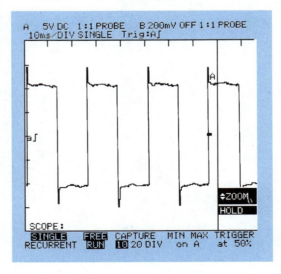

Figure 8-32 Command signals from the PCM to the IAC motor (Courtesy of Fluke Corporation)

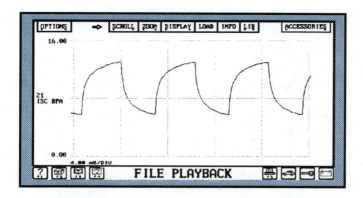

Figure 8-33 Duty cycle signal from the PCM to the IAC motor (Courtesy of EDGE Diagnostic Systems)

Some IAC motors have a single winding and the PCM operates the winding with a duty cycle to pulse the IAC motor plunger and control the amount of air bypassing the throttle. In some applications, this voltage signal from the PCM varies between battery voltage and 5 V, and the PCM never decreases this voltage to 0 V (Figure 8-33). If the A/C clutch is engaged, the duty cycle should change on the IAC motor. An increase in duty cycle causes a corresponding increase in engine rpm. If the duty cycle remains fixed, the PCM may be in the failure mode effects management (FMEM).

Some EFI systems use a stepper motor to control idle speed. This motor contains two windings. The PCM energizes one stepper motor winding and deenergizes the other winding to extend the motor pintle and close the air passage to decrease idle speed. The stepper motor pintle is retracted to open the air passage and increase idle speed when the PCM reverses the energizing and deenergizing of the stepper motor windings. The PCM operates the stepper motor windings in a series of steps. Each step is displayed on the DSO screen as an opposed square wave when the motor advances or retracts a step (Figure 8-34). When the ignition switch is shut off, the PCM positions the stepper motor pintle in the half-open position for restarting.

Failure mode effects management (FMEM) is a backup, or limp-in, mode in a Ford PCM.

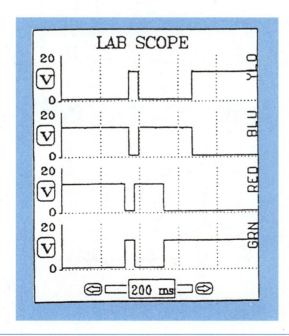

Figure 8-34 Idle speed control stepper motor waveforms (Courtesy of OTC Division, SPX Corporation)

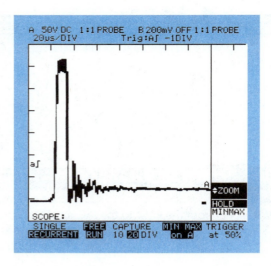

Figure 8-35 Normal DSO primary ignition waveform (Courtesy of Fluke Corporation)

The DSO must be capable of displaying four waveforms simultaneously to display the four square waves from the idle speed control stepper motor. Backprobe the four stepper motor connections, and connect the four DSO leads to these connections. When the idle speed changes, or when the ignition is turned off, the four square waves from the idle speed control stepper motor should be displayed on the DSO screen.

Ignition Waveforms

The DSO leads are connected from the negative primary coil terminal to ground to obtain a primary waveform. A normal primary waveform indicates the ignition module is opening and closing the primary circuit properly (Figure 8-35). The DSO leads may be connected to an inductive clamp attached to one of the secondary plug wires to obtain a secondary waveform (Figure 8-36).

Spark Advance Control

In many EI ignition systems the PCM receives a fixed square wave signal from the crankshaft position sensor. This signal informs the PCM regarding crankshaft position and rpm. The PCM scans the

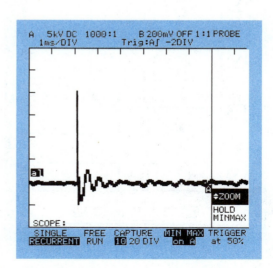

Figure 8-36 Normal DSO secondary ignition waveform (Courtesy of Fluke Corporation)

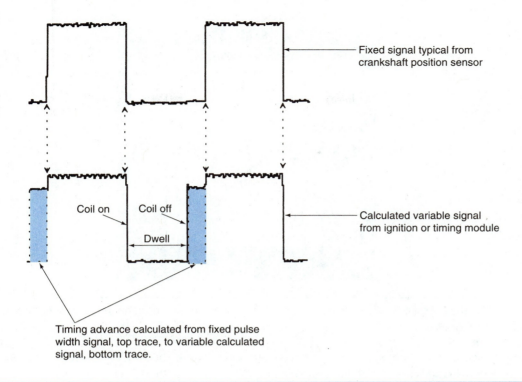

Figure 8-37 Waveforms indicating spark advance (Courtesy of Fluke Corporation)

input signal information and calculates the spark advance required by the engine. When this calculation is completed, the PCM sends a variable, calculated, square wave signal to the ignition module. This signal informs the module when to turn off the primary ignition circuit. The time difference between the leading edge of the crankshaft position sensor signal and the leading edge of the coil off signal indicates the spark advance (Figure 8-37). The DSO leads must be connected to the crankshaft position sensor signal wire and the electronic spark timing (EST) signal wire from the PCM to the ignition module to display these waveforms.

Diagnosis of Alternating Current Superimposed over Direct Current

Defects such as defective alternator diodes or a defective stator may result in the production of some alternating current (ac). This ac may be called ripple voltage or hash, and it can cause radio frequency interference (RFI). Some vehicle speed sensors and wheel speed sensors normally produce an ac voltage. If these sensors are defective, they may produce erratic ac voltage that is higher or lower than normal, resulting in RFI. The DSO may be connected to the radio power feed wire to determine if there is some ac voltage superimposed on top of the dc voltage supplied to the radio, resulting in RFI (Figure 8-38). The DSO may be connected to the alternator battery wire or the suspected sensor to determine the source of the ac voltage.

Classroom Manual Chapter 8, page 197

Diagnosing Repeated Component Failures

Shorted Output Actuators

Repeated computer failures may be caused by a shorted output actuator that allows excessive current flow through the output driver in the PCM. As was mentioned previously, some General Motors PCMs sense high current flow and shut the output driver off before it is damaged. Output drivers may also be damaged by output actuator wires that are shorted to a 12-V supply wire. To locate the cause of this problem, the technician may have to check all the output actuators and connecting wires for a

A quad driver is a group of transistors in a computer that switches certain components on and off.

217

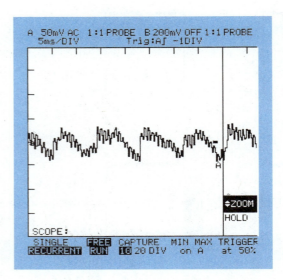

Figure 8-38 DSO waveform with ac superimposed over dc, resulting in RFI (Courtesy of Fluke Corporation)

shorted condition. If an output solenoid has a shorted winding, the winding has less resistance than specified when it is tested with an ohmmeter. In some cases, the shorted condition may cause a DTC to be stored in the PCM memory, which leads the technician to the circuit causing the problem. Some equipment manufacturers market a tester that connects to the General Motors PCM wiring connectors. This tester simulates the PCM and then monitors each output circuit. The tester energizes each output actuator and indicates defects such as high current draw caused by a shorted condition. Various PCM connectors and adapter cards are available with the tester (Figure 8-39).

Intermittent Open Circuits

An intermittent open circuit may result in a voltage spike that causes a repeated computer failure. For example, if the wire from the alternator battery terminal to the positive battery terminal develops an

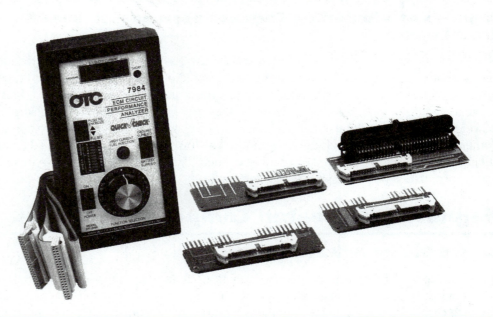

Figure 8-39 Tester for actuating General Motors PCM output actuators (Courtesy of OTC Division, SPX Corporation)

intermittent open circuit, a high voltage spike is produced each time the open circuit occurs. This voltage spike could damage the PCM and other modules and computers. Defective diodes or a defective stator in the alternator may result in excessive voltage fluctuations being supplied to the electrical system, including the PCM. These fluctuations may result in PCM damage.

Defective Voltage Suppression Diodes

Many solenoids have an internal voltage suppression diode connected across the solenoid winding. Each time the solenoid is deenergized, the magnetic field collapses across the winding, resulting in a high induced voltage spike in the winding. The voltage suppression diode reduces this voltage spike by completing the circuit between the two ends of the winding. When computer system solenoid windings are tested, the normal system polarity must be connected to the winding. If reversed polarity is connected to the winding, the voltage suppression diode may be damaged. A defective voltage suppression diode in a computer system solenoid, including the starter solenoid, results in high voltage spikes in the computer system, which may cause repeated and multiple computer failures.

> **SERVICE TIP:** A digital ohmmeter is not suitable for testing diodes because it may not supply enough voltage to forward bias the diode. Many digital meters have a special diode test position with an audible beeper to indicate if the diode is defective or satisfactory.

The computer system wiring diagram may indicate if the solenoids have a voltage suppression diode. Assuming a solenoid has a voltage suppression diode, connect the leads of an analog voltmeter across the suspected solenoid winding and then reverse the leads to check the suppression diode. If the diode is satisfactory, the ohmmeter reading will be slightly lower with one polarity connected to the winding compared to the reading when the meter leads are reversed. If the ohm reading on the solenoid winding is exactly the same with both meter connections, the diode is probably defective.

Excessive Radio Frequency Interference (RFI)

Repeated computer failures may be caused by excessive radio frequency interference (RFI). Additional electronic equipment such as high-powered stereo systems may result in excessive RFI, especially if some of the stereo wires are installed beside the voltage supply wires to the PCM. Installation of nonresistor spark plugs or spark plug wires also results in excessive RFI.

Inaccurate Diagnosis and Service

Repeated computer failures may be caused by inaccurate diagnosis. The PCM may be replaced to correct a problem that is really caused by a defective PCM ground wire.

Repeated input sensor or output actuator replacement may be caused by improper diagnosis. For example, an input sensor may be replaced to correct a problem when the problem is really an intermittent open circuit in the component wiring. A cylinder head coolant leak into a combustion chamber may contaminate the O_2 sensor and result in repeated sensor replacement.

Diagnosing Multiple Component Failures

> **CAUTION:** Battery boosting with reversed polarity or 24 V may result in battery explosions and personal injury. Always follow the battery boosting procedure in the vehicle manufacturer's service manual.

> **WARNING:** Battery boosting with reversed polarity or 24 V may result in severe damage to the vehicle electrical and electronic components.

Multiple component failures may be caused by improper service procedures. Connecting battery booster cables with reversed polarity may damage several computers or modules on the vehicle. Battery boosting with 24 V could have the same effect. These procedures will likely be evidenced by severe arcing damage on the battery terminals. If several computers or modules on a vehicle are damaged, always inspect the battery terminals for arcing damage.

Multiple computer failures may be caused by arc welding on the vehicle without disconnecting the battery terminals. When a vehicle has multiple component failures, always inspect the vehicle for recent signs of arc welding such as a recently installed trailer hitch.

Intermittent open circuits in the vehicle electrical system cause voltage spikes that may result in multiple electronic component failures. As was mentioned previously, RFI may damage electronic components causing multiple failures.

Internal shorts in a wiring harness or wires contacting the chassis ground may result in excessive current flow and multiple failures.

PCM Service, General Motors

WARNING: While servicing computers and chips and sitting in the vehicle, connect a ground wire from your clothing to a ground connection on the vehicle to prevent static charges and component damage. Some tool and equipment manufacturers sell a ground wire with a strap that fits around the technician's wrist. If these components are serviced on the workbench, connect the ground wire to a ground connection on the bench.

WARNING: Most automotive computer chips are shipped in a special container to prevent the chip from static damage. Do not remove this container until you are ready to install the chip in the PCM. Failure to observe this precaution may result in chip damage.

Service Bulletins

General Motors has a number of programmable read only memory (PROM) changes for their PCMs to correct various performance problems such as engine detonation and stalling. General Motors service bulletins provide the information regarding PROM changes. Several automotive test equipment suppliers and publishers also provide General Motors PROM identification (ID) and service bulletin books. This PROM ID and service bulletin information is absolutely necessary when diagnosing fuel injection problems on General Motors vehicles. Technicians must not have the idea they should immediately change a PROM to correct a problem. Most of the service bulletins relating to PROM changes provide a diagnostic procedure that directs the technician to test and eliminate all other causes of the problem before changing the PROM.

PROM Removal and Replacement

If a service bulletin or a fault code indicates that a programmable read only memory (PROM), calibration package (CALPAK), or memory calibrator (MEM-CAL) chip replacement is necessary, these chips are serviced separately from the PCM. A fault code representing one of these chips may indicate that the chip is defective or that the chip is improperly installed. The PROM and CALPAK chips are used in 160-baud PCMs, and the MEM-CAL chip is found in P4 PCMs. The MEM-CAL chip is a combined PROM, CALPAK, and electronic spark control (ESC) module. If PCM replacement is required, these chips are not supplied with the replacement PCM. Therefore, the chip or chips from the old PCM have to be installed in the new PCM. Always disconnect the negative battery cable before any chip replacement is attempted. Follow these steps for PROM replacement:

1. Connect a ground wire from your clothing to a vehicle ground, and touch a vehicle ground with your finger to dissipate static charge.

Baud rate refers to the rate at which a computer can sample input signals.

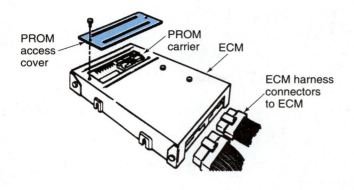

Figure 8-40 Removing PROM access cover (Courtesy of Chevrolet Motor Division, General Motors Corporation)

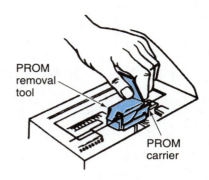

Figure 8-41 PROM removal tool (Courtesy of Chevrolet Motor Division, General Motors Corporation)

2. Remove the PROM access cover from the PCM (Figure 8-40).
3. Attach the PROM removal tool to the PROM and use a rocking action on alternate ends of the PROM carrier to remove the PROM (Figure 8-41).
4. Remove the replacement PROM from the antistatic bag in which it was shipped.

Special Tools

PROM removal and replacement tools

⚠️ **WARNING:** Prior to PROM installation, do not remove the replacement PROM from the antistatic bag in which it was shipped. Removing the PROM from the antistatic bag prior to installation may cause static charge damage to the PROM.

5. Align the notch on the PROM carrier with the notch in the PROM socket in the PCM. Press only on the PROM carrier and push the PROM and carrier into the socket (Figure 8-42). Never remove the PROM from the carrier.

MEM-CAL Removal and Replacement

1. Follow step 1 in the PROM removal and installation procedure.
2. Remove the MEM-CAL access cover on the ECM (Figure 8-43).
3. Use two fingers to push the retaining clips back away from the MEM-CAL, then grasp the MEM-CAL at both ends and lift it upward out of the socket.
4. Align the small notches on the MEM-CAL with the notches in the MEM-CAL socket in the PCM.

Special Tools

MEM-CAL removal and replacement tools

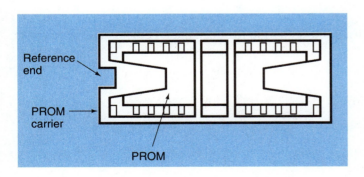

Figure 8-42 PROM installation (Courtesy of Chevrolet Motor Division, General Motors Corporation)

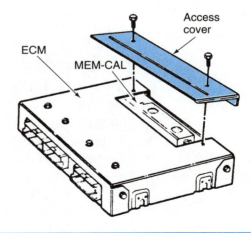

Figure 8-43 Removing MEM-CAL access cover (Courtesy of Oldsmobile Division, General Motors Corporation)

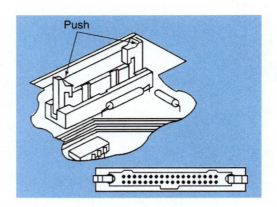

Figure 8-44 MEM-CAL installation (Courtesy of Oldsmobile Division, General Motors Corporation)

5. Press on the ends of the MEM-CAL and push it into the socket until the retaining clips push into the ends of the MEM-CAL (Figure 8-44).

OBD II Diagnosis

> Diagnostic systems previous to OBD II systems may be referred to as OBD I systems.

Standards Established by the Society of Automotive Engineers (SAE)

The SAE J1962 standards apply to the connectors, such as the DLC, and diagnostic repair tools. The J1962 standards also specify the location of the DLC under the instrument panel. The SAE J1930 standards apply to terms and abbreviations in automotive electronics. These standards provide universal terminology in this area. The SAE J1979 and J2190 standards apply to test modes in the automotive electronic system. The SAE J2012 standards define specific requirements for trouble codes.

Diagnostic Trouble Code (DTC) Interpretation

The SAE J2012 standards specify that all DTCs will have a five-digit alphanumeric numbering and lettering system. The following prefixes indicate the general area to which the DTC belongs:

1. P—powertrain
2. B—body
3. C—chassis

The first number in the DTC indicates who is responsible for the DTC definition.

1. 0—SAE
2. 1—Manufacturer

The third digit in the DTC indicates the subgroup to which the DTC belongs. The possible subgroups are:

0—Total system
1—Fuel-air control
2—Fuel-air control
3—Ignition system misfire
4—Auxiliary emission controls
5—Idle speed control
6—PCM and I/O
7—Transmission
8—Non-EEC powertrain

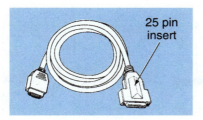

Figure 8-45 OBD II scan tester cable (Courtesy of OTC Division, SPX Corporation)

The fourth and fifth digits indicate the specific area where the trouble exists. Code P1711 has this interpretation:

1. P—Powertrain DTC
2. 1—Manufacturer-defined code
3. 7—Transmission subgroup
4. 11—transmission oil temperature (TOT) sensor and related circuit

Scan Tester Diagnosis

The scan tester must have the appropriate connector to fit the DLC on an OBD II system (Figure 8-45) and the proper software for the vehicle being tested. Different inserts are available for the OBD II scan tester cable depending on the vehicle being tested. Be sure the insert recommended by the scan tester manufacturer for the vehicle being tested is inserted in the OBD II cable. The vehicle make, model year, and engine size must be selected in the scan tester. Always follow the instructions in the manuals supplied by the scan tester manufacturer.

OBD II Datastream and DTC Trouble Code Diagnosis

After the proper vehicle year and VIN number are entered in the scan tester, OBD II must be selected. Following this selection, OBD II datastream can be selected (Figure 8-46). In this mode, the scan tester displays all the data from the inputs. This data may include engine data (Figure 8-47), EGR data (Figure 8-48), three-way catalyst data (Figure 8-49, page 226), HO_2S data (Figure 8-50, page 227), and misfire data (Figure 8-51, page 228). The actual data available varies depending on the vehicle and the scan tester. Always consult the vehicle manufacturer's service manual and the scan tester manufacturer's manual. These manuals provide an explanation of all displayed data.

Some OBD II systems provide flash DTCs with the MIL and OBD DTCs on the scan tester. On other OBD II systems, flash DTCs are not possible. Flash DTCs and OBD II DTCs, with the conditions and time required to set the DTCs in the PCM memory, are shown (Figure 8-52). When the option DTC trouble codes is selected, the scan tester may be used to read and erase DTCs. On many vehicles, OBD II codes are more extensive compared to previous systems. This provides additional diagnostic assistance for the technician. For example, on some OBD II vehicles, DTCs are available for

```
Function Menu:
1-GM
2-FORD
3-CHRYSLER
    → 1-OBDII
      2-OTHER
→ 4-OBDII
    → 1-OBDII Datastream
      2-DTC Trouble Codes
        1-READ DTC
        2-ERASE DTC
        3-FREEZE FRAME DATA
      3-Readiness Status
      4-O2 Tests
      5-Freeze Frame Data
      6-ECU Conflict Data
```

Figure 8-46 OBD II scan tester menu (Courtesy of OTC Division, SPX Corporation)

Tech 1 Parameter	Units Displayed	Typical Data Value	6E3 Reference Section
Engine Speed	RPM	± 100 RPM from Desired Idle	C1, C2, Chart C2-B, DTC P1508, DTC P1509
Desired Idle	RPM	PCM commanded idle speed (varies with temperature).	C1
ECT	°C, °F	85°C - 105°C / 185°F - 220°F (varies with temperature).	C1
IAT	°C, °F	Varies with ambient air temperature.	C1
IAC Position	Counts	10 - 40	C2, Chart C2-B, DTC P1508, DTC P1509
TP Angle	Percent	0%	C1, DTC P0121, DTC P0122 DTC P0123
TP Sensor	Volts	0.20 - 0.74	C1, DTC P0121, DTC P0122 DTC P0123
Throttle at Idle	Yes / No	Yes	C1, DTC P0121, DTC P0122 DTC P0123
Baro	kPa	65 - 110 (Depends on altitude and barometric pressure.)	C1, DTC P0106, DTC P0107, DTC P0108
MAP	kPa / Volts	29 - 48 kPa / 1 - 2 Volts (Depends on engine load and barometric pressure).	C1, DTC P0106, DTC P0107, DTC P0108
MAF Input Freq.	Hz	1200 - 3000 (Depends on engine load and barometric pressure).	C1, DTC P0101, DTC P0102, DTC P0103
MAF	gm/s	3 - 6 (Depends on engine load and barometric pressure).	C1, DTC P0101, DTC P0102, DTC P0103
Inj. Pulse Width	millisecond	Varies with engine load	C2
Air Fuel Ratio	Ratio	14.2 - 14.7	C2
Fuel Pump	On / Off	On	C2, Engine Fuel
VTD Fuel Disable	Active / Inactive	Inactive	DTC P1626, DTC P1629
ECT	°C, °F	85°C - 105°C / 185°F - 220°F (varies with temperature).	C1
Engine Run Time	Hr: Min: Sec	Depends on time since startup.	C1
Loop Status	Open / Closed	Closed	C2, DTC P0125 through DTC P0155
Fuel Trim Learn	Disabled / Enabled	Enabled	C2
H02S Bn1 Sen. 1	Not Ready / Ready	Ready	C2, DTC P0135
H02S Bn2 Sen. 1	Not Ready / Ready	Ready	C2, P0141
H02S Bn1 Sen. 1	millivolts	0 - 1000, constantly varying	C2
H02S Bn2 Sen. 1	millivolts	0 - 1000, constantly varying	C2
Rich/Lean Bn1	Rich / Lean	Constantly changing	C2
Rich/Lean Bn2	Rich / Lean	Constantly changing	C2
H02S Bn1 Sen. 2	millivolts	0 - 1000, constantly varying	C2
H02S Bn1 Sen. 3	millivolts	0 - 1000, varying	C2
Shrt Term FT Bn1	Percent	-10% - 10%	C2, DTC 171, DTC 172
Long Term FT Bn1	Percent	-10% - 10%	C2, DTC 171, DTC 172
Shrt Term FT Bn2	Percent	-10% - 10%	C2, DTC 174, DTC 175
Long Term FT Bn2	Percent	-10% - 10%	C2, DTC 171, DTC 175
Fuel Trim Cell	Cell #	0	C2
Engine Load	Percent	2% - 5% (varies)	C1
Desired EGR Pos.	Percent	0%	C7
Actual EGR Pos.	Percent	0%	C7
Vehicle Speed	MPH, Km/h	0	4L60-E Automatic Transmission Diagnosis
Engine Speed	RPM	100 RPM from Desired Idle	C1, C2, Chart C2-B, DTC P1508, DTC P1509
Commanded TCC	Engaged / Disengaged	Disengaged	4L60-E Automatic Transmission Diagnosis
Current Gear	1 / 2 /3 / 4	1	4L60-E Automatic Transmission Diagnosis
Ignition 1	Volts	13 (Varies)	C1
Commanded GEN	On / Off	On	C1, DTC P0560
Cruise Engaged	Yes / No	No	C1
Cruise Inhibited	Yes / No	Yes	C1
Engine Load	Percent	2% - 5% (varies)	C1
TWC Protection	Active / Inactive	Inactive	C2
Hot Open Loop	Active / Inactive	Inactive	C2
ECT	°C, °F	85°C - 105°C / 185°F - 220°F (varies with temperature).	C1
Decel Fuel Mode	Active / Inactive	Inactive	C2
Inj. Pulse Width	millisecond	Varies with engine load	C2
Power Enrichment	Active / Inactive	Inactive	C2
TP Angle	Percent	0%	C1, DTC P0121, DTC P0122 DTC P0123

Figure 8-47 Engine data (Courtesy of Chevrolet Motor Division, General Motors Corporation)

Tech 1 Parameter	Units Displayed	Typical Data Value	6E3 Reference Section
Desired EGR Pos.	Percent	0%	C7
Actual EGR Pos.	Percent	0%	C7
EGR Test Count	Counts	0	C7
EGR Feedback	Volts	0.14 - 1.0 volt	
EGR Duty Cycle	Percent	0%	C7
EGR Pos. Error	Percent	0%-9%	C7
EGR Closed Valve Pintle Position	Volts	0.14-1.0 volt	C7
Engine Speed	RPM	± 100 RPM from Desired Idle	C1, C2, Chart C2-B, DTC P1508, DTC P1509
Desired Idle	RPM	PCM commanded idle speed (varies with temperature).	C1
ECT	°C, °F	85°C - 105°C / 185°F - 220°F (varies with temperature).	C1
IAT	°C, °F	Varies with ambient air temperature.	C1
IAC Position	Counts	10 - 40	C2, Chart C2-B, DTC P1508, DTC P1509
TP Angle	Percent	0%	C1, DTC P0121, DTC P0122 DTC P0123
Baro	kPa	65 - 110 (Depends on altitude and barometric pressure).	C1, DTC P0106, DTC P0107, DTC P0108
MAP	kPa / Volts	29 - 48 kPa / 1 - 2 Volts (Depends on engine load and barometric pressure).	C1, DTC P0106, DTC P0107, DTC P0108
Loop Status	Open / Closed	Closed	C2, DTC P0125 through DTC P0155
Fuel Trim Learn	Disabled / Enabled	Enabled	C2
Rich/Lean Bn1	Rich / Lean	Constantly changing	C2
Rich/Lean Bn2	Rich / Lean	Constantly changing	C2
Fuel Trim Cell	Cell #	0	C2
Engine Load	Percent	2% - 5% (varies)	C1
Hot Open Loop	Active / Inactive	Inactive	C2
Ignition 1	Volts	13 (Varies)	C1
Knock Retard	Degrees	0°	C5, DTC P0325, DTC P0326, DTC P0327
KS Activity	Yes / No	No	C5, DTC P0325, DTC P0327
Current Gear	1 / 2 /3 / 4	1	4L60-E Automatic Transmission Diagnosis
Trans. Range	P/N / Reverse / Low / Drive 2 / Drive 3 / Drive 4	P/N	4L60-E Automatic Transmission Diagnosis
Commanded TCC	Engaged / Disengaged	Disengaged	4L60-E Automatic Transmission Diagnosis
Trans. Hot Mode	Active / Inactive	Inactive	4L60-E Automatic Transmission Diagnosis
Engine Load	Percent	2% - 5% (varies)	C1
TWC Protection	Active / Inactive	Inactive	C2
Decel Fuel Mode	Active / Inactive	Inactive	C2
Power Enrichment	Active / Inactive	Inactive	C2

Figure 8-48 EGR data (Courtesy of Chevrolet Motor Division, General Motors Corporation)

Tech 1 Parameter	Units Displayed	Typical Data Value	6E3 Reference Section
TWC Monitor Test Counter	Counts	0	DTC P0420
TWC Diagnostic	Enabled / Disabled	Enabled	DTC P0420
Engine Speed	RPM	± 100 RPM from Desired Idle	C1, C2, Chart C2-B, DTC P1508, DTC P1509
ECT	°C, °F	85°C - 105°C / 185°F - 220°F (varies with temperature).	C1
IAT	°C, °F	Varies with ambient air temperature.	C1
IAC Position	Counts	10 - 40	C2, Chart C2-B, DTC P1508, DTC P1509
TP Angle	Percent	0%	C1, DTC P0121, DTC P0122 DTC P0123
TP Sensor	Volts	0.20 - 0.74	C1, DTC P0121, DTC P0122 DTC P0123
Throttle at Idle	Yes / No	Yes	C1, DTC P0121, DTC P0122 DTC P0123
Engine Speed	RPM	± 100 RPM from Desired Idle	C1, C2, Chart C2-B, DTC P1508, DTC P1509
MAF	gm/s	3 - 6 (Depends on engine load and barometric pressure).	C1, DTC P0101, DTC P0102, DTC P0103
Engine Load	Percent	2% - 5% (varies)	C1
Air Fuel Ratio	Ratio	14.2 - 14.7	C2
ECT	°C, °F	85°C - 105°C / 185°F - 220°F (varies with temperature).	C1
Engine Run Time	HR: Min: Sec	Depends on time since startup.	C1
Loop Status	Open / Closed	Closed	C2, DTC P0125 through DTC P0155
Fuel Trim Learn	Disabled / Enabled	Enabled	C2
HO2S Bn1 Sen. 2	Millivolts	0 - 1000, constantly varying	C2
HO2S Bn1 Sen. 3	Millivolts	0 - 1000, varying	C2
Vehicle Speed	MPH, Km/h	0	4L60-E Automatic Transmission Diagnosis
MIL	On / Off	Off	On-Board Diagnostic System Check
Engine Load	Percent	2% - 5% (varies)	C1
TWC Protection	Active / Inactive	Inactive	C2
Hot Open Loop	Active / Inactive	Inactive	C2
ECT	°C, °F	85°C - 105°C / 185°F - 220°F (varies with temperature).	C1
Decel Fuel Mode	Active / Inactive	Inactive	C2
Inj. Pulse Width	millisecond	Varies with engine load	C2
Power Enrichment	Active / Inactive	Inactive	C2
TP Angle	Percent	0%	C1, DTC P0121, DTC P0122 DTC P0123

Figure 8-49 Three-way catalyst data (Courtesy of Chevrolet Motor Division, General Motors Corporation)

Tech 1 Parameter	Units Displayed	Typical Data Value	6E3 Reference Section
Start-up IAT	°C, °F	Depends on intake air temperature at time of startup	C1
Start-up ECT	°C, °F	Depends on engine coolant temperature at time of startup	C1
HO2S Bn1 Sen. 1	Millivolts	0 - 1000, constantly varying	C2
HO2S Bn2 Sen. 1	Millivolts	0 - 1000, constantly varying	C2
HO2S Bn1 Sen. 2	Millivolts	0 - 1000, constantly varying	C2
HO2S Bn1 Sen. 3	Millivolts	0 - 1000, varying	C2
HO2S X Counts Bn1	Counts	Varies	C2
HO2S X Counts Bn2	Counts	Varies	C2
HO2S Warm Up Time Bn1 Sen. 1	Hr: Min: Sec	Depends on startup intake air temperature, startup engine coolant temperature, and time to HO2S activity.	C2
HO2S Warm Up Bn2 SEN 1	HR: Min: Sec	Depends on startup intake air temperature, startup engine coolant temperature, and time to HO2S activity.	C2
HO2S Warm Up Bn1 SEN 2	HR: Min: Sec	Depends on startup intake air temperature, startup engine coolant temperature, and time to HO2S activity.	C2
HO2S Warm Up Bn1 SEN 3	HR: Min: Sec	Depends on startup intake air temperature, startup engine coolant temperature, and time to HO2S activity.	C2
Engine Speed	RPM	± 100 RPM from Desired Idle	C1, C2, Chart C2-B, DTC P1508, DTC P1509
ECT	°C, °F	85°C - 105°C / 185°F - 220°F (varies with temperature).	C1
MAF	gm/s	3 - 6 (Depends on engine load and barometric pressure).	C1, DTC P0101, DTC P0102, DTC P0103
EVAP Purge PWM	Percent	0% - 25% (varies)	C3
TP Angle	Percent	0%	C1, DTC P0121, DTC P0122 DTC P0123
Ignition 1	Volts	13 (Varies)	C1

Figure 8-50 HO$_2$S data (Courtesy of Chevrolet Motor Division, General Motors Corporation)

misfire detected on any cylinder or injector failure on any cylinder. Since the O$_2$ sensor heater is powered by the PCM on many OBD II systems, a DTC is provided for O$_2$ sensor heater malfunction. Refer to the vehicle manufacturer's service manual or the scan tester user's manual for complete lists of OBD II DTCs.

OBD II Freeze Frame and Failure Records Diagnosis

Freeze frame data is also available in the DTC trouble code function. OBD II systems freeze data into the PCM memory when an emissions-related fault is stored in the PCM memory and the MIL is illuminated. This freeze-frame data is very useful in diagnosing the cause of the code, especially with intermittent faults. On some General Motors vehicles, failure records information is also available on the

Tech 1 Parameter	Units Displayed	Typical Data Value	6E3 Reference Section
Misfire Cur. # 1	Counts	0	C1, C2, C4
Misfire Hist. #1	Counts	0	DTC P0300, C1, C2, C4
Misfire Cur. # 2	Counts	0	C1, C2, C4
Misfire Hist. #2	Counts	0	DTC P0300, C1, C2, C4
Misfire Cur. # 3	Counts	0	C1, C2, C4
Misfire Hist. #3	Counts	0	DTC P0300, C1, C2, C4
Misfire Cur. # 4	Counts	0	C1, C2, C4
Misfire Hist. #4	Counts	0	DTC P0300, C1, C2, C4
Misfire Cur. # 5	Counts	0	C1, C2, C4
Misfire Hist. #5	Counts	0	DTC P0300, C1, C2, C4
Misfire Cur. # 6	Counts	0	C1, C2, C4
Misfire Hist. #6	Counts	0	DTC P0300, C1, C2, C4
Misfire Failures Since First Fail	Counts	0	DTC P0300
Misfire Passes Since First Fail	Counts	0	DTC P0300
Total Misfire Current Count	Counts	0	DTC P0300
ECT	°C, °F	85°C - 105°C / 185°F - 220°F (varies with temperature).	C1
Hot Open Loop	Active / Inactive	Inactive	C2
Loop Status	Open / Closed	Closed	C2
Fuel Trim Learn	Disabled / Enabled	Enabled	C2
H02S Bn1 Sen. 1	Not Ready / Ready	Ready	C2
H02S Bn2 Sen. 1	Not Ready / Ready	Ready	C2
Rich/Lean Bn1	Rich / Lean	Constantly changing	C2
Rich/Lean Bn2	Rich / Lean	Constantly changing	C2
H02S X Counts Bn1	Counts	Varies	C2
H02S X Counts Bn2	Counts	Varies	C2
Shrt Term FT Bn1	Percent	-10% - 10%	C2
Long Term FT Bn1	Percent	-10% - 10%	C2
Shrt Term FT Bn2	Percent	-10% - 10%	C2
Long Term FT Bn2	Percent	-10% - 10%	C2
Fuel Trim Cell	Cell #	0	C2
Engine Load	Percent	2% - 5% (varies)	C1
Engine Speed	RPM	± 100 RPM from Desired Idle	C1
Ignition Mode	IC / Bypass	IC	C4
Spark	Degrees	16° (varies)	C4
Knock Retard	Degrees	0°	C5
Desired EGR Pos.	Percent	0%	DTC P1406, C7
Actual EGR Pos.	Percent	0%	DTC P1406, C7
Vehicle Speed	MPH, Km/h	0	4L60-E Transmission Diagnosis
Current Gear	1 / 2 /3 / 4	1	4L60-E Transmission Diagnosis
Engine Speed	RPM	± 100 RPM from Desired Idle	C1
TCC Enable	On / Off	Off	4L60-E Transmission Diagnosis
IAC Position	Counts	10 - 40	C2
Current Gear	1 / 2 /3 / 4	1	4L60-E Transmission Diagnosis
Engine Load	Percent	2% - 5% (varies)	C1
TWC Protection	Active / Inactive	Inactive	C2
Decel Fuel Mode	Active / Inactive	Inactive	C2
Power Enrichment	Active / Inactive	Inactive	C2
A/C Request	Yes / No	No (Yes with A/C "ON")	C10
Commanded A/C	On / Off	Off (On with A/C compressor engaged)	C10

Figure 8-51 Misfire data (Courtesy of Chevrolet Motor Division, General Motors Corporation)

INPUTS

COMPONENT	MALFUNC.	MIL	J2012 DTC	DIAGNOSTIC CRITERIA TO SET FAULT	MATURE TIME
Sync. (cam/crank)	Rationality	11	P1390	Cam/Crank reference Angle > threshold	25 counts
MAP sensor	Rationality	13	P1297	Baro-MAP < 4 in. of Hg at idle	2 sec.
	Short	14	P0107	Voltage signal < 0.02 Volts	2 sec.
	Open	14	P0108	Voltage signal > 4.667 Volts	2 sec.
Vehicle speed sensor	Rationality	15	P0500	No pulses during driving conditions	11 sec.
Coolant temp. sensor	Rationality	17	P0125	After start, engine temperature < 35°F	10 min.
	Short	22	P0117	Voltage signal < 0.51 Volts	3 sec.
	Open	22	P0118	Voltage signal > 4.96 Volts	3 sec.
O2 Sensor Upstream	Shorted Low	21	P0131	Voltage stays below .15 volts during O2 heater test	7 min.
O2 Sensor Upstream	Shorted High	21	P0132	Voltage signal > 1.22 Volts	3 sec.
O2 Sensor Upstream	Open	21	P0134	Voltage stays > .35 volts or < .59 volts	2 min.
O2 Sensor Downstream	Shorted Low	21	P0137	Voltage stays below .15 volts during O2 heater test	7 min.
O2 Sensor Downstream	Shorted High	21	P0138	Voltage signal > 1.22 Volts	3 sec.
	Rationality	21	P0139	At WOT Voltage signal > .61 volts and At decel fuel shutoff Voltage signal < .29 volts	15 sec.
Charge temp. sensor	Short	23	P0112	Voltage signal < 0.51 Volts	3 sec.
	Open	23	P0113	Voltage signal > 4.96 Volts	3 sec.
Throttle position sensor	Rationality	24	P0121	At idle Voltage signal > 4 Volts OR @ cruise Voltage signal < .51 Volts	3 sec.
	Short	24	P0122	Voltage signal < 0.157 Volts	0.7 sec.
	Open	24	P0123	Voltage signal > 4.706 Volts	0.7 sec.
PCM	ROM sum	53	P0605	At power down, bit sum not equal to calibration	Immediate
	SPI	53	P0605	No communication between devices	
Cam position sensor	Rationality	54	P0340	No pulses seen	5 sec.
P/S pressure switch	Rationality	65	P0551	At vehicle speed > 40 mph, high pressure	15 sec.

OUTPUTS

IAC motor	Open/Short	25	P0505	Voltage signal read is not equal to expected state	3 sec.
	Functionality	25	P1294	At idle, rpm not within ± 300 of target	20 sec.
Injector 1	Open/Short	27	P0201	Voltage signal read is not equal to expected state	3 sec.
Injector 2	Open/Short	27	P0202	Voltage signal read is not equal to expected state	3 sec.
Injector 3	Open/Short	27	P0203	Voltage signal read is not equal to expected state	3 sec.
Injector 4	Open/Short	27	P0204	Voltage signal read is not equal to expected state	3 sec.
EVAP solenoid	Functionality	31	P0441	Shift in O2 feedback or IAC or RPM from solenoid off to full on	25 sec.
	Open/Short	31	P0443	Voltage signal read is not equal to expected state	3 sec.
EGR solenoid	Open/Short	32	P0403	Voltage signal read is not equal to expected state	3 sec.
High speed fan relay	Open/Short	35	P1489	Voltage signal read is not equal to expected state	3 sec.
Low speed fan relay	Open/Short	35	P1490	Voltage signal read is not equal to expected state	3 sec.
Ignition coil 1 (DIS)	Open/Short	43	P0351	Maximum dwell current is not equal to expected state	3 sec.
Ignition coil 2 (DIS)	Open/Short	43	P0352	Maximum dwell current is not equal to expected state	3 sec.

Figure 8-52 Flash DTCs and OBD II DTCs, with the conditions and time required to set the DTCs in the PCM memory (Courtesy of Chrysler Corporation)

scan tester. This information is similar to freeze-frame data, but failure records include data related to any fault that set a code in the PCM memory and illuminated the MIL.

OBD II Readiness Status and O₂ Tests

OBD II readiness status indicates if the various monitors in the OBD II system have completed or not. This mode does not indicate whether the emission levels were excessive or not during each monitor. The O_2 sensor mode provides O_2 sensor data, which is also available in the OBD II datastream.

OBD II Conflicting Data

The OBD II conflicting data mode provides an identification number for each computer connected to the data links on the vehicle. The scan tester can be used to select each computer and display the

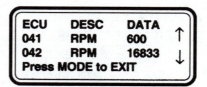

Figure 8-53 OBD II correct data and conflicting, incorrect data (Courtesy of OTC Division, SPX Corporation)

Classroom Manual
Chapter 8, page 204

data from that computer. The technician can then determine which computer is sending the correct data and which is sending the conflicting or incorrect data (Figure 8-53).

● **CUSTOMER CARE:** Never sell customers automotive service that is not required on their cars. Selling preventative maintenance is a sound business practice and may save customers some future problems. An example of preventative maintenance is selling a cooling system flush when the cooling system is not leaking but the manufacturer's recommended service interval has elapsed. If customers find out they were sold some unnecessary service, and some will find out, they will probably never return to the shop. They will likely tell their friends about their experience, and that kind of advertising the shop can do without.

Guidelines for Scan Tester, DSO, and OBD II Diagnosis

1. A scan tester may be connected to the DLC on many vehicles to obtain fault codes and perform many other diagnostic functions.
2. Serial data displayed on a scan tester may be used to diagnose input sensors, output actuators, and PCM control.
3. A breakout box may be connected in series with the PCM terminals on Ford products to obtain voltage readings from the input sensors.
4. Some scan testers now have the capability to read data from Ford computers.
5. On Chrysler products, the scan tester will extend and retract the IAC BPA motor in the ATM mode.
6. On some computer systems, the scan tester reads IAC BPA motor counts to indicate the motor pintle position.
7. The DSO provides more accurate diagnosis of automotive computer systems because it has a much faster signal sampling speed.
8. Repeated component failures in a computer system may be caused by shorted outputs, intermittent open circuits, defective voltage suppression diodes, excessive RFI, and inaccurate diagnosis and service.
9. Multiple component failures may be caused by battery boosting with reversed polarity or 24 V, arc welding on the vehicle without disconnecting the battery terminals, intermittent open circuits, and internal shorts or grounds in the wiring harness.
10. In General Motors PCMs, the PROM, CAL-PAK, and MEM-CAL chips may be replaced.

CASE STUDY

A customer complained about noise on the radio-stereo in his Mercedes. This noise occurred while driving the vehicle, but the noise disappeared with the engine idling and the vehicle stopped. When questioned about other problems, the customer replied the car operated normally except for the radio noise problem. Further questioning of the customer revealed two replacement radios had failed to correct the problem.

The technician obtained the vehicle wiring diagrams and studied them carefully, especially the electrical connections to the radio. The technician was surprised to discover the vehicle speed sensor (VSS) signal was connected through the radio. This signal gives the radio the capability to increase the volume slightly in relation to vehicle speed to compensate for wind and road noise.

The technician used the scan tester to obtain the serial data with the engine running and while driving the vehicle. All the serial data appeared to be normal. The technician connected the DSO to the vehicle speed sensor (VSS) signal wire at the radio. The VSS signal appeared ragged and erratic. The VSS was replaced and the DSO screen showed a normal waveform. During a road test, there was no radio or stereo noise.

Terms to Know

Block learn
Breakout box
Calibrator package (CAL-PAK)
Digital storage oscilloscope (DSO)
Integrator
Memory calibrator (MEM-CAL)
Programmable read only memory (PROM)
Quad driver
Radio frequency interference (RFI)
Scan tester
Serial data
Snapshot testing

ASE Style Review Questions

1. A customer complains about reduced fuel economy, and the following data is received on the scan tester with the engine operating at normal temperature and idle speed:

   MAP sensor 2.75 V   TPS sensor 0.6 V
   ECT sensor 0.55 V   IAT sensor 2.5 V
   HO_2S 0.3 V–0.8 V   VSS sensor 0 mph
   IAC counts 30   Engine rpm 850
   Bat V 14.2   EVAP solenoid off
   TCC off   EGR 0%
   Closed loop   MIL off
   DTCs none

   Technician A says there is a vacuum leak in the intake manifold.
   Technician B says the MAP sensor is defective.
   Who is correct?
   A. A only   **C.** Both A and B
   B. B only   **D.** Neither A nor B

2. While discussing block learn and integrator when the integrator number is 180 and the block learn number is 185:
   Technician A says these numbers indicate a normal condition.
   Technician B says these numbers indicate the PCM is trying to increase fuel delivery; therefore, the oxygen (O_2) sensor signal must be continually lean.
   Who is correct?
   A. A only
   C. Both A and B
   B. B only
   D. Neither A nor B

3. An engine idles faster than specified, and the following data is received on the scan tester at idle speed and normal engine temperature:

   MAP sensor 1.2 V
   ECT sensor 0.5 V
   HO_2S 0.4 V–0.7 V
   IAC counts 50
   Bat V 14.3
   TCC off
   Closed loop
   DTCs none
   TPS sensor 0.2 V
   IAT sensor 2.4 V
   VSS sensor 0 mph
   Engine rpm 1,050
   EVAP solenoid off
   EGR 0%
   MIL off

   Technician A says the engine has an intake manifold vacuum leak.
   Technician B says the IAC motor is defective.
   Who is correct?
   A. A only
   B. B only
   C. Both A and B
   D. Neither A nor B

4. A customer complains about no fast idle speed when the engine is cold and engine stalling. The following data is received on the scan tester with the engine operating at normal temperature and idle speed.

   MAP sensor 1.15 V
   ECT sensor 0.55 V
   HO_2S 0.3V–0.8V
   IAC counts 0
   Bat V 14.4
   TCC off
   Closed loop
   DTCs none
   TPS sensor 0.6 V
   IAT sensor 2.5 V
   VSS sensor 0 mph
   Engine rpm 650
   EVAP solenoid off
   EGR 0%
   MIL off

   Technician A says the IAC motor is stuck.
   Technician B says the IAC motor winding or connecting wires have an open circuit.
   Who is correct?
   A. A only
   B. B only
   C. Both A and B
   D. Neither A nor B

5. A customer complains about engine surging and stalling on a 1995 Ford F series truck with a 5.0-L engine. The scan tester displays this data with the engine operating at normal temperature and idle speed.

   MAF sensor 2.1 V
   ECT sensor 0.55 V
   HO_2S 0.3 V–0.8 V
   Knock sensor 0
   Bat V 14.4
   TCC off
   Closed loop
   DTCs 159, 337
   TPS sensor 0.6 V
   IAT sensor 2.5 V
   VSS sensor 0 mph
   Engine rpm 650–950
   EVAP solenoid off
   EGR EVP sensor 2.4 V
   MIL on

   Technician A says the MAF sensor is defective.
   Technician B says the EGR valve may be stuck open.
   Who is correct?
   A. A only
   B. B only
   C. Both A and B
   D. Neither A nor B

6. A technician is using a DSO to diagnose an OBD II system. The downstream HO_2S has the same voltage waveform as the upstream HO_2S.
   Technician A says the calalytic converter is defective.
   Technician B says the air-fuel ratio is too rich.
   Who is correct?
   A. A only
   B. B only
   C. Both A and B
   D. Neither A nor B

7. The following data is displayed on a scan tester with the engine operating at idle speed and normal temperature:

   MAP 1.1 V
   ECT sensor 2.1 V
   HO_2S 0.3 V–0.8 V
   IAC counts 32
   Bat V 14.2
   TCC off
   Open loop
   DTCs none
   TPS sensor 0.6 V
   IAT sensor 2.5 V
   VSS sensor 0 mph
   Engine rpm 750
   EVAP solenoid off
   Inj. pw 4 ms
   MIL off

   Technician A says the injectors should be purged.
   Technician B says the ECT sensor is defective.
   Who is correct?
   A. A only
   B. B only
   C. Both A and B
   D. Neither A nor B

8. While discussing injector diagnosis with a DSO, the voltage spike when the injector is turned off is much less on one injector compared to the other injectors:
   Technician A says this problem is caused by higher-than-specified resistance in the injector winding.
   Technician B says this problem is caused by a shorted injector winding.
   Who is correct?
   A. A only
   B. B only
   C. Both A and B
   D. Neither A nor B

9. While discussing repeated component failures:
   Technician A says repeated component failures may be caused by shorted output actuators.
   Technician B says repeated component failures may be caused by a defective voltage suppression diode in a relay.
   Who is correct?
   A. A only
   B. B only
   C. Both A and B
   D. Neither A nor B

10. While discussing PROM removal and replacement:
    Technician A says the new PROM should be kept in its protective shipping package until the PCM is prepared for PROM installation.
    Technician B says the technician should connect a ground wire from his clothing to a good ground on the vehicle while sitting on the car seat to install the new PROM.
    Who is correct?
    A. A only
    B. B only
    C. Both A and B
    D. Neither A nor B

Table 8-1 ASE Task

Interpret scan tool data to determine system condition.

Problem Area	Symptoms	Possible Causes	Classroom Manual	Shop Manual
INPUTS OR OUTPUTS	Not within specifications	Defective inputs, outputs, or connecting wires	193	191

Table 8-2 ASE Task

Establish relative importance of displayed serial data.

Problem Area	Symptoms	Possible Causes	Classroom Manual	Shop Manual
PERFORMANCE OR ECONOMY REDUCED	Serial data not within specifications	Defective inputs, outputs, or PCM	193	193

Table 8-3 ASE TASK

Test input sensor/sensor circuit using serial data.

Problem Area	Symptoms	Possible Causes	Classroom Manual	Shop Manual
PERFORMANCE OR ECONOMY REDUCED	Input sensor serial data not within specifications	Defective inputs or connecting wires	195	195

Table 8-4 ASE Task

Test output actuator/output circuit using serial data.

Problem Area	Symptoms	Possible Causes	Classroom Manual	Shop Manual
PERFORMANCE OR ECONOMY REDUCED	Input or output serial data not within specifications	Defective inputs, outputs, or connecting wires	195	198

Table 8-5 ASE Task

Test PCM control circuit using serial data.

Problem Area	Symptoms	Possible Causes	Classroom Manual	Shop Manual
PERFORMANCE OR ECONOMY REDUCED	Improper input or output data	Defective inputs, outputs, wires, or PCM	195	201

Table 8-6 ASE Task

Determine root cause of failures.

Problem Area	Symptoms	Possible Causes	Classroom Manual	Shop Manual
PERFORMANCE OR ECONOMY REDUCED	Input or output serial data not within specifications	Defective inputs, outputs, wires, or PCM	197	195

Table 8-7 ASE Task

Determine the root cause of multiple component failures.

Problem Area	Symptoms	Possible Causes	Classroom Manual	Shop Manual
SYSTEM DAMAGE	Damaged computer, inputs, outputs, or wires	1. Improper battery boost procedures	196	219
		2. Arc welding without disconnecting battery cables	196	219
		3. Intermittent circuits	196	219
		4. Internal shorts or grounds in the wiring	196	220

Table 8-8 ASE Task

Determine the root cause of repeated component failures.

Problem Area	Symptoms	Possible Causes	Classroom Manual	Shop Manual
SYSTEM DAMAGE	Repeated component failure	1. Shorted output actuators	193	217
		2. Intermittent open circuits	194	217
		3. Defective voltage suppression diodes	194	218
		4. Radio frequency interference	195	218
		5. Inaccurate diagnosis and service	196	218

CHAPTER 9

Emission Control Systems, Diagnosis and Service

Upon completion and review of this chapter, you should be able to:

❏ Inspect and test for missing, modified, or tampered emission components.

❏ Locate and utilize relevant emission service information.

❏ Determine appropriate diagnostic procedures based on available vehicle data and service information; determine if available information is adequate to proceed with effective diagnosis.

❏ Establish relative importance of displayed serial data.

❏ Differentiate between emission control systems' mechanical and electrical/electronic problems.

❏ Determine the need to diagnose emission control systems.

❏ Perform functional tests on emission control systems.

❏ Determine the effect on tailpipe emissions caused by a failure of an emission control component or subsystem.

❏ Diagnose hot or cold no-starting, hard starting, poor driveablility, incorrect idle speed, poor idle, flooding, hesitation, surging, engine misfire, power loss, stalling, poor mileage, and emission problems caused by a failure of emission control components or subsystems.

❏ Use a scan tester to diagnose an EGR system.

❏ Diagnose and service port EGR valves.

❏ Diagnose and service negative backpressure EGR valves.

❏ Diagnose and service positive backpressure EGR valves.

❏ Diagnose and service digital EGR valves.

❏ Diagnose and service linear EGR valves.

❏ Diagnose EGR vacuum regulator (EVR) solenoids.

❏ Diagnose exhaust gas temperature sensors.

❏ Diagnose EGR pressure transducers (EPT).

Locating Service Information

Basic Tools

Basic technician's tool set

Service manual

Wire brush

Lengths of vacuum hose

Heat-resistant water container

Thermometer

⚠ **WARNING:** Discarding, disconnecting, or tampering with any emission component is a serious federal offense for automotive technicians and repair facilities in the United States.

The first step in accurate emission system diagnosis is to locate the proper service information for the vehicle being diagnosed. The technician must have a wiring diagram, vacuum hose diagram, and specifications for the emission systems on the vehicle being serviced. It is very difficult and time consuming to attempt a diagnosis without this information. A vacuum hose diagram is mounted under the hood on many vehicles. The technician must have service bulletin information for the vehicle being serviced. If a service bulletin recommends a change-up on an emission component to correct the problem being diagnosed, the technician must have this information.

Service procedure information must be available in the vehicle manufacturer's service manual, generic service manual, or electronic data system. The diagnosis may indicate a problem in a specific emission system, but the diagnosis may not indicate whether the defect is in an emission system component or the connecting wires. The technician must have the proper service procedures available to locate the exact cause of the problem. A parts locator book often saves time by helping the technician find a component on the vehicle. Sometimes the diagnostic procedures inform the technician that a certain component is the problem, but the technician may spend time locating the component on today's complex electronic systems. A parts locator book allows the technician to find components quickly.

If the appropriate service information, such as service manuals, wiring diagrams, vacuum hose diagrams, service bulletin information, and specifications, is not available, it is probably advisable not to attempt a diagnosis of the vehicle.

Preliminary Emission System Inspection

All inspection, diagnosing, and servicing procedures in this chapter apply to the emission systems described in this chapter. Procedures related to other systems are explained in the next chapter. Before an emission system is diagnosed, the emission components and systems should be inspected for missing parts, modification, and tampering.

The exhaust system should be inspected to be sure the proper catalytic converter, or converters, is installed on the vehicle. Catalytic converter heat shields must be in place. If the catalytic converter has a hose connection from the air pump system, be sure this hose is securely connected.

The positive crankcase ventilation (PCV) system should be inspected for missing parts and hoses that are cracked, loose, oil-soaked, or deteriorated so they are spongy and soft. Remove the hose from the PCV valve to the intake manifold and inspect it for internal collapsing and deterioration. If the PCV valve is mounted in a rubber grommet in the rocker arm cover, be sure the valve fits snugly in this grommet. All hoses must be securely connected.

Inspect the exhaust gas recirculation (EGR) system for missing or modified components, and disconnected, improperly connected, leaking, or restricted vacuum hoses. Be sure that no restrictions such as ball bearings have been installed in vacuum hoses. Check all vacuum hoses for tight connections on the various EGR components. Inspect all vacuum hoses for oil soaking and deterioration.

Catalytic Converter Diagnosis and Service

CAUTION: Catalytic converters and other exhaust system components are extremely hot if the engine has been running. Wear protective gloves to avoid burns when servicing these components.

The most accurate method of testing a catalytic converter is to measure the tailpipe emissions with a four-gas infrared exhaust analyzer. A defective three-way catalytic converter may result in high emissions of HC, CO, and NO_x. Other systems, such as the fuel, ignition, and emission systems, also affect tailpipe emissions.

Special Tools

Digital pyrometer

The catalytic converter may be tested with a hand-held digital pyrometer (Figure 9-1). This electronic device measures heat wherever the probe is positioned. The meter probe should be

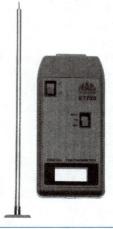

Figure 9-1 Digital pyrometer for measuring catalytic converter temperature (Courtesy of Mac Tools, Inc.)

positioned to measure the temperature at the converter inlet and outlet. If the catalytic converter is working properly, the outlet temperature should be a minimum of 100°F (38°C) higher than the inlet temperature. When the temperature difference between the converter inlet and outlet is less than 100°F (38°C), the converter is not working properly, and it should be replaced or repaired. If the converter is not working properly, always check the air pump system to be sure it is pumping air into the converter when the engine is at normal operating temperature. When this air flow is not present, converter operation is inefficient. Never install a piece of exhaust pipe in place of the catalytic converter.

> **SERVICE TIP:** The operating temperature of a defective catalytic converter is much lower than the temperature of a satisfactory catalytic converter. When driving short distances in cold climates, a defective catalytic converter may fill up with condensation. This condensation may freeze when the engine is not running, resulting in a completely restricted converter and a no-start problem.

Some catalytic converters have an inner and outer shell. If the outer shell is physically damaged on this type of converter, the outer shell may be cut apart to inspect the inner converter (Figure 9-2). If the inner part of the converter is damaged, replace the complete unit. If the damage is limited to the outer shell, a high-temperature sealant may be applied to the upper and lower shells. These shells are then retained with channels and clamps (Figure 9-3).

The pellets may be replaced in some pellet-type catalytic converters. Remove the plug in the bottom of the converter, and connect a vibrator tool to the plug opening. The vibrator tool shakes the pellets out of the converter into a storage can on the tool. Install the new pellets in the storage can on the vibrator tool, and connect a vacuum pump to the tailpipe. When a shop air hose is connected to the vacuum pump, the pellets are drawn from the storage can into the converter. If pellets come out of the tailpipe, replace the converter.

A pyrometer indicates the temperature of various components when the pickup is placed on the component.

Classroom Manual
Chapter 9, page 221

PCV System Service and Diagnosis

If the PCV valve is stuck in the open position, excessive air flow through the valve causes a lean air-fuel ratio and possible rough idle operation or engine stalling. When the PCV valve or hose is restricted, excessive crankcase pressure forces blow-by gases through the clean air hose and filter into the air cleaner. Worn rings or cylinders cause excessive blow-by gases and increased crankcase

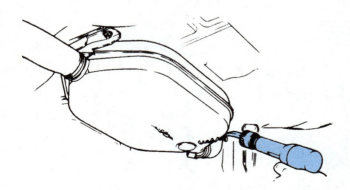

Figure 9-2 Cutting outer shell from a catalytic converter (Courtesy of Chevrolet Motor Division, General Motors Corporation)

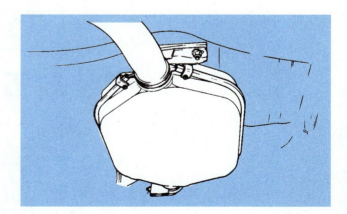

Figure 9-3 Catalytic converter outer shell reassembled with sealant, channels, and clamps (Courtesy of Chevrolet Motor Division, General Motors Corporation)

Figure 9-4 Checking engine gaskets, PCV hose, and clean air hose (Courtesy of Toyota Motor Corporation)

pressure, which forces blow-by gases through the clean air hose and filter into the air cleaner. A restricted PCV valve or hose may result in the accumulation of moisture and sludge in the engine and engine oil.

Leaks at engine gaskets, such as the rocker arm cover or crankcase gaskets, will result in oil leaks and the escape of blow-by gases to the atmosphere. However, the PCV system also draws unfiltered air through these leaks into the engine. This action could result in wear of engine components, especially when the vehicle is operating in dusty conditions.

When diagnosing a PCV system, the first step is to check all the engine gaskets for signs of oil leaks (Figure 9-4). Be sure the oil filler cap fits and seals properly. Check the clean air hose and the PCV hose for cracks, deterioration, loose connections, and restrictions.

Check the PCV clean air filter for contamination, and replace this filter if necessary (Figure 9-5). If there is evidence of oil in the air cleaner, check the PCV valve and hose for restriction. When the

> Blow-by is the escape of combustion chamber gases past the piston rings.

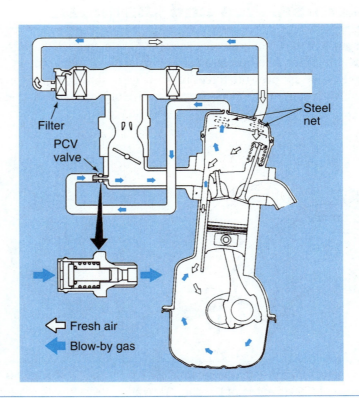

Figure 9-5 PCV air filter and complete system (Courtesy of Nissan Motor Co., Ltd.)

PCV valve and hose are in satisfactory condition, perform a cylinder compression test to check for worn cylinders and piston rings, which can cause excessive blow-by.

A restricted PCV valve or hose results in high CO emission levels on an emissions analyzer.

PCV Valve Diagnosis and Service

  CAUTION: Do not attempt to suck through a PCV valve with your mouth. Sludge and other deposits inside the valve are harmful to the human body.

Vehicle manufacturers recommend different PCV valve checking procedures. Always follow the procedure in the vehicle manufacturer's service manual. Some vehicle manufacturers recommend removing the PCV valve from the rocker arm cover and the hose. Connect a length of hose to the inlet side of the PCV valve, and blow air through the valve with your mouth while holding your finger near the valve outlet (Figure 9-6). Air should pass freely through the valve. If air does not pass freely through the valve, replace the valve.

Connect a length of hose to the outlet side of the PCV valve, and try to blow back through the valve (Figure 9-7). It should be difficult to blow air through the PCV valve in this direction. When air passes easily through the valve, replace the valve.

Some vehicle manufacturers recommend removing the PCV valve from the rocker arm cover and placing your finger over the valve with the engine idling. When there is no vacuum at the PCV valve, the valve, hose, or manifold inlet is restricted. Replace the restricted component or components.

Remove the PCV valve from the rocker arm cover and the hose. Shake the valve beside your ear, and listen for the tapered valve rattling inside the valve housing. If no rattle is heard, replace the PCV valve.

Classroom Manual
Chapter 9, page 224

Diagnosis of Exhaust Gas Recirculation (EGR) Systems

General Diagnosis

The engine must be at normal operating temperature before diagnosing the EGR system. If the EGR valve remains open at idle and low engine speed, the idle operation is rough and surging occurs at low speed. When this problem is present, the engine may hesitate on low-speed acceleration or stall after deceleration, or after a cold start. If the EGR valve does not open, engine detonation occurs, and NO_x emissions are high.

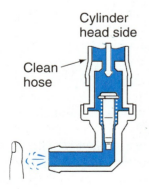

Figure 9-6 Blowing air through the PCV valve from the inlet side (Courtesy of Toyota Motor Corporation)

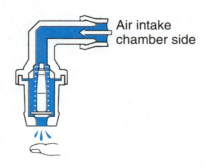

Figure 9-7 Blowing air through the PCV valve from the outlet side (Courtesy of Toyota Motor Corporation)

Since engine compression defects also result in rough idle operation, the compression should be verified with a compression test. Rough idle may be caused by vacuum leaks. Check for leaks at all vacuum hoses and vacuum-operated components. A vacuum leak causes a low steady reading on a vacuum gauge. Ignition defects may result in rough idle operation and detonation; therefore, the ignition system condition should be verified with an oscilloscope diagnosis before emission systems are diagnosed and serviced.

EGR Scan Tester Diagnostic Trouble Code Diagnosis

Special Tools

Scan tester

When an electrical defect occurs in the EGR system, a diagnostic trouble code (DTC) is usually set in the PCM memory. The actual defects required to set a DTC may vary depending on the vehicle make and model year.

When diagnosing any EGR system, the first step is to perform a preliminary emission system inspection. Repair any defects found during the inspection. In many EGR systems, the PCM uses inputs from the ECT, TPS, and MAP sensors to operate the EGR valve. Improper EGR operation may be caused by a defect in one of these sensors. A scan tester may be connected to the data link connector (DLC) to read PCM data and retrieve DTCs. Some DTCs indicate a fault in the EGR solenoid or solenoids, such as an open solenoid winding or connecting wires between the solenoid and the PCM. If a DTC indicates a defect in the EGR solenoid or connecting wires, proceed with the electrical tests on the EGR solenoid and connecting wires explained later in this chapter. In some systems, a fault code is set in the PCM memory if the EGR valve is open when the input sensor signals to the PCM indicate this valve should be closed. This type of DTC may be set in the PCM memory by a sticking EGR valve. If this type of code is present, diagnose the EGR valve as explained later in the chapter. Correct the cause of any DTCs before proceeding with EGR diagnosis.

EGR Scan Tester Data Diagnosis

✓ **SERVICE TIP:** In many computer systems when the EGR solenoid is off, the solenoid shuts off vacuum to the EGR valve. When the solenoid is energized or on, vacuum is supplied through the solenoid to the EGR valve. However, on some Chrysler products when the EGR solenoid is on, it shuts off vacuum to the EGR valve, and when this solenoid is off, it supplies vacuum through the solenoid to the EGR valve.

The EGR valve pintle refers to the movable, tapered end on the valve stem, which rests on a matching seat when the valve is closed.

The vehicle must be driven at various speeds on the road to determine if the EGR valve is working properly. While driving the vehicle, leave the EGR data displayed on the scan tester. If the vehicle has a digital EGR valve with three separate solenoids, the scan tester data indicates when each solenoid is on or off (Figure 9-8). In some applications, the scan tester data indicates the desired EGR position, actual EGR position, and the voltage signal from the EGR pintle position sensor to the PCM.

The EGR valve should open at the specified low vehicle speed, and it should close as the throttle opening exceeds a specific value. Since the exhaust flow through the EGR valve reduces cylinder temperature, high cylinder temperature and detonation may occur if the EGR valve is inoperative. This detonation occurs at a constant throttle opening.

If the EGR valve does not open, NO_x emissions will be higher than emission standards for that year of vehicle. The NO_x emission levels can be tested with a five-gas emissions analyzer.

When the EGR valve opens at a lower speed than specified, a hesitation may occur during acceleration at this speed. Remember the scan tester data only indicates if the EGR valve is opened electronically or not. Carbon buildup under the EGR pintle may hold the EGR valve open continually. This condition results in rough idle operation and engine stalling, plus a hesitation on low-speed acceleration. Since defective input sensor data may result in improper EGR system operation, always watch for the following input sensor problems when diagnosing an EGR system:

1. A defective O_2 sensor or ECT sensor signal may result in the PCM remaining in open loop with the engine at normal operating temperature. Under this condition, the EGR valve remains inoperative.

TYPICAL SCAN DATA VALUES
3.4L (VIN X) (SFI)

Idle / Upper Radiator Hose Hot / Closed Throttle / Park or Neutral / Closed Loop / Acc. off
Brake Not Applied

Tech 1 Parameter	Units Displayed	Typical Data Value	6E3 REFERENCE SECTION
Engine Speed	RPM	650–750	C1, C2
Desired Idle	RPM	675	C1
Eng Cool Temp	°C °F	Varies (85°–105°C) (185°–220°F)	C1
Int Air Temp	°C °F	Varies with air temp.	C1
Throt Position	Volts	.20–.74	C1
Throttle Angle	Percent	0%	C1
Mass Air Flow	Gm/Sec	4.9 (varies)	C1
Calc Eng Load	Counts	70–80	C1
HO2S	mV	Varies (100–1000)	C1, C2
Rich/Lean Status	Rich/Lean	Rich/Lean Changing Constantly	C1, C2
O2S Cross Counts	Counts	Always Changing	C1, C2
Loop Status	Open/Closed	Closed Loop	C2
Fuel Trim Cell	Cell #	0	C2
S.T. Fuel Trim	Counts	95–140*	C2
L.T. Fuel Trim	Counts	95–138*	C2
Injection Pulse Width	mSEC		
Spark Advance	Degrees	16° (Varies)	C4
Spark Retard	Degrees	0°	C5
Knock Signal	Yes/No	No	C5
Idle Air Control	Counts	16–20	C2
Ign. Cntrl Cam Sig	0 or 1	0 or 1, Constantly Changing	
Fuel Evap Purge	Percent Duty Cycle	Varies	C3
Power Steering	Normal/High Pressure	Normal	C1
EGR1	Off/On	Off	C7
EGR2	Off/On	Off	C7
EGR3	Off/On	Off	C7
Current Week Cyl.	Cylinder #/None	None	C4
A/C Request	Yes/No	No	C1, C10
A/C Clutch	On/Off	Off	C10
A/C Ref.Press.	PSI	OK	C12
Fan 1	On/Off	Off	C12
Fan 2	On/Off	Off	C12
PRNDL P	Low/High	Low	C1
PRNDL A	Low/High	Low	C1
PRNDL B	Low/High	High	C1
PRNDL C	Low/High	High	C1
Trans Range Switch	Low/Drive 2/Drive 3/ Drive 4/ Neutral/Reverse/Park	Park/Neutral	C1
TCC Brake Switch	Applied/Released	Released	C8
Commanded Gear	First/Second/Third/Fourth	1st	C8
P/N	Yes/No	Yes	SECTION 4T60-E
Vehicle Speed	mph, km/h	0	C1
TCC Apply Sol.	On/Apply/Off/Release	Off	C8
TCC pwm Sol.	Percent	0%	C8
TCC Slip Speed	RPM	+254	C8
Trans. Fluid Temp.	°C °F	Varies	C8
2nd GR Start Sw.	Off/On	Off	C8
SMCC Status	Off-Disengage/On-Engage	Off	C1
SMCC Inhibited	Yes/No	Yes	C1
QDM A	High/Low	Low	Refer to DTC P1640/P1650
QDM B	High/Low	Low	Chart in Section "6E3-A"
System Voltage	Volts	Varies	C1
PASS-KEY®II Fuel	Enabled/Disabled	Enabled	DTC P1626/P1629
Calibration ID	Prom ID #	Internal I.D. only	C1
Time From Start	Minutes	Varies	C1

NOTICE: IF ALL VALUES ARE WITHIN THE RANGE ILLUSTRATED, REFER TO "SYMPTOMS," SECTION "B".

* A poor PCM ground at the Transaxle Stud. could cause short term fuel trim and long term fuel trim to read around 150, make a careful physical inspection of this critical connection.

Figure 9-8 PCM data (Courtesy of Chevrolet Motor Division, General Motors Corporation)

2. A defective input sensor may cause the PCM to be in a limp-in or backup mode. The EGR valve is inoperative in this mode.
3. An ECT sensor signal that indicates the coolant temperature is lower than the actual coolant temperature may result in an inoperative EGR valve.
4. A TPS signal that is lower than specified may cause late opening of the EGR valve in relation to throttle opening and vehicle speed.
5. A TPS signal that is higher than specified may cause the EGR valve to open too soon in relation to throttle opening and vehicle speed.
6. A defective MAP sensor signal may result in improper EGR operation.

Since defective input sensor signals result in improper EGR valve operation, these signals must be diagnosed carefully when checking EGR data on the scan tester.

Diagnosis of Port EGR Valve

The EGR system diagnostic procedure varies depending on the vehicle make and model year. Always follow the diagnostic procedure in the vehicle manufacturer's service manual.

Special Tools

Vacuum hand pump

With the engine at normal operating temperature and operating at idle speed, disconnect the vacuum hose from the EGR valve. Supply 18 in. Hg of vacuum to the valve with a vacuum hand pump, and observe the EGR diaphragm movement. In some applications, a mirror may be held under the EGR valve to see the diaphragm movement. When the vacuum is applied, the EGR valve should open, and idle operation should become very rough. If the valve diaphragm does not hold the vacuum, replace the valve. When the valve does not open, remove the valve and check for carbon in the passages under the valve. Clean the passages as required.

 CAUTION: If the engine has been operating recently, the EGR valve may be very hot. Wear protective gloves when diagnosing or servicing this valve.

 WARNING: Do not wash an EGR valve in any type of solvent. This action will damage the valve diaphragm.

 WARNING: Sandblasting an EGR valve may damage valve components and plug orifices.

Carbon may be cleaned from the lower end of the EGR valve with a wire brush, but do not immerse the valve in solvent, and do not sandblast the valve.

Diagnosis of Negative Backpressure EGR Valve

With the engine at normal operating temperature and the ignition switch off, disconnect the vacuum hose from the EGR valve, and connect a hand vacuum pump to the vacuum fitting on the valve. Supply 18 in. Hg of vacuum to the EGR valve, and observe the valve operation and the vacuum gauge. The EGR valve should open and hold the vacuum for 20 seconds. When the valve does not operate properly, replace the valve.

With 18 in. Hg supplied to the EGR valve from the hand pump, start the engine. The vacuum should drop to zero, and the valve should close. If the valve does not operate properly, replace the valve.

Diagnosis of Positive Backpressure EGR Valve

With the engine at normal operating temperature and running at idle speed, disconnect the vacuum hose from the EGR valve. Connect a hand vacuum pump to the EGR valve vacuum fitting and operate the hand pump to supply vacuum to the valve. The vacuum should be bled off, and the EGR valve diaphragm and stem should not move. If the EGR valve does not operate properly, replace the valve.

Disconnect the EGR vacuum supply hose from the TBI unit, and connect a long hose from this port directly to the EGR valve. Accelerate the engine to 2,000 rpm and observe the EGR valve. The valve should open at this engine speed. Allow the engine to return to idle speed. The EGR valve should close. If the EGR valve does not open properly, remove the valve and check for a plugged or restricted exhaust passage under the valve. When these passages are not restricted, replace the valve.

Digital EGR Valve Diagnosis

A digital EGR valve is operated electronically with no vacuum connections. It contains one to three solenoids that are open or closed.

The digital EGR valve may be diagnosed with a scan tester. With the engine at normal operating temperature and the ignition switch off, connect the scan tester to the DLC. Start the engine, and allow the engine to operate at idle speed. Select EGR control on the scan tester, and then energize EGR solenoid #1 with the scan tester. When this action is taken, the engine rpm should decrease slightly. The engine rpm should drop slightly as each EGR solenoid is energized with the scan tester. When the EGR valve does not operate properly, check the following items before replacing the EGR valve:

1. Check for 12 V at the power supply wire on the EGR valve (Figure 9-9).
2. Check wires between the EGR valve and the PCM.
3. Remove the EGR valve, and check for plugged passages under the valve.

Linear EGR Valve Diagnosis

The linear EGR valve diagnostic procedure varies depending on the vehicle make and model year. Always follow the recommended procedure in the vehicle manufacturer's service manual. The scan tester may be used to diagnose a linear EGR valve. The engine should be at normal operating temperature prior to EGR valve diagnosis. Since the linear EGR valve has an EVP sensor, the actual pintle position may be checked on the scan tester. The pintle position should not exceed 3% at idle speed. The scan tester may be operated to command a specific pintle position, such as 75%, and this commanded position should be achieved within 2 seconds. With the engine idling, select various pintle positions and check the actual pintle position. The pintle position should always be within 10% of the commanded position. When the linear EGR valve does not operate properly:

A linear EGR valve contains one solenoid and a valve pintle that is operated electronically. The PCM operates this type of valve with a pulse width modulation principle.

1. Check the fuse in the 12-V supply wire to the EGR valve.
2. Check for open circuits, grounds, and shorts in the wires connected from the EGR valve winding to the PCM.
3. Use a digital voltmeter to check for 5 V on the reference wire to the EVP sensor.
4. Check for excessive resistance in the EVP sensor ground wire.
5. Leave the wiring harness connected to the valve, and remove the valve. Connect a digital voltmeter from the pintle position wire at the EGR valve to ground, and manually push the pintle upward (Figure 9-10). The voltmeter reading should change from approximately 1 V to 4.5 V.

If the EGR valve does not operate properly on the scan tester, and tests 1 through 5 are satisfactory, replace the valve.

SERVICE TIP: The same quad driver in a PCM may operate several outputs. For example, a quad driver may operate the EVR solenoid and the torque converter clutch solenoid. On General Motors computers, the quad drivers sense high current flow. If a solenoid winding is shorted and the quad driver senses high current flow, the quad driver shuts off all the outputs it controls, rather than being damaged by the high current flow. When the PCM fails to operate an output or outputs, always check the resistance of the solenoid windings in the outputs before replacing the PCM. A lower-than-specified resistance in a solenoid winding indicates a shorted condition, which may explain why the PCM quad driver stops operating the outputs.

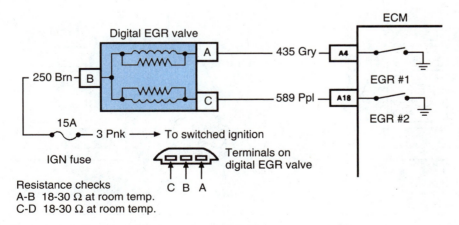

Figure 9-9 Digital EGR valve wiring diagram (Courtesy of Oldsmobile Division, General Motors Corporation)

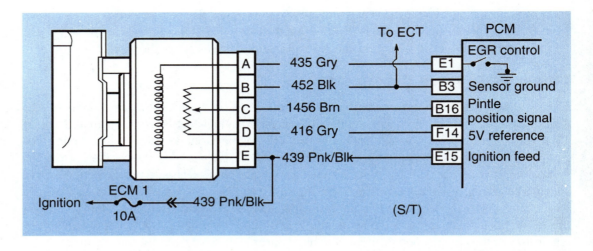

Figure 9-10 Linear EGR valve wiring diagram (Courtesy of Chevrolet Motor Division, General Motors Corporation)

EGR Vacuum Regulator (EVR) Tests

Special Tools

Digital multimeter

Connect a pair of ohmmeter leads to the EVR terminals to check the winding for open circuits and shorts (Figure 9-11). An infinite ohmmeter reading indicates an open circuit, whereas a lower-than-specified reading means the winding is shorted.

Connect the ohmmeter leads from one of the EVR solenoid terminals to ground on the solenoid case (Figure 9-12). A low ohmmeter reading indicates a grounded winding. An infinite reading indicates the winding is not grounded.

A scan tester may be used to diagnose the EVR solenoid operation. In the appropriate mode, the scan tester displays the EVR solenoid status as on or off. With the engine idling, the EVR solenoid should remain off. Road test the vehicle and drive the vehicle until the conditions required to open the EVR solenoid are present. When these conditions are present, the scan tester should indicate that the EVR solenoid is on.

Photo Sequence 7 shows a typical procedure for diagnosing an EGR vacuum regulator solenoid.

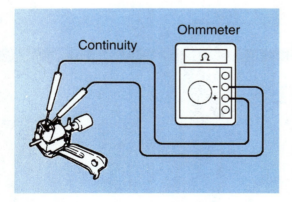

Figure 9-11 Ohmmeter connected to test the EVR solenoid winding for an open or shorted circuit (Courtesy of Toyota Motor Corporation)

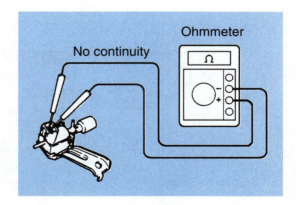

Figure 9-12 Ohmmeter connected to test the EVR solenoid winding for a grounded condition (Courtesy of Toyota Motor Corporation)

Photo Sequence 7
Typical Procedure for Diagnosing an EGR Vacuum Regulator Solenoid

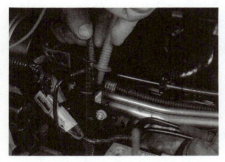

P7-1 Disconnect the EGR solenoid wiring connector and connect the ohmmeter leads to the solenoid terminals.

P7-2 Compare the ohmmeter reading to the vehicle manufacturer's specifications in the service manual.

P7-3 Connect the ohmmeter leads from one of the EGR solenoid terminals to ground. An infinite reading indicates the solenoid winding is not grounded, and a low reading indicates a grounded condition.

P7-4 Be sure the engine is at normal operating temperature and install the proper module in the scan tester for the vehicle being diagnosed.

P7-5 Plug the scan tester power cable into the cigarette lighter.

P7-6 Enter the vehicle model year and the VIN code in the scan tester.

P7-7 Connect the scan tester data cable to the DLC under the instrument panel.

P7-8 Start the engine and obtain the EGR data on the scan tester. The EGR valve should remain off.

P7-9 Drive the vehicle with the scan tester connected and the EGR data displayed. The EGR valve should be turned on at the vehicle manufacturer's specified speed.

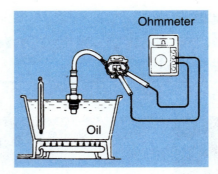

Figure 9-13 Testing the exhaust gas temperature sensor (Courtesy of Toyota Motor Corpora-

Exhaust Gas Temperature Sensor Diagnosis

Remove the exhaust gas temperature sensor, and place the sensor in a container of water. Place a thermometer in the water, and heat the container (Figure 9-13). Connect the ohmmeter leads to the exhaust gas temperature sensor terminals. The exhaust gas temperature sensor should have the specified resistance at various temperatures.

EGR Pressure Transducer (EPT) Diagnosis

CAUTION: If the engine has been running, EGR components, including the EPT, and especially the exhaust pressure supply pipe to the EPT, are hot. Wear protective gloves during diagnosis.

The EPT diagnosis varies depending on the vehicle make and model year. Always follow the instructions in the vehicle manufacturer's service manual. The following procedure is for a 1993 Toyota Camry:

1. Disconnect the vacuum hoses from ports P, Q, and R on the EPT.
2. With the ignition switch off, block ports P and R with your finger.
3. Connect a length of vacuum hose to port Q, and blow air into this port (Figure 9-14). Air should pass freely through the air filter on the side of the EPT.
4. Operate the engine at 2,500 rpm, and repeat step 3. There should be strong resistance to air flow (Figure 9-15). If the EPT does not operate properly, check for restrictions in the exhaust pressure tube to the EPT. If there are no restrictions in this tube, replace the EPT.

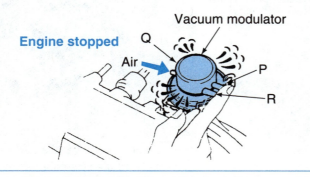

Figure 9-14 Testing EPT at idle speed (Courtesy of Toyota Motor Corporation)

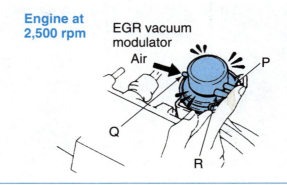

Figure 9-15 Testing EPT at 2,500 rpm (Courtesy of Toyota Motor Corporation)

Diagnosis of Specific Problems Related to Emission Systems, and Necessary Corrections

The diagnoses provided relate specifically to the emission systems explained in this chapter.

No-Start

1. Completely restricted catalytic converter—replace catalytic converter

Hard Starting

1. Low compression—perform compression test, repair engine as required
2. Lean air-fuel ratio, vacuum leak—test intake vacuum leaks, repair as necessary
3. EGR valve stuck open—replace EGR valve

Rough Idle

1. Low compression—perform compression test, repair engine as required
2. EGR valve stuck open—perform EGR valve test
3. Vacuum leak—test intake vacuum and EGR system leaks, repair as necessary

High Idle Speed

1. Vacuum leak—test and repair intake vacuum or EGR system leaks as required

Rich Air-Fuel Mixture, Low Fuel Economy, Excessive Catalytic Converter Odor

1. Low compression—test engine compression, perform engine repairs as required
2. Defective ignition—perform oscilloscope diagnosis, repair or replace components as necessary
3. PCV hose or valve restricted—replace hose or valve as required

Detonation

1. Lean air-fuel mixture—test intake vacuum leaks, repair as necessary
2. EGR valve not opening—diagnose EGR system, repair as required

Engine Stalling

EGR valve stuck open—test EGR valve and related components, repair as necessary

Engine Surging After Torque Converter Clutch Lockup

1. EGR valve opening and closing intermittently—test EGR valve and related system repair as necessary

Cylinder Misfiring

1. Low compression—test engine compression, perform engine repairs as necessary
2. Vacuum leak—test intake vacuum leaks, and correct as required

Engine Power Loss

1. Low compression—test engine compression, perform engine repairs as necessary
2. Improper EGR valve operation—test EGR valve and clean or replace as required

3. Restricted catalytic converter or exhaust system—evidenced by very low intake vacuum as engine is loaded, replace components as required

Hesitation on Acceleration

1. EGR valve opening too soon—diagnose EGR valve and system, repair as necessary

Classroom Manual
Chapter 9, page 228

CUSTOMER CARE: Customers should be advised that maintaining their vehicles according to the vehicle manufacturer's recommended maintenance schedule is one of the best ways to keep their vehicles meeting emission standards. Always advise customers when you find emission devices that are not operating properly. When customers know their vehicles are causing excessive emissions and air pollution, they are usually willing to have the necessary repairs completed. Familiarize yourself with state vehicle inspection and maintenance programs so you can advise customers regarding the necessary maintenance so their vehicles meet the standards of these programs.

Guidelines for Servicing and Diagnosing Catalytic Converters, PCV, and EGR Systems

1. Rough engine idle and surging at low speed occur if the EGR valve is open under these conditions.
2. If the EGR valve is open at idle and low speeds, the engine may hesitate on acceleration, stall after deceleration, or stall after a cold start.
3. The PCM uses information from the ECT, TPS, and MAP to operate the EGR valve.
4. Some EGR system faults will set a code in the PCM memory.
5. Scan tester data indicates whether or not the EGR valve is opened electrically.
6. When 18 in. Hg of vacuum is supplied to a port EGR valve with the engine idling, the valve should open, and the engine idle operation should become very rough.
7. When 18 in. Hg of vacuum is supplied to a negative backpressure EGR valve with the ignition switch off, the valve should open and hold the vacuum for 20 seconds. When the engine starts, the vacuum should drop to zero and the valve should close.
8. With the engine idling, if vacuum from a hand pump is supplied to a positive backpressure EGR valve, the vacuum should be bled off.
9. With manifold vacuum supplied to a positive backpressure EGR valve, the valve should open when the engine is accelerated to 2,000 rpm.
10. A scan tester may be used to command the PCM to operate each solenoid in a digital EGR valve, and the change in engine operation should be noticeable as each solenoid in the EGR valve is opened.
11. A scan tester may be used to command the PCM to provide a specific linear EGR valve opening, and the resulting EGR valve opening may be observed on the scan tester.
12. An EGR vacuum regulator (EVR) solenoid winding may be checked for open circuits, shorts, and grounds with an ohmmeter.
13. A scan tester will indicate whether the EVR solenoid is on or off during a road test.
14. An ohmmeter may be used to check an exhaust gas temperature sensor as it is heated to various temperatures in a container of water.
15. The vent port in an EGR pressure transducer (EPT) should be closed at 2,500 rpm.

CASE STUDY

A customer complained about a hesitation on acceleration on a Chevrolet truck with a 5.7-L engine. It was not necessary to road test the vehicle to experience the symptoms. Each time the engine was accelerated with the transmission in park, there was a very noticeable hesitation.

The technician removed the air cleaner and observed the injector spray pattern, which appeared normal. A fuel pressure test indicated the specified fuel pressure. The technician connected the scan tester to the DLC, and checked the PCM for DTCs. There were no DTCs in the PCM memory.

The technician checked all the sensor readings with the scan tester, and all readings were within specifications. When the engine was accelerated, the scan tester data indicated the EGR valve remained off, but the technician remembered the scan tester data only indicates whether or not the EGR valve is opened electronically.

Next the technician visually checked the operation of the EGR valve while accelerating the engine. The EGR valve remained closed at idle speed, but each time the engine was accelerated, the EGR valve moved to the wide-open position and remained there. With the hose removed from the EGR valve, the engine accelerated normally.

The technician checked the letters on top of the EGR valve and found it was a negative backpressure valve. A vacuum hand pump was connected to the EGR valve, and 18 in. Hg of vacuum were supplied to the valve with the ignition switch off. The valve opened and held the vacuum for 20 seconds. With 18 in. Hg of vacuum supplied to the EGR valve, the engine was started. The vacuum dropped slightly, but the valve remained open, indicating the exhaust pressure was not keeping the bleed valve open and the passages in the valve stem or under the valve were restricted.

The technician removed the EGR valve. Since there was no carbon in the passages under the valve, the technician replaced the valve. When the replacement EGR valve was installed, the engine accelerated normally.

Terms to Know

Diagnostic trouble code (DTC)
Digital EGR valve
EGR pressure transducer (EPT)
EGR vacuum regulator (EVR) solenoid
Exhaust gas recirculation (EGR) valve
Exhaust gas temperature sensor
Linear EGR valve
Negative backpressure EGR valve
Port EGR valve
Positive backpressure EGR valve
Positive crankcase ventilation (PCV) system

ASE Style Review Questions

1. While discussing PCV system diagnosis:
   Technician A says a restricted PCV valve may result in oil contamination of the air cleaner element.
   Technician B says a restricted PCV valve causes high NO_x emissions.
   Who is correct?
   A. A only   **C.** Both A and B
   B. B only   **D.** Neither A nor B

2. While discussing normal catalytic converter operation:
   Technician A says the temperature at the catalytic converter inlet should be higher than the temperature at the converter outlet.
   Technician B says the temperature at the catalytic converter outlet should be higher than the temperature at the converter inlet.
   Who is correct?
   A. A only   **C.** Both A and B
   B. B only   **D.** Neither A nor B

3. While discussing EGR system diagnosis:
   Technician A says if the EGR valve is open at idle and low speeds, the engine may surge during low-speed operation.
   Technician B says if the EGR valve is open at idle and low speeds, the engine may stall on deceleration.
   Who is correct?
   A. A only **C.** Both A and B
   B. B only **D.** Neither A nor B

4. While discussing EGR valve diagnosis:
   Technician A says if the EGR valve does not open, the engine may hesitate on acceleration.
   Technician B says if the EGR valve does not open, the engine may detonate on acceleration.
   Who is correct?
   A. A only **C.** Both A and B
   B. B only **D.** Neither A nor B

5. While discussing EGR valve diagnosis:
   Technician A says a defective throttle position sensor (TPS) may affect the EGR valve operation.
   Technician B says a defective engine coolant temperature (ECT) sensor may affect the EGR valve operation.
   Who is correct?
   A. A only **C.** Both A and B
   B. B only **D.** Neither A nor B

6. When discussing the diagnosis of a positive backpressure EGR valve:
   Technician A says with the engine running at idle speed, if a hand pump is used to supply vacuum to the EGR valve, the valve should open at 12 in. Hg of vacuum.
   Technician B says with the engine running at idle speed, any vacuum supplied to the EGR valve should be bled off, and the valve should not open.
   Who is correct?
   A. A only **C.** Both A and B
   B. B only **D.** Neither A nor B

7. While discussing digital EGR valve diagnosis:
   Technician A says a scan tester may be used to command the PCM to open each solenoid in the EGR valve.
   Technician B says the EGR valve should open when 18 in. Hg of vacuum are supplied to the valve at idle speed.
   Who is correct?
   A. A only **C.** Both A and B
   B. B only **D.** Neither A nor B

8. While discussing EGR vacuum regulator (EVR) diagnosis:
   Technician A says a scan tester will indicate whether the EVR is on or off.
   Technician B says the EVR winding is shorted if it has less resistance than specified.
   Who is correct?
   A. A only **C.** Both A and B
   B. B only **D.** Neither A nor B

9. While discussing exhaust gas temperature sensor diagnosis:
   Technician A says the resistance of this sensor should increase as the sensor temperature increases.
   Technician B says the resistance of this sensor should increase as the sensor temperature decreases.
   Who is correct?
   A. A only **C.** Both A and B
   B. B only **D.** Neither A nor B

10. While discussing EGR pressure transducer (EPT) diagnosis:
    Technician A says the vent in the EPT should be open at high engine rpm.
    Technician B says exhaust pressure is supplied to the top of the diaphragm in the EPT.
    Who is correct?
    A. A only **C.** Both A and B
    B. B only **D.** Neither A nor B

Table 9-1 ASE Task

Inspect and test for missing, modified, or tampered components.

Problem Area	Symptoms	Possible Causes	Classroom Manual	Shop Manual
FAILURE TO MEET STATE EMISSION TEST	Possible driveability problems such as rough idle, stalling, or hesitation on acceleration	Disconnected, modified, or tampered emission components	221	238

Table 9-2 ASE Task

Locate and utilize relevant service information.

Problem Area	Symptoms	Possible Causes	Classroom Manual	Shop Manual
DIAGNOSTIC TIME ACCURACY	Inaccurate, lengthy diagnosis	Service information not available or not utilized	224	237

Table 9-3 ASE Task

Determine appropriate diagnostic procedures based on available vehicle data and service information; determine if available information is adequate to proceed with effective diagnosis.

Problem Area	Symptoms	Possible Causes	Classroom Manual	Shop Manual
DIAGNOSTIC PROCEDURES	Improper, lengthy procedures	Service information, vehicle data not available	221	237

Table 9-4 ASE Task

Establish relative importance of displayed serial data.

Problem Area	Symptoms	Possible Causes	Classroom Manual	Shop Manual
PERFORMANCE OR ECONOMY REDUCED	Serial data not within specifications	Defective inputs or PCM-controlled emission system outputs	231	244

Table 9-5 ASE Task

Differentiate between emission control system's mechanical and electrical/electronic problems.

Problem Area	Symptoms	Possible Causes	Classroom Manual	Shop Manual
ENGINE PERFORMANCE	Rough idle, stalling	1. Low cylinder compression	234	245
		2. Vacuum leaks	234	245
		3. EGR valve open at idle	234	245

Table 9-6 ASE Task

Determine the need to diagnose emission control subsystems.

Problem Area	Symptoms	Possible Causes	Classroom Manual	Shop Manual
INOPERATIVE EMISSION SYSTEMS	EGR valve not opening, detonation	Defective EGR valve, solenoid, wires, or PCM	234	242

Table 9-7 ASE Task

Perform functional tests on emission control subsystems.

Problem Area	Symptoms	Possible Causes	Classroom Manual	Shop Manual
REDUCED PERFORMANCE	Rough idle, stalling	1. EGR valve stuck open	228	246
		2. PCV valve stuck open	224	246

Table 9-8 ASE Task

Determine the effect on tailpipe emissions caused by a failure of an emission control component or subsystem.

Problem Area	Symptoms	Possible Causes	Classroom Manual	Shop Manual
EXCESSIVE EMISSION LEVELS	High NO_x emissions	EGR valve inoperative	228	242
	High CO emissions	PCV valve or hose restricted	224	239

Table 9-9 ASE Task

Diagnose hot or cold no-starting, hard starting, poor driveability, incorrect idle speed, poor idle, flooding, hesitation, surging, engine misfire, power loss, stalling, poor mileage, dieseling, and emission problems caused by a failure of emission control components or subsystems.

Problem Area	Symptoms	Possible Causes	Classroom Manual	Shop Manual
REDUCED ECONOMY, PERFORMANCE	No-start	Completely restricted catalytic converter	221	249
	Hard starting	1. Low compression	228	249
		2. Vacuum leak, EGR stuck open	228	249
	Rough idle	1. Low compression	229	249
		2. Vacuum leak, EGR stuck open	229	249
	High idle speed	Vacuum leak	234	249
	Rich air-fuel ratio, low fuel economy, excessive catalytic converter odor	1. Low compression	224	249
		2. Defective ignition	224	249
		3. PCV valve restricted	224	249
	Engine surging after TCC lockup	EGR valve pulsing	234	249
	Cylinder misfiring	1. Low compression	235	249
		2. Vacuum leak	235	249
	Power loss	1. Low compression	236	249
		2. Improper EGR valve operation	236	249
		3. Restricted catalytic converter, exhaust system	221	249
			222	249
	Hesitation on acceleration	EGR valve opening too soon	230	250

Emission System Diagnosis, Part II: Five-Gas Emission Diagnosis and IM240 Failure Corrections

CHAPTER 10

Upon completion and review of this chapter, you should be able to:

❏ Inspect and test for missing, modified, or tampered components.

❏ Establish the relative importance of displayed serial data.

❏ Differentiate between emission control systems' mechanical and electrical/electronic problems.

❏ Determine the need to diagnose emission control subsystems.

❏ Perform functional tests on emission control subsystems.

❏ Determine the effect on tailpipe emissions caused by a failure of an emission control component or subsystem.

❏ Diagnose hot or cold no-starting, hard starting, poor driveability, incorrect idle speed, poor idle, flooding, hesitation, surging, engine misfire, power loss, stalling, poor mileage, dieseling, and emission problems caused by a failure of emission control components or subsystems.

❏ Determine the root cause of failures.

❏ Determine the root cause of multiple failures.

❏ Determine the root cause of repeated component failures.

❏ Utilize test instruments to observe, recognize, and interpret electrical/electronic signals.

❏ Evaluate HC, CO, NO_X, CO_2, and O_2 gas readings; determine the failure relationships.

❏ Analyze HC, CO, NO_X, CO_2, and O_2 readings; determine diagnostic test procedures.

❏ Diagnose the cause of HC emission failures.

❏ Diagnose the cause of CO emission failures.

❏ Diagnose the cause of NO_X emission failures.

❏ Diagnose the cause of IM240 HC failures.

❏ Diagnose the cause of IM240 CO failures.

❏ Diagnose the cause of IM240 NO_X failures.

❏ Evaluate emission readings obtained during an IM240 test to assist in emission failure diagnosis and repair.

❏ Verify the effectiveness of repairs in preparation for an IM240 retest.

❏ Diagnose the causes of evaporative emission system pressure test failures.

❏ Diagnose the causes of evaporative emission system purge flow test failures.

Preliminary Emission System Inspection

All inspection, diagnosing, and servicing procedures in this chapter apply to the emission systems described in this chapter. Procedures related to other systems are explained in Chapter 9. Before diagnosing an emission system, inspect the emission components and systems for missing parts, modification, and tampering.

Inspect all emission systems for vacuum hoses that are disconnected, restricted, improperly connected, leaking, cracked, deteriorated, or contacting hot components. Check the electrical terminals on all emission systems for corrosion-free, tight connections. Check all emission systems for modified or damaged components.

Check the pulsed secondary air injection system for burned or leaking pipes and hoses. A secondary air injection system has a belt-driven air pump; check the belt condition and tension.

Basic Tools

Basic technician's tool set

Service manual

Check all the vacuum hoses in this system for proper connections and hose condition. Be sure all electrical connections are tight and properly connected. Inspect all the air hoses and manifolds for proper connections, leaks, or indication of burning. Burned hoses are usually caused by defective one-way check valves. Be sure the pump inlet filter is in satisfactory condition.

Carefully inspect the condition and proper connection of all vacuum hoses and purge hoses in the EVAP control system. Be sure the electrical connections are secure on the purge solenoid. Inspect the vacuum and pressure valves in the gasoline filler cap. If the canister has a replaceable filter, be sure this filter is in satisfactory condition.

Inspect the spark control system for disconnected wires and clean, tight electrical connections. Check the vacuum decel valve for proper, tight vacuum hose and air hose connections. Inspect the throttle kicker diaphragm for proper vacuum hose connections, and be sure the electrical connections are tight on the idle stop solenoid.

If the engine is equipped with a heat riser valve, be sure this valve closes when the engine is cold and moves to the open position as the engine approaches normal operating temperature. Inspect all vacuum hoses and electrical connections in the system. When the engine is equipped with a heated air inlet system, be sure the air door closes the cold air inlet during cold engine operation. As the engine warms up, the air door should open the cold air inlet. Inspect all vacuum hoses in the heated air inlet system.

Pulsed Secondary Air Injection System Diagnosis

☑ SERVICE TIP: If the metal container or the clean air hose from the container show evidence of burning, some of the one-way check valves are allowing exhaust into this container and clean air hose.

Classroom Manual
Chapter 9, page 241

A defective pulsed air injection system causes high HC and CO emissions, particularly at lower engine speeds. Check all the hoses and pipes in the system for looseness and rusted or burned conditions. Remove the clean air hose from the air cleaner, and start the engine. With the engine idling, there should be steady audible pulses at the end of the hose. If these pulses are erratic, check for cylinder misfiring. When the cylinders are not misfiring, check for sticking one-way check valves or restricted exhaust inlet air tubes in the exhaust manifold.

Secondary Air Injection System Service and Diagnosis

General AIR System Diagnosis

If the AIR pump system does not pump air into the catalytic converter when the engine is at normal operating temperature, HC and CO emissions are higher than normal. If the AIR system does not pump air into the exhaust ports during engine warm-up, HC emissions are high during this mode, and the O_2 sensor or sensors takes longer to reach normal operating temperature. Under this condition, the PCM remains in open loop longer. Since the air-fuel ratio is richer in open loop, fuel economy is reduced.

When the AIR system pumps air into the exhaust ports with the engine at normal operating temperature, the additional air in the exhaust stream causes lean signals from the O_2 sensor or sensors. The PCM responds to these lean signals by providing a rich air-fuel ratio from the injectors, thereby increasing fuel consumption and CO emissions.

The first step in diagnosing a secondary air injection system is to check all vacuum hoses and electrical connections in the system. Many AIR system pumps have a centrifugal filter behind the pul-

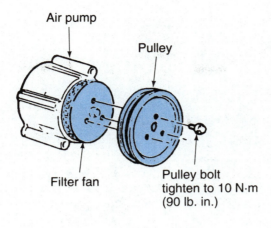

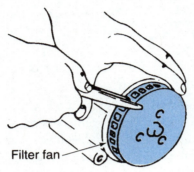

Figure 10-1 AIR pump pulley and centrifugal filter (Courtesy of Chevrolet Motor Division, General Motors Corporation)

ley. Air flows through this filter into the pump, and the filter keeps dirt out of the pump. The pulley and filter are bolted to the pump shaft, and they are serviced separately (Figure 10-1). If the pulley or filter is bent, worn, or damaged, it should be replaced. The pump assembly is usually not serviced.

The AIR pump belt must have the specified tension. A loose belt or a defective AIR system may result in high emission levels and/or excessive fuel consumption.

In some AIR systems, pressure relief valves are mounted in the AIRB and AIRD valves. Other AIR systems have a pressure relief valve in the pump. If the pressure relief valve is stuck open, air flow from the pump is continually exhausted through this valve, which causes high tailpipe emissions.

If the hoses in the AIR system show evidence of burning, the one-way check valves are leaking, which allows exhaust to enter the system. Leaking air manifolds and pipes result in exhaust leaks and excessive noise.

Some AIR systems will set DTCs in the PCM if there is a fault in the AIRB or AIRD solenoids and related wiring. In some AIR systems, DTCs are set in the PCM memory if the airflow from the pump is continually upstream or downstream. Always use a scan tester to check for any DTCs related to the AIR system. Correct the causes of these codes before proceeding with further system diagnosis.

AIR System Component Diagnosis

AIRB Solenoid and Valve Diagnosis. When the engine is started, listen for air being exhausted from the AIRB valve for a short period of time. If this air is not exhausted, remove the vacuum hose from the AIRB and start the engine. If air is now exhausted from the AIRB valve, check the AIRB solenoid and connecting wires. When air is still not exhausted from the AIRB valve, check the air supply from the pump to the valve. If the air supply is available, replace the AIRB valve.

Upstream air refers to airflow from the AIR pump to the exhaust ports.

Downstream air refers to airflow from the AIR pump to the catalytic converters.

Special Tools

Scan tester

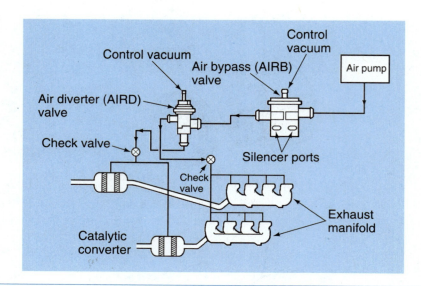

Figure 10-2 AIR system (Reprinted with the permission of Ford Motor Company)

During engine warm-up, remove the hose from the AIRD valve to the exhaust ports, and check for airflow from this hose (Figure 10-2). If airflow is present, the system is operating normally in this mode. When air is not flowing from this hose, remove the vacuum hose from the AIRD valve and connect a vacuum gauge to this hose. If vacuum is above 12 in. Hg, replace the AIRD valve. When the vacuum is zero, check the vacuum hoses to the AIRD solenoid, the AIRB solenoid, and the connecting wires.

SERVICE TIP: With the engine at normal operating temperature, the AIR system sometimes goes back into the upstream mode with the engine idling. It may be necessary to increase the engine speed to maintain the downstream mode.

With the engine at normal operating temperature, disconnect the air hose between the AIRD valve and the catalytic converters and check for airflow from this hose. When airflow is present, system operation in the downstream mode is normal. If there is no airflow from this hose, disconnect the vacuum hose from the AIRD valve and connect a vacuum gauge to the hose. When the vacuum gauge indicates zero vacuum, replace the AIRD valve. If some vacuum is indicated on the gauge, check the hose routing to the AIRD solenoid, the AIRB solenoid, and the connecting wires. When a defective secondary air injection system does not pump air into the exhaust ports during engine warm-up, HC emissions are high.

Classroom Manual
Chapter 9, page 242

Evaporative (EVAP) System Diagnosis and Service

General EVAP System Diagnosis

CAUTION: Gasoline vapors are extremely explosive. Do not smoke or allow any other sources of ignition near EVAP system components. A gasoline vapor explosion may result in personal injury and/or property damage.

CAUTION: If gasoline odor occurs in or around a vehicle, check the EVAP system for cracked or disconnected hoses, and check the fuel system for leaks. Gasoline leaks or escaping vapors may result in an explosion, causing personal injury and/or property damage. The cause of fuel leaks or fuel vapor leaks should be repaired immediately.

A leaking or inoperative EVAP system allows gasoline vapors to escape to the atmosphere. These vapors contain HC emissions. When the EVAP system allows vapor purging at idle or very low vehicle speeds, rough engine idle operation may be experienced. Since rough idle may be caused by low compression and intake vacuum leaks, always verify these items.

EVAP system diagnosis varies depending on the vehicle make and model year. Always follow the service and diagnostic procedure in the vehicle manufacturer's service manual. If the EVAP system is purging vapors from the charcoal canister when the engine is idling or operating at very low speed, rough engine operation will occur, especially at higher atmospheric temperatures. Cracked hoses or a canister saturated with gasoline may allow gasoline vapors to escape to the atmosphere, resulting in gasoline odor in and around the vehicle.

All the hoses in the EVAP system should be checked for leaks, restrictions, and loose connections. The electrical connections in the EVAP system should be checked for looseness, corroded terminals, and worn insulation. When a defect occurs in the canister purge solenoid and related circuit, a DTC is usually set in the PCM memory. If a DTC related to the EVAP system is set in the PCM memory, always correct the cause of this code before further EVAP system diagnosis.

A scan tester may be used to diagnose the EVAP system. In the appropriate scan tester mode, the tester indicates whether the purge solenoid is on or off. Connect the scan tester to the DLC and start the engine. With the engine idling, the purge solenoid should be off. Leave the scan tester connected and road test the vehicle. Be sure all the conditions required to energize the purge solenoid are present, and observe this solenoid status on the scan tester. The tester should indicate the purge solenoid is on when all the conditions are present for canister purge operation. If the purge solenoid is not on under the necessary conditions, check the power supply wire to the solenoid, solenoid winding, and the wire from the solenoid to the PCM. A defective input such as an ECT sensor may cause improper operation of the purge solenoid. Always verify proper input sensor operation from the serial data on the scan tester.

EVAP System Component Diagnosis

The canister purge solenoid winding may be checked with an ohmmeter in the same way as the EVR solenoid. With the tank pressure control valve removed, try to blow air through the valve with your mouth from the tank side of the valve (Figure 10-3). Some restriction to airflow should be felt until the air pressure opens the valve. Connect a vacuum hand pump to the vacuum fitting on

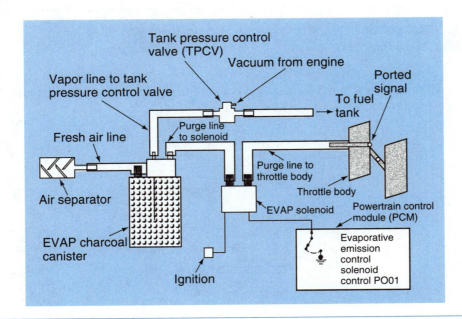

Figure 10-3 EVAP system (Courtesy of Cadillac Motor Car Division, General Motors Corporation)

Special Tools

Vacuum hand pump

the valve, and apply 10 in. Hg to the valve. Now try to blow air through the valve from the tank side. Under this condition, there should be no restriction to airflow. If the tank pressure control valve does not operate properly, replace the valve.

If the fuel tank has a pressure and vacuum valve in the filler cap, check these valves for dirt contamination and damage. The cap may be washed in clean solvent. When the valves are sticking or damaged, replace the cap.

When the charcoal canister has a replaceable filter, check the filter for dirt contamination. Replace the filter as required.

EVAP System Thermal Vacuum Valve (TVV) Diagnosis

Follow these steps for TVV diagnosis:

1. Drain the coolant from the radiator into a suitable container.
2. Remove the TVV from the water outlet.
3. Place the TVV and a thermometer in a container filled with water. The water temperature must be below 95°F (35°C).
4. Connect a length of vacuum hose to the upper TVV port, and try to blow air through the TVV with your mouth. The TVV should be closed, and no air should flow through the TVV (Figure 10-4). If the TVV allows airflow, it must be replaced.
5. Heat the water and observe the thermometer reading. When the water temperature is above 129°F (54°C), you should be able to blow air through the TVV. Replace the TVV if it does not open at the specified temperature.
6. Install thread sealant to two or three threads on the TVV, and install the TVV. Be sure this component is tightened to the specified torque.
7. Refill the radiator with coolant.

Special Tools

Heat-resistant container with thermometer

Classroom Manual
Chapter 9, page 244

Diagnosis of Knock Sensor and Knock Sensor Module

 WARNING: Operating an engine with a detonation problem for a sufficient number of miles may result in piston, ring, and cylinder wall damage.

If the spark control system does not reduce the spark advance, engine detonation occurs and NO_X emissions will increase. When the spark control system provides too much reduction in spark advance, engine performance and fuel economy are reduced.

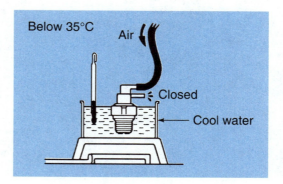

Figure 10-4 Testing TVV in EVAP system (Courtesy of Toyota Motor Corporation)

If the knock sensor system does not provide an engine detonation signal to the PCM, the engine detonates, especially on acceleration. When the knock sensor system provides excessive spark retard, fuel economy and engine performance are reduced.

The first step in diagnosing the knock sensor and knock sensor module is to check all the wires and connections in the system for loose connections, corroded terminals, and damage. With the ignition switch on, be sure 12 V are supplied through the fuse to terminal B on the knock sensor module. Repair or replace the wires, terminals, and fuse, as required.

The diagnostic procedure for the knock sensor system varies depending on the vehicle make and model year. Always follow the procedure in the vehicle manufacturer's service manual.

A quick test of the knock sensor may be performed with a scan tester connected to the DLC. Operate the engine at 2,000 rpm with the knock sensor data displayed on the tester. Tap gently on the engine block near the knock sensor with a small hammer. If the knock sensor is operational, the scan tester should indicate a knock sensor signal. On many scan testers, the word *yes* indicates a knock sensor signal.

Connect a scan tester to the DLC, and check for DTCs related to the knock sensor system. If DTCs are present, diagnose the cause of these codes. When no DTCs related to the knock sensor system are present, follow these steps to diagnose the system:

1. Connect the scan tester to the DLC, and be sure the engine is at normal operating temperature.

> **SERVICE TIP:** If the knock sensor torque is more than specified, the sensor may become too sensitive and provide an excessively high voltage signal, resulting in more spark retard than required. When the knock sensor torque is less than specified, the knock sensor signal is lower than normal, resulting in engine detonation.

2. Operate the engine at 1,500 rpm and observe the knock sensor signal on the scan tester. If a knock sensor signal is present, disconnect the wire from the knock sensor, and repeat the test at the same engine speed. If the knock sensor signal is no longer present, the engine has an internal knock, or the knock sensor is defective. When the knock sensor signal is still present on the scan tester, check the wire from the knock sensor to the knock sensor module for picking up false signals from an adjacent wire. Reroute the knock sensor wire as necessary.

3. If the knock sensor signal is not indicated on the scan tester in step 2, tap on the engine block near the knock sensor with a small hammer. When the knock sensor signal is present, the knock sensor system is satisfactory.

4. When a knock sensor signal is not present in step 3, turn the ignition switch off and disconnect the knock sensor module wiring connector. Connect a 12-V test light from 12 V to terminal D in this wiring connector (Figure 10-5). If the light is off, repair the wire connected from this terminal to ground. When the light is on, proceed to step 5.

5. Reconnect the knock sensor module wiring connector, and disconnect the knock sensor wire. Operate the engine at idle speed, and momentarily connect a 12-V test light from 12 V to the knock sensor wire. If a knock sensor signal is now generated on the scan tester, there is a faulty connection at the knock sensor, or the knock sensor is defective. When a knock sensor signal is not generated, check for faulty wires from the knock sensor to the module or from the module to the PCM. Check the wiring connections at the module. If the wires and connections are satisfactory, the knock sensor module is probably defective.

The knock sensor module may be called an electronic spark control (ESC) module.

Classroom Manual
Chapter 10, page 246

Vacuum-Operated Decel Valve Diagnosis

A defective vacuum-operated decel valve that does not open results in high CO and HC emissions on deceleration. If the vacuum-operated decel valve sticks in the open position, a rough idle problem is experienced.

A vacuum decel valve may be called a gulp valve.

263

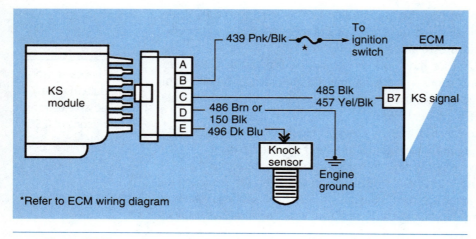

Figure 10-5 Knock sensor, knock sensor module, and wiring connections (Courtesy of Chevrolet Motor Division, General Motors Corporation)

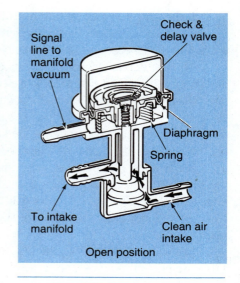

Figure 10-6 Vacuum-operated decel valve (Courtesy of Pontiac Motor Division, General Motors Corporation)

A vacuum-operated heat riser valve may be called an early fuel evaporation (EFE) system.

Classroom Manual Chapter 10, page 248

Before diagnosing the vacuum-operated decel valve, always inspect all the hoses connected to the valve for cracks, leaks, and loose connections. Disconnect the clean air hose from the valve and accelerate the engine to 2,500 rpm (Figure 10-6). Release the throttle suddenly, and listen for air intake at the clean air inlet on the valve. As the engine decelerates, there should be an audible rush of air through this clean air inlet for a few seconds. If this audible rush of air is present, the valve is operating normally. Be sure the clean air hose is not restricted before reinstalling the hose.

If an audible rush of airflow into the air inlet is not present on deceleration, connect a vacuum gauge to the vacuum signal hose on the valve. With the engine idling, there should be full manifold vacuum at the valve. If the manifold vacuum is normal at the valve signal hose, replace the valve. When the vacuum is lower than specified, check the vacuum signal hose for leaks. If this hose is satisfactory, check for intake manifold vacuum leaks and other causes of low manifold vacuum, such as late ignition timing and low engine compression.

Service and Diagnosis of Combination Throttle Kicker and Idle Stop Solenoid

A/C Idle Speed Check

The idle speed adjustment procedure varies with each vehicle, engine, or model year. Always follow the vehicle manufacturer's recommended procedure in the service manual. The following A/C idle rpm check applies to a carburetor with a combined vacuum and electric throttle kicker. The vacuum throttle kicker is activated when the driver selects the A/C mode, and the electric kicker solenoid is energized when the ignition switch is turned on. Prior to any idle speed check, the engine must be at normal operating temperature. Follow these steps for the A/C idle speed check:

1. Select the A/C mode, and set the temperature control to the coldest position.
2. Each time the A/C compressor cycles off and on, the kicker solenoid should be energized and the vacuum kicker plunger should move in and out. If the plunger reacts properly, the system is satisfactory. There is no adjustment on the vacuum kicker stem. When the vacuum kicker plunger does not react properly, check all the vacuum hoses and the solenoid, and test the vacuum kicker diaphragm for leaks.

Engine Idle Speed Check

If the idle stop solenoid is not working properly, a dieseling problem may be experienced.

Check the ignition timing, and adjust it as necessary before checking the idle speed. The engine idle speed adjustment procedure varies depending on the vehicle make and model year. Always follow the recommended procedure in the vehicle manufacturer's service manual. Follow these steps for a typical idle rpm check:

1. With the transaxle in neutral and the parking brake applied, turn all the accessories and lights off. Be sure the engine is at normal operating temperature, and connect a tachometer from the coil negative primary terminal to ground.
2. Disconnect the cooling fan motor connector, and connect 12 V to the motor terminal so the fan runs continually.
3. Remove the PCV valve from the crankcase vent module, and allow this valve to draw in underhood air.
4. Disconnect the O_2 feedback test connector on the left fender shield.
5. Disconnect the wiring connector from the kicker vacuum solenoid on the left fender shield.
6. If the idle rpm is not as specified on the underhood emission label, adjust the idle rpm with the screw on the kicker solenoid (Figure 10-7).
7. Reconnect the O_2 connector, PCV valve, and kicker solenoid connector on the left fender shield.
8. Increase the engine rpm to 2,500 for 15 seconds, and then allow the engine to idle. If the idle speed changes slightly, this is normal and a readjustment is not required.
9. Disconnect the jumper wire and reconnect the fan motor connector.

Antidieseling Adjustment

Follow these steps for a typical antidieseling adjustment:

1. Be sure the engine is at normal operating temperature and place the transaxle in neutral with the parking brake applied. Turn off all the lights and accessories.
2. Remove the red wire from the six-way carburetor connector on the carburetor side of the connector and disconnect the O_2 feedback test connector on the left fender shield.

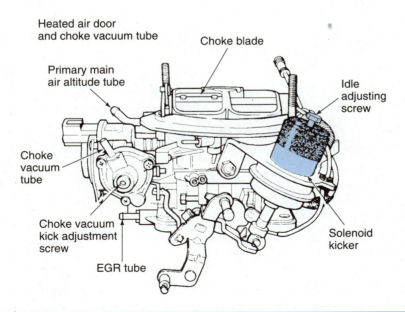

Figure 10-7 Idle speed screw on throttle kicker solenoid (Courtesy of Chrysler Corporation)

Classroom Manual
Chapter 10, page 249

3. Adjust the throttle stop screw to 700 rpm.
4. Reconnect the red wire in the six-way carburetor connector, and the O_2 feedback test connector.

Computer-Controlled Heat Riser Valve Diagnosis

Vacuum Diaphragm and Heat Riser Diagnosis

 CAUTION: If the engine has been running, heat riser valves and operating rods are extremely hot. Wear protective gloves to avoid burns when servicing these components.

If the heat riser valve does not close on a cold engine, HC and CO emissions are higher than normal during engine warm-up. If a defective heat riser remains closed with the engine at normal operating temperature, an engine power loss is experienced.

If the heat riser valve remains wide open at all engine temperatures, the engine may hesitate on acceleration during engine warm-up, and an increase in fuel consumption and emission levels may be experienced. When the heat riser valve remains closed at all engine temperatures, the engine has a loss of power on acceleration, and detonation may occur.

Disconnect the vacuum hose from the heat riser vacuum diaphragm, and connect a vacuum hand pump to the vacuum fitting on this diaphragm (Figure 10-8). Supply 18 in. Hg of vacuum to the heat riser diaphragm with the hand pump. If the diaphragm does not hold the vacuum, replace the diaphragm. If the heat riser valve does not move to the closed position, disconnect the diaphragm rod from the diaphragm and pull on the rod. If the heat riser valve is seized, squirt some penetrating oil on the ends of the valve shaft. When this action does not loosen the valve, replace the heat riser valve assembly.

Heat Riser Solenoid Diagnosis

Always check the vacuum hoses connected to the heat riser solenoid for cracks, leaks, kinks, loose connections, and deterioration. Check the wires and wiring terminals connected to this solenoid. Connect a scan tester to the DLC and check for a DTCs such as an engine coolant temperature sensor code, which could affect the operation of this system.

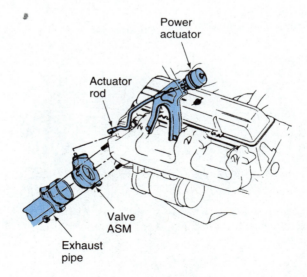

Figure 10-8 Vacuum diaphragm and heat riser valve (Courtesy of Pontiac Motor Division, General Motors Corporation)

Disconnect the heat riser solenoid wiring connector, and connect an ohmmeter to the solenoid terminals. A lower-than-specified reading indicates a shorted winding, whereas an infinite reading indicates an open winding. Connect the ohmmeter leads from one of the solenoid terminals to the solenoid case. An infinite reading indicates the winding is not grounded. A low reading indicates a grounded winding.

With the ignition switch on, use a digital voltmeter to check for 12 V at the solenoid input terminal. If 12 V are not available at this terminal, check the fuse connected in this circuit. When the fuse is satisfactory, check for an open wire between the ignition switch and the solenoid.

Remove the vacuum hoses from the heat riser solenoid, and connect a vacuum gauge to the vacuum hose connected to the intake manifold. With the engine idling, the vacuum should be above 16 in. Hg. When the vacuum is lower than 16 in. Hg, check the hose from the solenoid to the intake manifold for restrictions and leaks.

Be sure the coolant temperature is below 75°F (24°C), and connect a vacuum gauge to the vacuum fitting on the heat riser solenoid connected to the vacuum diaphragm. With the engine idling, the vacuum gauge should indicate over 16 in. Hg. If this vacuum is not present, use a jumper wire to ground the solenoid terminal connected to the PCM. When the specified vacuum is now available on the gauge, use an ohmmeter to check the wire from the solenoid to the PCM for an open circuit or a ground. If this wire is satisfactory, the PCM is probably faulty.

Special Tools
Multimeter

Special Tools
Vacuum gauge

Classroom Manual
Chapter 10, page 250

Heated Air Inlet System Diagnosis

If the air door in a heated air inlet system remains in the cold air position when the engine is cold, HC and CO emissions will be higher than normal during engine warm-up. When the air door remains in the hot air position with the engine at normal operating temperature, engine detonation may occur, resulting in an increase in NO_x emissions.

If the air door in a heated air inlet system remains in the cold air position while the engine is cold, an acceleration stumble may occur. When the engine is hot and the air door remains in the hot air position, the engine may detonate. Follow these steps to diagnose a heated air inlet system:

1. With the engine cold, install a thermometer inside the air cleaner near the bimetal sensor (Figure 10-9). Tape the thermometer to the air cleaner body to keep it from being drawn into the air intake. Install the air cleaner cover and wing nut.
2. Start the engine and observe the air door in the air cleaner snorkel. This door should be lifted upward to the hot air position.
3. Observe the air door as the engine warms up. When the air door begins to move downward to a modulated position, shut the engine off and remove the air cleaner cover. The temperature on the thermometer should be 75°F to 125°F (24°C to 52°C) if the heated air inlet system is working normally.
4. Install the air cleaner cover and observe the air door. This door should move to the cool air position as the engine approaches normal operating temperature.
5. If the air door does not move to the hot air position when the engine is cold, remove the vacuum hose from the air door vacuum motor diaphragm. Install a vacuum gauge on the

Special Tools
Thermometer

Figure 10-9 A thermometer may be installed in an air cleaner to check the heated air inlet system. (Courtesy of Mac Tools, Inc.)

267

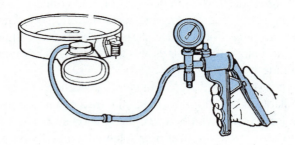

Figure 10-10 Hand vacuum pump connected to test air door vacuum motor diaphragm (Courtesy of Chrysler Corporation)

end of this hose with the engine idling and the engine cold. This vacuum reading should exceed 10 in. Hg (69 kPa). If the vacuum reading is normal, connect a hand vacuum pump to the vacuum nipple on the air door vacuum motor. Operate the hand pump to supply 10 in. Hg (69 kPa) to the vacuum motor (Figure 10-10). If this vacuum reading decreases in less than 120 seconds, replace the vacuum motor.

6. If 10 in. Hg (69 kPa) is not available at the air door vacuum motor with the engine idling, check all the system hoses for vacuum leaks and replace the hoses as necessary. To test each hose for leaks, connect one end of the hose to the vacuum hand pump and plug the opposite end of the hose. Operate the hand pump and supply 20 in. Hg (138 kPa) to the hose. If the gauge reading on the pump slowly decreases, the hose is leaking.

7. When the hoses are satisfactory, install a vacuum gauge in the vacuum hose from the intake manifold to the bimetal sensor. If the intake manifold vacuum is above 10 in. Hg, install the vacuum gauge on the bimetal sensor nipple connected to the air door vacuum motor. If the vacuum is less than 10 in. Hg (69 kPa), replace the bimetal sensor.

8. If the air door remains in the hot air position when the engine is at normal operating temperature, check all the system hoses for proper routing and connections. Check the air door to be sure it moves freely. Inspect the flexible hose and heat stove for damage and proper installation. If these system checks are satisfactory, replace the bimetal sensor.

Classroom Manual
Chapter 10, page 251

Diagnosis of Repeated Failures

Repeated emission system component failures may be caused by shorted output actuators such as a purge solenoid in the EVAP system. Intermittent opens cause sudden voltage spikes that may cause repeated failures. Some output solenoids contain voltage suppression diodes which reduce voltage spikes when the solenoid is deenergized. A defective voltage suppression diode results in voltage spikes and possible repeated component failures. Excessive radio frequency interference (RFI) may cause repeated component failures. Inaccurate diagnosis and service may cause repeated failures. For example, a PCM may fail because of a shorted purge solenoid winding. If the PCM is replaced without locating the cause of the failure in the purge solenoid, the PCM will fail again in a short time.

Diagnosis of Multiple Failures

Multiple component failures may result from improper service procedures such as reversing battery polarity with a booster battery. Arc welding on the vehicle without disconnecting the battery cables may result in multiple component failures. Intermittent open circuits or internal shorts in the wiring harness may cause multiple failures.

Diagnosis of Specific Emission-Related Problems and Necessary Corrections

The diagnoses of all the specific problems in this discussion apply to the emission systems explained in this chapter.

Hard Starting

1. Low compression—perform compression test, repair engine as required
2. Lean air-fuel ratio, vacuum leak, vacuum-operated decel valve stuck open—test intake vacuum leaks, vacuum-operated decel valve

Rough Idle

1. Low compression—perform compression test, repair engine as required
2. Vacuum leak—test intake vacuum leaks, repair as necessary
3. EVAP system purging at idle—test EVAP system with scan tester
4. Vacuum-operated decel valve stuck open—test valve, replace as required

High Idle Speed

1. Vacuum leak—test and repair intake vacuum leaks as required
2. Vacuum-operated decel valve stuck open—test valve and replace as necessary
3. Improper throttle kicker operation—test throttle kicker, repair or replace as required
4. Improper idle stop solenoid adjustment—adjust solenoid as required

Low Idle Speed

1. Improper idle stop solenoid operation or adjustment—adjust, test, and replace solenoid as required

Rich Air-Fuel Mixture, Low Fuel Economy, Excessive Catalytic Converter Odor

1. Low compression—test engine compression, perform engine repairs as required
2. Air pump air always upstream to exhaust ports with engine hot—test air pump system, repair or replace necessary components
3. EVAP system purging at idle and low speeds—test EVAP system with scan tester

Lean Air-Fuel Mixture

1. Vacuum leak—test intake manifold vacuum leaks, repair as required
2. Vacuum-operated decel valve stuck open—test and replace valve as required

Surging at Idle

1. Vacuum leak—test intake vacuum leaks, repair as necessary
2. Improper throttle kicker operation—test, repair, or replace throttle kicker

Detonation

1. Excessive spark advance—test spark advance knock sensor and connecting wires, repair or replace as required

2. Defective knock sensor or ESC module—test knock sensor module and connecting wires, repair or replace as necessary
3. Insufficient knock sensor torque—tighten knock sensor to the specified torque

Engine Stalling

1. Improper idle stop solenoid operation or adjustment—adjust, test, or replace idle stop solenoid
2. Improper throttle kicker operation—test throttle kicker operation, repair or replace as necessary
3. Vacuum-operated decel valve sticking open—test valve and replace as necessary

Engine Dieseling

1. Improper idle stop solenoid operation or adjustment—adjust, test, and replace solenoid as required

Cylinder Misfiring

1. Vacuum leak—test intake vacuum leaks, correct as required

Engine Power Loss

1. Low compression—test engine compression, perform engine repairs as necessary
2. Heat riser valve completely closed with hot engine—diagnose heat riser, repair or replace as necessary

Hesitation on Acceleration

1. Heat riser valve stuck open—diagnose, repair or replace as required
2. Heated air inlet system in cold air position with engine cold—diagnose heated air inlet system, repair as required
3. Vacuum leak—test and repair as necessary

Emissions Analyzer Testing

CAUTION: The emissions analyzer probe and sample hose may be very hot after the vehicle has been running. Wear protective gloves when handling these components.

WARNING: Regularly clean the water trap and change the filters to provide accurate analyzer readings and prevent analyzer damage.

WARNING: To dry moisture out of the tester and provide longer equipment life, operate the analyzer for 15 minutes after vehicle testing is completed.

WARNING: Do not use an emissions analyzer to sample exhaust on a diesel engine. This action will damage the analyzer.

WARNING: Remove the emissions analyzer probe from the tailpipe when using combustion chamber cleaners. If the probe is not removed under this condition, the analyzer will be damaged.

WARNING: Store the analyzer probe off the floor to avoid contamination and damage. Do not drop the analyzer probe.

Figure 10-11 Five-gas emissions analyzer (Courtesy of OTC Division, SPX Corporation)

 SERVICE TIP: Leaks in the exhaust system cause inaccurate emissions analyzer readings. Always inspect the exhaust system for leaks prior to infrared exhaust analysis.

Emissions Analyzer Calibration

Emissions analyzers have a warm-up and calibration period that is usually about 15 minutes. Modern emissions analyzers perform this calibration automatically (Figure 10-11). Always be sure the analyzer is calibrated properly so it provides accurate readings. Some older emissions analyzers had to be calibrated manually with calibration controls on the analyzer.

On most emissions analyzers, a warning light is illuminated if the exhaust flow through the analyzer is restricted because of a plugged filter, probe, or hose. This warning light must be off before proceeding with the exhaust emissions analysis.

When performing an exhaust emissions analysis, always follow the recommended procedure in the vehicle manufacturer's service manual or the equipment operator's manual. Exhaust emissions of HC, CO, and NO_X must conform to state and federal emission standards for the model year being tested.

Typical exhaust emission readings at idle speed on a catalytic converter-equipped vehicle with the AIR pump blocked off are: HC under 100 ppm, CO under 0.5%, CO_2 14.5 to 16.5%, O_2 0.5% to 3.0%, NO_X 500 to 800 ppm. On a vehicle without a catalytic converter and the AIR pump blocked off, typical emission levels are: HC under 300 ppm, CO under 3.0%, CO_2 12.5% to 14.5%, O_2 0.5% to 2.5%, NO_X 500 ppm to 800 ppm. These emission levels are not intended as accurate specifications on all vehicles. They are intended as a general guide that may help to baseline a vehicle.

Since a catalytic converter produces some CO_2, these emission levels are higher on catalytic converter-equipped vehicles than on noncatalytic converter-equipped vehicle.

Some exhaust gas analyzers read four or five emission levels plus a lambda reading. *Lambda* is a European word that is sometimes used for air-fuel ratio. On some exhaust gas analyzers, a lambda reading of 1 indicates the stoichiometric air-fuel ratio. A lambda reading above 1 indicates a lean air-

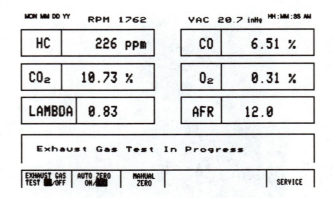

Figure 10-12 Four-gas analyzer display with lambda and air-fuel ratio readings (Courtesy of Snap-on Tools Corporation)

fuel ratio, and a reading below 1 represents a rich air-fuel ratio. Some exhaust gas analyzers also display air-fuel ratio (Figure 10-12).

Common Causes of Excessive Emissions

Excessive Hydrocarbon (HC) Emissions. Higher-than-normal HC emissions may be caused by one or more of the following conditions:

1. Ignition system misfiring—diagnose the ignition system
2. Improper ignition timing—adjust the timing
3. Excessively lean or rich air-fuel ratio—test the causes of lean or rich air-fuel ratio listed under high CO and high O_2 readings
4. Low cylinder compression—perform a compression test
5. Leaking head gasket—observe compression test results for low compression on two adjacent cylinders
6. Defective valves, guides, or lifters—observe compression test results for low compression with little increase on the four compression strokes on each cylinder
7. Defective rings, pistons, or cylinders—observe compression test results for low compression with some increase on each of the four compression strokes on each cylinder
8. Defective input sensor—diagnose the sensor input data
9. Defective heated air inlet system (if applicable, high HC during engine warm-up)—diagnose the heated air inlet system
10. Defective heat riser valve system (if applicable, high HC during engine warm-up)—diagnose the heat riser valve
11. Defective catalytic converter—test the catalytic converter
12. Defective PCM—diagnose the PCM

Excessive Carbon Monoxide (CO) Emissions. Higher-than-normal CO emissions may be caused by one of the following items:

1. Rich air-fuel ratio—test the fuel pressure; test input sensor data, air pump system for continual upstream air; and test the injectors
2. Dirty air filter—inspect the air filter, replace as necessary
3. Faulty injectors—test the injectors
4. Higher-than-specified fuel pressure—test the fuel pump pressure
5. Defective input sensor—test the input sensor data
6. Defective heated air inlet system (if applicable, high CO during engine warm-up)—test the heated air inlet system

7. Defective heat riser valve (if applicable, high CO during engine warm-up)—diagnose the heat riser valve
8. Defective secondary air injection system (always upstream)—diagnose the secondary air injection system
9. Defective catalytic converter—test the catalytic converter
10. Defective PCM—diagnose the PCM

Excessive HC and CO Emissions. When HC and CO emissions are higher than normal, check the following items:

1. Plugged PCV system—diagnose the PCV system
2. Heat riser valve stuck open (if applicable, high CO and HC during engine warm-up)—diagnose the heat riser system
3. AIR pump inoperative or disconnected—test the AIR pump system
4. Engine oil diluted with gasoline—visually inspect the engine oil.
5. Defective heated air inlet system (if applicable, high CO and HC during engine warm-up)—diagnose the heated air inlet system
6. Defective catalytic converter—test the catalytic converter
7. Defective PCM—test the PCM

Lower-than-Normal CO_2 Emissions. Lower-than-normal CO_2 levels may result from one of the following problems:

1. Exhaust gas sample dilution because of leaking exhaust system—check the exhaust system for audible leaks
2. Rich air-fuel ratio—diagnose the causes of rich air-fuel ratio listed under high CO emissions

Excessive NO_X Emissions. Higher-than-specified NO_X emissions may be caused by one of the following defects:

1. Improper EGR valve operation—diagnose the EGR valve and system
2. Detonation—check the engine compression and spark advance, diagnose the spark control system
3. Defective input sensor—diagnose the input sensor data
4. Carbon buildup in the combustion chambers—check the compression test results for higher-than-specified compression
5. Excessive spark advance—check the spark advance and spark control system
6. Excessive engine operating temperature—observe the engine temperature on the scan tester
7. Excessive air inlet temperature (if applicable)—diagnose the heated air inlet system
8. Defective catalytic converter—test the catalytic converter

Lower-than-Normal O_2 Readings and Higher-than-Normal CO Readings. Lower-than-normal O_2 readings and higher-than-normal CO readings may be caused by:

1. Rich air-fuel ratio—test the causes of rich air-fuel ratio listed under high CO emissions
2. Defective injectors, pintles not seating properly, dripping fuel—test the injectors
3. Higher-than-specified fuel pressure—test the fuel pump pressure
4. Defective input sensor—diagnose input sensor data on a scan tester
5. Restricted PCV system—diagnose the PCV system
6. Carbon canister purging at idle and low speeds—diagnose the EVAP system

Higher-than-Normal O_2 Readings and Lower-than-Normal CO Readings. Higher-than-normal O_2 readings and lower-than-normal CO readings are caused by:

1. Lean air-fuel ratio—test vacuum leaks, fuel pressure, injectors, and input sensor data

Legend
L = low, H = high, M = moderate

CO	CO$_2$	HC	O$_2$	POSSIBLE PROBLEM(S)
H	L	H	H	Rich mixture with ignition misfire.
H	L	H	L	Faulty thermostat or coolant sensor.
L	L	L	H	Exhaust leak after the converter.
L	H	L	H	Injector misfire, catalytic converter operating.
H	L	ML	L	Rich mixture.
H	H	H	H	Injector misfire, catalytic converter not working; combination of rich mixture and vacuum leak.
L	L	H	H	Ignition misfire; lean condition; vacuum or air leak between air flow sensor and throttle body (false air).
L	H	L	L	Good combustion efficiency and catalytic converter action.
L	H	L	M	All systems operating within tolerance; normal reading.

Figure 10-13 Causes of high emission levels (Courtesy of Snap-on Tools Corporation)

2. Vacuum leak—check for vacuum leaks visually and by squirting a small amount of oil on the suspected leak area
3. Lower-than-specified fuel pressure—test the fuel pump pressure
4. Defective injectors—test the injectors
5. Defective input sensor—test the input sensor data with a scan tester
6. AIR pump or pulsed secondary air injection system connected during infrared exhaust gas analysis—test the secondary air injection systems

The causes of high CO, HC, CO$_2$, and O$_2$ readings are summarized (Figure 10-13). To pinpoint the exact cause of the improper infrared analyzer readings, perform detailed tests of individual systems and components.

State Inspection Maintenance (I/M) Testing.

CAUTION: When performing any test with the engine running, be sure the parking brake is applied, with the automatic transmission in park or the manual transmission in neutral. Ignoring these precautions may cause personal injury and/or property damage.

CAUTION: If a vehicle is equipped with a vacuum parking brake release, disconnect and plug the vacuum hose to the release diaphragm before applying the parking brake before performing engine running tests. Personal injury and/or property damage may occur if this precaution is not followed.

The emissions analyzer may be used to check emission levels for state I/M testing. Infrared analyzers must meet specific standards such as California BAR-80 or ETI-80 to be used in state I/M testing. Some analyzers may require a software change or the addition of a printer to comply with these programs. During the emissions test, the vehicle must be operated under the conditions specified in the state emission standards.

Catalytic Converter Efficiency

The catalytic converter may be tested with the four- or five-gas analyzer. Before performing the catalytic converter test, tap on the converter with a soft hammer to check for loose internal components. Catalytic converter replacement is necessary if the internal components are loose. Check and repair any exhaust system leaks. Follow these steps to test the catalytic converter:

1. Disable the AIR pump.
2. Be sure the engine is at normal operating temperature.
3. Operate the engine at 2,000 rpm.
4. On fuel injected or computer-controlled carburetor engines, the O_2 must be 0.0 or very low, and some CO must be present. Add propane to the air intake if necessary with a propane enrichment kit to obtain some CO. On non-computer-controlled fuel systems, some O_2 is acceptable.
5. Suddenly accelerate the engine and observe the O_2. If the O_2 goes above 1.2%, the catalytic converter is defective.

Special Tools

Five-gas emissions analyzer

Secondary Air Injection (AIR) System Efficiency

Operate the engine at normal temperature and idle speed. Observe the O_2 readings, and then disconnect the AIR system. Record the O_2 level, and compare the reading to the level when the AIR system was operational. Most AIR or PAIR systems should increase the O_2 level by 2% to 5% when they are operational.

Positive Crankcase Ventilation (PCV) Test

With the engine at normal operating temperature and running at idle speed, observe the O_2 and CO readings. Remove the PCV valve from the rocker arm cover, and allow this valve to pull in fresh air. The O_2 reading should increase, and the CO reading should decrease. If the O_2 reading does not change, the PCV system is restricted. When the O_2 reading increases more than 1% or the CO reading decreases more than 1%, there are excessive blow-by gases, or the oil is contaminated with fuel.

Cylinder Misfiring and Vacuum Leaks

If a cylinder misfires, there is a large increase in HC emissions. Since CO is a by-product of combustion, it usually decreases slightly when a cylinder misfires. Many engine analyzers have an infrared emission analyzer combined with an oscilloscope and other test equipment. The engine analyzer has the capability to stop each spark plug from firing momentarily, and record the rpm drop. If a cylinder is not contributing to engine power, there is very little rpm drop when the spark plug stops firing. Some of these analyzers also record the HC change when each spark plug stops firing. When a cylinder is misfiring and not contributing to engine power, the HC emissions from that cylinder are high. Therefore, when the spark plug stops firing, there is not much change in HC emissions. A vacuum leak results in higher-than-normal O_2 levels.

Head Gasket and Combustion Chamber Leaks

Remove the radiator cap when the engine is cold. Place the infrared analyzer probe above the radiator filler neck and run the engine. Do not immerse the probe in the coolant. Head gasket or combustion chamber leaks into the cooling system result in an HC reading (Figure 10-14).

Evaporative (EVAP) System Leaks

With the engine at normal operating temperature, place the analyzer probe near any component in the EVAP system where a gasoline vapor leak is suspected. When the HC reading increases, the probe is near the gasoline vapor leak. Photo Sequence 8 shows a typical procedure for five-gas emission analysis.

Photo Sequence 8
Typical Procedure for Five-Gas Emission Analysis

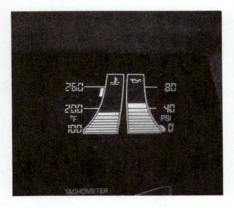

P8-1 Be sure the engine is at normal operating temperature.

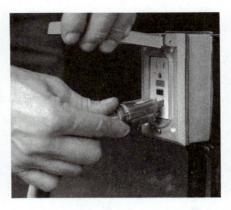

P8-2 Connect the emissions analyzer to the proper voltage supply.

P8-3 Allow the emissions analyzer to perform the automatic calibration function.

P8-4 Connect the emissions analyzer probe to the vehicle tailpipe.

P8-5 Start the engine and record the HC, CO, CO_2, O_2, and NO_x readings with the engine idling.

P8-6 Hold the engine speed at 2,500 rpm and record the HC, CO, CO_2, O_2, and NO_x readings.

P8-7 Compare the vehicle emissions to state or federal emission standards.

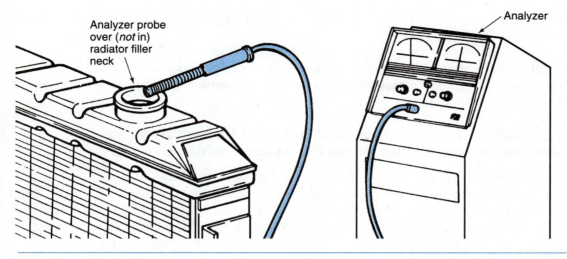

Figure 10-14 Testing for combustion chamber leaks with an emissions analyzer (Courtesy of Chrysler Corporation)

Infrared Emission Analyzer with Diagnostic Assistance

Some infrared emission analyzers provide diagnostic assistance for the technician in the tester software program. If one or more of the emission levels are high, the technician may press the appropriate tester button and request assistance in diagnosing the excessive emission level.

Classroom Manual
Chapter 10, page 253

IM240 Emission Testing

IM240 Emission Standards

Final IM240 allowable emission standards in grams per mile (gpm) are:

	HC	CO	NO_x
1983 and later	0.8	15.0	2.0
1981–1982	0.8	30.0	2.0
1980	0.8	30.0	4.0
1977–1979	3.0	65.0	4.0
1968–1976	3.0	65.0	6.0

Phase-in IM240 allowable emission standards in gpm are:

	HC	CO	NO_x
1989–1992	1.2	20.0	2.5
1983–1988	2.0	30.0	3.0
1981–1982	2.0	60.0	6.0
1980	2.0	60.0	6.0
1977–1979	7.5	130.0	6.0
1968–1975	7.5	130.0	9.5

Importance of Accurate Diagnosis

IM240 failures must be diagnosed accurately to avoid unnecessary repairs and expense. For example, replacing the catalytic converter on some vehicles without accurate diagnosis will allow the vehicle to pass the IM240 test, because the new converter is somewhat more efficient compared to the old converter. However, once the car is driven for awhile, it will fail the IM240 test because the real cause of the high emissions was not corrected.

If IM240 emission test failures are not diagnosed accurately, customers will waste millions of dollars, and customer reaction will be negative. Conversely, if IM240 diagnosis is accurate the first time, customer reaction will be positive toward enhanced emission testing, and we will all benefit from reduced air pollution.

Evaluation of O_2 Sensor and Injector Pulse Width to Determine the Cause of an Emission Failure

Special Tools
Digital storage oscilloscope (DSO)

Evaluation of O_2 Sensor. The O_2 sensor signal is the best indicator of the computer system being in proper control of the air-fuel ratio. The first step in evaluating the O_2 sensor signal is to prove this sensor is in good condition. Test the O_2 sensor switching capability with a DSO. Enrich the air-fuel ratio with propane, and create a lean air-fuel ratio with a large vacuum leak. Evaluate the capability of the O_2 sensor to switch from rich to lean, and lean to rich, in the proper time.

Evaluation of O_2 Sensor and Injector Pulse Width. If the O_2 sensor signal is valid, remove the propane cylinder and hose and install the vacuum line on the intake manifold connection. Observe the O_2 sensor signal on the DSO with the engine idling and operating at 2,500 rpm with the system in closed loop. At 2,500 rpm, the O_2 sensor should have 10 to 40 rich-lean oscillations in 10 seconds (Figure 10-15).

If the O_2 sensor shows a continually high voltage signal, the air-fuel ratio is rich. Under this condition, check the injector pulse width on the DSO. When the pulse width is normal or less than specified, check for mechanical engine problems or high fuel pump pressure. If the injector pulse width is more than specified, the rich problem is caused by a defective input or PCM. Check the input sensor data with a scan tester.

When the O_2 sensor signal is continually low, indicating a lean air-fuel ratio, check the injector pulse width on the DSO. If the injector pulse width is more than specified, there is a mechanical problem such as a vacuum leak creating a lean air-fuel ratio. Under this condition, the PCM is responding to the vacuum leak and increasing the injector pulse width in an attempt to bring the air-fuel ratio back to stoichiometric. Under this condition, always verify fuel pump pressure, which may be responsible for the lean air-fuel ratio. When the injector pulse width is normal, there is an input sensor problem or a defective PCM. For example, the O_2 sensor signal wire may be open and this signal is not reaching the PCM.

When the O_2 sensor signal displays excessive rich-to-lean oscillations, it usually indicates the cylinders have different air-fuel ratios. This condition may result from dripping injectors, a vacuum leak, or a cylinder misfiring from low compression or a defective ignition.

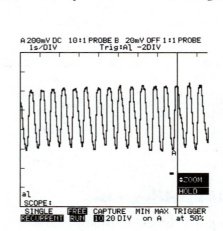

Figure 10-15 Normal O_2 sensor waveform at 2,000 rpm (Courtesy of Fluke Corporation)

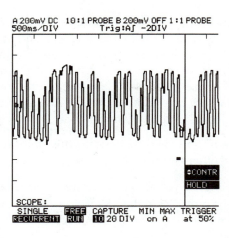

Figure 10-16 O₂ sensor waveform caused by a shortened spark plug (Courtesy of Fluke Corporation)

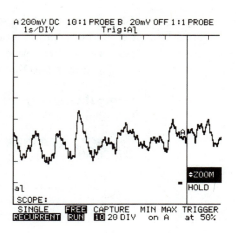

Figure 10-17 O₂ sensor waveform caused by an open secondary spark plug wire (Courtesy of Fluke Corporation)

A shorted spark plug causes excessive O₂ sensor lean-to-rich transitions on a DSO waveform (Figure 10-16). Excessive O₂ sensor lean-to-rich transitions may also be caused by an open secondary spark plug wire (Figure 10-17). A leaking injector or an inoperative injector also results in excessive lean-to-rich O₂ sensor transitions (Figures 10-18 and 10-19). When a cylinder misfires, the extra oxygen in the exhaust stream causes the excessive lean-to-rich O₂ sensor transitions. It does not matter whether the cylinder misfire is caused by an injector, ignition, or mechanical defect. With some experience, the technician should be able to recognize specific defects from the O₂ sensor waveform.

SERVICE TIP: A five-gas analyzer used with the DSO may be very helpful in diagnosing specific defects. For example, if the O₂ sensor waveform has excessive lean-to-rich transitions, and the HC emissions are considerably higher than normal, the cylinder misfire is probably caused by an ignition misfire or a mechanical defect. Since the fuel is entering the cylinder and not being burned, all the unburned HCs are flowing out of the exhaust system. When the O₂ sensor waveform has excessive transitions and the HC emissions are satisfactory, the cylinder misfire is probably caused by an injector problem. Since the fuel is not entering the cylinder with an injector defect, HC emissions do not increase.

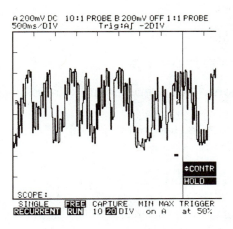

Figure 10-18 O₂ sensor waveform caused by a leaking injector (Courtesy of Fluke Corporation)

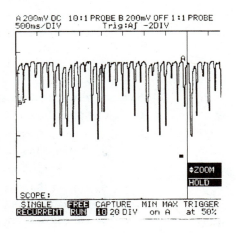

Figure 10-19 O₂ sensor waveform caused by a defective injector (Courtesy of Fluke Corporation)

When the O_2 sensor has the required number of lean-rich transitions at idle and 2,500 rpm, the PCM is in proper control of air-fuel ratio and the intake system, compression, and ignition systems are not causing any cylinder misfiring. The required number of rich-to-lean O_2 sensor transitions also indicates that the injectors are not dripping.

General Diagnosis of an IM240 Emission Failure for HC and/or CO. A general diagnostic procedure for a vehicle that failed an IM240 test for HC and/or CO is as follows:

> **SERVICE TIP:** Diagnosis of an emission failure is basically the same whether the vehicle failed an IM240 test or some other emission test procedure.

1. Baseline the vehicle emission levels using the vehicle manufacturer's specifications or the general emission level specifications provided in this chapter.
2. Evaluate the O_2 sensor condition, and replace the sensor if required.
3. Evaluate the O_2 sensor waveform. If this sensor does not have the required number of lean-to-rich transitions, check the injectors and fuel system, vacuum leaks, compression, ignition, and emission systems such as the AIR pump always pumping upstream.
4. If the O_2 sensor has the required number of lean-to-rich transitions, test the catalytic converter and the other emission systems.
5. After the necessary repairs are completed, be sure the O_2 sensor has the required number of lean-to-rich transitions. Be sure the baseline CO and HC readings are obtained on a gas analyzer.

> Baselining a vehicle means determining the maximum allowable emission levels and other specifications for that particular vehicle during a repair procedure.
>
> IM240 emission testing may be referred to as enhanced emission testing.

Excessive Hydrocarbon (HC) Emissions. An IM240 test should be run with the engine at normal operating temperature. Higher-than-normal HC emissions may be caused by one or more of the following conditions:

1. Excessively lean or rich air-fuel ratio—evaluate the O_2 sensor and injector pulse width
2. Defective input sensor—test inputs with DSO, or input data
3. Ignition system misfiring—diagnose the ignition system
4. Improper ignition timing—adjust the timing
5. Low cylinder compression—perform a compression test
6. Leaking head gasket—observe the compression test results for low compression on two adjacent cylinders
7. Defective valves, guides, or lifters—observe the compression test results for low compression with little increase on the four compression strokes on each cylinder
8. Defective rings, pistons, or cylinders—observe the compression test results for low compression with some increase on each of the four compression strokes on each cylinder
9. Defective emission systems, PCV, AIR, decel valve, and EVAP system—test the emission systems
10. Defective catalytic converter—test the catalytic converter

Excessive Carbon Monoxide (CO) Emissions. Higher-than-normal CO emissions may be caused by one of the following items:

1. Rich air-fuel ratio—evaluate the O_2 sensor and injector pulse width
2. Dirty air filter—inspect the air filter, replace as necessary
3. Faulty injectors—test the injectors
4. Higher-than-specified fuel pressure—test the fuel pump pressure
5. Defective input sensor—test inputs with DSO, or input data
6. Defective secondary air injection system (always upstream)—diagnose the secondary air injection system
7. Defective emission systems, PCV, decel valve, EVAP system—test the emission systems
8. Defective catalytic converter—test the catalytic converter

General Diagnosis of an IM240 Emission Failure for NO_X. A general diagnostic procedure for a vehicle that failed an IM240 emissions test for NO_X follows:

> ✓ **SERVICE TIP:** When testing NO_X emissions on a vehicle with a computer-controlled EGR system, be sure to operate the vehicle under the conditions required to open the EGR valve. These conditions include engine at or near normal operating temperature and a specific vehicle speed or throttle opening. This action requires raising the vehicle on a lift.

1. Baseline the vehicle for NO_X emission levels. NO_X emissions should be less than 100 ppm with the engine idling. When a vehicle is operating in the normal cruising speed range, NO_X emissions should be 100 ppm to 600 ppm. If the engine is operating at wide open throttle, or under heavy load, NO_X emissions should not exceed 1,000 ppm.
2. Operate the vehicle under the conditions necessary to open the EGR valve, and test the NO_X emission levels with a five-gas emission analyzer. If the NO_X emissions are excessive, proceed with steps 3 through 11.
3. Defective input sensor—evaluate the O_2 sensor and injector pulse width to locate the cause of the problem.
4. Improper EGR valve operation—diagnose the EGR valve and system.
5. Plugged EGR passages—remove the EGR valve and clean the passages.
6. Detonation—check engine compression and spark advance, and diagnose the spark control system.
7. Carbon buildup in the combustion chambers—check compression test results for higher-than-specified compression.
8. Excessive spark advance—check the spark advance and knock sensor system.
9. Excessive engine operating temperature—observe the engine temperature on the scan tester.
10. Excessive air inlet temperature—diagnose the heated air inlet system.
11. Defective catalytic converter—test the catalytic converter.

Diagnosing EVAP System Purge Flow and Pressure Test Failures

If the EVAP system fails the purge flow test during an IM240 test, check the EVAP system hoses for restrictions. Inspect the canister filter for restrictions. Diagnose the purge solenoid with a scan tester as explained previously.

When the EVAP system fails the pressure test, inspect the EVAP system for leaking or disconnected hoses and components. Inspect the pressure valve in the gasoline filler cap. If the EVAP system fails the IM240 test, repair or replace the necessary system components and hoses before the vehicle is retested.

Classroom Manual
Chapter 10, page 260

> ● **CUSTOMER CARE:** Like everyone else, individuals involved in the automotive service industry make mistakes. If you make a mistake that results in a customer complaint, always be willing to admit your mistake and correct it. Do not try to cover up the mistake or blame someone else. Customers are usually willing to live with an occasional mistake that is corrected quickly and efficiently.

Guidelines for Servicing and Diagnosing Emission Systems and Testing Vehicle Emissions

1. Steady audible pulses should be heard at the clean air hose in a pulsed secondary air injection system when this hose is removed from the air cleaner.
2. Some secondary air injection (AIR) systems exhaust air to the atmosphere for a brief interval when the engine is started.

3. During engine warm-up, many AIR systems deliver airflow to the exhaust ports.
4. When the engine is at normal operating temperature, many AIR systems deliver airflow to the catalytic converters.
5. In many EVAP systems, the PCM energizes the purge solenoid and provides canister purging when specific input signals such as vehicle speed above 20 mph (32 km/h) are present.
6. In some EVAP systems, vacuum is supplied to the charcoal canister by a thermal vacuum valve (TVV).
7. When the engine is operating at 1,500 rpm, if the engine is tapped near the knock sensor, a signal from this sensor should be indicated on the scan tester.
8. A vacuum-operated decel valve should pull air into the intake manifold when the engine is decelerated from 2,500 rpm.
9. Many throttle kickers provide additional throttle opening when the A/C is on.
10. The vacuum-operated heat riser valve should be observed visually to be sure it is closed when the engine is cold and open when the engine is at normal temperature.
11. The air door in the heated air inlet system should be observed to be sure it is in the hot air position when the engine is cold. This door should move to the cold air position as the engine warms up.
12. Tailpipe emission levels of HC, CO, CO_2, O_2, and NO_x are tested with a five-gas emissions analyzer.
13. IM240 emission testing is a test required by the EPA in the Clean Air Act Amendments of 1990. This is a computer-controlled test run under various operating conditions on a dynamometer.

CASE STUDY

A 1994 Ford Taurus with a 3.0-L engine failed a state emission test for HC and CO. The emission readings indicated higher-than-normal CO and HC, and lower-than-normal O_2. CO_2 was also lower than normal. The technician concluded these emission readings indicated a rich air-fuel ratio.

An O_2 sensor test with a DSO indicated this sensor was in satisfactory condition. However, the O_2 sensor did have excessive lean-to-rich transitions on the DSO waveform. The technician tested the fuel pressure and found it was within specifications. When the technician tested the PCV system, it was in normal condition. When the scan tester was connected to the DLC, all the input sensor serial data appeared normal. The output serial data also appeared to be normal. The technician realized the rich air-fuel ratio was not caused by high fuel pressure or defective inputs or outputs.

The technician performed an injector balance test and discovered three out of the six injectors had an excessive pressure drop, indicating they were not seating properly. The injectors were cleaned and the balance test repeated. All six injectors now had the same pressure drop. The O_2 sensor waveform on the DSO now indicated a normal amount of lean-to-rich transitions. When the CO, HC, and O_2 were retested, they were normal.

Terms to Know

AIR bypass (AIRB) solenoid
AIR diverter (AIRD) solenoid
Bimetal sensor
Canister purge solenoid
Downstream air
Evaporative (EVAP) system
Idle stop solenoid
IM240 emission testing
Knock sensor
Knock sensor module, or electronic spark control module
Pulsed secondary air injection system
Secondary air injection (AIR) system
Thermal vacuum valve (TVV)
Throttle kicker
Upstream air
Vacuum-operated decel valve

ASE Style Review Questions

1. A vehicle has 1,200 ppm NO_x at 2,000 rpm:
   Technician A says the MAF sensor may be defective.
   Technician B says the EGR valve passages may be restricted.
   Who is correct?
   A. A only
   B. B only
   C. Both A and B
   D. Neither A nor B

2. A vehicle fails an emissions test for high CO and HC. When the vehicle is checked in the shop, the readings obtained on a five-gas analyzer at idle speed and normal engine temperature are: HC 200 ppm, CO 2%, O_2 0.2%, CO_2 11%, and NO_x 500 ppm.
   Technician A says the AIR pump may be pumping air upstream all the time.
   Technician B says the fuel pressure may be lower than specified.
   Who is correct?
   A. A only
   B. B only
   C. Both A and B
   D. Neither A nor B

3. A vehicle has the following scan tester data at idle speed and normal engine temperature:

   ECT sensor—0.38 V   IAT sensor—2.00 V
   Engine rpm—750 HO_2S—0.8 V
   VSS sensor—0 mph   Bat voltage—14.2
   TPS sensor—0.6 V   MAP sensor—1.2 V
   IAC counts—30   TCC—off
   EGR—off   Can. purge—on
   MIL—on   loop—open
   Fuel pump relay—on

   Technician A says the O_2 may be defective.
   Technician B says the ground wire from the canister purge solenoid to the PCM may be grounded to the chassis.
   Who is correct?
   A. A only
   B. B only
   C. Both A and B
   D. Neither A nor B

4. A catalytic converter-equipped vehicle provides the following readings on a five-gas analyzer at 2,000 rpm and normal engine temperature: HC 500 ppm, CO 0.2%, CO_2 10%, O_2 0.2%, NO_x 200 ppm.
   Technician A says one cylinder may be misfiring.
   Technician B says some of the fuel injectors may be partially clogged.
   Who is correct?
   A. A only
   B. B only
   C. Both A and B
   D. Neither A nor B

5. A vehicle fails an emissions test for CO and HC. When tested in the shop with the engine at idle speed and normal temperature, the following scan tester and five-gas analyzer readings are obtained:

   HC 150 ppm, CO 2%, CO_2 11%, O_2 0.2%, NO_x 750 ppm
   ECT sensor—0.40 V   IAT sensor—1.8 V
   Engine rpm—850 HO_2S—0.8 V
   VSS sensor—0 mph   Bat voltage—14.2
   TPS sensor—0.6 V   MAP sensor—4.5 V
   IAC counts—30   TCC—off
   EGR—off   Can. purge—off
   MIL—on   Loop—open
   Fuel pump relay—on

   Technician A says the ECT sensor is defective.
   Technician B says the air cleaner is restricted.
   Who is correct?
   A. A only
   B. B only
   C. Both A and B
   D. Neither A nor B

6. A catalytic converter-equipped vehicle fails an emissions test for high NO_x, and the engine stumbles on acceleration. The scan tester data and five-gas analyzer readings obtained with the engine running at a steady 2,000 rpm and normal engine temperature are:

   HC 200 ppm, CO 0.1%, CO_2 11.5%, O_2 4%, NO_x 1100

   ECT sensor—0.45 V   Eng. knock —yes
   Engine rpm—800 HO_2S—0.4 V
   VSS sensor—0 mph   Bat voltage—14.4
   TPS sensor—1.8 V   MAP sensor—2 V
   IAC counts—50   TCC—off
   EGR—off   Can. purge—off
   MIL—off   loop—closed
   Fuel pump relay—on

   Technician A says the fuel pressure is much lower than specified.
   Technician B says the knock sensor is defective.
   Who is correct?
   A. A only   **C.** Both A and B
   B. B only   **D.** Neither A nor B

7. A catalytic converter-equipped vehicle fails an IM240 test for high CO. The CO trace printed out at the conclusion of the test indicates that CO was high only at speeds above 20 mph.
   Technician A says the intake manifold has a vacuum leak.
   Technician B says the charcoal canister is saturated with fuel.
   Who is correct?
   A. A only   **C.** Both A and B
   B. B only   **D.** Neither A nor B

8. A catalytic converter-equipped vehicle fails an emissions test for HC and CO. There are no performance complaints. When the vehicle is tested in the shop, the following emission levels and scan tester data are obtained with the engine running at 2,000 rpm and normal engine temperature:
   HC 240 ppm, CO 2.8%, CO_2 13.5%, O_2 0.4%
   ECT sensor—0.40 V   IAT sensor—1.8 V
   Engine rpm—2000 HO_2S—0.6 V
   VSS sensor—0 mph   Bat voltage—14.0
   TPS sensor—1.5 V   MAP sensor—1.8 V
   IAC counts—50   TCC—off
   EGR—off   Can. purge—off

   Technician A says the injectors are dripping.
   Technician B says the AIR pump may be not be pumping air into the catalytic converter.
   Who is correct?
   A. A only
   B. B only
   C. Both A and B
   D. Neither A nor B

9. A TBI engine fails an emissions test for NO_x, and the engine detonates on acceleration.
   Technician A says the thermostatic air cleaner is always in the hot air position.
   Technician B says there is carbon buildup in the combustion chambers.
   Who is correct?
   A. A only
   B. B only
   C. Both A and B
   D. Neither A nor B

10. The O_2 sensor voltage is low, indicating a lean air-fuel ratio, and the injector pulse width is higher than specified.
    Technician A says the PCM may be defective.
    Technician B says there may be a vacuum leak in the intake manifold.
    Who is correct?
    A. A only
    B. B only
    C. Both A and B
    D. Neither A nor B

Table 10-1 ASE Task

Inspect and test for missing, modified, or tampered components.

Problem Area	Symptoms	Possible Causes	Classroom Manual	Shop Manual
FAILURE TO MEET STATE EMISSION TEST	Possible driveability problems such as engine hesitation or reduced fuel economy	Disconnected, modified, or tampered emission and fuel system components	241	257

Table 10-2 ASE Task

Establish relative importance of displayed serial data.

Problem Area	Symptoms	Possible Causes	Classroom Manual	Shop Manual
PERFORMANCE OR ECONOMY REDUCED	Serial data not within specifications	Defective inputs, outputs, or PCM	247	263

Table 10-3 ASE Task

Differentiate between emission control systems' mechanical and electrical/electronic problems.

Problem Area	Symptoms	Possible Causes	Classroom Manual	Shop Manual
ENGINE PERFORMANCE	Rough idle	1. Low compression	242	269
		2. Vacuum leaks	242	269
		3. EVAP system purging at idle	242	269

Table 10-4 ASE Task

Determine the need to diagnose emission control subsystems.

Problem Area	Symptoms	Possible Causes	Classroom Manual	Shop Manual
INOPERATIVE EMISSION SYSTEMS	EVAP system not purging	1. Defective purge solenoid, wires, or PCM	244	260
		2. Defective input sensors	244	260

Table 10-5 ASE Task

Perform functional tests on emission control subsystems.

Problem Area	Symptoms	Possible Causes	Classroom Manual	Shop Manual
REDUCED PERFORMANCE	Rough idle	1. Defective purge solenoid	244	261
		2. Defective wires purge solenoid to PCM	244	261
		3. Defective PCM	244	261

Table 10-6 ASE Task

Determine the effect on tailpipe emissions caused by a failure of an emission contro component or subsystem.

Problem Area	Symptoms	Possible Causes	Classroom Manual	Shop Manual
EXCESSIVE EMISSIONS	High HC and CO emissions	Defective AIR, EVAP, decel, heat riser, heated air inlet system, or catalytic converter	254	273
	High NO_x emissions	Defective EGR system, spark control system, or catalytic converter	254	273

Table 10-7 ASE Task

Determine the root cause of failures.

Problem Area	Symptoms	Possible Causes	Classroom Manual	Shop Manual
PERFORMANCE OR ECONOMY REDUCED	Improper operation of an emission system	Defective inputs or emission system components	256	263

Table 10-8 ASE Task

Determine the root cause of repeated component failures.

Problem Area	Symptoms	Possible Causes	Classroom Manual	Shop Manual
SYSTEM DAMAGE	Repeated component failures	1. Shorted output actuators	256	268
		2. Intermittent open circuits	256	268
		3. Defective voltage suppression diode	256	268
		4. Radio frequency interference	256	268
		5. Inaccurate diagnosis and service	256	268

Table 10-9 ASE Task

Determine the root cause of multiple component failures.

Problem Area	Symptoms	Possible Causes	Classroom Manual	Shop Manual
SYSTEM DAMAGE	Damaged computers, inputs, outputs, or wires	1. Improper battery boost procedures	242	268
		2. Arc welding without disconnecting battery cables	246	268
		3. Intermittent open circuits	246	268
		4. Internal shorts or grounds in the wiring	246	268

Table 10-10 ASE Task

Diagnose the cause of HC emission failures.

Problem Area	Symptoms	Possible Causes	Classroom Manual	Shop Manual
HC EMISSION TEST FAILED	Engine performance, economy reduced	Defective ignition, compression, lean or rich air-fuel ratio, defective inputs	254	280

Table 10-11 ASE Task

Diagnose the cause of CO emission failures.

Problem Area	Symptoms	Possible Causes	Classroom Manual	Shop Manual
CO EMISSION TEST FAILED	Engine performance, economy reduced	Rich air-fuel ratio, high fuel pressure, defective inputs	256	280

Table 10-12 ASE Task

Diagnose the cause of NO_x emission failures.

Problem Area	Symptoms	Possible Causes	Classroom Manual	Shop Manual
NO_x EMISSION TEST FAILED	Detonation, loss of power	Improper EGR system operation, defective input	257	281

Table 10-13 ASE Task

Analyze HC, CO, NO$_x$, CO$_2$, and O$_2$ readings; determine the failure relationships.

Problem Area	Symptoms	Possible Causes	Classroom Manual	Shop Manual
TWO OR MORE EMISSIONS IMPROPER	High O$_2$ with low CO	1. Lean air-fuel ratio, vacuum leak, low pump pressure, defective injectors	254	273
		2. Defective input sensor	254	273
	High CO and HC emissions	AIR pump inoperative; PCV valve, hose plugged; defective catalytic converter	254	273

Table 10-14 ASE Task

Evaluate emission readings obtained during an IM240 test to assist in emission failure diagnosis and repair.

Problem Area	Symptoms	Possible Causes	Classroom Manual	Shop Manual
IM240 FAILURE	Possible reduced economy and performance	Improper O$_2$ sensor and injector pulse width caused by defective input, improper air-fuel ratio, or PCM	260	278

Table 10-15 ASE Task

Verify effectiveness of repairs in preparation for IM240 retest.

Problem Area	Symptoms	Possible Causes	Classroom Manual	Shop Manual
IM240 FAILURE	Corrections have been performed	Verify corrections by observing the O$_2$ sensor waveform	260	280

Table 10-16 ASE Task

Diagnose causes of evaporative emission system pressure test failures.

Problem Area	Symptoms	Possible Causes	Classroom Manual	Shop Manual
EVAP HC EMISSIONS	EVAP system pressure test failure, gasoline odor	Leaking, disconnected EVAP system hoses, components	244	281

Table 10-17 ASE Task

Diagnose causes of evaporative emission system purge flow test failures.

Problem Area	Symptoms	Possible Causes	Classroom Manual	Shop Manual
EVAP HC EMISSIONS	EVAP system purge flow test failure	Restricted EVAP system hoses, components, canister, filter	244	281

Table 10-18 ASE Task

Diagnose hard starting, rough idle, high idle speed, low idle speed, poor mileage, lean air-fuel ratio, surging, stalling, dieseling, misfiring, power loss, and hesitation and emission problems caused by a failure of emission control components or subsystems.

Problem Area	Symptoms	Possible Causes	Classroom Manual	Shop Manual
ENGINE PERFORMANCE	Hard starting	Low compression, lean air-fuel ratio	244	269
	Rough idle	Low compression, vacuum leak, defective EVAP system or decel valve	244	269
	High idle speed	Vacuum leak, improper throttle kicker operation, idle stop solenoid, or IAC motor	246	269

Table 10-18 ASE Task (continued)

Problem Area	Symptoms	Possible Causes	Classroom Manual	Shop Manual
	Low idle speed	Improper idle stop solenoid operation or IAC motor	249	269
	Rich air-fuel ratio, poor fuel economy	Low compression, AIR pump upstream, defective EVAP system	242	269
	Lean air-fuel ratio	Vacuum leak, decel valve stuck open	248	269
	Stalling	Improper idle stop solenoid, throttle kicker, or decel valve	249	270
	Dieseling	Improper idle stop solenoid or decel valve operation	249	270
	Misfiring	Vacuum leak	248	270
	Power loss	Low compression, heat riser valve stuck shut	250	270
	Hesitation on acceleration	Heat riser valve stuck open, defective heated air inlet system, vacuum leak	250	270

Table 10-19 ASE Task

Evaluate HC, CO, NO_x, CO_2, and O_2 gas readings; determine the diagnostic test sequence.

Problem Area	Symptoms	Possible Causes	Classroom Manual	Shop Manual
EMISSION TEST FAILURE	High HC emissions	Ignition or cylinder misfiring	254	272
	High CO emissions	Rich air-fuel ratio, defective input sensors	256	272
	High NO_x emissions	Defective EGR or input sensors	257	273
	High O_2 emissions	Lean air-fuel ratio, low pump pressure, defective injectors, input sensors	257	273

Table 10-20 ASE Task

Diagnose the cause of IM240 HC failures.

Problem Area	Symptoms	Possible Causes	Classroom Manual	Shop Manual
IM240 HC FAILURE	Reduced fuel economy and performance	Rich or lean air-fuel ratio, low compression, ignition misfiring, defective input or catalytic converter	260	280

Table 10-21 ASE Task

Diagnose the cause of IM240 CO failures.

Problem Area	Symptoms	Possible Causes	Classroom Manual	Shop Manual
IM240 CO FAILURE	Reduced fuel economy or performance	Rich air-fuel ratio, defective input, AIR system, or catalytic converter	260	280

Table 10-22 ASE Task

Diagnose the cause of IM240 NO_x failures.

Problem Area	Symptoms	Possible Causes	Classroom Manual	Shop Manual
IM240 NO_x FAILURE	Possible detonation	Defective input, EGR system, excessive spark advance, or catalytic converter	261	281

Servicing and Diagnosing Body Computer Systems

Upon completion and review of this chapter, you should be able to:

- Perform a preliminary diagnostic procedure on a body computer system.
- Connect a scan tester to the DLC and select body computer from the initial selections.
- Use the scan tester to select and perform system tests on the body computer menu.
- Select and read faults with the scan tester on the body computer menu.
- Select and perform actuator tests with the scan tester on the body computer menu.
- Select and perform adjustments with the scan tester on the body computer menu.
- Select and display module information with the scan tester on the body computer state menu.
- Select and test sensors with the scan tester on the body computer state menu.
- Select and test inputs and outputs with the scan tester on the body computer state menu.
- Use system monitors with the scan tester on the body computer state menu to test the C^2D bus.
- Perform a self-diagnostic test on the overhead travel information system (OTIS).
- Perform a compass variance adjustment.
- Program the remote keyless entry remote control and module.
- Diagnose power door lock systems, switches, and motors.

Preliminary Diagnostic Procedure

Basic Tools

Basic technician's tool set
Service manual
Scan tester manual
Jumper wire

Prior to any body computer system diagnosis, the engine must be at normal operating temperature. Inspect all wiring connectors for proper, tight connections, and check for wiring harness damage. Inspect all vacuum hoses for proper connections, leaks, restrictions, and deterioration. A general diagnostic procedure follows:

1. Verify the complaint; road test the vehicle if necessary.
2. Verify related symptoms.
3. Isolate the problem to a specific system or systems.
4. Diagnose the causes of the isolated problem.
5. Repair the causes of the isolated problem.
6. Verify the correction of the complaint; road test the vehicle if necessary.

Scan Tester Diagnosis of Body Computer Faults

Initial Scan Tester Selections

 WARNING: Installing an improper module in the scan tester for the vehicle or system being tested may result in electronic component or scan tester damage.

WARNING: When using the scan tester, the scan tester manufacturer's and the vehicle manufacturer's recommended service procedures must be followed. The use of other service procedures may result in electronic component or scan tester damage.

293

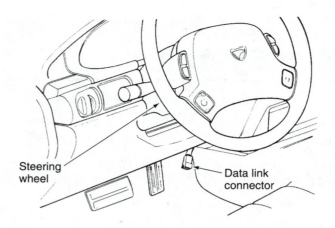

Figure 11-1 Data link connector for computer systems connected to the C²D bus (Courtesy of Chrysler Corporation)

Special Tools

Scan tester

The Chrysler collision detection (C²D) multiplex system is a twisted pair of data links interconnecting many of the onboard computers.

The proper module for body computer diagnosis must be installed in the scan tester. The scan tester is connected to the data link connector (DLC) under the instrument panel (Figure 11-1) to diagnose the body computer system and other computer systems connected to the C²D bus. Since body computer systems vary widely depending on the vehicle model and year, always follow the service procedures in the vehicle manufacturer's appropriate service manual. Scan tester diagnostic procedures may vary depending on the scan tester manufacturer. Always follow the scan tester manufacturer's recommended procedures.

SERVICE TIP: Some Chrysler products previous to 1993 have a separate DLC for the PCM in the engine compartment, and a DLC for the transmission computer and the body computer is located under the dash. These DLCs have different shapes and require different scan tester connectors.

The scan tester must be programmed for the vehicle being tested. The initial selections on the the scan tester are:

1. Engine
2. Transmission
3. Body
4. Suspension
5. ABS
6. Passive restraint
7. Theft alarm
8. System monitors
9. LH climate control

The technician presses number 3 on the scan tester to select body computer. When this selection is made, an automatic test of the C²D bus is performed, and the scan tester displays BUS TEST IN PROGRESS. If the C²D bus is satisfactory, the scan tester displays BUS OPERATIONAL. The scan tester then displays the following selections:

1. Body computer
2. Electronic cluster
3. Elect/Mech cluster
4. Engine node
5. Auto temp control
6. Info systems
7. System tests

SERVICE TIP: Some Chrysler products previous to 1993 have an engine node that is usually mounted behind the grill. This node sends compass and outside air temperature readings to the BCM. Since the compass is now contained in the overhead travel information system (OTIS), this node is no longer required. If engine node is selected on the scan tester on 1993 or later models, the scan tester displays N/A NO RESPONSE.

When the technician selects number 1 for body computer, the scan tester displays the body computer menu with the following options:

1. System tests
2. Read faults
3. Actuator tests
4. Adjustments
5. State display

Classroom Manual
Chapter 11, page 267

Body Computer Menu Tests

System Tests

If the technician presses number 1 on the scan tester to select system tests, communication between the BCM and PCM on the C²D bus is tested. If this communication is satisfactory, the scan tester displays ENGINE CONTROLLER ACTIVE ON THE BUS. When this communication is unsatisfactory, the scan tester displays NO RESPONSE.

Read Faults

When the technician presses number 2 on the scan tester to select read faults, the scan tester displays any fault messages in the BCM memory. If there are no faults in the BCM memory, the scan tester displays NO FAULTS DETECTED. Fault messages in the BCM vary depending on the vehicle model and year. The possible fault messages in the BCM memory are:

1. Ambient temperature sensor
2. ATC blend door stall failure
3. ATC mode door stall failure
4. ATC recirculation door stall failure
5. ATC head communications test
6. ATC mode door feedback failure
7. ATC blend door feedback failure
8. Battery power to BCM
9. EATX PRNDL message
10. EEPROM constant checksum
11. Engine temperature message test
12. Evaporator temperature sensor
13. In-car sensor failure
14. Internal module tests
15. Sun load sensor failure
16. Fuel level sender failure

In some body computer systems, fault messages are provided in place of two- or three-digit codes.

In the electronic automatic transaxle (EATX), all shifting is controlled by a separate EATX computer or by the PCM.

Ambient Temperature Sensor Failure. When the ambient temperature sensor input voltage is above or below a specific value for 10 seconds indicating an open or shorted sensor, the ambient temperature failure message is stored in the BCM memory. The ambient temperature sensor is located to the left of center behind the front bumper on some models.

ATC Blend Door Stall Failure. A motor stall is determined by the amount of current draw through the motor when the blend door reaches the end of its travel in either direction. This failure may be caused by the motor being disconnected electrically or by an improper mechanical connection between the motor and the door. A feedback signal is sent from each of the mode doors to the BCM. When the proper stall current is not sensed by the BCM, the fault message is set in the BCM memory.

ATC Mode Door Stall Failure. The BCM senses a mode door stall failure in the same way as it senses a blend door stall failure. The mode door stall failure message is stored in the BCM memory if the proper stall current is not available when the mode door reaches the end of its travel in either direction.

ATC Recirculation Door Stall Failure. The BCM also senses a recirculation door motor stall failure from the motor stall current at either end of the motor travel (Figure 11-2).

ATC Head Communications Test. On vehicles with an ATC system, the ATC head communications test message is set in the BCM memory if the ATC head is not transmitting messages to the BCM on the C^2D bus for 8 to 16 seconds.

ATC Mode Door Feedback Failure. When the feedback signal difference exceeds a predetermined value at the two ends of the mode door travel, or if this feedback signal is not within the proper range during normal operation, the ATC mode door feedback failure message is set in the BCM memory.

ATC Blend Door Feedback Failure. The same conditions to set the mode door feedback failure message are also required to set the blend door feedback failure message in the BCM memory.

Battery Power to Module. When the battery voltage to the BCM decreases below 5 V, the battery power to module message is set in the BCM memory.

EATX PRNDL Message. If the engine rpm is above 450 rpm and a message has not been received by the BCM from the electronic automatic transaxle (EATX) computer on the C^2D bus for 8 to 16 seconds, a possible fault condition is noted by the BCM. If this message is not received in another 10 seconds, the EATX PRNDL message is stored in the BCM memory.

EEPROM Constant Checksum. The BCM checks the EEPROM information on demand to determine if this component contains valid information. When valid information is not available from the EEPROM, the EEPROM constant checksum message is set in the BCM memory.

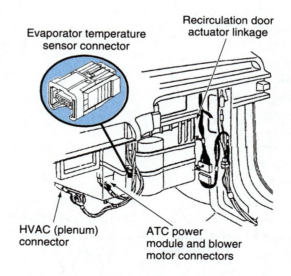

Figure 11-2 Recirculation door actuator and linkage location (Courtesy of Chrysler Corporation)

Engine Temperature Message Test. On vehicles equipped with ATC, if the engine temperature message is not received by the BCM from the PCM within 4 seconds after the ignition switch is turned on, a fault condition is noted.

Evaporator Temperature Sensor. If the evaporator temperature sensor input voltage signal is above or below a predetermined value for 10 seconds indicating an open or shorted sensor, the evaporator temperature sensor message is set in the BCM memory (Figure 11-3).

In-Car Sensor Failure. If the in-car sensor input voltage is above or below a predetermined value for 10 seconds indicating an open or shorted sensor, the in-car sensor failure message is set in the BCM memory (Figure 11-4).

Internal Module Tests. The BCM performs internal self-tests on demand to determine if the BCM microprocessor is operating properly. If this fault message is displayed, the BCM has an internal fault.

Panel Lamps Failure. When the instrument panel lamps are shorted for 10 seconds with the headlamp switch in any panel lamps position, this fault message is set in the BCM memory.

Sun Load Sensor Failure. When the sun load sensor input voltage is above or below a specific value indicating an open or shorted sensor, the sun load sensor failure message is set in the BCM memory.

Actuator Tests

If the technician presses the appropriate number on the body computer menu, the following actuators may be activated and deactivated with the scan tester:

1. Chime—on/off
2. Low washer fluid lamp—on/off
3. Wiper low speed—on/off
4. Wiper high speed—on/off
5. Door lock relay—on/off
6. Door unlock relay—on/off
7. Headlamp relay—on/off
8. Courtesy lamps—on/off
9. Key-in lamp—on/off
10. More actuators—on/off

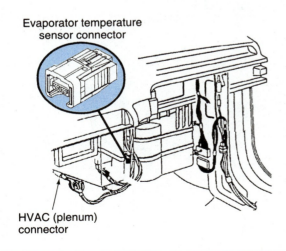

Figure 11-3 Evaporator temperature sensor connector (Courtesy of Chrysler Corporation)

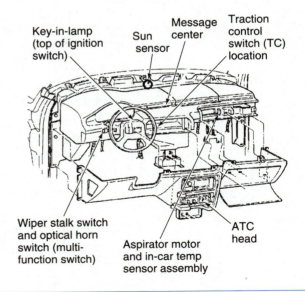

Figure 11-4 In-car temperature sensor location (Courtesy of Chrysler Corporation)

When number 10 is pressed to select more actuators, the ACT high blower relay, in-car aspirator motor, mode door, temperature air door, and recirculation door may be activated and deactivated. If any actuator cannot be activated by the scan tester, voltmeter and ohmmeter tests must be performed in that actuator system to locate the exact cause of the problem. These voltmeter and ohmmeter test procedures are in the vehicle manufacturer's service information.

Adjustments

When the technician selects adjustments on the body computer menu, the following selections appear on the scan tester display:

1. Auto door locks
2. Speed chime
3. Recalibrate ATC doors

If any of these selections are made, the BCM performs an internal calibration check of its timer function related to that output.

Body Computer State Displays

When the technician selects state displays, the following menu choices are displayed:

1. Module info
2. Sensors
3. Inputs/outputs
4. Monitors
5. Custom display

Module Info

If the technician presses number 1 on the scan tester body computer state menu to select module info, the scan tester displays information regarding the BCM such as the vehicle model year, model letter designation, and BCM number.

Sensors

When 2 is pressed on the scan tester to select sensors on the body computer state menu, the scan tester displays voltage signals from the following inputs:

1. Intermittent wipers
2. Panel lamps
3. Battery voltage
4. Fuel level
5. Ignition voltage
6. Door lock switch
7. Evaporator temperature sensor
8. In-car temperature sensor
9. Outside temperature sensor
10. Sun load temperature sensor
11. Common wire
12. Mode door
13. Temperature door
14. Door stall

Each input voltage signal should equal the vehicle manufacturer's specifications.

Inputs/Outputs

If number 3 on the body computer state menu is pressed to select inputs/outputs, the following items may be tested:

1. L/F door ajar—open/closed
2. Door ajar—open/closed
3. Key in ignition—open/closed
4. Ignition run/start—open/closed
5. Seat belt—open/closed
6. Washer fluid—low/OK
7. Windshield washer switch—open/closed
8. Wiper park switch—open/closed
9. Headlamps—on/off
10. A/C—on/off
11. Illuminated entry—on/off
12. Courtesy switch—on/off
13. Park lamps—on/off

When any of these switches or systems are activated and deactivated, the reading should change on the scan tester indicating that particular input to the BCM is satisfactory.

Monitors

If the technician selects monitors by pressing number 4 on the body computer state menu, the C^2D bus is tested for proper operation.

Custom Display

When custom display is selected by pressing 5 on the body computer state menu, the technician can view and modify certain scan tester displays.

Individual System and Component Diagnosis

Overhead Travel Information System (OTIS) Self-Diagnostic Procedure

These steps may be followed to perform an OTIS self-diagnostic procedure:

1. Be sure the ignition switch is off and press the OTIS C/T and STEP buttons simultaneously.
2. All OTIS segments must be illuminated for 2 to 4 seconds.
3. If the OTIS displays PASS after the segment illumination, the OTIS module is satisfactory.
4. If the OTIS displays FAIL after the segment illumination, the OTIS module must be replaced.
5. When the OTIS displays CCD after the segment illumination, there is an open or shorted circuit in the C^2D bus.
6. Press the C/T and STEP buttons simultaneously to exit the diagnostic mode.

Adjusting Compass Variance

The compass in the OTIS is self-calibrating and does not require any adjustment. The word *CAL* may be displayed in the OTIS, indicating the compass is in a fast calibration mode. This display is turned off when the vehicle has gone in three complete circles without stopping in an area free of

> The compass variance adjustment compensates for the difference between the geographic north pole and the magnetic north pole.

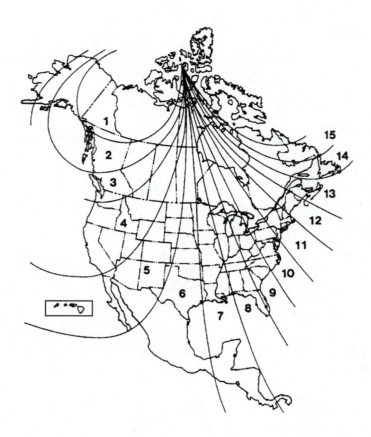

Figure 11-5 Compass variance zone numbers (Courtesy of Chrysler Corporation)

magnetic disturbance. However, the variance between the magnetic north pole and geographic north pole may cause inaccurate compass readings depending on the area in which the vehicle is driven. The procedure for setting the compass variance follows:

1. Select the proper variance number from the variance chart (Figure 11-5).
2. Turn on the ignition switch and press the OTIS C/T button to select the compass/temperature display.
3. Press and hold the reset button for 5 seconds. OTIS now displays the present variance zone entered in the OTIS module with the letters *VAR*.
4. Press the STEP button until the proper variance number appears in the display.
5. Press the RESET button to set this proper number in the OTIS memory.

Programming the Remote Keyless Entry Transmitter and Receiver Module

The remote keyless entry module may receive signals from two remote control transmitters. Each remote control has its own code, and this code must be programmed into the remote keyless entry module. When a remote control is added or replaced, the remote control code must be programmed into the module. If the module is replaced, the codes from the remote control or controls must be programmed into the new module. Each remote control contains two 3-V batteries that can be easily replaced with Duracell DL 2016 or its equivalent. The remote control should operate the door locks within 30 ft. (9 m) of the vehicle. The procedure for programming the remote control codes into the module follows:

WARNING: Never ground or short across any terminals in a computer system unless instructed to do so in the vehicle manufacturer's service manual. This action may result in component or multiple component damage.

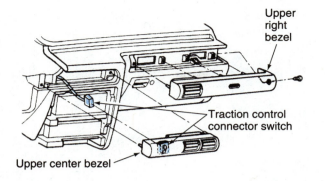

Figure 11-6 Removing center vent bezel from the instrument panel (Courtesy of Chrysler Corporation)

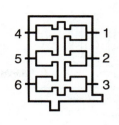

1	Ground
2	ABS module
3	Panel lamps
4	Remote keyless program wire
5	Not used
6	Not used

Figure 11-7 Remote keyless entry module programming connection (Courtesy of Chrysler Corporation)

1. Remove the center vent bezel from the instrument panel.
2. Locate the 2 × 3 traction control black connector from the traction control switch or positioned behind the bezel housing (Figure 11-6).
3. Turn on the ignition switch.
4. Connect a jumper wire from pin 4 of the traction control connector to a ground (Figure 11-7). Pin 4 is the programming line for the remote keyless entry system. The door locks will lock and unlock to indicate the module is prepared for the new code.
5. Press any button on the remote control to set the code. The door locks will cycle to verify code programming is complete. If there is a second remote control, press any button on it to complete the programming.
6. Disconnect the ground wire, and install the traction control wiring connector on the switch or on the back of the bezel.
7. Install the center vent bezel and verify the remote keyless entry system operation.

Door Lock System Tests

Switch Power Supply Test. To test the power supply to the left door power lock switch:

1. Remove the left power door lock switch from the door trim panel.
2. Remove the wiring harness connector from the switch.
3. Connect a pair of digital voltmeter leads from the red wire in the switch wiring harness to ground.
4. If the voltmeter indicates over 11 V, the power supply is satisfactory. When the voltage is below 11 V, check the 30-ampere circuit breaker in the power lock system and the connecting wires.

Special Tools

Multimeter

Power Door Lock Switch Test

The procedure for testing the power door lock switch follows:

1. Remove the switch from the door panel.
2. Connect a pair of ohmmeter leads from terminal 1 to 4 on the switch (Figure 11-8).
3. Hold the switch in the lock position; the ohmmeter should read 2,700 Ω.
4. Move the switch to the unlock position; the ohmmeter should indicate 620 Ω.

If the readings in steps 3 and 4 do not equal the specified ohm values, replace the switch.

Power Door Lock Motor Tests

When none of the power door lock motors operate, there may be a shorted motor in the system. Disconnecting the defective motor allows the other motors to operate. Test individual door lock motors as follows:

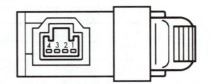

Switch position	Continuity between	Resistance value
Lock	1 and 4	2700 Ω ± 10%
Unlock	1 and 4	620 Ω ± 10%

Figure 11-8 Power door lock switch test connections (Courtesy of Chrysler Corporation)

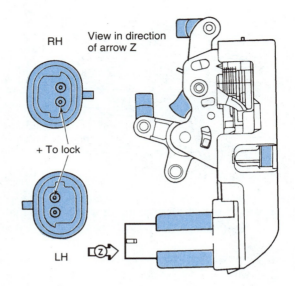

Figure 11-9 Power door lock test connections (Courtesy of Chrysler Corporation)

1. Remove the wiring connector from the door lock motor.
2. Connect a 12-V power wire to the positive terminal on the door lock motor, and ground the other terminal in the motor connector (Figure 11-9). This action should move the motor to the lock position.
3. Reverse the connections in step 2. When this connection is completed, the motor should move to the unlocked position.

Classroom Manual
Chapter 11, page 273

If the door lock motor does not perform as described in steps 2 and 3, replace the motor.

CAUTION: Before removing an electric or electronic component on an air bag-equipped vehicle, always disconnect the negative battery cable and wait the length of time specified by the vehicle manufacturer. This time is usually 1 to 2 minutes. Failure to observe this precaution may cause an accidental air bag deployment resulting in personal injury and an expensive air bag replacement.

Photo Sequence 9 shows a typical procedure for diagnosing body computer systems.

Photo Sequence 9
Typical Procedure for Diagnosing Body Computer Systems

P9-1 Be sure the engine is at normal operating temperature.

P9-2 Install the proper module in the scan tester for body computer diagnosis and the proper vehicle model and year.

P9-3 Connect the scan tester power supply wires to the battery terminals or cigarette lighter.

Photo Sequence 9
Typical Procedure for Diagnosing Body Computer Systems (continued)

P9-4 Program the scan tester for the vehicle model and year.

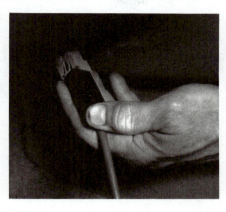

P9-5 Connect the scan tester data wire to the DLC with the proper connector.

P9-6 Select body computer and actuator test on the scan tester.

P9-7 Using the scan tester to activate and deactivate the low wiper speed.

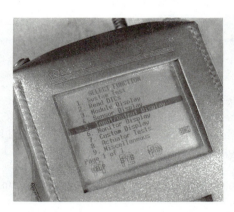

P9-8 Select body computer state tests and inputs/outputs on the scan tester.

P9-9 Display the door ajar functions on the scan tester.

P9-10 Open and close the appropriate door and the scan tester should display open or closed if the door ajar switch is operating normally.

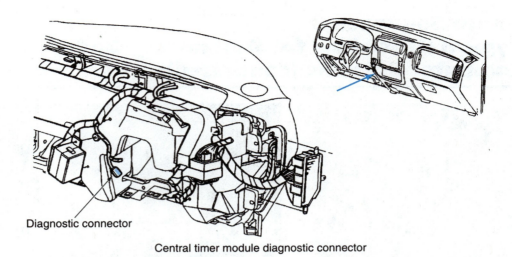

Central timer module diagnostic connector

Figure 11-10 Central timer module diagnostic connector (Reprinted with the permission of Ford Motor Company)

Diagnosis of Central Timer Module (CTM) System

Tone and Door Ajar Light Test

The tone and door ajar light test is entered by grounding the single wire in the CTM diagnostic connector under the instrument panel (Figure 11-10). The CTM immediately enters the tone and door ajar light test when this diagnostic lead is grounded without the key in the ignition switch. When this wire is grounded, the technician must observe the door ajar light and listen to the CTM tone to diagnose the door ajar light and tone circuit (Figure 11-11). The CTM completes the tone and door ajar light test in about 3 seconds. After a 1-second pause, the CTM enters the diagnostic trouble code (DTC) test.

Diagnostic Trouble Code (DTC) Test

With the CTM diagnostic lead grounded, DTCs are flashed by the door ajar light. These DTCs are two-digit codes (Figure 11-12). Code 31 is indicated by three light flashes followed by a one-second pause and one light flash. DTCs are erased if the ignition switch is turned to the start position with the CTM diagnostic connector grounded. Disconnecting the CTM wiring connector with the ignition switch off also erases DTCs.

Door Ajar Lamp	Tone	Meaning
Lights	Sounds	Door Ajar Lamp and Chime are Functioning Correctly.
Doesn't Light	Sounds	One of the Following Problems May Exist: • Door Ajar Lamp Is Burned Out. • Door Ajar Lamp Connection Is Faulty. • Door Ajar Lamp Output on CTM Is Bad, Use the Trouble Code Output Mode to Diagnose.
Lights	Does Not Sound	Chime Circuitry is Faulty.
Doesn't Light	Does Not Sound	Connection to the Diagnostic Input Is Probably Faulty. Check Connections and Try Again.

Figure 11-11 Test results, tone and door ajar light test (Reprinted with the permission of Ford Motor Company)

Code	Description
11	No trouble codes are currently stored.
12	The wiper/washer mode switch was open at least 30 seconds.
13	The wiper delay interval switch was open at least 30 seconds.
14	An internal software failure was detected.
15	An internal hardware failure was detected.
21	The wiper/high low speed relay coil shorted.
22	The wiper/brake run relay coil shorted.
23	The battery saver relay coil shorted.
24	The courtesy lamp relay coil shorted.
31	The washer pump relay coil shorted.
32	The safety belt lamp relay coil shorted.
34	The door ajar/liftgate relay coil shorted.

Figure 11-12 Central timer module diagnostic trouble codes (DTCs) (Reprinted with the permission of Ford Motor Company)

Input Test

With the CTM diagnostic lead grounded, the key must be placed in the ignition switch to enter the input test mode. In this mode, the technician turns each output on and off. When this action is taken, the door ajar light should flash once and the CTM tone should sound once if the input is received by the CTM.

Generic Electronic Module (GEM) System Diagnosis

On-Demand Self-Test

The on-demand self-test includes testing of the GEM and the related inputs and outputs. The GEM performs this self-test when the following conditions are present:

1. The ignition switch is in the run position with all the doors closed.
2. The windshield wiper multifunction switch is in the off position with the wipers parked.
3. The headlights are off and all the windows are in the up position.
4. The liftgate is closed.
5. The seat belts are unbuckled.

Diagnostic Trouble Codes (DTCs)

A scan tester must be connected to the 16-terminal data link connector (DLC) under the instrument panel to retrieve the DTCs (Figure 11-13). The GEM communicates DTC information through the ISO-9141 data links to the DLC. The DTCs are displayed on the scan tester.

Functional Tests, Active Test Modes

The scan tester can command the GEM to cycle the speed relay, wiper relay, and washer relay. When commanded by the scan tester, the GEM will cycle the door ajar light, seat belt light, and tone. The scan tester may be used to cycle the battery saver relay and illuminated entry relay. When the appropriate command is received from the scan tester, the GEM cycles the rear defrost

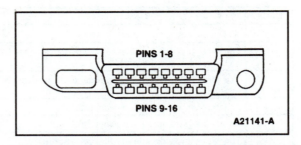

Cavity	General Assignment
1	Ignition Control
2	BUS (+) SCP
3	Discretionary (Not Used)
4	Chassis Ground
5	Signal Ground (SIG RTN)
6	Discretionary (Not Used)
7	K Line of ISO 9141
8	Discretionary (Not Used)
9	Discretionary (Not Used)
10	BUS (-) SCP
11	Discretionary (Not Used)
12	Discretionary (Not Used)
13	FEPS (Flash EEPROM)
14	Discretionary (Not Used)
15	L Line of ISO 9141
16	Battery Power

Figure 11-13 Data link connector (DLC) terminals (Reprinted with the permission of Ford Motor Company)

relay. This action causes the light in the rear defrost switch to blink on and off. The scan tester can command the GEM to cycle the driver's side window relay and the accessory delay relay.

● **CUSTOMER CARE:** When talking to customers, always remember the two *P*s, pleasant and polite. There may be many days when you do not feel like being pleasant and polite. Perhaps there are several problem vehicles in the shop with symptoms that are difficult to diagnose and correct. Some service work may be behind schedule, and customers may be irate because their vehicles are not ready on time. However, always remain pleasant and polite with customers. Your attitude does much to make customers feel better and to realize their business is appreciated. A customer may not feel very happy about an expensive repair bill, but a pleasant attitude on your part may help to improve the situation. When the two *P*s are remembered by service personnel, customer relations are enhanced, and the customers return to the shop. Conversely, if service personnel have a grouchy, indifferent attitude, customers may be turned off and take their business to another shop.

Guidelines for Diagnosing and Servicing Body Computer Systems

1. Before diagnosing a body computer system a preliminary diagnostic procedure should be performed.
2. The scan tester is connected to a DLC under the instrument panel to diagnose the body computer system.

3. Body computer is selected on the initial scan tester menu to enter the body computer diagnostics.
4. The technician may select system tests, read faults, actuator tests, adjustments, or state display on the body computer menu.
5. System tests allows the technician to test the C²D bus communication between the BCM and PCM.
6. The read faults selection on the body computer menu displays any faults in the BCM memory.
7. The actuator tests selection allows the technician to actuate many of the outputs in the BCM system.
8. Selecting adjustments on the body computer menu allows the body computer to perform a calibration of some timed functions.
9. When body computer state display is selected on the body computer menu, the technician can then select module information, sensors, inputs/outputs, monitors, or custom display.
10. If module information is selected on the body computer state menu, certain information is provided regarding the vehicle and body computer.
11. When sensors is selected on the body computer state menu, the technician can read the voltage from specific sensors.
12. If inputs/outputs is selected on the body computer state menu, the technician can obtain an open or closed signal from many of the inputs, or outputs.
13. When the technician selects monitors on the body computer state menu, a C²D bus test is performed.
14. If custom display is selected on the body computer state menu, the technician may change specific scan tester readings.
15. A self-diagnostic test may be performed on the OTIS to determine the OTIS module condition.
16. A compass variance test may be required to compensate for the difference between the geographic and magnetic north poles.
17. If the remote keyless entry module is replaced or another remote control is added, the module must be programmed to the remote control code.

CASE STUDY

A customer stated the power door locks operated intermittently on her 1995 Chrysler New Yorker LHS. The technician checked the door lock operation with the door lock switches, and then checked the automatic door lock operation while driving the vehicle. The door locks worked perfectly.

A preliminary diagnostic procedure did not reveal any loose wiring connections or vacuum hose problems. The technician connected the scan tester to the DLC and selected the body computer menu and actuator tests. When door lock relay and door lock unlock relay were selected the door locks operated properly.

The technician disconnected the scan tester, and tapped the doors near the door lock switch while operating the switch. On the driver's side the door locks did not work when the switch area was tapped. The technician removed the door lock switch from the door, and found the switch harness was loose on the switch. It appeared the harness had never been installed properly when the car was manufactured. The technician installed the wiring securely on the door lock switch and re-installed the switch. When the driver's door was tapped near the switch, the door locks continued to operate properly.

Terms to Know

Body computer menu
Body computer state menu
Chrysler collision detection (C^2D) multiplex system
Compass variance
Electronic automatic transaxle (EATX)
Overhead travel information system (OTIS)

ASE Style Review Questions

1. While discussing data link connectors (DLCs):
   Technician A says on some vehicles the DLC under the instrument panel is connected to the C^2D bus, which interconnects many of the on-board computers.
   Technician B says some older vehicles have an engine computer DLC in the engine compartment and a body computer DLC under the instrument panel.
   Who is correct?
   A. A only
   B. B only
   C. Both A and B
   D. Neither A nor B

2. While discussing body computer system diagnosis:
   Technician A says the scan tester performs an automatic test of the C^2D bus when body computer is selected from the initial menu.
   Technician B says remote keyless entry systems appears on the initial selections menu.
   Who is correct?
   A. A only
   B. B only
   C. Both A and B
   D. Neither A nor B

3. While discussing an ATC blend door stall failure fault message on the scan tester:
   Technician A says there is no feedback signal from the blend door to the BCM.
   Technician B says this message may be caused by improper mechanical connection between the blend door and the motor.
   Who is correct?
   A. A only
   B. B only
   C. Both A and B
   D. Neither A nor B

4. While discussing body computer system diagnosis:
   Technician A says the ATC head communications test message is set in the BCM memory if the ATC head is not sending messages to the BCM for a period of 8 to 16 seconds.
   Technician B says the ATC head communications test message may be caused by a defective power module in the ATC system.
   Who is correct?
   A. A only
   B. B only
   C. Both A and B
   D. Neither A nor B

5. While discussing actuator tests in a body computer system:
   Technician A says the actuator tests allow the technician to activate and deactivate the wiper high speed mode.
   Technician B says the actuator tests allow the technician to activate and deactivate the chime function in the body computer.
   Who is correct?
   A. A only
   B. B only
   C. Both A and B
   D. Neither A nor B

6. While discussing sensor tests in the body computer state menu:
   Technician A says the sensor tests allow the technician to read the voltage signals from specific sensors.
   Technician B says the sensor tests allow the technician to read the ohms resistance in specific sensors.
   Who is correct?
   A. A only
   B. B only
   C. Both A and B
   D. Neither A nor B

7. While discussing inputs and outputs selected from the body computer state menu:
   Technician A says the scan tester displays the voltage supplied from specific inputs and outputs to the BCM.
   Technician B says the scan tester displays fault messages from specific inputs and outputs.
   Who is correct?
   A. A only
   B. B only
   C. Both A and B
   D. Neither A nor B

8. While discussing the OTIS self-test:
   Technician A says the C/T and STEP buttons are pressed simultaneously to enter the self-test mode.
   Technician B says the STEP and RESET buttons are pressed simultaneously to enter the self-test mode.
   Who is correct?
   A. A only
   B. B only
   C. Both A and B
   D. Neither A nor B

9. While discussing compass variance setting:
   Technician A says the variance setting varies depending on the geographic location of the vehicle.
   Technician B says the C/T button is pressed followed by the RESET button for 5 seconds to set the compass variance.
   Who is correct?
   A. A only
   B. B only
   C. Both A and B
   D. Neither A nor B

10. While discussing remote keyless entry module programming:
    Technician A says if this module is replaced the new module must be programmed to the remote control.
    Technician B says the scan tester is used to program the remote control to the module.
    Who is correct?
    A. A only
    B. B only
    C. Both A and B
    D. Neither A nor B

Table 11-1 ASE Task

Diagnose the cause of constant, intermittent, or no operation of audible warning devices.

Problem Area	Symptoms	Possible Causes	Classroom Manual	Shop Manual
AUDIBLE WARNING TONES, CHIMES	Inoperative, constant, or intermittent chime operation	Defective inputs, wires, or body computer	282	304

Table 11-2 ASE Task

Diagnose the cause of poor, intermittent, or no operation of electric door locks.

Problem Area	Symptoms	Possible Causes	Classroom Manual	Shop Manual
POWER DOOR LOCKS	Intermittent or no power door lock operation	Switches, wires, circuit breaker, motors, or body computer	273	301

Table 11-3 ASE Task

Diagnose the cause of poor, intermittent, or no operation of keyless and remote lock/unlock devices.

Problem Area	Symptoms	Possible Causes	Classroom Manual	Shop Manual
REMOTE KEYLESS ENTRY SYSTEM	Intermittent, or no operation	1. Low batteries in remote control	274	300
		2. Defective remote control module or wiring	274	300

Table 11-4 ASE Task

Inspect, test, and repair or replace components, connectors, and wires of keyless and remote lock/unlock device circuits.

Problem Area	Symptoms	Possible Causes	Classroom Manual	Shop Manual
REMOTE KEYLESS ENTRY SYSTEM	Intermittent, or no operation	Defective remote control, module, or wiring	274	300

Electronic Instrument Cluster and Vehicle Theft Security System Diagnosis and Service

CHAPTER 12

Upon completion and review of this chapter, you should be able to:

❏ Diagnose a prove-out display on an electronic instrument cluster (EIC).
❏ Perform a function diagnostic mode on an EIC.
❏ Perform a special test mode on an EIC.
❏ Perform a preprogrammed signal check on an EIC.
❏ Diagnose the power unit on an EIC.
❏ Test the speed sensor and wiring in an EIC system.
❏ Diagnose head-up displays (HUDs).
❏ Diagnose vehicle theft security systems.

Diagnosis of a Typical Electronic Instrument Cluster

Prove-Out Display

Most EICs have a prove-out display each time the ignition switch is turned on. During this display, all the EIC segments are illuminated momentarily and then turned off momentarily (Figure 12-1). The EIC returns to normal displays after the prove-out display. If the EIC is not illuminated during the prove-out display, check the power supply to the EIC. When some of the EIC segments are not illuminated during the prove-out display, the EIC is defective and must be replaced. During the prove-out mode, the turn signal indicators and high beam indicator are not illuminated. Other indicator lights remain on when the EIC display is turned off momentarily in the prove-out mode. These indicator lights go out shortly after the EIC returns to normal displays after the prove-out mode is completed.

Function Diagnostic Mode

The diagnostic procedure for electronic instrument clusters (EICs) varies depending on the vehicle make and model year. Always follow the diagnostic procedures in the vehicle manufacturer's service manual. Some EICs have a function diagnostic mode which provides diagnostic information in the display readings if certain defects occur in the system. If the coolant temperature sender has a shorted circuit, the two top and bottom bars are illuminated in the temperature gauge, and the ISO symbol is extinguished (Figure 12-2). If the engine coolant never reaches normal operating temperature or the coolant temperature sender circuit has an open circuit, the bottom bar in the temperature gauge is illuminated with the ISO symbol.

If the fuel gauge sender develops a shorted or open circuit, the two top and bottom bars in the fuel gauge are illuminated, and the ISO symbol is not illuminated. A shorted fuel gauge sender causes *CS* to be displayed in the fuel remaining or distance to empty displays. If the fuel gauge sender has an open circuit, *CO* is displayed in the fuel remaining and distance to empty displays. When the function diagnostic mode indicates shorted or open circuits in the inputs, the cause of the problem must be located by performing voltmeter and ohmmeter tests in the circuit with the indicated problem. These voltmeter and ohmmeter tests are included in the vehicle manufacturer's service information.

Basic Tools

Basic technician's tool set
Service manual
Scan tester manual
Glass cleaner

Each time the ignition switch is turned on, all EIC segments are illuminated in a prove-out display.

International standards organization (ISO) symbols are universal symbols recognized in many countries to help overcome the language barrier.

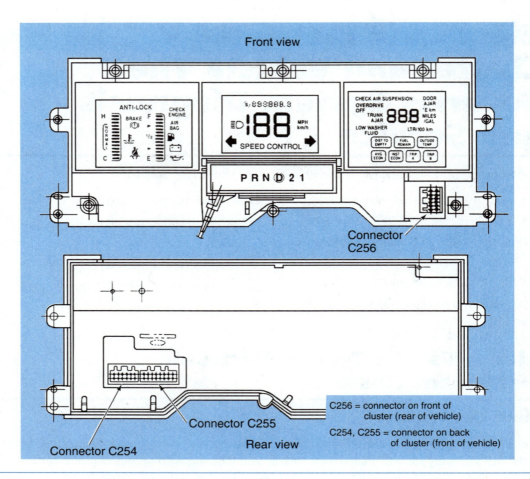

Figure 12-1 All EIC segments are illuminated during the prove-out display. (Reprinted with the permission of Ford Motor Company)

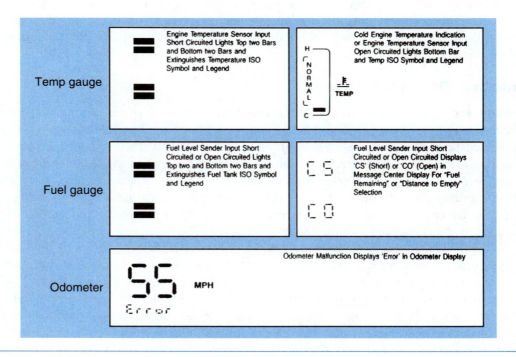

Figure 12-2 EIC function diagnostic mode (Reprinted with the permission of Ford Motor Company)

When the word *ERROR* appears in the odometer display, the EIC computer cannot read valid odometer information from the nonvolatile memory chip.

Special Test Mode

Most EICs have a self-test procedure to determine if the display is working properly. To enter the special test mode, press the E/M and SELECT buttons simultaneously, and turn the ignition switch from the off to the run position. When this action is completed, a number appears in the speedometer display, and two numbers are illuminated in the odometer display. The gauges and message center displays are not illuminated. If any of the numbers are flashing in the speedometer or odometer displays, the EIC is defective and must be replaced.

Photo Sequence 10 shows a typical procedure for diagnosing electronic instrument clusters.

The special test mode may be referred to as a self-test mode.

Photo Sequence 10
Typical Procedure for Diagnosing Electronic Instrument Clusters

P10-1 Turn on the ignition switch to enter the EIC prove-out mode.

P10-2 Observe the EIC segments; all segments must be illuminated. If all the segments are not illuminated, the EIC is defective.

P10-3 With the ignition switch off, press and hold the E/M and SELECT buttons simultaneously to enter the special test mode.

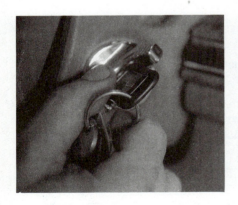

P10-4 Turn on the ignition switch.

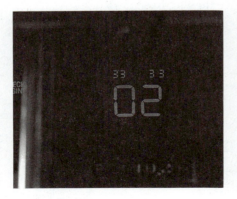

P10-5 Observe the speedometer and odometer displays. A number should appear in the speedometer display, and two numbers should be displayed in the odometer. If these numbers are flashing, the EIC is defective.

P10-6 Turn off the ignition switch to exit the special test mode.

Diagnosis of a Typical Import Electronic Instrument Cluster

Display Check

The display check tests for an open circuit in each display segment and shorts between the segments. Press and hold reset switch A, and turn the ignition switch from off to on to initiate the display check (Figure 12-3). After this action is taken, all the display segments should illuminate, one after the other. If any segment is not illuminated, the EIC must be replaced.

Preprogrammed Signal Check

 WARNING: VFD displays are easily damaged by physical shock. When handling EICs, do not drop or jar them.

 WARNING: When servicing EICs, follow all service precautions related to static discharge in the vehicle manufacturer's service manual to avoid EIC damage.

All EICs are sensitive to static electricity damage, and EIC cartons usually have a static electricity warning label. When servicing EICs:

1. Do not open the EIC carton until you are ready to install the component.
2. Ground the carton to a known good ground before opening the package.
3. Always touch a known good ground before handling the component.
4. Do not touch EIC terminals with your fingers.

Follow all service precautions and procedures in the vehicle manufacturer's service manual.

The preprogrammed signal check tests for defects in various displays. To complete the preprogrammed signal check:

1. Disconnect the negative battery cable. If the vehicle is equipped with an air bag, wait the specified time recommended by the vehicle manufacturer.
2. Remove the EIC power unit.
3. Remove the retaining nuts on the EIC switches.
4. Remove the EIC switches (Figure 12-4).
5. Remove cluster lid A.
6. Remove the EIC assembly.

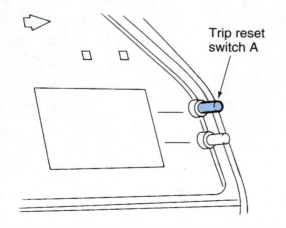

Figure 12-3 EIC reset button (Courtesy of Nissan Motor Co., Ltd.)

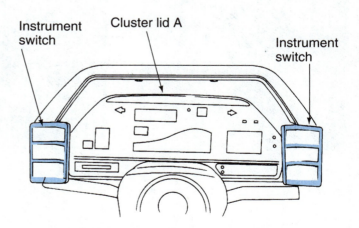

Figure 12-4 EIC switches (Courtesy of Nissan Motor Co., Ltd.)

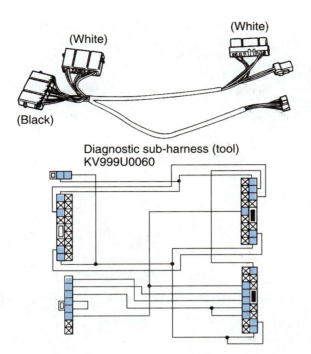

Figure 12-5 Special self-checking wiring harness (Courtesy of Nissan Motor Co., Ltd.)

7. Obtain a special self-checking wiring harness (Figure 12-5).
8. Connect the special self-checking wiring harness to the EIC terminals (Figure 12-6).
9. Connect the negative battery cable, turn on the ignition switch, and observe the EIC displays. Each display should change to a specific reading (Figure 12-7). If each display changes as specified by the vehicle manufacturer, the EIC is satisfactory. When some of the displays do not change as specified, voltmeter and ohmmeter tests are required to locate the exact cause of the problem.
10. Turn off the ignition switch, and disconnect the negative battery cable. If the vehicle is equipped with an air bag, wait the length of time specified by the vehicle manufacturer.
11. Disconnect the special self-checking wiring harness, and connect all EIC connectors securely.
12. Install the EIC and cluster lid.

Special Tools

EIC self-checking wiring harness

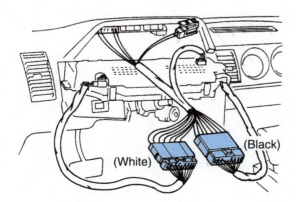

Figure 12-6 Special self-checking wiring harness connected to EIC terminals (Courtesy of Nissan Motor Co., Ltd.)

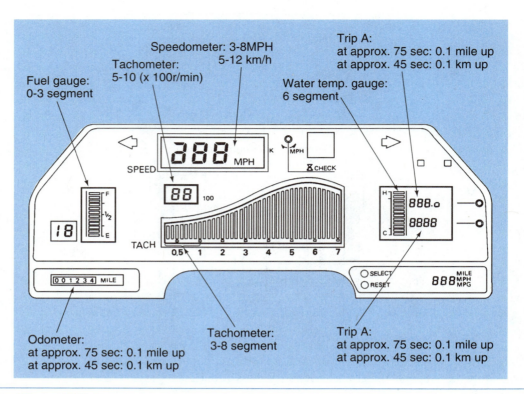

Figure 12-7 Display changes during preprogrammed signal check (Courtesy of Nissan Motor Co., Ltd.)

13. Install the EIC switches and tighten the retaining nuts to the specified torque.
14. Install the power unit.
15. Connect the negative battery cable and check the EIC for proper operation.

CAUTION: If the odometer has been repaired or replaced, and the odometer cannot indicate the same mileage indicated before it was removed, the law in most areas requires that an odometer mileage label must be attached to the left front door frame. Failure to comply with this procedure could lead to court action.

Power Unit Check

A defective power unit may cause the EIC displays to be inoperative. Remove the power unit and leave the wiring harness connected to the unit (Figure 12-8). The power unit supplies different

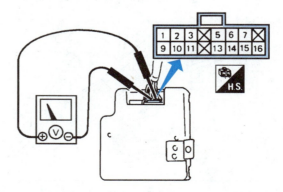

Figure 12-8 Power unit with wiring harness connected (Courtesy of Nissan Motor Co., Ltd.)

Voltmeter terminal ⊕	Voltmeter terminal ⊖	Voltage [V]	Remarks
②	⑨	Approx. 12	Check when no display appears.
③	⑨	Approx. 0	
⑤	⑨	Approx. 22	
⑥	⑨	Approx. 26	
	⑦	Approx. 23	
⑨	⑬ ⑭	Approx. 14	For speedometer, fuel, information, tachometer
⑨	⑮ ⑯	Approx. 19	For temp., trip

Figure 12-9 Voltage at various power unit terminals (Courtesy of Nissan Motor Co., Ltd.)

voltages to various EIC displays. Therefore, it is possible for a defective power unit to cause the failure of specific EIC displays to illuminate. With the ignition switch on, test the voltage at the power unit terminals. Each power unit terminal should have the voltage specified by the vehicle manufacturer (Figure 12-9). The power unit ground wire is connected from terminal 9 to ground. With the ignition switch off, connect a pair of ohmmeter leads from power unit terminal 9 to ground. If the meter reading is above 0.5 Ω, repair the ground wire. If the power unit does not have the specified voltage at some of the terminals, replace the unit.

Special Tools

Multimeter

Speed Sensor Check

A defective speed sensor may cause an inoperative or erratic speedometer reading. The speed sensor testing procedure follows:

1. Remove cluster lid A.
2. Connect a pair of voltmeter leads to terminals 11 and 1 on the EIC with the wiring harness connected (Figure 12-10).

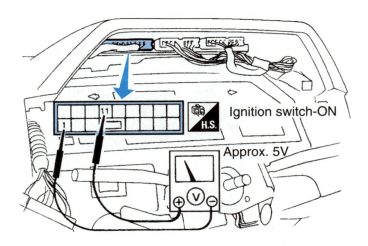

Figure 12-10 Voltmeter connections to EIC terminals (Courtesy of Nissan Motor Co., Ltd.)

3. Turn on the ignition switch and check the voltmeter reading. If the voltage is zero, check the power unit.
4. Turn off the ignition switch.
5. Disconnect the speedometer cable from the speed sensor and remove the speed sensor with the wiring harness connected.
6. Remove the wiring harness connector containing terminals 1 and 12 from the EIC, and connect a pair of analog voltmeter leads to terminals 1 and 12 in the wiring harness connector.

SERVICE TIP: While testing the speed sensor, slowly turn the sensor with a small screwdriver. The sensor produces 24 signals per revolution, which are difficult to read during fast sensor rotation.

7. Use a small screwdriver to slowly rotate the speed sensor (Figure 12-11). If the voltmeter pointer does not deflect, the speed sensor is defective or the connecting wires are defective.
8. Connect the voltmeter leads directly to the speed sensor terminals and repeat step 7. If the voltmeter pointer deflects, repair the wires between the speed sensor and EIC terminals 1 and 12. When the voltmeter pointer does not deflect, replace the speed sensor.

Speedometer Diagnosis

If all speedometer segments fail to illuminate properly while driving the car, test the power unit. Replace the power unit if it is unsatisfactory. Perform the display check. If the display check is satisfactory, replace the control unit. When the speedometer display is not satisfactory in the display check, replace the display unit (Figure 12-12).

Classroom Manual
Chapter 12, page 298

Head-Up Display (HUD) Diagnosis

HUD Display Prove-Out

To check the HUD display, turn the HUD brilliance control switch to maximum intensity, and turn the ignition switch from the off to the run position. When this action is taken, all segments and indicators in the HUD display should be illuminated for 4 seconds. After this time, the speedometer should read 0 with the mph or km/h indicator illuminated, and other indicators should remain off.

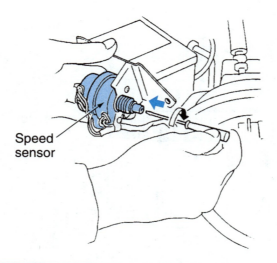

Figure 12-11 Testing the speed sensor (Courtesy of Nissan Motor Co., Ltd.)

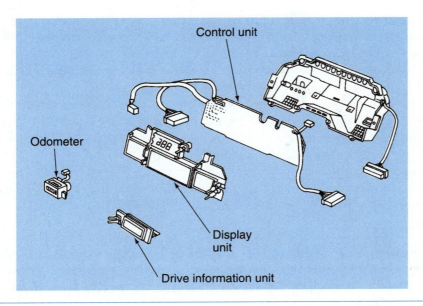

Figure 12-12 Display unit and control unit (Courtesy of Nissan Motor Co., Ltd.)

HUD Window and Windshield Cleaning

WARNING: Do not spray glass cleaner directly on the HUD window, because this action may damage HUD module internal components.

If the display is not sharp and clear, clean the windshield and the module window, and be sure there is nothing obstructing the HUD module window or the windshield. Spray glass cleaner on a soft clean cloth, and then use this cloth to clean the HUD module window. Do not spray glass cleaner directly on the HUD module window. The windshield may be cleaned with a chamois and a soft clean cloth.

HUD Diagnosis

If the HUD display is inoperative, follow these steps to diagnose the system:

1. Disconnect the HUD control switch connector and connect a pair of digital voltmeter leads from terminal 5 in the wiring harness connector to ground (Figure 12-13).

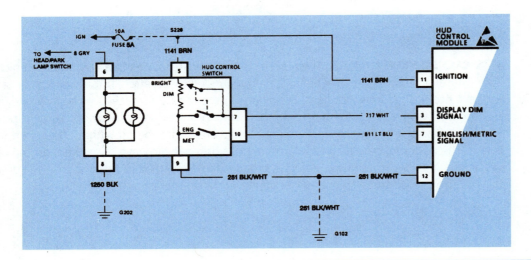

Figure 12-13 HUD wiring diagram (Courtesy of Pontiac Division, General Motors Corporation)

319

2. Turn on the ignition switch. If the voltage is less than 11 V, check for an open circuit in fuse 5A or the circuit from the ignition switch to HUD switch terminal 5. When the voltage is above 11 V, proceed to step 3.
3. Turn off the ignition switch, and check for loose wiring connections on the HUD module. Disconnect the HUD module connector.
4. Turn on the ignition switch, and connect a pair of digital voltmeter leads from terminal 11 in the HUD wiring harness to ground. If the voltage is less than 11 V, repair the open circuit in fuse 5A or the wire from the ignition switch to HUD module terminal 11. If the voltage is above 11 V, proceed to step 5.
5. Connect a pair of digital voltmeter leads from terminal 11 to terminal 12 in the HUD module connector. If the reading is less than 11 V, repair the open circuit in the HUD ground wire. When the reading is over 11 V, replace the HUD module.

Classroom Manual
Chapter 12, page 300

Diagnosis of Vehicle Theft Security Systems

Vehicle Theft Security System Diagnosis with Ignition Switch Cycling

Diagnostic procedure for the vehicle theft security system varies depending on the vehicle model and year. Always follow the diagnostic procedure in the vehicle manufacturer's service manual. Some vehicle theft security systems enter the diagnostic mode when the ignition switch is cycled three times from the off to the accessory position. When the vehicle theft security system is in the diagnostic mode, the horn should sound twice and the park and tail lamps should flash. If the horn does not sound or the lights do not flash, voltmeter and ohmmeter tests are required to locate the cause of the problem (Figures 12-14 and 12-15). During the vehicle theft alarm system diagnosis, some voltmeter tests are required at the module connectors (Figure 12-16, page 323, and Figure 12-17, page 324).

The scan tester may be used to diagnose 1993 and later model vehicle theft alarm systems. Follow the scan tester manufacturer's recommended procedure to enter the vehicle theft alarm system diagnostic mode. When this diagnostic mode is entered, the horn sounds twice to indicate the trunk lock cylinder is in its proper position. When the key is placed in the ignition switch, the park and tail lamps should begin flashing.

Special Tools
Scan tester

The following procedures should cause the horn to sound once if the system is operating normally:

1. Activate the power door locks to the locked and unlocked positions.
2. Use the key to lock and unlock each front door.
3. Turn on the ignition switch.

When the ignition switch is turned on in step 3, the diagnostic mode is exited.

Classroom Manual
Chapter 12, page 302

● **CUSTOMER CARE:** Always check the indicator lights in a customer's vehicle. These indicator lights may be indicating a dangerous situation, but the customer may not have noticed the indicator light. For example, the vehicle theft security system set light may not be flashing when the normal system arming procedure is followed. This indicates an inoperative security system, and someone could break into the car without triggering the alarms. The customer paid a considerable amount of money to have this system on the car, and it should be working. If this defect is brought to the customer's attention, he will probably have you repair the system and will appreciate your interest in the vehicle.

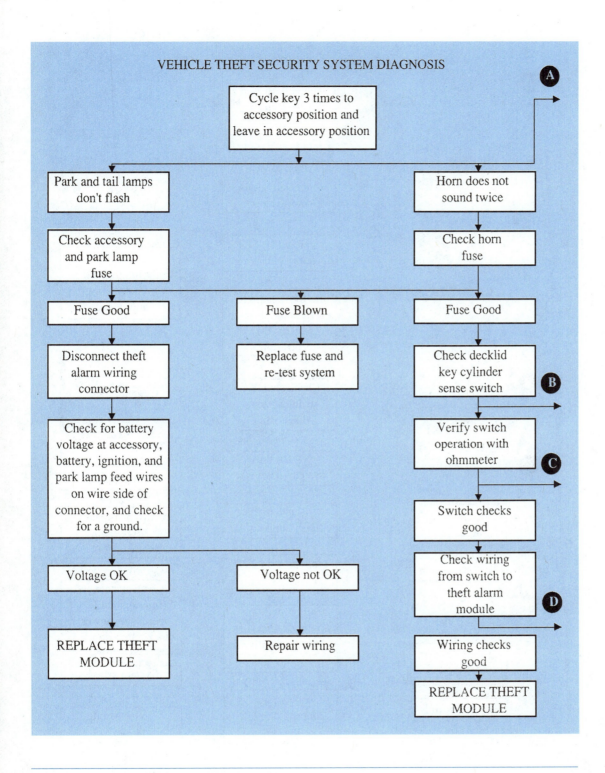

Figure 12-14 Vehicle theft alarm system diagnosis (Courtesy of Chrysler Corporation)

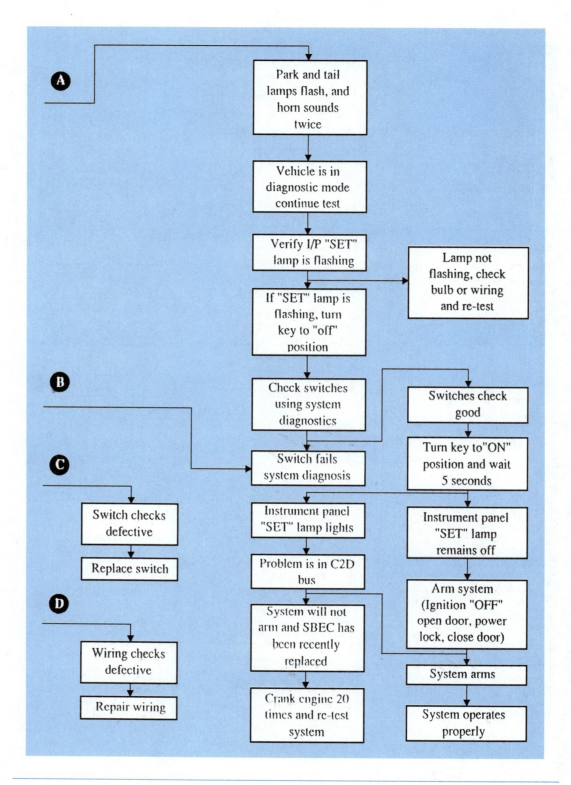

Figure 12-15 Vehicle theft alarm system diagnosis continued (Courtesy of Chrysler Corporation)

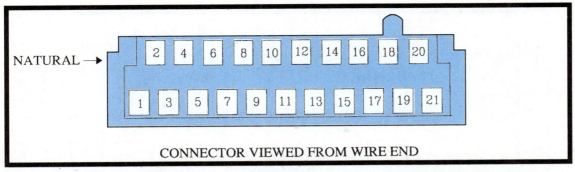

CONNECTOR VIEWED FROM WIRE END

SECURITY ALARM MODULE CONNECTOR

CAV	CIRCUIT	GAUGE	COLOR	FUNCTION	VOLTAGE WITH IGN. ON
1	H40	20	BK/LG*	Electronic ground	0 volts
2	MX2	20	WT*	Serial data bus (-)	2.5 volts
3	M6	20	TN/BK*	Hood ajar	8.5 volts closed 0 volts open
4	MX1	20	BK	Serial data bus (+)	2.5 volts
5	M3	22	TN	LF door ajar	5 volts closed 0 volts open
6	G5	20	DB*	Ign. feed (run)	12 volts key on (fuse 8)
7	M5	22	OR*	LR door ajar	5 volts closed 0 volts open
8	M32	22	DG/OR*	RF door key cylinder (unarm)	12 volts at rest 0 volts to unlock
9	SA1	22	BK/OR*	Ground signal to security alarm lamp	12 volts off 0 volts on
10	H40	20	BK/LG*	Electronic ground	0 volts

Figure 12-16 Vehicle theft alarm system module terminals and voltage tests (Courtesy of Chrysler Corporation)

11	P36	18	PK/VT*	Power door locks (unlock)	12 volts unlock (from relay)
12	H3	20	BK/RD*	Ground signal to horn relay	12 volts off 0 volts on
13	DL1	20	OR/WT*	Power door locks (lock)	12 volts lock (from relay)
14	L7	18	BK/YL*	Tail/park Lamps	12 volts to turn on lamps
15	M33	22	LG/OR*	LF Door key (unarm)	12 volts at rest 0 volts to unlock
16	M13	22	TN/RD*	RF door ajar	5 volts closed 0 volts open
17	M41	20	VT/YL*	Trunk key cylinder ground	12 volts at rest 0 volts when removed
18	X1	20	RD	Battery feed	12 volts
19	V1	18	DB	Ign feed (run/acc)	12 volts key on or acc.
20	M15	22	TN/YL*	RR door ajar	5 volts closed 0 volts open
21	L8	18	PK/BK*	Tail/park lamp feed	12 volts

MECHANICAL INSTRUMENT CLUSTER

If the vehicle is equipped with a Mechanical Instrument Cluster, the connector will NOT use cavities 7, 16, and 20. Cavity 5 will change to the following:

5	M3	22	WT/TN*	Door (ALL)	5 volts closed 0 volts open

Figure 12-17 Vehicle theft alarm system voltage tests continued (Courtesy of Chrysler Corporation)

Guidelines for Servicing EICs, HUDs, and Vehicle Theft Security Systems

1. Each time the ignition switch is turned on, a prove-out display occurs in which all the EIC segments are illuminated for a few seconds.
2. Some EICs have a function diagnostic mode, which provides specific displays if certain defects occur in the EIC system.
3. Some EICs have a special test mode in which two buttons are pressed simultaneously and specific EIC displays occur if these displays are working normally.

4. Other EICs have a preprogrammed signal check in which a special wiring harness is connected to the EIC terminals and specific gauge readings must occur if the EIC is operating normally.
5. Some EICs have a power unit that supplies different voltages to specific EIC terminals. A defective power unit may cause inoperative EIC displays.
6. EICs are easily damaged by physical shock or static charges.
7. The HUD has a prove-out display in which all segments and indicators are illuminated when the ignition switch is turned on.
8. The HUD module window may be cleaned with glass cleaner sprayed on a soft cloth.
9. Some vehicle theft security systems enter the diagnostic mode when the ignition switch is cycled three times from the off to the accessory position.
10. Other vehicle theft security systems may be diagnosed with a scan tester.

CASE STUDY

A customer brought a 1995 Chrysler Concorde into the shop because the vehicle theft security system would not arm. The technician checked the system and found the security system indicator light was inoperative when the proper arming procedure was followed.

The technician attempted to enter the security system diagnostic mode with the scan tester and found this mode to be inoperative. The technician checked the accessory and park lamp fuse, which supplies voltage to the security system module. This fuse was satisfactory. Digital voltmeter readings at each of the module terminals indicated the specified voltage at each terminal. The technician connected the digital voltmeter leads from ground to terminals 1 and 10 in the module connector and found the reading on each ground wire to be 0.2 V with the ignition switch on. The technician noticed the ground wire in terminal 10 was pushed back so it was partly out of the wiring connector, preventing it from making proper contact with the module terminal. This ground wire was pushed securely into the wiring connector until it snapped into place.

The technician armed the system with the normal procedure and the set light indicated the system armed properly. The diagnostic mode was entered with the scan tester. During this mode, the park and tail lamps flashed normally and the horn sounded as specified by the vehicle manufacturer.

Terms to Know

Control unit
Display check
Display unit
Electronic instrument cluster (EIC)
Function diagnostic mode

Head-up display
International standards organization (ISO)
Power unit
Preprogrammed signal check

Prove-out display
Special test mode
Vehicle theft security system

ASE Style Review Questions

1. While discussing an EIC function diagnostic mode:
   Technician A says if the fuel gauge sender circuit is open, the two top and bottom bars in the fuel gauge are illuminated.
   Technician B says if the fuel gauge sender circuit is open, the letters *CO* appear in the fuel remaining and distance to empty displays.
   Who is correct?
   A. A only
   B. B only
   C. Both A and B
   D. Neither A nor B

2. While discussing an EIC function diagnostic mode:
   Technician A says if the temperature gauge sender circuit is shorted, all the bars in the temperature gauge are illuminated.
   Technician B says if the temperature gauge sender circuit is shorted, the ISO symbol is flashing.
   Who is correct?
   A. A only
   B. B only
   C. Both A and B
   D. Neither A nor B

3. While discussing an EIC preprogrammed signal check:
   Technician A says when the special wiring harness is connected to the EIC, the display readings should change as indicated in the service manual.
   Technician B says when the special wiring harness is connected to the EIC, all the display readings should begin flashing.
   Who is correct?
   A. A only
   B. B only
   C. Both A and B
   D. Neither A nor B

4. While discussing EIC power units:
   Technician A says the power unit supplies the same voltage to all EIC displays.
   Technician B says an open power unit ground affects the EIC displays.
   Who is correct?
   A. A only
   B. B only
   C. Both A and B
   D. Neither A nor B

5. While discussing the speed sensor:
   Technician A says a defective speed sensor may cause an erratic or inoperative speedometer display.
   Technician B says a defective speed sensor may cause inaccurate tachometer operation.
   Who is correct?
   A. A only
   B. B only
   C. Both A and B
   D. Neither A nor B

6. While discussing HUD diagnosis:
   Technician A says each time the ignition switch is turned on the HUD enters a prove-out mode in which all the displays and indicators are illuminated.
   Technician B says the prove-out mode lasts for 4 seconds.
   Who is correct?
   A. A only
   B. B only
   C. Both A and B
   D. Neither A nor B

7. While discussing HUD service:
   Technician A says the HUD module window may be cleaned by spraying glass cleaner directly on the window.
   Technician B says a dirty windshield may affect the HUD displays.
   Who is correct?
   A. A only
   B. B only
   C. Both A and B
   D. Neither A nor B

8. While discussing HUD diagnosis:
   Technician A says after the HUD prove-out the speedometer reading should read 88 with the engine idling and the car stopped.
   Technician B says the mph or kmh indicator is illuminated with the speedometer reading.
   Who is correct?
   A. A only
   B. B only
   C. Both A and B
   D. Neither A nor B

9. While discussing vehicle theft security system diagnosis:
Technician A says on some security systems the diagnostic mode is entered by cycling the ignition switch three times from the off to the accessory position.
Technician B says the security system diagnostic mode is entered by cycling the ignition switch three times from the off to the on position.
Who is correct?
A. A only
B. B only
C. Both A and B
D. Neither A nor B

10. While discussing vehicle theft security systems:
Technician A says during the diagnostic mode, the lights should flash and the horn should sound.
Technician B says a scan tester may be used to diagnose some security systems.
Who is correct?
A. A only
B. B only
C. Both A and B
D. Neither A nor B

Table 12-1 ASE Task

Diagnose the cause of intermittent, high, or no readings on electronic digital instrument clusters.

Problem Area	Symptoms	Possible Causes	Classroom Manual	Shop Manual
ELECTRONIC INSTRUMENT CLUSTER	Intermittent, high, or no displays	1. Defective power supply 2. Defective inputs 3. Defective EIC	291 293 293	311 311 311

Table 12-2 ASE Task

Inspect, test, and repair or replace sensors, sending units, connectors, and wires of electronic instrument clusters.

Problem Area	Symptoms	Possible Causes	Classroom Manual	Shop Manual
ELECTRONIC INSTRUMENT CLUSTER	Intermittent, inaccurate displays	1. Defective inputs 2. Defective wiring	293 293	316 316

Table 12-3 ASE Task

Diagnose the cause of false, intermittent, unintended, or no operation of antitheft systems.

Problem Area	Symptoms	Possible Causes	Classroom Manual	Shop Manual
ANTI-THEFT SYSTEM	Intermittent, unintended, or no operation	1. Defective inputs 2. Defective wiring 3. Defective module	302 302 303	320 320 320

Table 12-4 ASE Task

Inspect, test, and repair or replace components, switches, relays, connectors, sensor, and wires of antitheft systems.

Problem Area	Symptoms	Possible Causes	Classroom Manual	Shop Manual
ANTI-THEFT SYSTEM	Intermittent, unintended, or no operation	Defective inputs, wiring or module	303	320

Air Bag System Diagnosis and Service

CHAPTER 13

Upon completion and review of this chapter, you should be able to:

- ❏ Observe all the necessary precautions when servicing and diagnosing air bag systems.
- ❏ Perform an air bag system diagnostic system check to determine if the system is operational.
- ❏ Obtain flash codes from an air bag system.
- ❏ Connect the test harness for an air bag voltmeter diagnosis.
- ❏ Perform a scan tester diagnosis of an air bag system.
- ❏ Disable an air bag system before servicing the system.
- ❏ Remove and replace inflator modules.
- ❏ Center the clock spring electrical connector.
- ❏ Repair air bag wiring.
- ❏ Perform an air bag deployment procedure before vehicle scrapping.

Diagnostic System Check

Basic Tools

Basic technician's tool set
Service manuals
Jumper wires

CAUTION: When servicing any air bag system component, always disconnect the battery negative cable, isolate the cable end, and wait for the amount of time specified by the vehicle manufacturer before proceeding with the necessary diagnosis or service. The average waiting period is 2 minutes, but some vehicle manufacturers specify up to 10 minutes. Failure to observe this precaution may cause accidental air bag deployment and personal injury.

CAUTION: Replacement air bag system parts must have the same part number as the original part. Replacement parts of lesser quality or questionable quality must not be used. Improper or inferior components may result in improper air bag deployment and injury to the vehicle occupants.

CAUTION: Do not strike or jar a sensor or an air bag system diagnostic monitor (ASDM) because this may cause air bag deployment or make the sensor inoperative. Accidental air bag deployment may cause personal injury, and an inoperative sensor may result in air bag deployment failure, which could cause personal injury to the vehicle occupants.

CAUTION: All sensors and mounting brackets must be properly torqued before an air bag system is powered up to ensure correct sensor operation. If sensor fasteners do not have the proper torque, improper air bag deployment may result in injury to the vehicle occupants.

An air bag system diagnostic module (ASDM) may be called a diagnostic energy reserve module (DERM) or air bag diagnostic monitor. Since there are several terms used for the air bag computer, we use the term *ASDM* in this chapter to avoid confusion.

Before an air bag system is diagnosed, a diagnostic system check must be performed to avoid diagnostic errors. The diagnostic system check may vary depending on the vehicle. Always consult the manufacturer's specific information. The diagnostic system check involves observing the air bag warning light to determine if it is operating normally. A typical diagnostic system check follows:

1. Turn on the ignition switch and observe the air bag warning light. On some General Motors systems, this light should flash 7 to 9 times and then go out. On other vehicles, the air bag warning light should be illuminated continually for 6 to 8 seconds and then go out. If the air bag warning light does not operate properly, further system diagnosis is necessary.

> Current codes are diagnostic trouble codes (DTCs) that are present at the time of testing.
>
> History codes are DTCs resulting from an intermittent problem that occurred long enough to be recorded in the ASDM memory.

2. Observe the air bag warning light while cranking the engine. On many General Motors vehicles, this light should be illuminated continually while cranking the engine. Always refer to the vehicle manufacturer's service manual. During engine cranking if the air bag warning light does not operate as specified by the vehicle manufacturer, complete system diagnosis is required.
3. Observe the air bag warning light after the engine starts. This light should flash 6 to 9 times and then remain off. If the air bag warning light remains off, there are no current diagnostic trouble codes (DTCs) in the air bag system diagnostic module (ASDM). When the air bag warning light remains on, obtain the DTCs with a scan tester or flash code method. History DTCs on General Motors vehicles can only be obtained with a scan tester. Some vehicles such as Chrysler products do not provide air bag flash codes. The scan tester must be used to diagnose these systems.

Air Bag System Flash Code and Voltmeter Diagnosis

Ford Air Bag System Flash Code Diagnosis

✓ **SERVICE TIP:** On some air bag systems such as those on Ford vehicles, if the air bag warning light is illuminated continually with the ignition switch on or the engine running, the ASDM is defective or disconnected. The air bag warning light may not be illuminated at full brilliance.

✓ **SERVICE TIP:** On some Ford vehicles, the ASDM provides an audible tone if the air bag warning light is not operating properly. Under this condition, the ASDM provides 5 sets of 5 beeps every half-hour.

✓ **SERVICE TIP:** On Ford vehicles, the air bag deployment loop is connected to the battery positive terminal even with the ignition switch off. Therefore, air bag deployment is possible with the ignition switch off.

On some air bag systems, the ASDM disarms the system if a fault exists that could result in unwarranted air bag deployment. The ASDM disarms the system by opening a thermal fuse inside the ASDM. This fuse is not replaceable.

On many Ford vehicles, the air bag warning light begins flashing a DTC when a defect occurs in the air bag system. On many Ford systems, the air bag warning light priorizes the DTCs and flashes the highest priority DTC if there is more than one fault in the system. When the fault represented by the flashing air bag warning light is corrected, the light flashes the DTC with the next highest priority if a second fault exists. Since DTCs vary depending on the model year, the technician must have the DTC list for the model year being diagnosed.

General Motors Air Bag System Flash Code Diagnosis

✓ **SERVICE TIP:** The air bag systems on some General Motors vehicles must be diagnosed with a scan tester. Always consult the proper vehicle manufacturer's service manual.

✓ **SERVICE TIP:** If the air bag warning light is illuminated continually on a General Motors vehicle, always check the ASDM wiring connector. When this connector is disconnected, shorting bars in the connector illuminate the air bag warning light.

On many General Motors vehicles, air bag flash codes are obtained by connecting a jumper wire between terminals A and K in the data link connector (DLC) (Figure 13-1). When this connection is completed and the ignition switch is on, the air bag warning light flashes current DTCs. Fault codes are flashed by the air bag light in much the same way as the malfunction indicator light (MIL) flashes engine codes. The air bag warning light only flashes current codes representing faults present at the time of testing. Current codes are stored in the ASDMs random access memory (RAM). When the fault is corrected, the current code is erased, but the same code may be recorded as a history code. History codes represent all faults detected since the last time the EEPROM in the ASDM was cleared. History codes represent intermittent or current faults. These codes can only be obtained and erased with a scan tester connected to the DLC.

When the flash code diagnostic mode is entered, code 12 indicates that the air bag system is in the diagnostic mode, whereas code 13 indicates that history codes are present in the ASDM memory. Codes 51 and 52 indicate accident detection. If these codes exist as history codes, they cannot be erased even with a scan tester. Each code is flashed once, and the codes are given in numerical order. The code sequence is repeated until the jumper wire is removed or the ignition switch is turned off. The fault codes may vary depending on the year and make of General Motors vehicle. Always consult the manufacturer's information for the exact fault codes and code diagnosis.

Toyota Air Bag System Flash Code Diagnosis

SERVICE TIP: On Toyota vehicles, if the air bag warning light flashes a DTC that is not on the fault code list for that model year, the ASDM is defective.

On Toyota vehicles, air bag flash codes may be obtained by cycling the ignition switch on and off five times. Each time the ignition switch is cycled on or off, the technician must wait 20 seconds. Toyota air bag DTCs may also be obtained by connecting a special jumper wire supplied by the vehicle manufacturer between terminals TC and E1 in the DLC2. Some vehicles have a round DLC2

The Society of Automotive Engineers (SAE) J1930 terminology is an attempt to standardize electronics terminology in the automotive industry.

In the SAE J1930 terminology, the term *data link connector (DLC)* replaces previous terms such as assembly line diagnostic link (ALDL).

A random access memory (RAM) is a memory chip in a computer. The computer can write information into this chip or read data from the chip.

An electronically erasable programmable read only memory (EEPROM) is a computer chip that can be erased and reprogrammed with special equipment.

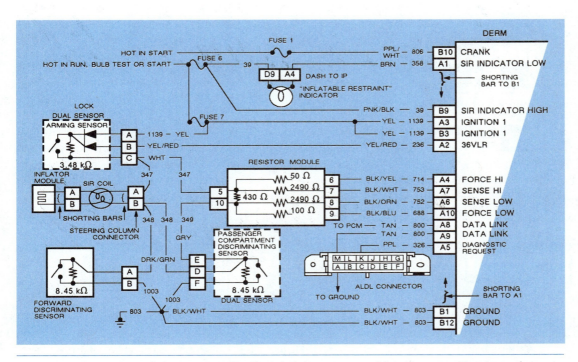

Figure 13-1 Data link connector (DLC) and air bag system wiring diagram (Courtesy of Oldsmobile Division, General Motors Corporation)

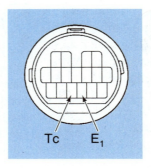

Figure 13-2 Round-shaped DLC2 (Courtesy of Toyota Motor Corporation)

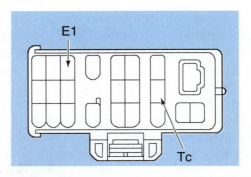

Figure 13-3 Rectangular-shaped DLC2 (Courtesy of Toyota Motor Corporation)

(Figure 13-2), whereas a rectangular-shaped DLC2 is found on other vehicles (Figure 13-3). Before connecting this jumper wire, make sure the ignition switch is in the ACC or ON position. After the ignition switch is in one of these positions, wait 30 seconds. If there are no DTCs in the ASDM memory, the air bag warning light flashes two times per second (Figure 13-4). When DTCs are present in the ASDM memory, the air bag warning light flashes these codes in numerical order (Figure 13-5). A DTC indicates a fault in a certain area such as a specific air bag sensor. Voltmeter or ohmmeter tests recommended in the vehicle manufacturer's service manual are usually necessary to locate the exact cause of the problem.

Honda Air Bag System Voltmeter Diagnosis

CAUTION: Use only the vehicle manufacturer's recommended tools and equipment for air bag system service and diagnosis. Failure to observe this precaution may result in unwarranted air bag deployment and personal injury.

CAUTION: Do not use battery-powered or A/C-powered voltmeters or ohmmeters except those meters specified by the vehicle manufacturer. Failure to observe this precaution may result in unwarranted air bag deployment and personal injury.

CAUTION: Do not use nonpowered probe-type test lights or self-powered test lights to diagnose the air bag system. Unwarranted air bag deployment and personal injury may result from this procedure.

CAUTION: Follow the vehicle manufacturer's service and diagnostic procedures. Failure to observe this precaution may cause inaccurate diagnosis and unnecessary repairs, or unwarranted air bag deployment, resulting in personal injury.

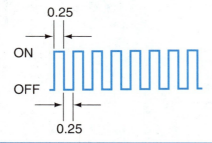

Figure 13-4 The air bag warning light flashes two times per second if there are no DTCs in the ASDM memory. (Courtesy of Toyota Motor Corporation)

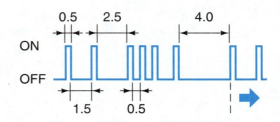

Figure 13-5 The air bag warning light flashes the two-digit DTCs in numerical order. (Courtesy of Toyota Motor Corporation)

Honda recommends testing the voltage at specific terminals on the ASDM, inflator modules, and sensor to diagnose the air bag system. When voltage tests are performed at these terminals, special jumper wires are connected to the terminals to allow the necessary voltmeter connections without damaging the terminals. Before connecting the special wiring harness, remove and isolate the battery negative cable. Then wait for the time period specified by the vehicle manufacturer.

Wiring harness A is connected to a terminal on the ASDM, and wiring harness B is connected in series between the large ASDM wiring connector and the matching terminals on the ASDM (Figure 13-6). Wiring harness C is connected in series at the inflator module connector, and harness D is connected in series at the dash sensor (Figure 13-7). After the special wiring harness is connected, the voltmeter tests provided in the vehicle manufacturer's service manual may be performed to diagnose the system.

Special Tools

Air bag system diagnostic wiring harness

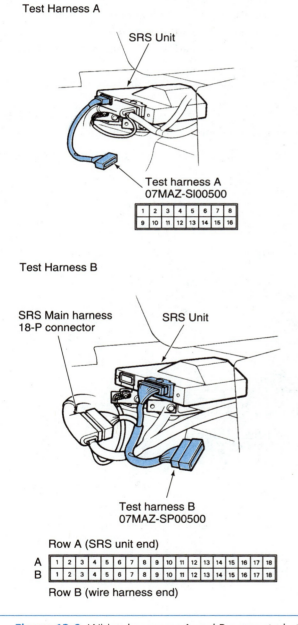

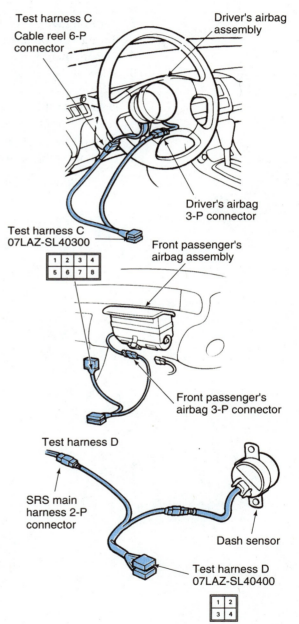

Figure 13-6 Wiring harnesses A and B connected at the ASDM terminals (Courtesy of American Honda Motor Co., Inc.)

Figure 13-7 Wiring harnesses C and D connected at the inflator module and dash sensor terminals (Courtesy of American Honda Motor Co., Inc.)

Scan Tester Diagnosis of Air Bag Systems

Special Tools

Scan tester

Many air bag systems must be diagnosed with a scan tester, because they do not provide flash codes. The scan tester must be compatible with the air bag system being diagnosed, and the proper test module must be installed in the tester for the vehicle make, model year, and type of ASDM. Some vehicles such as Chrysler LH cars had an ASDM change during the 1994 model year. An air bag system scan tester diagnostic procedure follows:

1. Disconnect and isolate the negative battery cable and wait 2 minutes or the time specified by the vehicle manufacturer.
2. Use the proper connector to connect the scan tester to the DLC on the vehicle (Figure 13-8).
3. Turn on the ignition switch. Roll down the driver's window and move the scan tester through the open window. Use the scan tester while standing outside the vehicle.
4. Be sure there is no one in the vehicle, then reconnect the negative battery cable.
5. Select air bag system on the scan tester menu and read the DTC messages on the tester (Figure 13-9).
6. Disconnect the battery negative cable and wait 2 minutes or the time specified by the vehicle manufacturer.
7. Turn off the ignition switch.
8. Disconnect the scan tester and reconnect the battery negative cable.

When the air bag system repairs are completed, the DTCs may be erased with the scan tester.

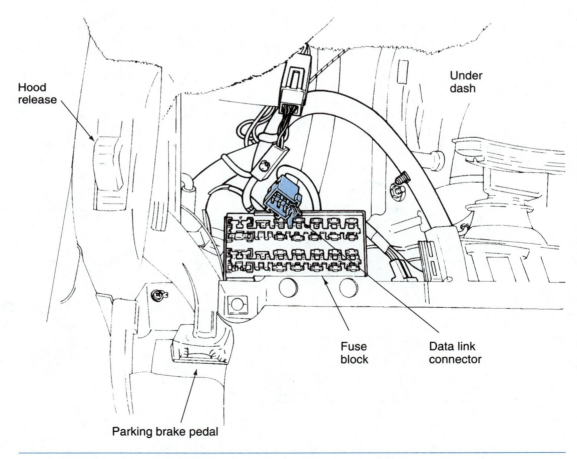

Figure 13-8 Data link connector (DLC) for air bag system and diagnosis of other systems (Courtesy of Chrysler Corporation)

Diagnostic Trouble Code	Set Condition
Initiator Open	Resistance greater than 10 Ohms
Safing Sensor Open	Internal measurement
Safing Sensor Short	Internal measurement
Front Sensor Short	Resistance under 1000 Ohms
Two Front Sensors Open	Resistance above 30 K Ohms
One Front Sensor Open	Resistance above 10.2 K Ohms
Initiator Short	Resistance under 1.6 Ohms
Low Stored Energy	Internal measurement
Ignition 'Run/Start' Low	Internal measurement
Internal Diagnostic Module Codes	Internal malfunction
Ignition 'Run Only' Low	Voltage less than 2.86v
Warning Lamp Open	Internal Measurement

Figure 13-9 Air bag system DTC messages (Courtesy of Chrysler Corporation)

Disabling the Air Bag System

Many vehicle manufacturers recommend disabling the air bag system to prevent unwarranted air bag deployment while servicing the system components. A typical General Motors air bag disabling procedure follows:

1. Turn off the ignition switch.
2. Disconnect and isolate the negative battery cable, and wait the time specified by the vehicle manufacturer.
3. Disconnect the yellow two-wire SIR system connector at the base of the steering column.
4. Install the J 38715 load tool to the yellow two-wire connector as per the load tool instructions (Figure 13-10).
5. Perform the necessary service work or diagnosis. Connect the negative battery cable and turn on the ignition switch as requested by the vehicle manufacturer.
6. When service work is completed, be sure that the ignition switch is off, disconnect and isolate the negative battery cable, and wait the specified time. Disconnect the load tool.
7. Reconnect the two-wire yellow connector.
8. Check the air bag warning light for proper operation.

Special Tools

Air bag system load tool

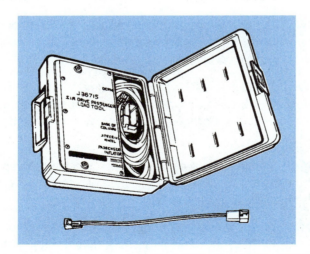

Figure 13-10 Air bag system load tool (Courtesy of Chevrolet Motor Division, General Motors Corporation)

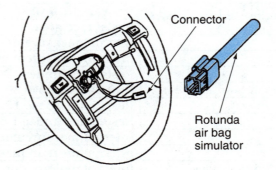

Figure 13-11 Air bag simulator (Reprinted with the permission of Ford Motor Company)

A similar air bag system disabling procedure is recommended on Ford vehicles. However, on these vehicles, the driver's side and passenger's side inflator modules must be removed and an air bag simulator tool installed in the inflator module wiring connectors (Figure 13-11). The air bag simulator contains a resistor that simulates the inflator module during air bag system service.

Special Tools

Inflator module simulator tool

Removing and Replacing the Inflator Module

CAUTION: Always carry live inflator modules with the trim cover pointing away from you to minimize personal injury from accidental air bag deployment.

CAUTION: When inflator modules or steering wheels containing these modules are placed on a bench, position the trim cover away from the bench surface. If this precaution is not observed, the inflator module could become a dangerous projectile, causing personal injury if the air bag deploys.

WARNING: Do not carry any air bag system component by the lead wires. This action may cause component or wiring damage.

CAUTION: Wear safety goggles, gloves, and a long-sleeved shirt or coveralls when handling deployed air bags to avoid skin irritation from sodium hydroxide.

CAUTION: Do not attempt to repair sensors, inflator modules, ASDMs, or other system components. This action may cause improper air bag deployment resulting in personal injury to the vehicle occupants.

CAUTION: If sensors or ASDMs have arrows marked forward, the arrows must face toward the front of the vehicle when these components are mounted. Failure to follow this procedure may prevent proper air bag deployment, causing injury to the vehicle occupants.

WARNING: Inflator modules must not be subjected to temperatures above 150°F (65°C). If the inflator module encounters temperatures above this value, an unwarranted air bag deployment may occur.

CAUTION: Never connect an ohmmeter across the inflator module terminals to check the resistance of this unit. This action may result in an unwarranted air bag deployment and personal injury.

Photo Sequence 11 shows a typical procedure for removing and replacing an inflator module.

Photo Sequence 11
Typical Procedure for Removing and Replacing an Inflator Module

P11-1 Disconnect the negative battery cable, isolate the cable end, and wait the length of time specified by the vehicle manufacturer.

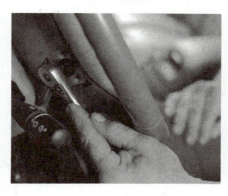

P11-2 Remove the four inflator module retaining nuts under the steering wheel.

P11-3 Lift the inflator module up from the steering wheel and disconnect the inflator module wiring connector.

P11-4 Carry the inflator module with the trim cover facing away from your body.

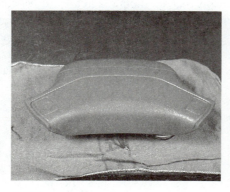

P11-5 Place the inflator module on the workbench with the trim cover facing upward.

P11-6 Hold the inflator module above the steering wheel and connect the wiring connector to the inflator module.

P11-7 Properly position the inflator module on the steering wheel. Be careful not to jam the wiring harness between the module and the steering wheel.

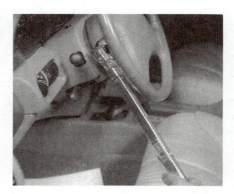

P11-8 Tighten the four inflator module retaining nuts to the specified torque.

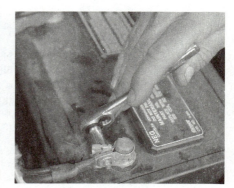

P11-9 Reconnect and tighten the negative battery cable.

If the air bag has been deployed recently, allow 20 minutes for the inflator module to cool before replacing this component. A typical inflator module removal and replacement procedure follows:

1. Be sure the ignition switch is off. Disconnect the negative battery cable and isolate the cable end. Wait the specified time recommended in the vehicle manufacturer's service manual.
2. Remove the four inflator module retaining nuts under the steering wheel (Figure 13-12).
3. Lift the inflator module from the steering wheel and disconnect the inflator module wiring connector (Figure 13-13).
4. Check the inflator module wiring attached to the clock spring electrical connector. If this wiring is burned, replace the clock spring electrical connector.
5. Hold the new inflator module near the steering wheel and connect the clock spring electrical connector to the inflator module.
6. Properly position the inflator module on the steering wheel. Be sure the inflator module wiring is not jammed against other components.
7. Install the four inflator module retaining nuts under the steering wheel, and tighten these nuts to the specified torque.
8. Reconnect the negative battery cable and check the air bag warning light for proper operation.

After a collision involving air bag deployment, most vehicle manufacturer's recommend replacing any sensors that are mounted in the area of collision damage. If the vehicle is involved in a collision without air bag deployment, all air bag system components, component mounting surfaces, and wiring harnesses must be inspected for damage. Live or undeployed inflator modules must be shipped in special containers with correct hazardous material identification. Deployed inflator modules may be treated as normal scrap material and disposed of in the usual manner.

Classroom Manual
Chapter 13, page 309

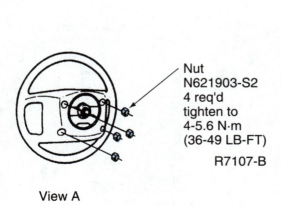

Figure 13-12 Removing the four inflator module retaining nuts under the steering wheel (Reprinted with the permission of Ford Motor Company)

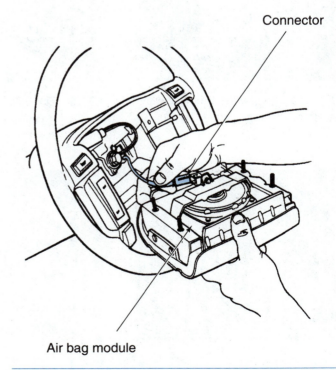

Figure 13-13 Disconnecting the inflator module electrical connector (Reprinted with the permission of Ford Motor Company)

Centering the Clock Spring Electrical Connector

If the steering gear is removed from the steering column, the front wheels must be centered in the straight-ahead position. After this action is taken, the ignition key should be removed to lock the steering wheel in the straight-ahead position. Before the steering gear is removed from the column, the stub shaft and the rack should be clearly marked in the straight-ahead position so they can be reinstalled in the same position. The clock spring electrical connector becomes uncentered if the steering gear is separated from the column and the steering wheel is allowed to rotate. On some systems, this uncentered clock spring electrical connector condition also occurs if the clock spring electrical connector is removed and the centering spring is depressed, allowing the coil hub to rotate. *If the clock spring electrical connector is uncentered in relation to the steering column, the conductive ribbon inside the unit will probably break when the steering wheel is turned fully in one direction.* Replacement clock spring electrical connectors are supplied in the centered position. They must be installed with the front wheels in the straight-ahead position.

The clock spring electrical connector centering procedure varies depending on the vehicle. Always follow the procedure in the vehicle manufacturer's service manual. The following procedure applies to Honda vehicles:

1. Rotate the clock spring electrical connector clockwise until it stops.
2. Rotate the clock spring electrical connector counterclockwise for approximately two turns until the yellow gear tooth lines up with the alignment mark on the cover (Figure 13-14).
3. Observe the arrow mark on the cover to be sure it points straight up.

Air Bag System Wiring Repairs

If a wiring pigtail is damaged on any air bag system component, the component must be replaced complete with new pigtails. Wire, connector, or terminal repairs must not be attempted on the sensors, inflator module, or clock spring electrical connector. If any of the air bag wiring is damaged

Special Tools

Crimp and seal splice sleeves

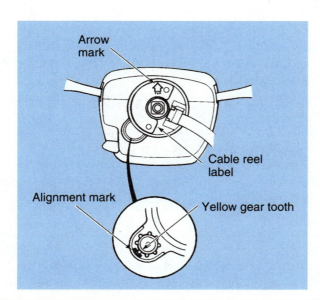

Figure 13-14 Centering the clock spring electrical connector (Courtesy of American Honda Motor Co., Inc.)

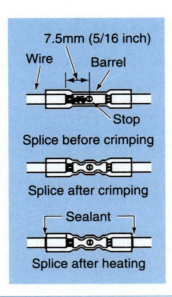

Figure 13-15 Crimp and seal wiring connectors for air bag wiring repairs (Courtesy of Pontiac Division, General Motors Corporation)

other than pigtails, use only crimp and seal splice sleeves recommended by the vehicle manufacturer (Figure 13-15). When connectors in the air bag wiring harness are damaged, complete connector repair kits are available. These connectors and pigtails should be installed using the crimp and seal sleeves for each wire.

Air Bag Deployment Before Vehicle Scrapping

 CAUTION: If the air bag is not deployed before vehicle scrapping, an air bag deployment during this operation may result in personal injury.

If the air bag has not been deployed and the vehicle is being scrapped, follow these steps to deploy the air bag:

1. Be sure the ignition switch is off. Disconnect the negative battery cable and isolate the cable end. Wait 2 minutes or the time specified by the vehicle manufacturer.
2. Disconnect the two-wire connector at the base of the steering column.
3. Cut the two disconnected wires 6 inches from the connector on the wiring harness side of the connector.
4. Attach two 20-foot wires to the 6-inch wires connected to the connector, and leave the outer ends of the 20-foot wires shorted together.
5. Reconnect the two-wire connector at the base of the steering column.
6. Check that the inflator module is securely attached to the steering wheel.
7. Remove any loose objects from the front seat.
8. Be sure that no one is in the vehicle.
9. Straighten the two wires so that the wire ends are as far away from the vehicle as possible.
10. Disconnect the outer ends of the 20-foot wires from each other, and connect the two wires to the terminals of a fully charged 12-V battery to deploy the air bag (Figure 13-16).
11. The inflator module is hot after deployment; do not touch this component for 20 minutes.
12. When handling the air bag, wear gloves, safety goggles, and a long-sleeved shirt or coveralls. Wash your hands with mild soap and water after handling the air bag.

Special Tools

Twenty-foot jumper wires for air bag deployment

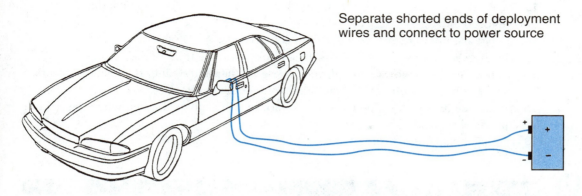

Figure 13-16 Two 20-foot wires and a fully charged battery used to deploy an inflator module before vehicle scrapping (Courtesy of Pontiac Division, General Motors Corporation)

If the inflator module is removed from the vehicle, it may be deployed using the same procedure. Prior to inflator module deployment, place the inflator module in a parking lot, and be sure there is nothing within 20 feet of the inflator module.

● **CUSTOMER CARE:** During many automotive service operations such as air bag system service, the technician literally has the customer's life in her hands! Always perform automotive service, including air bag system repairs, carefully and thoroughly. Always watch for unsafe conditions such as improper air bag system warning light operation, and report these problems to the customer. When you prove to the customer that you are concerned about vehicle safety, you will probably have a steady customer.

Classroom Manual
Chapter 13, page 317

Guidelines for Diagnosing and Servicing Air Bag Systems

1. When diagnosing or servicing any air bag system component, always disconnect the negative battery cable, isolate the cable end, and wait for the amount of time specified by the vehicle manufacturer before proceeding with the required service.
2. Air bag system components must not be replaced with components of lesser quality.
3. Do not strike or jar air bag system sensors or ASDMs.
4. All sensors and brackets must be tightened to the specified torque.
5. Before diagnosing an air bag system, observe the air bag warning light for proper operation during an air bag system diagnostic check.
6. On Ford products, an audible tone is heard if the air bag system warning light does not provide proper operation.
7. On some air bag systems such as those on Ford products, the air bag warning light begins flashing a DTC if a fault occurs in the system.
8. On other air bag systems, a jumper wire may be connected to the specified terminals on the DLC to obtain flash codes on the air bag system warning light.
9. A scan tester must be used to obtain DTCs on some air bag systems.
10. Some manufacturers recommend the installation of a load tool or inflator module simulator tool while servicing air bag system components.
11. Always carry live inflator modules with the trim cover facing away from your body.
12. When placing live inflator modules on the workbench, always position them with the trim cover facing upward.

13. Wear safety gloves, eye protection, and long-sleeved coveralls when handling deployed air bags.
14. Do not attempt to repair air bag system components.
15. The arrows on sensors and ASDMs must face toward the front of the vehicle.
16. Do not attempt to remove a deployed air bag for 20 minutes after deployment occurs.
17. Do not connect an ohmmeter across inflator module terminals.
18. An uncentered clock spring electrical connector will likely be broken when the steering wheel is turned.
19. Deploy the air bags before scrapping a vehicle.

CASE STUDY

A customer brought a 1995 Monte Carlo into the shop with the air bag warning light illuminated continually. The technician checked the electrical connections on the sensing and diagnostic module (SDM) and the inflator modules. These connectors appeared to be in satisfactory condition. The technician connected the scan tester to the DLC under the instrument panel. The scan tester displayed code 71 indicating an internal (SDM) problem. The technician realized this code could result from a full crash data memory in the SDM. An inspection of the vehicle did not reveal any indication of collision repairs.

The technician followed all the necessary precautions and replaced the SDM. During a diagnostic system check, the air bag warning light flashed 7 times and then remained off. The technician connected the scan tester to the DLC, and no DTCs appeared on the tester.

Terms to Know

Air bag system diagnostic module (ASDM)
Clock spring electrical connector
Current DTC
Data link connector (DLC)
Diagnostic trouble code (DTC)
Flash code
History DTC
Inflator module

ASE Style Review Questions

1. While discussing air bag system service:
   Technician A says before an air bag system component is replaced, the negative battery cable should be disconnected, and the technician should wait 2 minutes.
   Technician B says this waiting period is necessary to dissipate the reserve energy in the air bag system computer.
   Who is correct?
   A. A only
   B. B only
   C. Both A and B
   D. Neither A nor B

2. While discussing air bag sensor service:
   Technician A says incorrect torque on air bag sensor fasteners may cause improper air bag deployment.
   Technician B says the arrow on an air bag sensor must face toward the driver's side of the vehicle.
   Who is correct?
   A. A only
   B. B only
   C. Both A and B
   D. Neither A nor B

3. While discussing air bag warning light operation:
   Technician A says on some vehicles if the air bag system is working normally, the air bag warning light should flash 7 to 9 times when the engine is started and then remain on.
   Technician B says on some vehicles if the air bag system is working normally, the air bag warning light remains on for 6 to 8 seconds after the engine is started and then remains off.
   Who is correct?
   A. A only
   B. B only
   C. Both A and B
   D. Neither A nor B

4. While discussing air bag system flash code diagnosis:
   Technician A says on some Ford products, the air bag computer priorizes faults and flashes the code representing the highest priority fault.
   Technician B says on some air bag systems, the air bag computer disarms the system if a fault occurs that could result in an unwarranted air bag deployment.
   Who is correct?
   A. A only
   B. B only
   C. Both A and B
   D. Neither A nor B

5. While discussing flash code diagnosis of air bag systems:
   Technician A says the air bag system warning light flashes codes in random order.
   Technician B says on General Motors vehicles, the air bag system warning light flashes current and history codes.
   Who is correct?
   A. A only
   B. B only
   C. Both A and B
   D. Neither A nor B

6. While discussing air bag system flash code diagnosis:
   Technician A says on Toyota vehicles if there are no faults in the air bag system, the air bag warning light flashes 4 times per second in the diagnostic mode.
   Technician B says on Toyota vehicles a jumper wire must be connected between terminals TC and E1 in the DLC to obtain air bag system codes.
   Who is correct?
   A. A only
   B. B only
   C. Both A and B
   D. Neither A nor B

7. While discussing air bag system service and diagnosis:
   Technician A says A/C-powered voltmeters may be used to diagnose air bag systems.
   Technician B says probe-type 12-V test lights may be used to diagnose air bag systems.
   Who is correct?
   A. A only
   B. B only
   C. Both A and B
   D. Neither A nor B

8. While discussing scan tester diagnosis of air bag systems:
   Technician A says some air bag systems do not provide flash codes and the scan tester must be used to diagnose these systems.
   Technician B says the scan tester should be used to diagnose the air bag system while sitting in the vehicle.
   Who is correct?
   A. A only
   B. B only
   C. Both A and B
   D. Neither A nor B

9. While discussing air bag system disabling:
   Technician A says some manufacturers recommend disabling the air bag system by connecting a special load tool to the inflator module connector near the bottom of the steering column.
   Technician B says one manufacturer recommends disabling the air bag system by removing the inflator modules and connecting a simulator tool to the wiring harness side of the inflator module connector.
   Who is correct?
   A. A only
   B. B only
   C. Both A and B
   D. Neither A nor B

10. While discussing air bag system service:
    Technician A says a live inflator module should be carried with the trim cover facing your body.
    Technician B says an inflator module may be tested by connecting an ohmmeter to the module terminals.
    Who is correct?
    A. A only
    B. B only
    C. Both A and B
    D. Neither A nor B

Table 13-1 ASE Task

Diagnose the cause(s) of the air bag warning light staying on or flashing.

Problem Area	Symptoms	Possible Causes	Classroom Manual	Shop Manual
Air bag system	Air bag warning light stays on or flashes	Defective air bag system component or wiring	309	329

Table 13-2 ASE Task

Inspect, test, repair, or replace the air bag, air bag module, sensors, connectors, and wires of the air bag system circuit(s).

Problem Area	Symptoms	Possible Causes	Classroom Manual	Shop Manual
Air bag system	Fault codes present in air bag computer	Defective air bag system component or wiring	310	330

Servicing and Diagnosing Computer-Controlled Transmissions

CHAPTER 14

Upon completion and review of this chapter, you should be able to:

- ❏ Define customer complaints related to transmission or transaxle operation.
- ❏ Check transaxle fluid level and condition.
- ❏ Perform a throttle cable adjustment.
- ❏ Perform a shift control cable adjustment.
- ❏ Complete a manual valve lever position sensor (MVLPS) adjustment.
- ❏ Perform a scan tester diagnosis of transmission or transaxle electronics.
- ❏ Perform a key on engine off (KOEO) test and obtain diagnostic trouble codes (DTCs) related to transaxle problems.
- ❏ Perform a wiggle test.
- ❏ Complete an output cycling test.
- ❏ Perform a key on engine running (KOER) test and obtain DTCs related to transaxle problems.
- ❏ Obtain flash DTCs related to transaxle problems.
- ❏ Diagnose transaxle and transmission problems using a transmission tester.
- ❏ Perform a DTC diagnosis on a transaxle with a pattern select switch.
- ❏ Diagnose the throttle position sensor and brake switch on a transaxle with a pattern select switch.

Basic Tools

Basic technician's tool set

Service manual

Jumper wires

A computer-controlled transmission may be defined as an electrohydraulic, mechanical device that provides various gear ratios. The transmission is mounted longitudinally behind the engine in most rear wheel drive vehicles.

General Diagnosis of Computer-Controlled Transaxles and Transmissions

Complaint Definition

When transmission problems occur, always define the complaint and the exact conditions when the malfunction occurs. For example, determine if the problem occurs when the engine and transmission are hot, cold, or at any temperature. If necessary, road test the vehicle with the transmission at various temperatures. *During our discussions in this chapter the words* transmission *and* transaxle *are interchangeable.*

Find out the exact operating conditions when the malfunction is present. Does the problem occur when accelerating, decelerating, upshifting, downshifting, or cornering? Does the problem occur at a specific vehicle speed?

Determine if the problem is affected by road or weather conditions. Does the problem occur during severely hot or cold conditions? Does the problem occur on slippery road surfaces? During a road test, remember any unusual operating problems or conditions. When the customer complaint is carefully identified, the diagnosis of the problem will likely be faster and more accurate.

Reduced Engine Performance

Engine performance affects transmission operation. Reduced engine performance may result in a wider throttle opening at any specific engine speed. This wider throttle opening may result in improper transmission shifting. When diagnosing a transmission problem, one of the first checks is to verify the engine performance. Road test the vehicle and from your experience determine if the engine performance is normal.

A transaxle may be defined as a transmission and differential combined in one unit. A transaxle is used in front wheel drive (FWD) vehicles. Many transaxles are mounted transversely with the engine, but some transaxles and engines are mounted longitudinally.

345

Fluid Condition and Level Check

Contaminated fluid or incorrect fluid level causes improper transmission operation. For example, low fluid level may result in clutch slipping or erratic shifting. When the fluid level is higher than specified, foaming of the fluid may cause some of the fluid to be forced from the transaxle vent.

> **CAUTION:** If the vehicle has been driven recently, the transaxle fluid may be very hot. When changing transmission fluid, wear protective gloves and face protection to avoid burns.

Remove the transmission dipstick and check the fluid on the dipstick for discoloration and a burned smell. If the fluid is darker than normal or smells burned, change the fluid and filter. When the oil pan is removed to change the oil and filter, check the oil pan for accumulation of clutch material, engine coolant, or metallic and copper filings. If there is an accumulation of clutch material or metal and copper filings in the oil pan, transmission overhaul may be required. When engine coolant is present in the oil pan, the transmission cooler in the radiator is leaking.

To perform a typical transmission fluid level check:

1. Drive the vehicle until the transmission fluid is at the normal operating temperature of 180°F (82°C).
2. Park the vehicle on a level surface and apply the parking brake.
3. With the engine idling, move the gear selector lever through each gear position and return this selector to the park position.
4. Remove the transmission dipstick and wipe it clean with a shop towel.
5. Fully insert the dipstick into the transaxle filler tube.
6. Remove the dipstick and observe the fluid level on the dipstick. The fluid level should be in the hot range (Figure 14-1). Add the vehicle manufacturer's specified fluid as required to bring the fluid level within the hot range on the dipstick.

> **SERVICE TIP:** Always use the vehicle manufacturer's recommended transaxle fluid. The friction qualities vary in different transaxle fluids. A fluid other than the one recommended by the vehicle manufacturer may result in improper transmission operation such as harsh shifting.

> **SERVICE TIP:** The transmission and differential sumps are separate in some automatic transaxles, and each sump must be filled separately with the vehicle manufacturer's specified fluid. In other automatic transaxles, the transmission and differential have a common sump. Always refer to the vehicle manufacturer's draining and refilling procedures. Be sure both the transmission and differential are filled to the specified level.

Improper Adjustments

Throttle Cable Adjustment. If the throttle cable connected from the accelerator pedal to the throttle linkage is not adjusted properly, the throttle valve may not be in the wide-open position when the accelerator pedal is pushed fully downward. Under this condition, engine power is reduced and transmission shifting may be improper. The throttle cable adjustment procedure varies depending on the vehicle model and year. Always follow the instructions in the vehicle manufacturer's service manual. The following throttle cable adjustment procedure is for a 1993 or later model Toyota Camry:

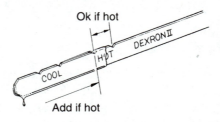

Figure 14-1 Transaxle dipstick (Courtesy of Toyota Motor Corporation)

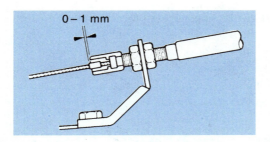

Figure 14-2 Throttle cable adjusting nuts (Courtesy of Toyota Motor Corporation)

1. Fully depress the accelerator pedal.
2. Loosen the throttle cable adjusting nuts (Figure 14-2).
3. Adjust the outer cable so the distance between the end of the boot and the stop on the cable is 0–0.040 in. (0–1 mm).
4. Tighten the adjusting nuts to the specified torque.
5. Recheck the throttle opening with the accelerator pedal fully depressed.

Shift Control Cable Adjustment. An improper shift control cable adjustment may result in the transmission being in a different gear from the one selected by the driver. The shift control cable adjustment varies depending on the vehicle model and year. Always follow the adjustment procedure in the vehicle manufacturer's service manual. Following is a typical shift control cable adjustment procedure on a 1993 or later model Toyota Camry:

1. Loosen the shift lever swivel nut at the transaxle (Figure 14-3).
2. Push the manual valve fully toward the right side of the vehicle.
3. Return the lever two notches to the neutral position.
4. Position the shift lever in the passenger compartment in the neutral position.
5. Lightly hold the shift lever on the transaxle toward the reverse position and tighten the swivel nut to the specified torque.

Manual Valve Lever Position Sensor. Since the manual valve lever position sensor (MVLPS) informs the transmission control module (TCM) regarding the gear selector lever position selected by the driver, improper MVLPS adjustment may result in improper transaxle shifting. If the engine starts in any gear selector position except neutral or park, an MVLPS adjustment is required. A typical MVLPS adjustment follows:

1. Loosen the MVLPS mounting bolts and position the gear selector lever in the neutral position.
2. Align the neutral basic line on the MVLPS with the groove on the transaxle shift linkage (Figure 14-4).
3. Hold the MVLPS in this position and tighten the mounting bolts to the specified torque.

A manual valve lever position sensor (MVLPS) may be called a manual lever position (MLP) sensor or a park/neutral position switch.

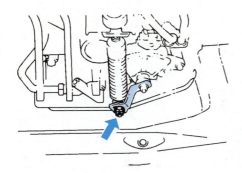

Figure 14-3 Shift lever swivel on the transaxle (Courtesy of Toyota Motor Corporation)

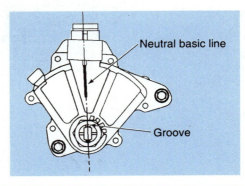

Figure 14-4 Manual valve lever position sensor (MVLPS) adjustment (Courtesy of Toyota Motor Corporation)

Shop Manual
Chapter 13, page 331

Idle Speed Adjustment Check. Specified engine idle speed is essential for proper transmission operation. For example, if the idle speed is faster than specified, the shifting may be harsh when the gear selector is moved from park or neutral to drive or reverse. Improper downshifting may occur if the idle speed is faster than specified. Since the idle speed is computer controlled on most fuel injected engines, it is not adjustable. Improper idle speed may be caused by an intake manifold vacuum leak or a defective input sensor.

Hydraulic and Mechanical Malfunctions

Since this chapter is a discussion regarding computer-controlled transmissions and transaxles, we will concentrate on the electronic diagnosis. However, it is sometimes necessary to differentiate between mechanical, hydraulic, and electronic problems. Therefore, a summary of mechanical and hydraulic problems and their causes is provided (Figure 14-5).

POSSIBLE CAUSE \ CONDITION	HARSH ENGAGEMENT FROM NEUTRAL TO D	R	DELAYED ENGAGEMENT FROM NEUTRAL TO D	R	POOR SHIFT QUALITY	SHIFTS ERRATIC	DRIVES IN NEUTRAL	DRAGS OR LOCKS	GRATING, SCRAPING, GROWLING NOISE	ENGINE MISFIRE	BUZZING NOISE	BUZZING NOISE DURING SHIFTS ONLY	HARD TO FILL OIL BLOWS OUT FILLER TUBE	TRANSAXLE OVERHEATS	HARSH UPSHIFT	NO UPSHIFT INTO OVERDRIVE	NO TORQUE CONVERTER CLUTCH	HARSH DOWNSHIFTS	HIGH SHIFT EFFORTS	HARSH CONVERTER CLUTCH
Engine Performance	X	X			X										X			X		
Worn or faulty clutch(es)	X	X	X	X	X	X	X								X	X		X		
— Underdrive clutch	X		X		X	X	X											X		
— Overdrive clutch					X	X	X								X	X				
— Reverse clutch		X		X	X	X														
— 2/4 clutch					X		X								X			X		
— Low/reverse clutch	X	X			X		X											X		
Clutch(es) dragging						X														
Insufficient clutch plate clearance						X								X						
Damaged clutch seals			X	X														X		
Worn or damaged accumulator seal ring(s)	X	X	X	X														X		
Faulty cooling system														X						
Engine coolant temp. too low																X	X			
Incorrect gearshift control linkage adjustment			X	X	X	X								X						
Shift linkage damaged																			X	
Chipped or damaged gear teeth								X	X											
Planetary gear sets broken or seized								X	X											
Bearings worn or damaged								X	X											
Driveshaft(s) bushing(s) worn or damaged									X											
Worn or broken reaction shaft support seal rings			X	X	X	X										X				
Worn or damaged input shaft seal rings			X	X											X					
Valve body malfunction or leakage	X	X	X	X	X	X	X				X				X	X	X			
Hydraulic pressures too low			X	X	X	X								X	X	X				
Hydraulic pressures too high	X	X													X			X		
Faulty oil pump			X	X		X									X			X		
Oil filter clogged			X	X	X	X						X								
Low fluid level			X	X	X	X		X						X				X	X	
High fluid level														X	X					
Aerated fluid			X	X	X	X		X						X	X			X	X	
Engine idle speed too low			X	X																
Engine idle speed too high	X	X													X			X		
Normal solenoid operation											X									
Solenoid sound cover loose											X									
Sticking torque converter clutch position																				X
Torque Converter Failure	X														X		X			X
Drive Plate cracked or bent								X	X											

Figure 14-5 Transaxle mechanical and hydraulic problems and their causes (Courtesy of Chrysler Corporation)

Electronic Diagnosis

Scan Tester Diagnosis

The proper module must be installed in the scan tester for the model year and transaxle being tested. Drive the vehicle until the engine and transaxle reach normal operating temperature before an electronic transaxle diagnosis. A transaxle electronic diagnostic procedure follows:

1. Be sure the transaxle is at normal operating temperature.
2. Check the transaxle fluid level; if necessary, add fluid until the level is at the hot mark on the dipstick.
3. Check all the electrical connections on the transaxle for loose, corroded, or damaged connections.
4. Install the proper module in the scan tester for the vehicle model year and transaxle being tested.
5. Connect the scan tester voltage supply wires to the battery terminals with the proper polarity. This voltage supply wire may also be connected to the cigarette lighter.
6. Enter the model year and vehicle identification number (VIN) in the scan tester plus any other information requested by the scan tester.
7. Connect the scan tester data lead to the data link connector (DLC) under the instrument panel (Figures 14-6 and 14-7).
8. Select transaxle diagnosis on the scan tester.
9. Perform a shift lever status test with the scan tester. The shift lever position displayed on the scan tester should match the actual shift lever position in each gear. If the actual and displayed shift lever positions do not match, perform the gear shift linkage and manual valve lever position sensor (MVLPS) adjustments. Do not proceed with the fault code diagnosis until the actual and displayed shift lever positions are the same.

WARNING: Inaccurate diagnosis may occur if the actual and displayed gear shift positions do not match.

10. Record all the diagnostic trouble codes (DTCs) displayed on the scan tester.

Special Tools

Scan tester

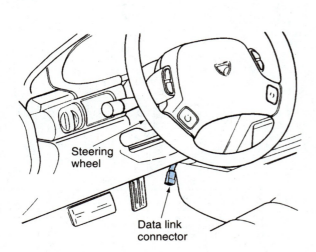

Figure 14-6 Data link connector (DLC) under the instrument panel (Courtesy of Chrysler Corporation)

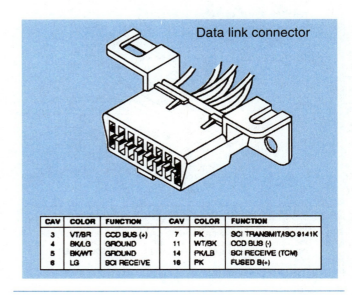

Figure 14-7 Data link connector (DLC), 16 terminal OBD II (Courtesy of Chrysler Corporation)

An intermittent diagnostic trouble code (DTC) may be called a memory code.

Classroom Manual
Chapter 14, page 333

On Chrysler 42LE transaxles in LH cars, a hard fault is a DTC that comes back within three engine starts. A counter in the TCM records the ignition switch cycles. A soft fault is an intermittent defect that is not present every time the TCM checks the electronic system. Soft faults are usually caused by wiring and electrical connector problems.

If a code is present, further diagnosis with a voltmeter or ohmmeter will probably be necessary to locate the exact cause of the problem. For example, if code 22 is present indicating a problem with low 2-4 clutch pressure, the vehicle manufacturer's service procedure may direct the technician to check the voltage supply to the solenoid pack with the ignition switch on. This voltage is supplied on terminal 7 of the transaxle eight-way harness side connector (Figure 14-8). Next, the technician may be instructed to check the resistance of the 2-4 solenoid, which is connected to terminals 2 and 7 on the transaxle side eight-way connector (Figure 14-9). Detailed voltmeter and ohmmeter diagnoses are provided in the vehicle manufacturer's service information. This service information also lists the DTCs and possible causes (Figure 14-10). Photo Sequence 12 shows a typical procedure for performing a scan tester diagnosis of a computer-controlled transaxle.

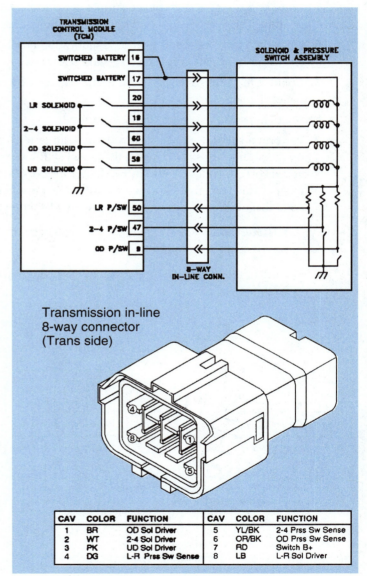

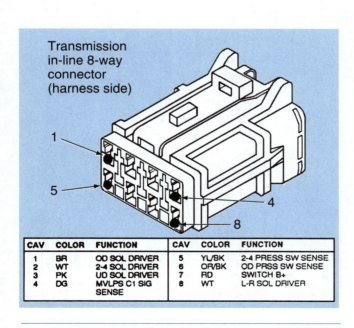

Figure 14-8 Harness side of the transaxle eight-way connector (Courtesy of Chrysler Corporation)

Figure 14-9 Transaxle side of the transaxle eight-way connector (Courtesy of Chrysler Corporation)

Fault Code Number	Condition	Planetary gear sets broken or seized	Faulty cooling system	Torque converter clutch failure	Internal solenoid leak	Pressures too high	Valve body leakage	Regulator valve	Torque converter control valve	Solenoid switch valve	Stuck/sticky valves	Plugged filter	Worn or damaged accumulator seal rings	Damaged clutch seals	L/R clutch	2/4 clutch	Reverse clutch	OD clutch	UD clutch	Damaged or failed clutches	Worn pump	Worn or damaged input shaft seal rings	Worn or damaged reaction shaft support seal rings	Aeroled fluid (high fluid level)	Low fluid level
21	OD clutch—pressure too low	x	x	x	x	x	x			x	x	x	x			x		x		x		x		x	x
22	2/4 clutch—pressure too low	x	x	x	x	x	x			x	x	x	x		x			x		x				x	x
23	2/4 clutch and OD clutch—pressures too low	x	x	x	x	x	x			x	x	x	x					x		x				x	x
24	L/R clutch—pressure too low	x	x	x	x	x	x			x	x	x	x	x				x		x				x	x
25	L/R clutch and OD clutches—pressures too low	x	x	x	x	x	x			x	x	x	x					x		x				x	x
26	L/R clutch and 2/4 clutches—pressures too low	x	x	x	x	x	x			x	x	x	x					x		x				x	x
27	OD, 2/4, and L/R clutches—pressures too low	x	x	x	x	x	x			x	x	x	x					x		x				x	x
31	OD clutch pressure switch response failure			x	x		x											x		x				x	x
32	2/4 pressure switch response failure				x		x						x					x		x				x	x
33	2/4 and O/D clutch pressure response failures				x		x				x							x		x				x	x
37	Solenoid switch valve stuck in the LO position				x		x			x	x							x		x					
38	Partial torque converter clutch out of range			x	x	x	x		x	x	x									x	x				
47	Solenoid switch valve stuck in the LR position				x		x			x	x							x		x					
50	Speed ratio default in reverse	x	x		x	x	x	x			x	x	x	x	x		x			x	x	x			x
51	Speed ratio default in 1st	x	x		x	x	x	x			x	x	x	x	x				x	x	x	x			x
52	Speed ratio default in 2nd	x	x		x	x	x	x			x		x	x		x		x		x	x				x
53	Speed ratio default in 3rd	x	x		x	x	x	x			x		x	x				x		x	x				x
54	Speed ratio default in 4th	x	x		x	x	x	x			x		x	x		x		x		x	x				x
60	Inadequate LR element volume	x					x						x	x	x										
61	Inadequate 2/4 element volume	x					x						x	x		x									
62	Inadequate OD element volume	x					x						x	x				x							

NOTE: Code 36 is not stored alone. It is stored if a speed error (codes 50 through 58) is detected immediately after a shift. Look at the possible causes associated with the speed error code.

Figure 14-10 Diagnostic trouble codes (DTCs) and possible causes (Courtesy of Chrysler Corporation)

Photo Sequence 12
Typical Procedure for Performing a Scan Tester Diagnosis of a Computer-Controlled Transaxle

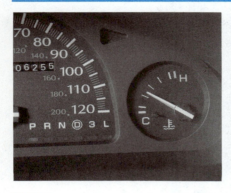

P12-1 Be sure the transaxle and engine are at normal operating temperature.

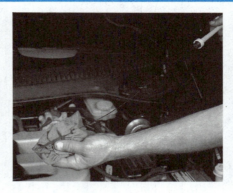

P12-2 Check the transaxle fluid level; if necessary, add fluid until the level is at the hot mark on the dipstick.

P12-3 Check all the electrical connections on the TCM and transaxle for loose, corroded, or damaged connections.

P12-4 Install the proper module in the scan tester for the vehicle model year and transaxle being tested.

P12-5 Connect the scan tester voltage supply wires to the battery terminals with the proper polarity. This voltage supply wire may also be connected to the cigarette lighter.

P12-6 Enter the model year and vehicle identification number (VIN) in the scan tester plus any other information requested by the scan tester.

P12-7 Connect the scan tester data lead to the data link connector (DLC) under the instrument panel and select transaxle diagnosis on the scan tester.

P12-8 Perform a shift lever status test with the scan tester. The shift lever position displayed on the scan tester should match the actual shift lever position in each gear.

P12-9 Record all the diagnostic trouble codes (DTCs) displayed on the scan tester.

Electronic Diagnosis of Ford Computer-Controlled Transaxles and Transaxles with Electronic Pressure Control Solenoids

Preliminary Tests

As was explained previously in this chapter, the following items must be checked first during transaxle diagnosis:

1. Complaint definition
2. Engine performance
3. Fluid level and condition
4. External adjustments
5. Idle speed check

Key On Engine Off Test

Prior to electronic diagnosis, the transmission must be at normal operating temperature. Since the same PCM controls transmission and engine functions, the codes related to engine and transmission defects are obtained simultaneously. The key on engine off (KOEO) test procedure follows:

1. Be sure the ignition switch is off.
2. Install the proper module in the scan tester for the vehicle make and year.
3. Connect the scan tester voltage supply wires to the battery terminals with the proper polarity. Enter the model year and VIN in the scan tester.
4. Connect the scan tester data lead to the DLC (Figure 14-11).
5. Turn on the ignition switch and observe the scan tester display. In the KOEO test, hard fault DTCs are displayed followed by a separator code 10 and DTCs representing intermittent faults (Figure 14-12).
6. Record all the DTCs displayed on the scan tester.

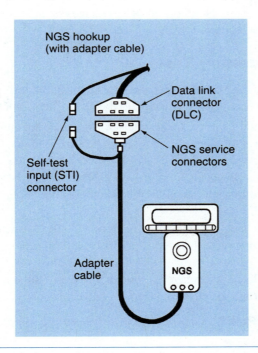

Figure 14-11 Scan tester connected to the data link connector (DLC) (Reprinted with the permission of Ford Motor Company)

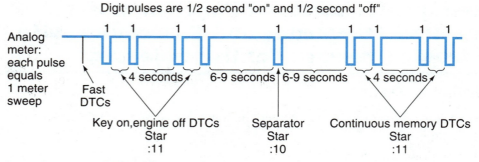

Figure 14-12 Key on engine off (KOEO) test format (Reprinted with the permission of Ford Motor Company)

When a DTC is displayed, voltmeter and ohmmeter tests may be necessary to pinpoint the exact cause of the problem. Pinpoint tests for most DTCs are included in the vehicle manufacturer's service information.

Wiggle Test

At the conclusion of the KOEO test, the wiggle test may be performed. Tap, wiggle, or move the suspected connector or sensor. If a malfunction such as an open circuit is detected, the scan tester provides continuous tone and the malfunction indicator light (MIL) is illuminated.

Output Cycling Test

When the KOEO test is completed, all the PCM outputs related to transmission and engine functions may be turned on by depressing the accelerator pedal to the wide-open position. If the accelerator pedal is depressed again to the wide-open position, all the outputs are turned off. The scan tester indicates whether the outputs are on or off. A voltage test may be performed at a suspected output during the output cycling test.

Key On Engine Running Test

If other tests have preceded the key on engine running (KOER) test, turn off the ignition switch and wait 10 seconds. Leave the scan tester connected to the DLC. The KOER test procedure follows:

1. Start the engine and observe the scan tester. An engine identification code should be displayed representing one-half the engine cylinders. On a V8 engine, code 40 should be displayed.
2. If the vehicle has a power steering pressure switch (PSPS), turn the steering wheel at least one-half turn in each direction. If this action is not taken, a DTC representing the PSPS is displayed.
3. When the vehicle is equipped with a brake on/off (BOO) switch, step on the brake pedal. Failure to complete this step causes a DTC representing the BOO switch to be displayed.
4. If a separator code 10 is displayed, momentarily depress the accelerator pedal to the wide-open position and release it (Figure 14-13).

✓ **SERVICE TIP:** When a separator code 10 is not displayed, do not touch the accelerator pedal.

5. Record the DTCs displayed on the scan tester. The DTCs in the KOER test represent hard faults that are present during the test.

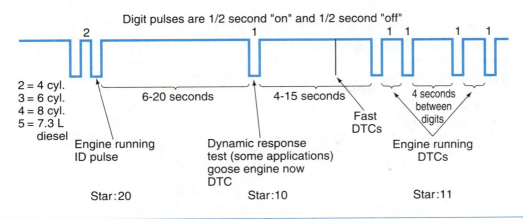

Figure 14-13 Key on engine running (KOER) test format (Reprinted with the permission of Ford Motor Company)

After the necessary repairs are completed to correct the problems, the DTCs may be erased with the scan tester.

Alternate Test Method

SERVICE TIP: Since many Ford vehicles have three-digit DTCs, it is inconvenient to read codes such as 638 with the MIL flashes. This code is displayed as six flashes followed by a pause, and then three flashes followed by a pause and eight flashes. It is easier to read the codes in digital form on the scan tester.

Rather than using the scan tester to perform the various tests, a jumper wire can be connected from the single self-test input wire near the DLC to the appropriate DLC terminal (Figure 14-14). After this connection is completed, the ignition switch can be turned on to enter the KOEO test. When the jumper wire is connected, the engine can be started to enter the KOER test. During the KOEO and KOER tests, the MIL flashes the DTCs. If the jumper wire is disconnected during the KOEO test, DTCs are erased.

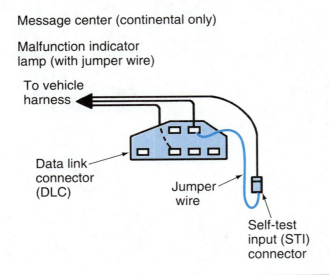

Figure 14-14 Jumper wire connected from the self-test input wire to the appropriate wire in the DLC to obtain DTCs (Reprinted with the permission of Ford Motor Company)

Computer-Controlled Transmission and Transaxle Diagnosis with Transmission Tester

Transmission Tester Connecting Procedure

Special Tools

Transmission tester

A transmission tester is recommended by some vehicle manufacturers. The vehicle wiring harness is disconnected from the transmission, and the transmission tester wiring connector is plugged into the transmission wiring connector (Figure 14-15). The vehicle wiring harness connector remains disconnected. The use of the transmission tester separates the vehicle electronics and the transmission electronics. The transmission tester replaces the PCM. If the fault is present with the transmission tester connected, the defect is in the transmission. When the defect is present with the PCM wiring connected to the transmission, but the problem disappears with the transmission tester connected, the problem is in the vehicle wiring, sensors, or PCM.

Disconnecting the transmission wiring connector sets DTCs in the PCM memory. After the diagnosis and service is completed, these codes must be erased. When the transmission wiring connector is disconnected, the transmission defaults to maximum pressure. This action results in firm shifts. Various overlays are available for the transmission tester depending on the transmission or transaxle being tested. These overlays fit over the tester switches, indicator lights, and jacks. Always be sure the proper overlay is positioned on the tester for the transmission or transaxle being tested (Figure 14-16). Before the transmission tester is connected, check all the items mentioned previously under General Diagnosis. The procedure for connecting the transmission tester follows:

WARNING: Do not attempt to pry off connectors with a screwdriver. This action could damage the connector, resulting in high-resistance problems in the terminals.

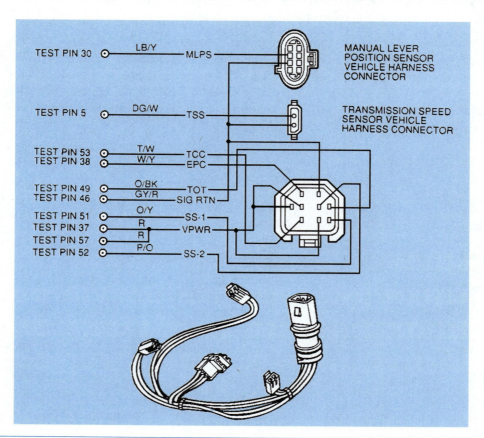

Figure 14-15 Transmission wiring connector and internal wiring (Reprinted with the permission of Ford Motor Company)

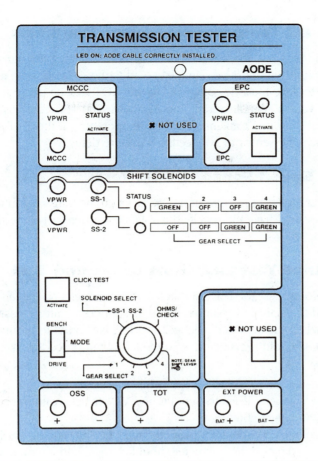

Figure 14-16 Overlay for transmission tester (Reprinted with the permission of Ford Motor Company)

1. Disconnect the wiring harness at the transmission connector.
2. Rotate the tester solenoid selector switch to the ohms/check position.
3. Install the proper tester interface cable and place the correct overlay on the tester.
4. Install a line pressure gauge into the line pressure tap on the transmission.

 WARNING: Route all cables and hoses away from heat sources such as exhaust pipes and manifolds to prevent cable and hose damage.

5. Install the transmission tester power supply cord into the cigarette lighter. All the tester LEDs should be illuminated for a brief time and then turn off. This action proves the tester is operating properly.
6. Place the bench/drive switch in the bench position.

Resistance Tests

If the DTCs obtained in the KOEO or KOER tests indicate a problem in a specific solenoid or sensor, the technician may test that component first. To test the resistance in the shift solenoids or modulated converter clutch control (MCCC) solenoid, connect a pair of ohmmeter leads from the VPWR terminal to the solenoid terminal. For example, to test the MCCC solenoid, connect the ohmmeter leads to the MCCC solenoid terminal and the VPWR terminal adjacent to this solenoid terminal. The shift solenoids, MCCC solenoid, and EPC solenoid must have the specified resistance. If resistance of any solenoid is less than specified, the solenoid winding is shorted. An infinite ohmmeter reading indicates an open circuit in the solenoid winding or wiring connections. An ohmmeter reading above the specified value indicates a resistance problem in the solenoid winding or wiring harness.

Special Tools

Digital multimeter

Transmission Oil Temperature Sensor Diagnosis

The ohmmeter should be set to the x1,000 ohm scale before testing the transmission oil temperature (TOT) sensor. Connect the positive ohmmeter lead to the +TOT jack on the transmission tester. The negative ohmmeter lead must be connected to the -TOT jack on the tester. Compare the ohmmeter reading to the vehicle manufacturer's specifications. The TOT sensor should have the specified resistance in relation to the transmission fluid temperature. If the TOT sensor does not have the specified resistance, the sensor or connecting wires are defective.

Output Shaft Speed Sensor Static Test

Connect the ohmmeter leads to the +OSS and -OSS jacks on the transmission tester with the proper polarity. If the OSS does not have the specified resistance, the sensor is defective or the connecting wires are defective.

Solenoid Voltage Test and Short to Ground Test

When any solenoid is activated, the LED for that solenoid turns green. The LED goes out if the solenoid is deactivated. If the activated solenoid wiring is shorted to battery positive, the LED for that solenoid turns red. When the activated solenoid winding is shorted to ground or has an open circuit, the LED for that solenoid remains off. Use the following procedure to test the EPC solenoid and connecting wires:

1. Leave the tester control switch in the bench position.
2. Connect the voltmeter leads to the EPC and VPWR jacks with the proper polarity.
3. Press the EPC activate button.
4. If the EPC solenoid is operating properly, the EPC LED should be illuminated green, the voltage reading on the voltmeter should change, and an audible click should be heard from the solenoid.

EPC Solenoid Dynamic Testing

The procedure for EPC solenoid dynamic testing follows:

WARNING: Do not depress the EPC switch while accelerating the engine with the brake applied and the transmission in drive. This action may cause transmission damage.

1. Leave the tester control switch in the bench position.
2. Place the gear selector in park and start the engine.
3. Observe and record the line pressure. This pressure should be at maximum specification.
4. Depress the EPC switch. Line pressure should drop to the minimum specification.
5. Set the control switch to the drive position.
6. Rotate the gear select switch on the transmission tester to the first gear position and depress the EPC switch. Line pressure should decrease to idle pressure.
7. Move the gear selector on the vehicle to reverse. This shift should occur normally.
8. Move the gear selector on the vehicle from reverse to drive. This shift action should be normal.

Upshift and Downshift Diagnosis

The car must be driven on the road for the upshift and downshift diagnosis, and the mode switch on the transmission tester must be in the drive position. When a shift solenoid is activated, the appropriate LED on the transmission tester is illuminated. Shifts are firm during this diagnosis because the transmission tester is controlling the transmission shifting and line pressure is maximum.

The vehicle manufacturer's service information instructs the technician to drive the vehicle at a specific speed and then rotate the gear selector switch on the transmission tester to a certain position. The shift selected on the tester should occur and the specified shift solenoid LED should be illuminated for the energized solenoid or solenoids. Follow the vehicle manufacturer's specified procedure to check all the upshifts and downshifts.

Converter Clutch Engagement

 WARNING: Do not depress the MCCC switch with the transmission in gear and the engine idling. This action may cause converter clutch damage.

This test should be performed on the road and the tester mode switch must be in the drive position. Drive the vehicle in third gear as determined by the transmission tester gear selector switch position. Press the MCCC switch and check the vehicle operation for converter clutch engagement. A slight momentary shock should be felt as this clutch engages.

Output Shaft Speed Sensor Dynamic Test

Connect the voltmeter leads to the +OSS and -OSS jacks on the transmission tester with the proper polarity. Slowly accelerate the vehicle and observe the voltmeter. The voltage should increase gradually in relation to vehicle speed if the OSS is working normally.

When the transmission testing is completed, disconnect the transmission tester wiring harness and reconnect the vehicle wiring harness to the transmission connector. Disconnect the transmission tester power lead from the cigarette lighter. Erase all the DTCs with the scan tester or with the procedure mentioned in the KOEO test. Repeat the KOEO and KOER tests to be sure all the faults have been corrected.

Classroom Manual
Chapter 14, page 342

Diagnosis of General Motors 4L80-E Transmission

Fault Code Diagnosis

The PCM continually monitors the inputs and outputs, and each of these components is assigned a normal operating range in the PCM software. If an input or output goes out of the normal, assigned range, a code is set in the PCM memory. When the fault applies to both engine and transmission operation, such as code 21 TPS, the MIL is illuminated. On some vehicles, if the fault relates to transmission control only, the MIL is not illuminated. Prior to diagnosis, the engine and transmission should be at normal operating temperature. A jumper wire must be connected between terminals A and B in the data link connector (DLC) to obtain the fault codes. After this jumper connection is completed, turn on the ignition switch and observe the MIL. The diagnostic trouble codes (DTCs) are flashed in numerical order, and each code is repeated three times.

Scan Tester Serial Data Diagnosis

A scan tester may be connected to the DLC under the instrument panel to obtain DTCs and transmission data. Select transmission data on the scan tester. The engine and transmission must be at normal operating temperature before the serial data is obtained. Many of the serial data items related to such items as transmission shifting and torque converter clutch operation must be obtained by lifting the vehicle and operating the transmission in various gears at different engine and vehicle speeds.

CAUTION: Be sure the vehicle is securely supported on safety stands with the drive wheels several inches off the floor before operating the engine with the transmission in gear. Failure to observe this precaution may result in personal injury and vehicle damage.

The technician must place the gear selector lever in various positions and observe the results on the scan tester to test the transmission range pressure switches. The scan tester indicates whether the A, B, and C circuits in these pressure switches are on or off (Figure 14-17).

Voltmeter and Ohmmeter Tests

Ohmmeter tests may be performed at the transmission connector to determine if an electrical problem is inside the transmission. The technician must disconnect the wiring harness from the transmission connector and identify the terminal location in the transmission side of this connector. For example, terminal E is a battery voltage supply wire and terminal A is connected from shift sole-

SCAN Position	Units Displayed	Typical Data Value
Engine Speed	RPM	± 50 RPM from Desired
Trans Output Speed	RPM	0 RPM
Eng Cool Tamp	C°/F°	85°C - 105°C (185°F - 221°F)
Trans Fluid Temp	C°/F°	82°C - 94°C (180°F - 200°F)
Throt Position	Volts	0.3 - 0.9V
Throttle Angle	Percentage	0%
# A/B/C RNG	Off/On	Off/On/Off
* A/B/C RNG	Off/On	On/Off/On
Trans Range Sw	Invalid, Rev, Drive 4, 3, 2, Low, Park/Neut	Park/Neut
Commanded Gear	1-4	1
# Adaptable Shift	No, Yes	No
1-2 Sol, 2-3 Sol	Off/On	On/On
CTR FDBK 1/2 2/3	Off/On	On/On
* Trans Input Speed	RPM	± 50 RPM of Engine Speed
# 3-2 Control Sol	Percentage	0%
# 3-2 Control FDBK	Off/On	Off
Hot Mode	No, Yes	No
TCC PWM Solenoid	Percentage	0%
TCC Slip Speed	RPM	± 50 RPM from Engine Speed
# TCC Solenoid	Off/On	Off
# CTR FDBK TCC Sol	Off/On	Off
Desired PCS	Amps	1.01 Amps
Actual PCS	Amps	1.01 Amps
PCS Duty Cycle	Percentage	40% - 60%
MPH Km/h	0-255	0
4WD Low Switch	No, Yes	No
Cruise Engaged	No, Yes	No
TCC Brake Switch	Open/Closed	Closed
Kickdown Enabled	No, Yes	No
* Trans Gear Ratio	Ratio	0.00
* Turbine Speed	RPM	± 50 RPM of Engine Speed
1-2 Shift Time	Seconds	0
2-3 Shift Time	Seconds	0
# 1-2 Shift Time Error	Seconds	0
# Curr Adapt Cell	% TP	Not Used
Trans Calib ID	0-65535	Internal ID
System Voltage	Volts	12.0 - 14.5V

4L60-E   * 4L80-E

Figure 14-17 Transmission serial data (Courtesy of Chevrolet Motor Division, General Motors Corporation)

noid A to the PCM (Figure 14-18). The ohmmeter leads should be connected to terminals A and E in the transmission connector to test shift solenoid A for open circuits and shorts. The resistance of shift solenoid A or B should be 20 Ω with the solenoid temperature at 68°F (20°C) (Figure 14-19). When the ohmmeter leads are connected from terminals A or E to ground on the transmission case, an infinite reading indicates the solenoid winding and connecting wires are not grounded. With this ohmmeter connection, a low ohmmeter reading indicates the winding in shift solenoid A is grounded or the connecting wires are grounded.

The pressure control solenoid resistance should be 3.5–8 Ω with the solenoid at 68°F (20°C.) The resistance of the TCC PWM solenoid winding should be 10–15 Ω at 68°F (20°C.) The maximum current flow through the TCC PWM solenoid should be 1.5 amperes.

Classroom Manual
Chapter 14, page 347

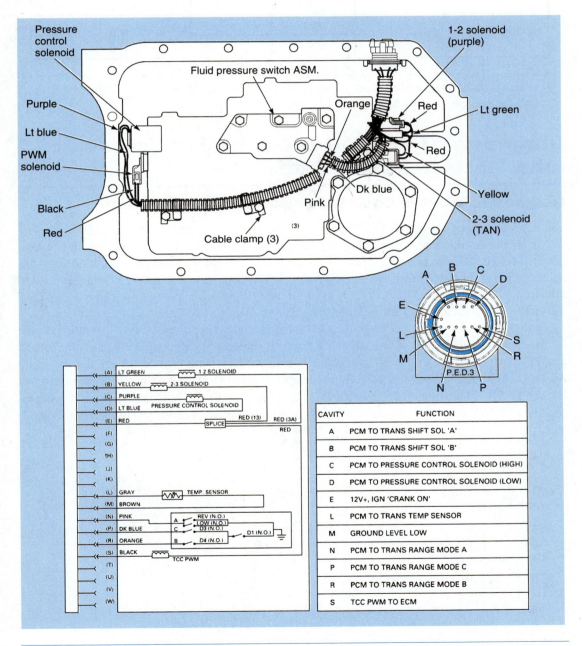

Figure 14-18 Transmission wiring connector terminal identification (Courtesy of Chevrolet Motor Division, General Motors Corporation)

4L80-E COMPONENT RESISTANCE CHART

COMPONENT	WIRE COLOR	PASS-THRU PIN	RESISTANCE @ 20°C	CKT #
1-2 SHIFT SOLENOID	RED	E *	20 - 40 Ohms	1149A
	LT GREEN	A		1222
2-3 SHIFT SOLENOID	RED	E *	20 - 40 Ohms	1149B
	YELLOW	B		1223
PRESSURE CONTROL SOLENOID	PURPLE	C	3.5 - 8 Ohms	1228
	LT BLUE	D		1229
TCC SOLENOID	RED	E *	10 - 15 Ohms	1149C
	BLACK	S		1350

Figure 14-19 Transmission solenoid wire colors and ohm specifications (Courtesy of Chevrolet Motor Division, General Motors Corporation)

Diagnosis of Computer-Controlled Transaxle with Pattern Select Switch

Diagnostic Trouble Code Diagnosis

The items explained previously in this chapter under General Diagnosis must be checked before the diagnostic trouble code (DTC) diagnosis. The transaxle must be at normal operating temperature before the DTC diagnosis. Since the DTCs are flashed out on the overdrive (O/D) off light, this light should be checked before proceeding with the DTC diagnosis. When the O/D light is pressed inward, the O/D is on, and the O/D light is off. If the O/D switch is pressed so it moves outward, the O/D is off, and the O/D light is illuminated (Figure 14-20).

The DTCs can only be read with the O/D switch in the on position. If the O/D button is in the off position, the O/D light is illuminated continually. Connect a jumper wire between terminals E1 and TE1 in the DLC under the left side of the instrument panel and turn on the ignition switch to read the DTCs (Figure 14-21). The DTCs are indicated by the O/D off light in the instrument panel (Figure 14-22). If there are no DTCs in the PCM memory, the O/D off light flashes twice per second (Figure 14-23). When the O/D light flashes four times followed by a pause and two flashes, code 42

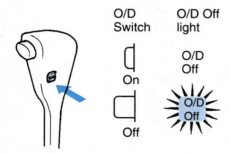

Figure 14-20 Overdrive (O/D) off light operation (Courtesy of Toyota Motor Corporation)

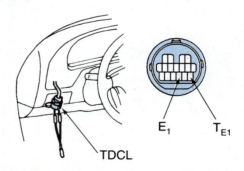

Figure 14-21 Connect a jumper wire between terminals E1 and TE1 in the DLC to read the DTCs. (Courtesy of Toyota Motor Corporation)

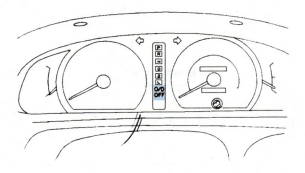

Figure 14-22 The overdrive (O/D) off light in the instrument panel flashes the DTCs. (Courtesy of Toyota Motor Corporation)

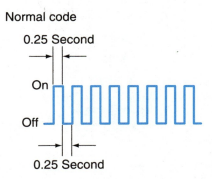

Figure 14-23 If there are no DTCs in the PCM memory, the O/D off light flashes twice per second. (Courtesy of Toyota Motor Corporation)

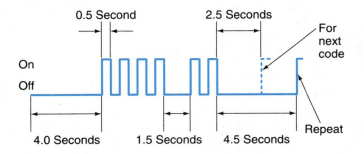

Figure 14-24 When the O/D light flashes four times followed by a pause and 2 flashes, code 42 is indicated. (Courtesy of Toyota Motor Corporation)

representing the number 1 speed sensor is indicated (Figure 14-24). Other transaxle DTCs are available representing the number 2 speed sensor, shift solenoids, and lockup solenoids (Figure 14-25). If there is more than one code in the PCM memory, the DTCs are displayed in numerical order. The DTCs are erased by turning off the ignition switch and removing the EFI fuse for 10 seconds or more (Figure 14-26).

Code No.	Light Pattern	Diagnostic System
–	⎍⎍⎍⎍⎍⎍⎍	Normal
42	⎍⎍⎍⎍ ⎍⎍	Defective No. 1 speed sensor (in combination meter) – severed wire harness or short circuit
61	⎍⎍⎍⎍⎍⎍ ⎍	Defective No. 2 speed sensor (in ATM) – severed wire harness or short circuit
62	⎍⎍⎍⎍⎍⎍ ⎍⎍	Severed No. 1 solenoid or short circuit – severed wire harness or short circuit
63	⎍⎍⎍⎍⎍⎍ ⎍⎍⎍	Severed No. 2 solenoid or short circuit – severed wire harness or short circuit
64	⎍⎍⎍⎍⎍⎍ ⎍⎍⎍⎍	Severed lock-up solenoid or short circuit – severed wire harness or short circuit

Figure 14-25 Other transaxle DTCs are available representing the number 2 speed sensor, shift solenoids, and lockup solenoid. (Courtesy of Toyota Motor Corporation)

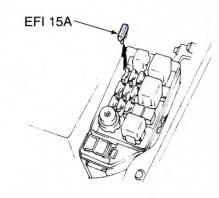

Figure 14-26 The DTCs are erased by turning off the ignition switch and removing the EFI fuse for 10 seconds or more. (Courtesy of Toyota Motor Corporation)

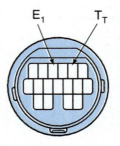

Figure 14-27 Voltmeter connection from the TT to the E1 terminals in the DLC to test the TPS (Courtesy of Toyota Motor Corporation)

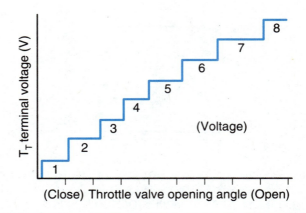

Figure 14-28 Voltmeter reading between the TT and E1 terminals in the DLC as the throttle is moved to the wide-open position (Courtesy of Toyota Motor Corporation)

Diagnosis of Throttle Position Sensor

The throttle position sensor (TPS) signal may be tested by connecting a voltmeter from the TT to the E1 terminals in the DLC (Figure 14-27). When this voltmeter connection is completed, slowly depress the accelerator pedal to the wide-open position. The voltmeter reading should gradually increase to 8 V (Figure 14-28). If the voltmeter reading is erratic, or not within specifications, the TPS or connecting wires are defective.

Brake Signal Test

With the voltmeter connected to the TT and E1 terminals in the DLC, depress and release the brake pedal. When the brake pedal is released, the voltmeter should indicate 8 V. Depressing the brake pedal should provide a 0-V reading. If the voltmeter reading is not within specifications, the brake switch is defective or the connecting wires are defective.

Upshift Voltage Test

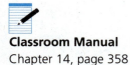

Classroom Manual
Chapter 14, page 358

Connect a voltmeter between the TT and E1 terminals in the DLC, and road test the vehicle. Stop the vehicle and then accelerate slowly while observing the voltmeter. If the transaxle electronic control system is operating normally, the specified voltage reading is available after each upshift (Figure 14-29).

● **CUSTOMER CARE:** Always be willing to spend a few minutes explaining problems, including safety concerns, regarding the customer's vehicle. When a customer understands why certain repairs are necessary, he feels better about spending the money. For

ECT Terminal (V)	Gear Position
0	1st
2	2nd
3	2nd Lock-up
4	3rd
5	3rd Lock-up
6	O/D
7	O/D Lock-up

Figure 14-29 Voltage readings between the TT and E1 terminals in the DLC after each upshift (Courtesy of Toyota Motor Corporation)

example, if you explain that an excessive amount of clutch material in the transaxle oil pan indicates a transaxle failure in the near future, the customer appreciates this warning and is willing to spend the money for the necessary repairs.

Guidelines for Diagnosing and Servicing Computer-Controlled Transmissions and Transaxles

1. Prior to any transmission diagnosis, the customer complaint related to transmission operation must be defined.
2. Reduced engine performance causes improper transmission operation.
3. Before any transmission diagnosis, the fluid level and condition must be checked.
4. The throttle cable and shift control cable must be properly adjusted before any transmission diagnosis.
5. Improper manual valve lever position sensor (MVLPS) adjustment will affect transmission operation.
6. The engine must be idling at the specified speed to obtain proper transmission operation.
7. A scan tester may be connected to the DLC to obtain transmission DTCs and perform other transmission diagnosis.
8. A KOEO and KOER test may be performed to obtain transmission DTCs on some vehicles.
9. On some vehicles, a wiggle test may be performed with the scan tester to test for defective electrical connections.
10. On some vehicles, an output cycling test may be performed with the scan tester to turn the PCM outputs on and off.
11. On some computer-controlled transmissions, a jumper wire may be connected to the proper terminals on the DLC to obtain flash codes related to transmission problems.
12. Some vehicle manufacturers recommend the use of a transmission tester. The use of this tester separates the vehicle electronics and the transmission electronics for diagnostic purposes.

CASE STUDY

A customer complained about an inoperative torque converter clutch on a 1995 Lincoln Town Car. When asked when the problem occurred, the customer replied that she could never feel the torque converter locking up. A road test indicated that the customer's description of the problem was accurate.

The technician performed the KOEO and KOER tests with a scan tester. In both tests, code 652 appeared as a hard fault indicating a problem in the torque converter clutch electrical circuit. The technician connected the transmission tester. When an ohmmeter was connected across the MCCC and VPWR jacks on the transmission tester, the ohmmeter provided an infinite reading, indicating an open circuit in the MCCC solenoid winding or connecting wires inside the transmission.

The technician disconnected the transmission tester and carefully examined the MCCC solenoid terminals in the connector. This examination indicated the terminal connected from the MCCC solenoid to the PCM was pushed back out of place in the transmission connector. The transmission fluid was drained and the pan was removed from the transmission. A replacement internal wiring assembly was installed. A check of the pan did not reveal any sign of clutch material, metal and copper cuttings, or engine coolant. Since the car had low mileage and the fluid and filter were clean, the technician did not change the filter. The transmission pan was installed with a new gasket, and the transmission was refilled with the car manufacturer's specified fluid.

The technician checked the wiring harness side of the transmission connector. The MCCC solenoid terminals were in satisfactory condition in this terminal. The transmission tester was reconnected. When an ohmmeter was connected to the MCCC and VPWR jacks on the tester, the ohmmeter indicated the specified MCCC solenoid resistance.

A voltmeter was connected across the MCCC and VPWR terminals in the transmission tester. When the MCCC solenoid switch on the tester was pressed, the voltmeter reading changed and the MCCC status LED came on. Simultaneously, an audible click was heard in the transmission. The technician disconnected the transmission tester and reconnected the transmission wiring harness. The DTC was erased from the PCM memory. During a road test, the transmission, including the torque converter clutch, operated normally.

Terms to Know

Diagnostic trouble code (DTC)
Electronic pressure control (EPC) solenoid
Key on engine off (KOEO) test
Key on engine running (KOER) test
Malfunction indicator light (MIL)
Manual valve lever position sensor (MVLPS)

Output cycling test
Output shaft speed (OSS) sensor
Powertrain control module (PCM)
Shift control cable adjustment
Throttle cable adjustment
Transaxle
Transmission

Transmission control module (TCM)
Transmission oil temperature (TOT) sensor
Transmission tester
Transmission tester overlay
Wiggle test

ASE Style Review Questions

1. While discussing general transaxle diagnosis:
   Technician A says reduced engine performance may affect transaxle operation.
   Technician B says an improper manual valve lever position sensor adjustment may affect transaxle operation.
   Who is correct?
   A. A only
   B. B only
   C. Both A and B
   D. Neither A nor B

2. While discussing a transaxle fluid level check:
   Technician A says the transaxle fluid should be checked with the fluid cold.
   Technician B says the transaxle fluid should be changed if it is darker than normal.
   Who is correct?
   A. A only
   B. B only
   C. Both A and B
   D. Neither A nor B

3. While discussing transaxle fluid and fluid changing:
   Technician A says various types of transaxle fluid have different friction qualities.
   Technician B says some transaxles have separate transmission and differential sumps.
   Who is correct?
   A. A only
   B. B only
   C. Both A and B
   D. Neither A nor B

4. While discussing a key on engine off (KOEO) test:
   Technician A says the DTCs displayed before the separator code represent hard faults.
   Technician B says the DTCs displayed after the separator code represent hard faults.
   Who is correct?
   A. A only
   B. B only
   C. Both A and B
   D. Neither A nor B

5. While discussing a wiggle test:
   Technician A says if a connection provides an open circuit during the wiggle test, the words *open circuit* are displayed on the scan tester.
   Technician B says if a connection provides an open circuit during the wiggle test, the scan tester provides a continuous tone.
   Who is correct?
   A. A only   C. Both A and B
   B. B only   D. Neither A nor B

6. While discussing a key on engine running test:
   Technician A says if the engine identification code is 4, the vehicle being tested has a 4-cylinder engine.
   Technician B says if a separator code 10 is provided, the brake pedal should be depressed at that time.
   Who is correct?
   A. A only   C. Both A and B
   B. B only   D. Neither A nor B

7. While discussing the use of the transmission tester:
   Technician A says the PCM controls transmission shifting while the transmission tester is connected.
   Technician B says the transmission tester controls transmission shifting when it is connected.
   Who is correct?
   A. A only   C. Both A and B
   B. B only   D. Neither A nor B

8. While discussing the use of the transmission tester:
   Technician A says if a shift occurs with the transmission tester connected but the same shift does not occur with the vehicle wiring harness connected to the transmission, the problem may be in the vehicle wiring.
   Technician B says if an ohmmeter is connected across the EPC and VPWR jacks on the transmission tester and the ohmmeter reading is less than specified, the EPC solenoid has an open circuit.
   Who is correct?
   A. A only   C. Both A and B
   B. B only   D. Neither A nor B

9. While discussing the use of the transmission tester:
   Technician A says if the MCCC solenoid LED turns red when this solenoid is activated with the MCCC switch, the MCCC wiring inside the transmission is shorted to ground.
   Technician B says if the MCCC solenoid LED turns red when this solenoid is activated with the MCCC switch, the MCCC wiring inside the transmission is shorted to battery positive.
   Who is correct?
   A. A only   C. Both A and B
   B. B only   D. Neither A nor B

10. While discussing the diagnosis of a transaxle equipped with a pattern select switch:
    Technician A says the overdrive switch must be in the off position before the DTCs are obtained.
    Technician B says the DTCs are flashed out by the check engine light.
    Who is correct?
    A. A only   C. Both A and B
    B. B only   D. Neither A nor B

Table 14-1 ASE Task

Inspect, adjust, and replace electronic sensors, wires, and connectors.

Problem Area	Symptoms	Possible Causes	Classroom Manual	Shop Manual
IMPROPER TRANSMISSION OPERATION	Improper shifting	1. Inaccurate manual valve lever position sensor adjustment	335	347
		2. Defective turbine speed or output speed sensors	333	349

Table 14-2 ASE Task

Inspect, test, adjust, or replace electrical/electronic components, including computers, solenoids, sensors, relays, switches, and harnesses.

Problem Area	Symptoms	Possible Causes	Classroom Manual	Shop Manual
IMPROPER TRANSMISSION OPERATION	Improper shifting	1. Defective input sensor or connecting wires	343	356
		2. Inaccurate mechanical adjustments	343	347
		3. Defective outputs in transmission or connecting wires	347	356
		4. Defective transmission computer	348	359

Antilock Brake and Traction Control System Diagnosis and Service

Upon completion and review of this chapter, you should be able to:

- ❏ Perform a preliminary antilock brake inspection.
- ❏ Diagnose rear wheel antilock (RWAL) brake systems.
- ❏ Inspect wheel speed sensors and measure sensor gap.
- ❏ Diagnose an antilock brake system with an antilock brake tester.
- ❏ Obtain flash codes on an ABS from the flashes of the amber ABS warning light.
- ❏ Relieve accumulator pressure on antilock brake systems.
- ❏ Follow the proper procedure for accumulator disposal.
- ❏ Bleed air from an ABS modulator with an antilock brake tester.
- ❏ Check the fluid level on an integral ABS with high-pressure accumulator.
- ❏ Use a pressure bleeder to bleed the brakes on an integral ABS with high-pressure accumulator.
- ❏ Use a fully charged accumulator to bleed the rear brakes on an integral ABS with high-pressure accumulator.
- ❏ Use a scan tester to diagnose an integral ABS with a high-pressure accumulator.
- ❏ Perform a scan tester diagnosis of a Delco Moraine ABS VI.
- ❏ Use a scan tester to diagnose an ABS with wheel spin traction control.
- ❏ Diagnose an ABS with spark advance reduction and transmission upshift traction control.
- ❏ Diagnose an ABS with throttle control, wheel spin control, and spark advance reduction traction control.

Diagnosis of Rear Wheel Antilock (RWAL) Systems

Preliminary Inspection

CAUTION: The size and type of tires recommended by the vehicle manufacturer must be maintained on a vehicle equipped with antilock brakes. If tires of a different size are installed, braking efficiency may be reduced, resulting in vehicle damage and personal injury.

Before performing an antilock brake system (ABS) diagnosis, a preliminary inspection should be performed. During a preliminary inspection, the following items should be checked:

1. Check the brake fluid level in the master cylinder reservoir. If a separate reservoir is mounted in the hydraulic control unit, check the level in both reservoirs. Fill these reservoirs to the specified level with the manufacturer's specified brake fluid.
2. Inspect the ABS hydraulic system for fluid leaks.
3. Inspect the ABS for worn mechanical parts such as brake linings on pads or shoes, rotors, and drums.
4. Inspect all the tires. Be sure all the tires are the proper size and rating specified by the manufacturer.

Basic Tools

Basic technician's tool set
Service manual
Feeler gauge
Jumper wire

5. Inspect all wiring connections in the ABS for loose, corroded connections and damaged wires.
6. Inspect the notched rings on the wheel speed sensors for damage.
7. If the wheel speed sensors are adjustable, check these adjustments according to the vehicle manufacturer's recommended procedure.
8. Check all ABS fuses and fuse links.

Diagnostic Trouble Code (DTC) Diagnosis

The DTCs on RWAL systems are obtained by grounding a diagnostic connector. On many RWAL systems, the ignition switch must be on and the diagnostic connector should be grounded for 1 to 2 seconds and then disconnected from ground.

On General Motors trucks with a separate electronic brake control module (EBCM), connect terminal H to A in the data link connector (DLC) to obtain the DTCs. When a General Motors truck has the EBCM combined in the vehicle control module (VCM), continually connect terminals A and B in the DLC to obtain the DTCs. The engine DTCs are indicated by the flashes of the MIL, followed by the RWAL DTCs flashed out by the RWAL warning light. The RWAL DTCs are indicated by one long flash followed by some short flashes. Count the long flash as number one and the first short flash as number two. For example, if one long flash is followed by eight short flashes, code 9 is indicated.

The diagnostic connector is located near the EBCM under the instrument panel on many Dodge trucks.

On some Ford trucks, the diagnostic connector is attached to the wiring harness near the bulkhead connector. On some later model Ford vehicles, the diagnostic connector is located at the forward end of the power distribution box in the engine compartment. These vehicles have an in-line diagnostic connector that must be disconnected, and the black wire with an orange stripe in the connector must be grounded to obtain the RWAL DTCs (Figure 15-1). On late-model Ford trucks, count the long flash as the last in the sequence. For example, five short RWAL light flashes followed by a long flash is code 6.

SERVICE TIP: Some RWAL modules only store one DTC at a time. After one fault is repaired, always drive the vehicle to be sure another fault is not present.

Always consult the vehicle manufacturer's service manual for the erase code procedure. One of the following methods may be recommended to erase codes:

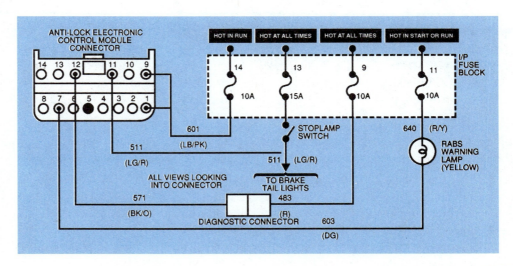

Figure 15-1 RWAL diagnostic connector for late-model Ford trucks (Reprinted with the permission of Ford Motor Company)

1. With the ignition switch off, disconnect the wiring harness from the RWAL module for 10 seconds.
2. On General Motors vehicles with an RWAL system, disconnect the stop fuse for 5 seconds with the ignition switch off.
3. Use the scan tester to erase DTCs on General Motors vehicles with the EBCM in the VCM.
4. On late-model Ford trucks with an in-line diagnostic connector, obtain the DTCs and then turn off the ignition switch with the diagnostic connector disconnected.

On General Motors zero pressure rear wheel antilock (ZPRWAL) systems, follow these steps to erase the DTCs:

1. Be sure the ignition switch is off.
2. Remove the stop/hazard fuse.
3. Turn on the ignition switch.
4. Connect terminals A to H in the DLC for 1 second.
5. Remove the A to H terminal connection for 1 second.
6. Connect terminals A to H for 1 second.
7. Turn off the ignition switch and replace the stop/hazard fuse.
8. Turn on the ignition switch and connect terminals A to H to verify the DTCs are erased.

Scan Tester Diagnosis

On some General Motors RWAL systems, a scan tester may be connected to the DLC to test the RWAL system. The scan tester must have the proper module for the system being tested. When RWAL system is selected on the scan tester, the technician may choose the following modes:

1. Functional test—during this test the scan tester signals the EBCM to cycle the system components. If a fault is present, a DTC is set in the module memory (Figure 15-2).
2. Trouble codes—the scan tester displays the DTCs and supplies some related diagnostic information.
3. Miscellaneous test—this test provides specialized testing of the VSS buffer module and brake switch.
4. VSS monitor—the scan tester displays the VSS signal in mph or km/h.

Classroom Manual
Chapter 15, page 366

Diagnosis of Nonintegral ABS with High-Pressure Accumulator

Wheel Speed Sensor Gap Measurement

SERVICE TIP: Some antilock brake systems, especially integral systems with high-pressure accumulators, require a special procedure for checking the master cylinder fluid reservoir. Always follow the recommended procedure in the vehicle manufacturer's service manual.

SERVICE TIP: Wheel speed sensor gap measurement or adjustment procedures vary depending on the vehicle make and model year. Always follow the instructions in the vehicle manufacturer's service manual.

1. Raise the vehicle on a lift.
2. Check the notched ring at each wheel speed sensor for chipped or damaged teeth. Replace the notched rings if necessary.
3. Use the specified feeler gauge to measure the air gap between the notched ring teeth and the wheel speed sensor tip while rotating the wheel for one revolution (Figure 15-3). Repeat this procedure at each wheel.

A nonintegral ABS has some of the components such as the master cylinder and the modulator mounted separately from each other.

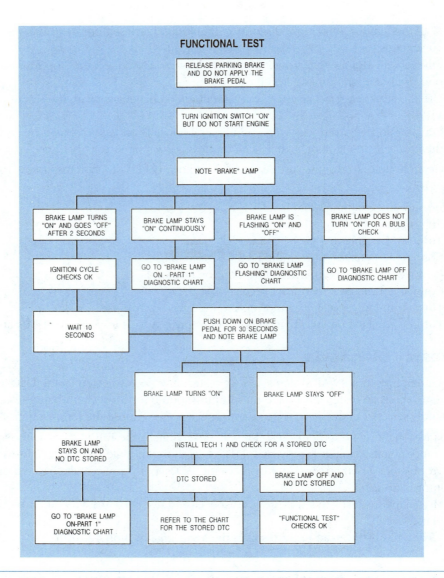

Figure 15-2 Functional test procedure (Courtesy of Chevrolet Motor Division, General Motors Corporation)

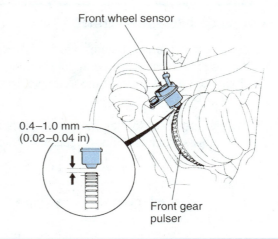

Figure 15-3 Wheel speed sensor air gap measurement (Courtesy of American Honda Motor Co., Inc.)

4. If any wheel speed sensor gap is not within specifications, check the wheel speed sensor retaining bolt for proper torque. When a wheel speed sensor air gap is not within specifications at certain locations around the notched ring, check for distorted components such as wheel hubs.

Antilock Brake System Diagnosis with Antilock Brake Tester

SERVICE TIP: Do not use the antilock brake tester if the ABS warning light in the instrument panel indicates a problem with the ABS. If the ABS warning light is illuminated with the engine running, proceed with the diagnostic trouble code (DTC) diagnosis.

CAUTION: Never drive the car with the antilock brake tester connected. This action may result in a partial or complete loss of braking ability resulting in personal injury and vehicle damage.

Some vehicle manufacturers provide an antilock brake (ALB) tester to check for proper operation of the ABS. This tester checks each system function for satisfactory operation. Follow these steps when using the ALB tester:

1. Place the vehicle on a level shop floor with the wheels blocked. Place the transmission selector in park for automatic transmissions or neutral for manual transmissions.
2. With the ignition switch off, remove the cover from the ABS inspection connector under the passenger's seat (Figure 15-4).
3. Connect the ALB tester to the ABS inspection connector.
4. Start the engine and be sure the parking brake is released.
5. Turn the ALB tester mode selector switch to position #1 and push the start test switch (Figure 15-5). The test-in-progress light on the tester should be illuminated, and in a few seconds, all four monitor lights on the tester should come on. The ABS warning light on the instrument panel should not come on.
6. Rotate the mode selector switch to position #2. Firmly depress the brake pedal and push the start test switch on the tester. The test-in-progress light should come on and the ABS warning light on the instrument panel should remain off. Kickback should be felt on the brake pedal.
7. Repeat step 6 with the mode selector switch in positions 3, 4, and 5. In each of these positions, kickback should be felt on the brake pedal, and the test-in-progress light should come on while the ABS warning light remains off.

Explanation of Test Modes in ALB Tester. During test mode #1, a simulated driving speed signal of 0 mph to 113 mph (180 km/h) and then from this speed back to 0 mph is sent from each wheel speed sensor to the ABS control unit. There should be no pedal kickback in this mode.

Special Tools

Antilock brake (ALB) tester

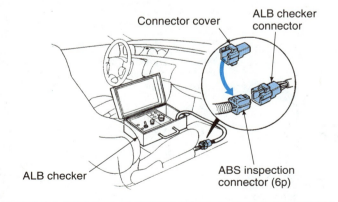

Figure 15-4 ABS inspection connector (Courtesy of American Honda Motor Co., Inc.)

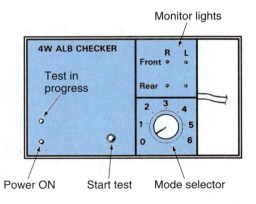

Figure 15-5 Antilock brake (ALB) tester (Courtesy of American Honda Motor Co., Inc.)

In test mode #2, a driving signal from each wheel followed by a lockup signal from the left rear wheel is sent to the ABS control unit. There should be pedal kickback in this mode.

During test mode #3, a driving signal from each wheel followed by a lockup signal from the right rear wheel is sent to the ABS control unit. This lockup signal should result in pedal kickback.

In test mode #4, a driving signal from each wheel followed by a lockup signal from the left front wheel is sent to the ABS control unit. There should be pedal kickback in this mode.

In test mode #5, a driving signal from each wheel followed by a lockup signal from the right front wheel is sent to the ABS control units. This mode should provide pedal kickback.

Interpretation of Test Results from ALB Tester. When the ABS warning light in the instrument panel is illuminated during any of the test modes, the ABS system has an electrical defect. Under this condition, proceed with the DTC diagnosis. If there is brake pedal kickback in modes 2 through 5 and the ABS warning light does not come on, check for air in the high-pressure lines, a restricted high-pressure line, or a faulty modulator unit.

Wheel Speed Sensor Test with ALB Tester

The procedure for testing the wheel speed sensor signals with the ALB tester follows:

1. Leave the ALB tester connected to the ABS inspection connector.
2. Raise the vehicle on a lift.
3. Turn on the ignition switch, and position the ALB tester mode selector switch in the 0 position.
4. Briskly rotate each wheel by hand at one revolution per second. If the wheel speed sensor signal is satisfactory on the wheel being rotated, the monitor light flashes on the ALB tester for that wheel. When any monitor light does not flash, that wheel speed sensor signal is not received by the ABS control unit. If a wheel speed sensor signal is not provided during this test, check the sensor air gap, and check the sensor resistance with an ohmmeter. When these items are satisfactory, check the wires from the ABS control unit to the sensor.

Diagnostic Trouble Code Diagnosis

When the ABS control unit senses a defect in the system, the control unit illuminates the amber ABS warning light and disables the ABS system. Normal power-assisted braking is still available. When the ABS warning light in the instrument panel is illuminated with the engine running, the ABS control unit should be checked for DTCs. The procedure to obtain the DTCs follows:

1. Remove the service check connector from the connector cover mounted behind the front of the center console (Figure 15-6).

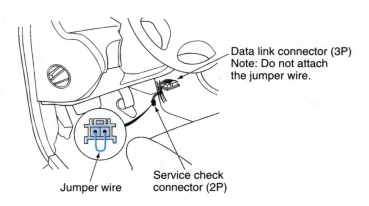

Figure 15-6 Service check connector (Courtesy of American Honda Motor Co., Inc.)

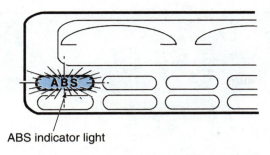

Figure 15-7 The ABS warning light flashes the system DTCs. (Courtesy of American Honda Motor Co., Inc.)

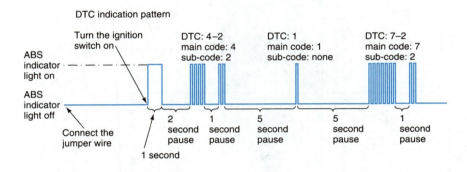

Figure 15-8 The ABS warning light flashes the DTCs as a main code that is usually followed by a sub-code. (Courtesy of American Honda Motor Co., Inc.)

2. Connect the two terminals in the service check connector with a jumper wire.
3. Turn on the ignition switch, but do not start the engine.
4. Record the flashes displayed by the ABS warning light (Figure 15-7).

The ABS warning light flashes the DTCs as a main code, which is usually followed by a sub-code (Figure 15-8). The ABS control unit is capable of storing three DTCs. Turn the ignition switch off and on to repeat the DTC display. DTCs are available for most of the ABS system components (Figure 15-9). A DTC indicates a problem in a specific area. Voltmeter or ohmmeter tests are usually required to locate the exact cause of the problem.

After the necessary repairs are completed, the DTCs may be erased in the control unit memory by turning the ignition switch off and disconnecting the ABS B2 15-A fuse in the underhood fuse/relay box for 3 seconds. The DTCs are also erased if the control unit connector is disconnected with the ignition switch off. The ABS control unit may be removed from the vehicle body with the ignition switch off to erase the DTCs.

Relieving Accumulator Pressure

CAUTION: Always relieve the accumulator pressure before loosening the accumulator or the hose from the accumulator to the modulator. Failure to observe this precaution may release high-pressure brake fluid, resulting in personal injury.

WARNING: Brake fluid is a paint remover! If brake fluid is spilled on the vehicle paint, serious paint damage will result. When brake fluid is spilled on the vehicle paint, immediately wash the spill area with generous amounts of clean water. Even with this action, paint damage may occur.

DIAGNOSTIC TROUBLE CODE (DTC)		PROBLEMATIC COMPONENT/ SYSTEM	AFFECTED			
MAIN CODE	SUB- CODE		FRONT RIGHT	FRONT LEFT	REAR RIGHT	REAR LEFT
①	—	ABS pump motor over-run	—	—	—	—
	②	ABS pump motor circuit problem	—	—	—	—
	③	High pressure leakage	—	—	—	—
	④	Pressure switch	—	—	—	—
	⑧	Accumulator gas leakage	—	—	—	—
②	①	Parking brake switch-related problem	—	—	—	—
③	①	Pulser(s)	○			
	②			○		
	④				○	○
④	①	Wheel sensor	○			
	②			○		
	④				○	
	⑧					○
⑤	—	Wheel sensor(s)			○	○
	④				○	
	⑧					○
⑥	—	Fail-safe relay		○		○
	①			○		
	④					○
⑦	①	Solenoid related problem	○			
	②			○		
	④				○	○

Figure 15-9 DTCs for an ABS (Courtesy of American Honda Motor Co., Inc.)

A high-pressure accumulator in an ABS is a round metal container with a heavy diaphragm in the center. Pressurized nitrogen gas is sealed on one side of the diaphragm, and pressurized brake fluid is pumped into the opposite side of the diaphragm.

Follow these steps to relieve the accumulator pressure:

1. Remove the red cap from the bleeder screw on the modulator body (Figure 15-10).
2. Install the special bleeder tool on the bleed screw and slowly turn this tool for 90° to release the high-pressure brake fluid from the accumulator into the reservoir on the special tool.
3. Rotate the special tool one complete turn to release all the accumulator pressure.
4. Retighten the bleeder screw and install the red cap on this screw.
5. Discard the brake fluid from the special tool.

Accumulator Disposal

CAUTION: Never puncture or attempt to disassemble an accumulator. Since the accumulator contains high-pressure nitrogen gas, this action may result in personal injury.

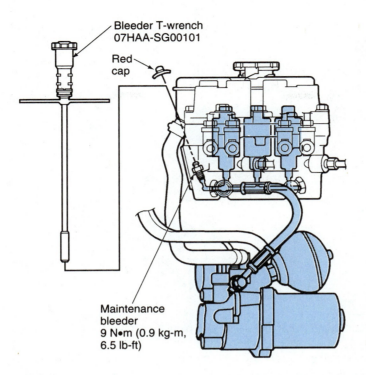

Figure 15-10 Bleeder screw and cap on modulator body (Courtesy of American Honda Motor Co., Inc.)

CAUTION: Never expose an accumulator to excessive heat or flame. Since the accumulator contains high-pressure nitrogen gas, this action may cause an explosion resulting in personal injury.

CAUTION: Never throw an accumulator containing gas pressure in an incinerator. This action could result in an explosion causing personal injury.

Follow these steps for accumulator disposal:

1. Relieve the accumulator pressure.
2. Secure the pump and motor assembly in a vise and loosen the accumulator with the proper wrench. Turn the accumulator to thread it out of the pump assembly (Figure 15-11).

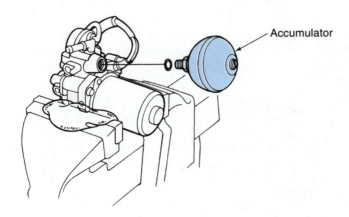

Figure 15-11 Removing the accumulator from the pump and motor assembly (Courtesy of American Honda Motor Co., Inc.)

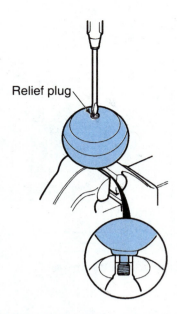

Figure 15-12 Loosening the accumulator relief plug (Courtesy of American Honda Motor Co., Inc.)

3. Clamp the hex-shaped area on the lower end of the accumulator in a vise with the relief plug pointing upward (Figure 15-12).
4. Slowly turn the relief plug 3.5 turns counterclockwise and wait 3 minutes for the pressure to escape.
5. Completely remove the relief plug and discard the accumulator.

Modulator Air Bleeding with the Antilock Brake Tester

SERVICE TIP: Air in the modulator may result in longer-than-normal pump motor run time and improper antilock brake operation. When air in the modulator causes longer-than-normal pump motor run time, the pump runs with a constant soft sound.

The modulator air bleeding procedure with the antilock brake (ALB) tester follows:

1. Park the vehicle on a level shop floor and block the wheels. Place the transmission in park for automatic transmissions, or neutral for manual transmissions. Release the parking brake.
2. Connect the ALB tester to the ABS inspection connector under the passenger's seat.
3. Fill the modulator reservoir to the maximum fill line with the vehicle manufacturer's specified brake fluid (Figure 15-13). Install the reservoir cap.
4. Start the engine and allow the engine to idle for a few minutes. Shut off the engine and check the fluid level in the modulator reservoir. If necessary, add fluid to the reservoir until the level is at the maximum fill line.

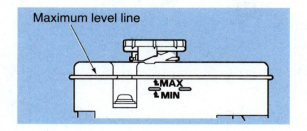

Figure 15-13 Maximum fill line on modulator reservoir (Courtesy of American Honda Motor Co., Inc.)

5. Remove the cap from the accumulator bleeder screw and install the special bleeder tool. Loosen the bleeder screw with the tool and allow the high-pressure fluid into the tool reservoir. Close the bleeder screw.
6. Repeat step 4.
7. Turn the mode selector on the ALB tester to position #2. Firmly depress the brake pedal and push the start test button. Kickback should be felt on the brake pedal. If no kickback is felt, repeat steps 5 through 7. When kickback is felt, proceed to step 8.
8. Repeat step 7 with the mode selector in positions 3, 4, and 5. Refill the modulator reservoir to the maximum fill line if required.

Special Tools
Brake bleeder tool

Classroom Manual
Chapter 15, page 372

Diagnosis of Integral ABS with High-Pressure Accumulator

Antilock Brake Warning Lights

The amber antilock brake warning light is located in the roof console or in the instrument panel. When the ignition switch is turned on, the electronic controller performs a check of all the ABS components. This check requires 3 to 4 seconds, and during this time the amber ABS warning light is illuminated. When the electronic controller completes the system check and no defects are located, the controller shuts off the amber ABS light. If the amber ABS warning light is illuminated with the engine running, the electronic controller has sensed a defect in the ABS. Under this condition, the electronic controller shuts down the antilock function and a DTC is set in the controller memory. Normal power-assisted braking is still available.

The red brake warning light may be illuminated by a parking brake application, low fluid level in the master cylinder, or the electronic controller.

Checking and Filling the Master Cylinder Reservoir

CAUTION: Before any hydraulic component or line is disconnected, the accumulator pressure must be relieved. This is done by pumping the brake pedal 25 times with the ignition switch off. Failure to observe this procedure may result in personal injury.

The brake fluid level may be above the maximum fill line on the master cylinder reservoir depending on the accumulator state of charge. If the master cylinder reservoir is overfilled, the reservoir may overflow when the accumulator discharges. To check and fill the master cylinder reservoir:

1. Turn on the ignition switch and pump the brake pedal until the hydraulic pump motor starts in the master cylinder.
2. Wait until the pump motor shuts off. The accumulator is now fully charged.
3. The brake fluid should be at the maximum fill line on the master cylinder reservoir (Figure 15-14). If the fluid level is below this line, add the vehicle manufacturer's specified brake fluid to bring the brake fluid up to the maximum fill line. When the fluid level is above the maximum fill line, use a suction gun to remove fluid until the level is at the maximum fill line.

Brake Bleeding

SERVICE TIP: When the brakes are applied, air in the wheel calipers or lines causes a soft, spongy feeling on the brake pedal.

The front or rear brakes may be bled with a pressure bleeder. If a pressure bleeder is attached to the master cylinder for bleeding purposes, maintain the bleeder pressure at 35 psi (240 kPa). Be sure the ignition switch is off and the brake pedal released. Connect a hose from the bleeder screw into a plastic container, and open the bleeder screw for 10 seconds. Tighten the bleeder screw and

Special Tools
Pressure brake bleeder

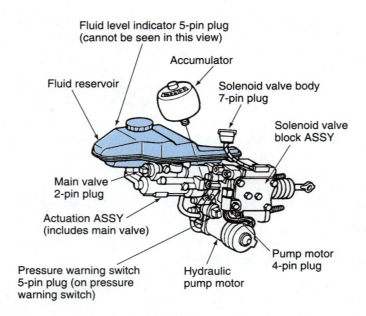

Figure 15-14 Master cylinder reservoir (Reprinted with the permission of Ford Motor Company)

repeat the procedure at each wheel. Be sure the bleeder pressure is maintained. Adjust the master cylinder fluid level to the maximum fill line when the bleeding procedure is completed.

Since accumulator pressure is supplied to the rear wheels, the rear brakes may be bled with a fully charged accumulator. The rear brake bleeding procedure with a fully charged accumulator follows:

> **CAUTION:** When bleeding the rear brakes with a fully charged accumulator, the accumulator pressure is considerably higher than pressure bleeder pressure or conventional master cylinder pressure. Use caution when loosening the rear bleeder screws to avoid personal injury.

1. Turn on the ignition switch and pump the brake pedal until the hydraulic pump motor starts.
2. Wait until the pump motor shuts off.
3. With the ignition switch on, have an assistant apply the brake pedal.
4. Connect a hose from one of the rear bleeder screws into a plastic container. Loosen the bleeder screw until a clear stream of brake fluid is released into the container. Repeat steps 1 through 4 until the clear stream of brake fluid is obtained. Add brake fluid to the master cylinder reservoir as required during and after the bleeding procedure. Repeat the procedure on the opposite rear wheel.

Wheel Speed Sensor Adjustments

On some models, the wheel speed sensors are not adjustable. When the sensor is installed, the retaining bolt must be tightened to the specified torque (Figure 15-15). Some early models required a wheel speed sensor adjustment. On these models, a paper shim had to be installed on the sensor tip. Install the sensor until the paper shim lightly touches the notched ring. With the sensor held in this position, tighten the sensor retaining bolt to the specified torque.

Diagnostic Trouble Code Diagnosis

Before performing the diagnostic trouble code (DTC) diagnosis, always perform the preliminary inspection mentioned previously in this chapter. The DTC diagnostic procedure follows:

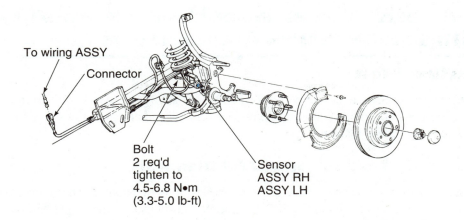

Figure 15-15 Wheel speed sensor mounting (Reprinted with the permission of Ford Motor Company)

 SERVICE TIP: Some ABS DLCs are located in the engine compartment.

1. Be sure the ignition switch is off.
2. Locate the ABS data link connector (DLC) in the trunk right-rear quarter panel area (Figure 15-16).
3. Install the proper module in the scan tester for ABS diagnosis and the model year being diagnosed. Connect the scan tester to the ABS DLC (Figure 15-17).
4. Operate the scan tester according to the tester manufacturer's recommended procedure. Select ABS on the scan tester.
5. Turn on the ignition switch and record the DTCs.

When a DTC in the 20s is received, no other codes are displayed. If a 20 series code is received, repair the problem causing this code and then repeat the DTC diagnostic procedure. A DTC indicates a fault in a specific area. Voltmeter and ohmmeter tests are usually required to locate the exact cause of the problem. After all the faults are corrected, the DTCs are erased when the vehicle is driven above 25 mph (40 km/h).

Special Tools

Scan tester

Classroom Manual
Chapter 15, page 379

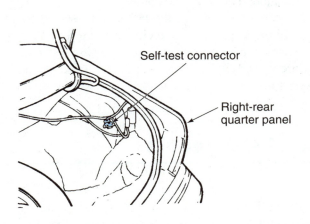

Figure 15-16 ABS data link connector (DLC) located in the trunk (Reprinted with the permission of Ford Motor Company)

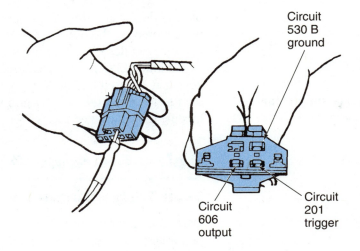

Figure 15-17 Scan tester connected to the DLC (Reprinted with the permission of Ford Motor Company)

381

Diagnosis of Four Wheel Nonintegral ABS with Low-Pressure Accumulators

Functional Test

A preliminary inspection should be performed before any ABS diagnosis. When the preliminary inspection does not indicate any problems, perform the functional test described in the vehicle manufacturer's service manual.

Diagnostic Trouble Code (DTC) Diagnosis

When terminals A and H are connected in the DLC with the ignition switch on, the antilock warning light flashes codes stored in the EBCM memory. The antilock warning light flashes the codes in same way that the MIL flashes engine function codes. To erase DTCs:

1. Turn on the ignition switch.
2. Connect a jumper wire from terminal H to A in the DLC for 2 seconds.
3. Disconnect the jumper wire for 1 second.
4. Reconnect the jumper wire for 2 seconds.
5. When the DTCs are erased, the antilock and brake warning lights turn on and then off.

Scan Tester Diagnosis

A scan tester may be connected to the DLC to diagnose this ABS. Be sure the proper module for the system being tested is installed in the scan tester. The following tests may be performed with the scan tester:

1. Data display—displays input sensor and output function data in the ABS.
2. Function test—allows the scan tester to command the EBCM to cycle the outputs, including the pump motor.
3. Diagnostic trouble codes—displays the DTCs on the scan tester.
4. Snapshot—allows the technician to record ABS data in the scan tester during a road test. The technician may select any code, a specific code, manual trigger, intermittent fault, or ABS brake application to trigger the record function.
5. Miscellaneous tests—in this test mode, PROM ID, history data, history codes, and tire size calibration tests are available.

Classroom Manual
Chapter 15, page 382

In the PROM ID test, the scan tester displays the software version in the EBCM. The history data mode displays the ABS activation and ignition switch cycle counts in relation to when DTCs were stored. In the history code test, the scan tester displays the DTCs in the stored codes and enhanced codes blocks. In the tire size calibration mode, the technician can read the tire size stored in the EBCM. If the technician selects new tire size calibration on the scan tester, the arrow keys may be used to scan the available tire sizes. When the proper new tire size appears on the scan tester, press the enter button to select this tire size. This action recalibrates the VSS buffer module to the new tire size.

Delco Moraine ABS VI Diagnosis

Scan Tester Diagnosis

Data List. Prior to any ABS diagnosis, the preliminary inspection explained previously must be performed. The next step in the diagnostic procedure is to complete the diagnostic system check in the vehicle manufacturer's service manual. Since flash codes are not available on a Delco Moraine VI system, a compatible scan tester must be used to diagnose the system (Figure 15-18). Be sure the proper module is installed in the scan tester for ABS diagnosis and the model year being diag-

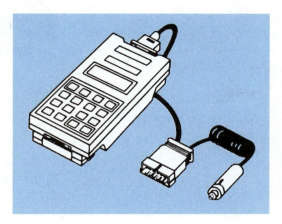

Figure 15-18 Scan tester for ABS diagnosis (Courtesy of Oldsmobile Division, General Motors Corporation)

nosed. The scan tester is connected to the DLC under the instrument panel. When data list is selected on the scan tester, the scan tester continually monitors all the inputs and outputs and displays the appropriate readings from each input and output (Figure 15-19).

DTC History. In the code history mode, the scan tester displays how many ignition cycles have occurred since the DTC was stored. This code history is available for up to five DTCs.

Show DTCs. In the trouble code mode, the scan tester displays both current and history DTCs. The scan tester may be used to clear the DTCs in this mode.

A current DTC represents a hard fault that is present continually.

A history DTC represents an intermittent fault that occurred long enough to be set in the computer memory.

MSC CHASSIS APPLICATION SELECTION	TECH 1 SCREEN DISPLAY
STEP 1 SELECT VEHICLE MODEL YEAR	MODEL YEAR . 19??
STEP 2 SELECT VEHICLE TYPE	"?" CAR
STEP 3 SELECT MODE (FIRST SCREEN)	

- F0: DATA LIST → FRONT WHL SPEEDS / REAR WHL SPEEDS / VEHICLE SPEED / ABS WARNING LAMP / LF ABS MOTOR CMD / LF ABS MOTOR FDBK / RF ABS MOTOR CMD / RF ABS MOTOR FDBK / REAR ABS MOTOR CMD / REAR ABS MOTOR FBK / LF SOLENOID CMD / LEFT FRONT EMB / RF SOLENOID CMD / RIGHT FRONT EMB / BRAKE SWITCH / ABS RELAY CMD / BRK WARN LP CMD / BRK FLUID LEVEL / EBCM BAT VOLTAGE / EBCM IGN VOLTAGE
- F1: DTC HISTORY → INFORMATION ABOUT PREVIOUS DTCs STORED
- F2: SHOW DTC(s) → DTCs STORED AND CLEAR DTCs
- F3: SNAPSHOT → SNAPSHOT OPTIONS
  - F0: REPLAY DATA
  - F1: MANUAL TRIG.
  - F2: AUTO TRIG.
  - F3: ANY DTC
  - F4: SINGLE DTC
  - F9: TRIGGER POINT
    - F0: BEGINNING
    - F1: CENTER
    - F2: END OF DATA
- F4: MISC. TESTS → CONTINUED ON FOLLOWING CHART (PAGE 2 OF 3)
- F5: MOTOR REHOME → REHOMES MOTORS PRIOR TO BLEEDING THE BRAKE SYSTEM.

Figure 15-19 Delco Moraine ABS VI diagnostic modes (Courtesy of Chevrolet Motor Division, General Motors Corporation)

Snapshot. The snapshot mode may be used to store data in the tester memory for future reference. For example, if a problem occurs at a specific speed, the vehicle can be driven at this speed and the input or output data can be recorded in the tester memory. The vehicle can be returned to the shop and the data played back to display the problem.

Miscellaneous Tests. In the ABS tests mode, the scan tester performs functional tests on the hydraulic modulator to assist in problem isolation (Figures 15-20 and 15-21).

Motor Rehome. In this mode, the scan tester can be used to command the EBCM to move the motors, ball screws, and pistons fully upward prior to brake bleeding.

Photo Sequence 13 shows a typical procedure for performing a scan tester diagnosis of an antilock brake system.

> The fully upward piston position in the hydraulic modulator may be referred to as the home position.

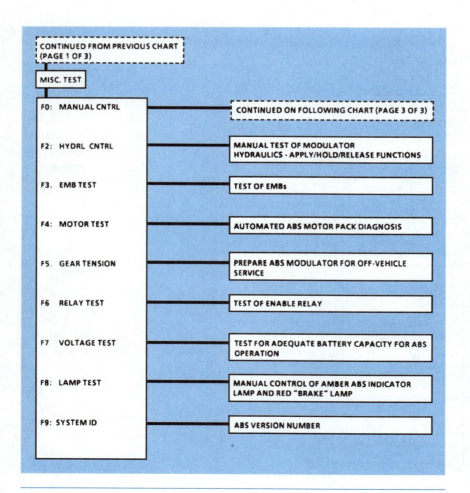

Figure 15-20 Miscellaneous tests in Delco Moraine ABS VI diagnosis (Courtesy of Chevrolet Motor Division, General Motors Corporation)

Figure 15-21 Miscellaneous tests continued, Delco Moraine ABS VI diagnosis (Courtesy of Chevrolet Motor Division, General Motors Corporation)

Photo Sequence 13
Typical Procedure for Performing a Scan Tester Diagnosis of an Antilock Brake System

P13-1 Be sure the ignition switch is off.

P13-2 Install the proper module in the scan tester for the vehicle make, model year, and ABS diagnosis.

P13-3 Connect the scan tester power lead to the cigarette lighter socket.

P13-4 Enter the model year and engine VIN in the scan tester.

P13-5 Connect the scan tester data lead to the DLC under the dash.

P13-6 Select ABS diagnosis from the initial menu.

P13-7 Select data list to display input and output data.

P13-8 Select trouble codes to display codes in the ABS computer memory.

P13-9 With the scan tester connected, drive the vehicle on a road test, and press the proper scan tester button to record data in the tester memory.

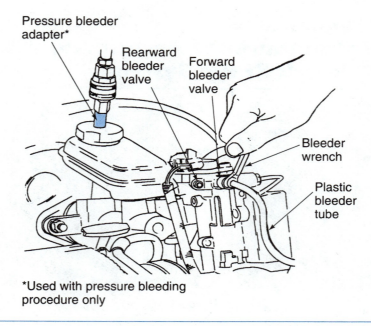

Figure 15-22 Hydraulic modulator bleeding (Courtesy of Chevrolet Motor Division, General Motors Corporation)

Pressure Brake Bleeding

Hydraulic Modulator Bleeding. Before performing a brake bleeding procedure, make sure all of the pistons in the hydraulic modulator are in the fully upward position. The procedure for assuring fully upward piston position follows:

1. Start the engine and let it run for 10 seconds with the brake pedal released.
2. Be sure the ABS warning light is off.
3. Turn off the ignition switch and repeat steps 1 and 2.
4. Be sure the proper module is installed in the scan tester and connect the scan tester to the DLC.
5. Select the motor rehome mode on the scan tester and use this tester to command the EBCM to move the pistons to the fully upward position.
6. Proceed with the pressure or manual brake bleeding procedure.

SERVICE TIP: Diaphragm-type pressure bleeding equipment must be used with a diaphragm between the air supply and the brake fluid to prevent air and moisture from contaminating the brake fluid.

The proper adapter must be used to connect the pressure bleeder to the master cylinder reservoir. Use the air supply to increase pressure in the bleeding equipment to 5 to 10 psi (35 to 70 kPa), and check for leaks in the hose or master cylinder reservoir. When no leaks are present, increase the bleeder pressure to 30 to 35 psi (205 to 240 kPa). Use the following procedure to bleed air from the modulator:

1. Connect a clear plastic hose from the right rear bleeder valve to a clean container partially filled with brake fluid. Submerge the end of the hose in the brake fluid.
2. Slowly open the bleeder valve and allow fluid to flow into the container until there is no air in the fluid.
3. Close the bleeder valve and repeat step 2 if necessary.
4. Repeat steps 2 and 3 on the forward bleeder valve (Figure 15-22).
5. Place a shop towel under the modulator assembly tubing connections.

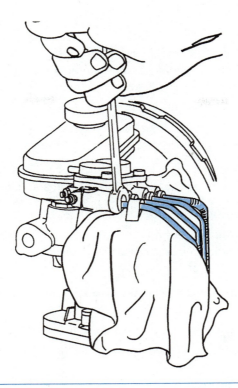

Figure 15-23 Hydraulic modulator pipe bleeding (Courtesy of Chevrolet Motor Division, General Motors Corporation)

6. Slightly loosen the forward brake tubing nut (Figure 15-23). When air stops escaping, tighten the tubing nut to the specified torque.
7. Repeat steps 4 and 5 on the remaining three brake tubes.

Wheel Brake Bleeding. The proper sequence for wheel brake bleeding is right rear, left rear, right front, and left front. The vehicle must be lifted and properly supported to bleed the wheel brakes. The wheel brake bleeding procedure follows:

1. Connect a clear plastic bleeder hose from the rearward bleeder valve, and submerge the other end of the hose in a clean container partially filled with brake fluid.
2. Slowly open the bleeder valve and observe the fluid movement through the hose into the container.
3. Close the bleeder valve when there is no airflow through the bleeder hose.
4. Repeat steps 1, 2, and 3 on the left rear, right front, and left front wheels.

SERVICE TIP: The brake caliper may be tapped lightly with a rubber mallet to dislodge any trapped air bubbles.

Manual Brake Bleeding

Modulator Priming

1. Fill the master cylinder reservoir to the proper level with DOT 3 brake fluid.
2. Connect a clear plastic bleeder hose from the rearward bleeder valve, and submerge the other end of the hose in a clean container partially filled with brake fluid.
3. Slowly open the rearward bleeder valve one-half turn, and depress the brake pedal. Hold the pedal until fluid begins to flow through the bleeder hose.
4. Close the bleeder valve and release the brake pedal. Repeat steps 3 and 4 until there is no air flowing through the bleeder hose. Maintain the proper brake fluid level in the master cylinder as required.

Wheel Brake Bleeding

5. Lift the vehicle and provide proper vehicle support.
6. Connect a clear plastic bleeder hose from the right rear bleeder screw, and submerge the opposite end of the hose in a clean container partially filled with brake fluid.
7. Open the right rear bleeder valve and slowly depress the brake pedal.
8. Close the bleeder valve and release the brake pedal and wait 5 seconds. Repeat steps 6, 7, and 8 until there is no air moving through the bleeder hose.
9. Maintain the brake fluid level in the master cylinder reservoir as required.
10. Repeat steps 6, 7, and 8 on the left rear, right front, and left front wheels.

Hydraulic Modulator Bleeding. The procedure for bleeding the hydraulic modulator follows:

1. Connect a clear plastic bleeder hose to the rearward bleeder valve on the modulator, and submerge the other end of the hose in a container partially filled with brake fluid.
2. Use moderate force to apply the brake pedal.
3. Slowly open the rearward bleeder valve one-half turn and allow the brake fluid to flow into the container.
4. Close the bleeder valve, release the brake pedal, and wait 5 seconds.
5. Repeat steps 2, 3, and 4 until there is no airflow through the bleeder hose.
6. Repeat steps 2, 3, and 4 on the forward bleeder valve.

Classroom Manual
Chapter 15, page 384

Diagnosis of ABS with Wheel Spin Traction Control System

Initialization and Dynamic Tests

Each time the ignition switch is turned on, the controller antilock brake (CAB) performs an initialization test. During this test, the CAB checks all the electrical components in the ABS and traction control system. The CAB momentarily cycles all the build pressure, decay pressure, and isolation solenoids. If the driver has the brake pedal depressed during this test, pedal pulsations may be felt. This is a normal condition.

Each time the engine is started and the vehicle is accelerated to 5 to 10 mph (8 to 16 km/h), the CAB momentarily turns on the pump motor. When the CAB detects a fault during the initialization or dynamic tests, it illuminates the amber ABS warning light and the amber traction control off light. Under this condition, the CAB disables the ABS and the traction control system.

An initialization test is a test completed by the controller antilock brake (CAB) each time the ignition switch is turned on. During this test, the CAB checks all the electrical components in the ABS system.

Diagnostic Mode

A compatible scan tester must be connected to the DLC under the instrument panel to enter the diagnostic mode (Figure 15-24). During the diagnostic mode, both the red and amber brake warning lights flash. If a current fault is detected by the CAB, these lights remain illuminated without flashing. The vehicle speed must be below 10 mph (16 km/h) to enter the diagnostic mode. If the vehicle speed is above this value, the scan tester displays NO RESPONSE. The solenoid valves in the valve body cannot be actuated above 5 mph (8 km/h). When an attempt is made to energize the solenoid valves above this speed, a VEHICLE IN MOTION message is displayed.

A dynamic test is completed by the CAB each time the engine is started and the vehicle is accelerated to a specific speed. During this test, the CAB momentarily cycles the pump motor.

Latching Faults. Many of the possible faults that may be detected by the CAB are latching faults. When the CAB detects a latching fault, the ABS and traction control system are disabled and the amber ABS and traction control off warning lights are illuminated. These systems remain disabled until the ignition switch is cycled, even if the fault disappears. When the ignition switch is turned

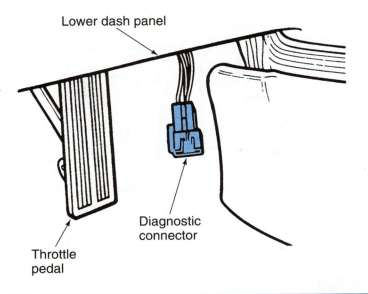

Figure 15-24 ABS data link connector (DLC) under the instrument panel (Courtesy of Chrysler Corporation)

off and on, the systems are operational and the amber ABS and traction control off warning lights remain off until the CAB detects the fault again. Following are latching faults:

1. Main relay or power circuit failure
2. Pump motor circuit not operating properly
3. Pump motor running without command
4. Pedal travel sensor circuit
5. Solenoid valve fault (any one of ten solenoids with traction control)
6. Fluid level switch #2 signal not being processed
7. Pressure switch or brake switch circuits
8. Wheel speed sensor circuit failure (any one of four sensors)
9. Signal missing, wheel speed sensor
10. Wheel speed sensor signals improper comparison
11. Wheel speed sensor continuity below 25 mph
12. Wheel speed sensor continuity above 25 mph

Nonlatching Faults. When the CAB detects a nonlatching fault, the amber ABS and traction control off warning lights are illuminated, and the ABS and traction control system are disabled. If the defective condition disappears, both warning lights go off, and the ABS and traction control system are operational, but a DTC remains in the CAB memory. FLUID LEVEL SWITCH #2 OPEN is a nonlatching fault.

Locked Fault. If the CAB detects a locked fault, the amber ABS and traction control off warning lights are illuminated and the ABS and traction control system are disabled. These lights remain illuminated until the fault is erased from the CAB memory. HYDRAULIC FAILURE is a locked fault; it can only be erased with the erase faults mode in a scan tester.

Noneraseable Fault. CONTROLLER FAILURE is a noneraseable fault. When this fault message is displayed on the scan tester, the CAB must be replaced.

A latching fault causes the CAB to turn on the amber ABS and traction control off lights and disable these systems until the next ignition cycle, even when the fault disappears.

A nonlatching fault causes the CAB to turn on the amber ABS and traction control lights and disable these systems. If the fault disappears, the warning lights go off and these systems are returned to operation.

A locked fault causes the CAB to illuminate the amber ABS and traction control off warning lights and disable these systems. The lights remain illuminated and the systems are disabled until the locked fault code is erased from the CAB memory.

A nonerasable fault cannot be erased with a scan tester or by disconnecting battery voltage from the CAB.

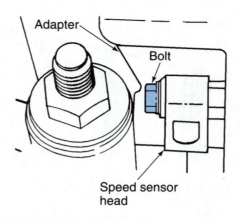

Figure 15-25 Wheel speed sensor mounting bolt (Courtesy of Chrysler Corporation)

Brake Bleeding

When bleeding the brakes on the combined ABS and traction control system, the first step is to bleed the brakes at each wheel with a pressure bleeder. The second step in the brake bleeding procedure is to connect the scan tester to the ABS DLC under the instrument panel and use the scan tester to perform a hydraulic control unit bleed procedure. The third step is to repeat the brake bleeding procedure at each wheel using the pressure bleeder.

Since this system does not have a high-pressure accumulator, there is no special procedure for checking the brake fluid level in the master cylinder reservoir.

Wheel Speed Sensor Service

 WARNING: If a wheel speed sensor is seized in the opening, do not pull on it with a pair of pliers. This action may damage the sensor.

Wheel speed sensor adjustments are not required on these systems. To remove a wheel speed sensor, first remove the sensor mounting bolt (Figure 15-25). If the sensor is seized in the mounting hole, lightly tap the sensor ear with a hammer and punch (Figure 15-26). Lightly tap the sensor on alternate sides of the sensor ear until it comes out of the opening. After the wheel speed sensor is installed, tighten the mounting bolt to the specified torque. Always be sure to position the wheel

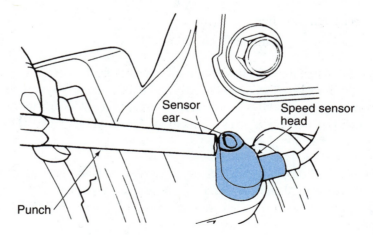

Figure 15-26 Tap lightly on the wheel speed sensor ear with a hammer and punch to loosen the sensor in the mounting hole. (Courtesy of Chrysler Corporation)

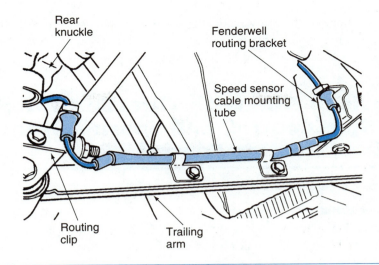

Figure 15-27 Wheel speed sensor wires must be positioned securely in the original clamps. (Courtesy of Chrysler Corporation)

speed sensor wires in the original clamps to prevent these wires from rubbing on other components (Figure 15-27).

Diagnosis of ABS with Spark Advance Reduction and Transmission Upshift Traction Control

Scan Tester Diagnosis

**WARNING:** Use only DOT 3 brake fluid in this system. Do not use DOT 5 brake fluid. The use of brake fluid other than DOT 3 may adversely affect system operation, resulting in reduced braking efficiency and a possible collision.

There is no provision for obtaining flash codes on this system. Prior to system diagnosis, complete the preliminary inspection procedure. Then perform the diagnostic system test provided in the vehicle manufacturer's service manual.

The scan tester must be connected to the DLC under the instrument panel. Before connecting the scan tester, always insert the proper module in the scan tester for the system being tested. The following tests may be performed with a scan tester:

1. **Data list**—When data list is selected on the scan tester, input and output data are displayed, including wheel speed data, brake switch status, and enhanced traction system (ETS) switch and lamp status.
2. **DTC history**—This data includes the number of ignition cycles since a DTC occurred. The first five DTCs and the last DTC recorded are displayed in the order in which they occurred. This information may be useful in determining if a previous defect is linked to the last DTC set in the EBCM memory.
3. **Show DTCs**—In this mode, the scan tester displays history and current DTCs. The scan tester also clears DTCs in this mode. DTCs are also cleared by 100 drive cycles after the fault is corrected. A drive cycle is an ignition on followed by a vehicle speed of 10 mph (16 km/h).
4. **Snapshot**—This mode allows the technician to road test the vehicle and record data into the scan tester memory. This data may be played back to pinpoint intermittent problems.

Classroom Manual Chapter 15, page 391

Classroom Manual
Chapter 15, page 401

5. **Motor rehome**—In this test mode, the scan tester commands the EBCM to move the motors to the home position (fully upward) in the ABS motor pack assembly. This position is necessary before brake bleeding.

Diagnosis of ABS with Throttle Control, Wheel Spin Control, and Spark Advance Reduction Traction Control

Scan Tester Diagnosis

WARNING: Use only DOT 3 brake fluid in this system. Do not use DOT 5 brake fluid. The use of brake fluid other than DOT 3 may adversely affect system operation, resulting in reduced braking efficiency and a possible collision.

Flash codes are not available on this system. Scan tester diagnosis of this system is similar to diagnosis of the ABS system with spark advance reduction and transmission upshift traction control. The test modes include data list, DTC history, DTCs, snapshot, miscellaneous tests, and bleed preparation. Since this system contains traction control motors in the hydraulic modulator and a throttle adjuster motor, additional traction control tests are available to test these components with the scan tester in the data list mode (Figure 15-28). In the miscellaneous test mode, the tests are divided into ABS tests (Figures 15-29 and 15-30) and TCS tests (Figures 15-31 and 15-32). These tests vary depending on the vehicle model year and the type of scan tester. Always follow the instructions in the vehicle manufacturer's service manual and scan tester manufacturer's manual.

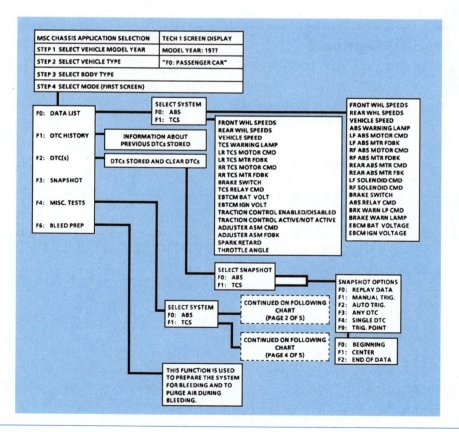

Figure 15-28 ABS and TCS diagnosis (Courtesy of Chevrolet Motor Division, General Motors Corporation)

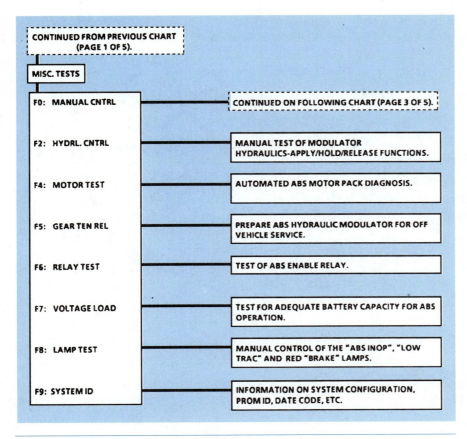

Figure 15-29 ABS system miscellaneous tests (Courtesy of Chevrolet Motor Division, General Motors Corporation)

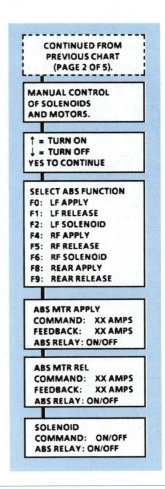

Figure 15-30 ABS system miscellaneous tests continued (Courtesy of Chevrolet Motor Division, General Motors Corporation)

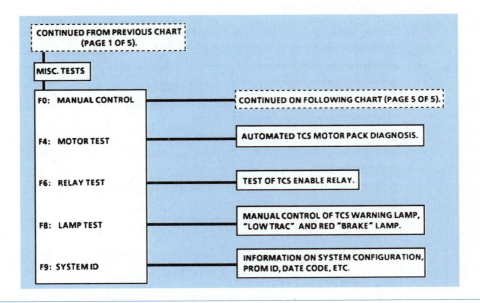

Figure 15-31 TCS miscellaneous tests (Courtesy of Chevrolet Motor Division, General Motors Corporation)

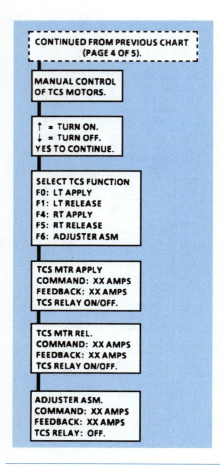

Figure 15-32 TCS miscellaneous tests continued (Courtesy of Chevrolet Motor Division, General Motors Corporation)

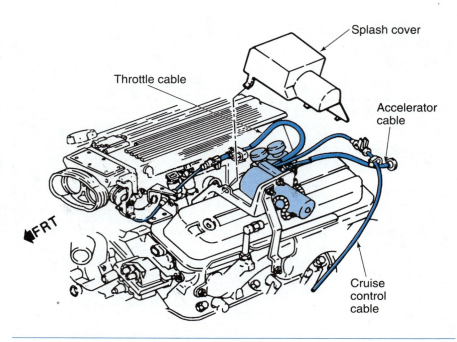

Figure 15-33 Throttle body and cruise control cables on throttle adjuster (Courtesy of Chevrolet Motor Division, General Motors Corporation)

Adjustment of Throttle Body and Cruise Control Cables on Throttle Adjuster

If the throttle adjuster assembly is removed and replaced or the cables are removed and replaced, the throttle body and cruise control cables must be adjusted (Figure 15-33). To adjust the throttle body and cruise control cables:

1. Remove the cover on the throttle adjuster assembly.
2. Pull the tabs upward on the cruise control and throttle body cable adjusters to release these mechanisms (Figure 15-34).
3. Firmly hold the throttle in the idle position.
4. Grasp the throttle body cable just behind the adjuster lock and remove slack from the cable.
5. Push the lock tab down on the throttle body cable adjuster to lock this mechanism.
6. Continue holding the throttle lever in the idle position.
7. Grasp the cruise control cable just behind the cruise control cable adjuster and remove slack from the cable.
8. Lock the cruise control cable adjuster by pushing downward on the lock tab.

SERVICE TIP: If the cruise control or throttle body cables are not adjusted properly, the throttle adjuster may hold the throttle open, resulting in higher-than-specified idle speed.

Classroom Manual
Chapter 15, page 402

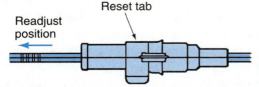
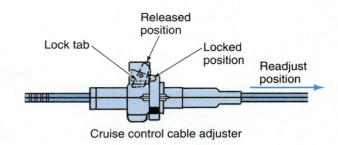

Figure 15-34 Throttle body and cruise control cable adjusters (Courtesy of Chevrolet Motor Division, General Motors Corporation)

● **CUSTOMER CARE:** Never make negative statements about a customer's car even though the car has some defects. The customer probably worked hard to earn the money to buy the car, and it is entirely possible it is the best car he can afford. This type of statement only turns the customer off and may encourage him to take his car to another shop for repairs. When a customer asks about the condition of any part of his car, always be honest with the customer.

Guidelines for Diagnosing and Servicing Antilock Brake and Traction Control Systems

1. Prior to any ABS diagnosis or service, a preliminary ABS inspection should be completed.
2. Tire sizes other than the size recommended by the vehicle manufacturer must not be installed on a vehicle with ABS.
3. A wheel speed sensor gap measurement is recommended on some vehicles even though the sensor gap is not adjustable.
4. Some vehicle manufacturers provide an antilock brake (ALB) tester for ABS diagnosis.
5. On some ABS systems, flash codes may be obtained on the amber ABS warning light when a jumper wire is connected across the proper terminals in a service check connector.
6. On ABS systems with a high-pressure accumulator, the accumulator pressure must be relieved with the manufacturer's recommended procedure before disconnecting any brake line to which accumulator pressure can be supplied.
7. Before disposing of an accumulator, the nitrogen gas pressure must be relieved in the accumulator.
8. In most ABS systems, if a fault occurs in the system, the ABS computer turns on the amber ABS warning light and disables the ABS system.
9. On an integral ABS system with a high-pressure accumulator, the brake fluid level in the master cylinder reservoir must be checked with a fully charged accumulator.
10. On an integral ABS system with a high-pressure accumulator, the brakes may be bled with a pressure bleeder, or the rear brakes may be bled with a fully charged accumulator.
11. On an integral ABS system with a high-pressure accumulator, a scan tester may be connected to the DLC to obtain DTCs.
12. On a Delco Moraine ABS VI system, a scan tester may be connected to the DLC for data tests, code history test, obtain trouble codes, and ABS test.

13. On a combined antilock brake and traction control system, the CAB performs an initialization test each time the ignition switch is turned on. The CAB also performs a dynamic test each time the engine is started and the vehicle is accelerated to a specific speed.
14. On a combined antilock brake and traction control system in the diagnostic mode, the faults may be classified as latching, nonlatching, locked, or nonerasable.
15. Wheel speed sensor wires must be securely positioned in their original clamps to prevent them from rubbing on other components.

CASE STUDY

A customer complained about the amber ABS warning light being illuminated on a 1994 Honda Prelude. A preliminary inspection of the ABS system did not reveal any problems. The technician checked the ABS system for flash codes and discovered a code 1 in the control unit memory. This code indicated a pump motor over-run time.

The technician suspected the excessive pump motor run time was caused by air in the modulator, because the pump motor had a soft constant sound when it was running. The special bleeder tool was used to bleed air from the modulator, and the modulator reservoir was filled to the proper level with the car manufacturer's recommended brake fluid. The technician disconnected the ABS B2 fuse for 3 seconds with the ignition switch off to erase the DTC.

The test with the antilock brake tester was completed and the proper results were obtained in each test mode. The amber ABS light remained off with the engine running, and a road test indicated normal brake operation.

Terms to Know

ABS inspection connector
Current code
Diagnostic trouble code (DTC)
Dynamic test
High-pressure accumulator

History code
Initialization test
Latching faults
Locked fault
Nonerasable fault

Nonintegral ABS system
Nonlatching faults
Service check connector
Snapshot mode
Wheel speed sensor gap

ASE Style Review Questions

1. While discussing a preliminary ABS inspection:
   Technician A says the tire size recommended by the vehicle manufacturer must be maintained on an ABS-equipped vehicle.
   Technician B says a different tire size than the size recommended by the vehicle manufacturer may be installed without affecting the ABS operation.
   Who is correct?
   A. A only
   B. B only
   C. Both A and B
   D. Neither A nor B

2. While discussing wheel speed sensor gap measurement:
   Technician A says the wheel speed sensor gap should be measured at one location on the notched ring.
   Technician B says if the wheel speed sensor gap varies at different locations around the notched ring, the wheel hub may be distorted.
   Who is correct?
   A. A only
   B. B only
   C. Both A and B
   D. Neither A nor B

3. While discussing the use of an antilock brake tester on a nonintegral ABS with high-pressure accumulator:
   Technician A says in modes 2 through 5 a lockup signal is sent from one of the wheels to the ABS control unit.
   Technician B says in modes 2 through 5 some kickback should be felt on the brake pedal.
   Who is correct?
   A. A only   C. Both A and B
   B. B only   D. Neither A nor B

4. While discussing diagnostic trouble codes (DTCs) on a nonintegral ABS with high-pressure accumulator:
   Technician A says the amber ABS warning light begins flashing a DTC when a fault occurs in the ABS.
   Technician B says the amber ABS warning light flashes three-digit fault codes.
   Who is correct?
   A. A only   C. Both A and B
   B. B only   D. Neither A nor B

5. While discussing high-pressure accumulators:
   Technician A says the accumulator brake fluid pressure must be relieved before disconnecting any line to which accumulator pressure may be supplied.
   Technician B says the gas pressure in an accumulator must be relieved before disposing of the accumulator.
   Who is correct?
   A. A only   C. Both A and B
   B. B only   D. Neither A nor B

6. While discussing an integral ABS with a high-pressure accumulator:
   Technician A says the fluid level in the master cylinder reservoir must be checked with a discharged accumulator.
   Technician B says the fluid level in the master cylinder reservoir must be checked with a fully charged accumulator.
   Who is correct?
   A. A only   C. Both A and B
   B. B only   D. Neither A nor B

7. While discussing rear brake bleeding on an integral ABS with a high-pressure accumulator:
   Technician A says the rear brakes may be bled with a fully charged accumulator.
   Technician B says the rear brakes may be bled with a pressure bleeder.
   Who is correct?
   A. A only   C. Both A and B
   B. B only   D. Neither A nor B

8. While discussing combined ABS and traction control system diagnosis:
   Technician A says during the initialization test, the CAB cycles all the solenoids in the hydraulic control unit (HCU).
   Technician B says during the initialization test, the driver may feel pulsations on the brake pedal.
   Who is correct?
   A. A only   C. Both A and B
   B. B only   D. Neither A nor B

9. While discussing combined ABS and traction control diagnosis:
   Technician A says when a latching fault is present, the amber ABS and traction control off warning lights go off if the fault disappears.
   Technician B says when a latching fault is present, the amber ABS and traction control off warning lights remain on until the ignition switch is cycled even if the fault disappears.
   Who is correct?
   A. A only   C. Both A and B
   B. B only   D. Neither A nor B

10. While discussing combined ABS and traction control system diagnosis:
    Technician A says if a locked fault is present, the amber ABS and traction control lights remain illuminated until the fault is erased from the CAB memory.
    Technician B says a locked fault code can only be erased with a scan tester.
    Who is correct?
    A. A only   C. Both A and B
    B. B only   D. Neither A nor B

Table 15-1 ASE Task

Follow accepted service and safety precautions during inspection, testing, and servicing of ABS hydraulic, electrical, and mechanical components.

Problem Area	Symptoms	Possible Causes	Classroom Manual	Shop Manual
IMPROPER BRAKE OPERATION	Amber and/or red brake warning lights on	Defect in ABS electrical, hydraulic, or mechanical system	378	375

Table 15-2 ASE Task

Diagnose poor stopping, wheel lockup, pedal feel, pulsation, and noise problems caused by the ABS; determine needed repairs.

Problem Area	Symptoms	Possible Causes	Classroom Manual	Shop Manual
IMPROPER BRAKE OPERATION	Poor stopping, wheel lockup, squealing, scrapping	Worn out, contaminated linings	372	373
	Spongy pedal	Air in hydraulic system	372	378
	Pedal pulsations non-antilock stop	Excessive runout on rotors or drums	372	378

Table 15-3 ASE Task

Observe ABS warning light(s) at startup; determine if further diagnosis is needed.

Problem Area	Symptoms	Possible Causes	Classroom Manual	Shop Manual
DEFECTIVE ABS OPERATION	Amber ABS light on with engine running	Defect in ABS electrical system	378	375

Table 15-4 ASE Task

Diagnose ABS electronic control(s) and components using self-diagnosis and/or recommended test equipment; determine needed repairs.

Problem Area	Symptoms	Possible Causes	Classroom Manual	Shop Manual
IMPROPER OR NO ANTILOCK FUNCTION	Amber ABS light on with engine running	Defect in ABS wiring, components, or computer	372	382

Table 15-5 ASE Task

Depressurize integral (high-pressure) components of the ABS following the manufacturer's recommended safety procedures.

Problem Area	Symptoms	Possible Causes	Classroom Manual	Shop Manual
IMPROPER POWER ASSIST, ABS OPERATION	Lack of power assist, no antilock function	Defective accumulator, pump, motor, switches, or wiring	379	379

Table 15-6 ASE Task

Fill the ABS master cylinder with the recommended fluid to the proper level following the manufacturer's procedures; inspect the system for leaks.

Problem Area	Symptoms	Possible Causes	Classroom Manual	Shop Manual
IMPROPER ABS OPERATION	Red brake warning light on with engine running	Low master cylinder fluid level	379	379
	Spongy brake pedal	Low master cylinder fluid level	378	379

Table 15-7 ASE Task

Bleed the ABS front and rear hydraulic circuits following the manufacturer's procedures.

Problem Area	Symptoms	Possible Causes	Classroom Manual	Shop Manual
IMPROPER PEDAL FEEL	Spongy pedal	Air in brake hydraulic system	379	379

Table 15-8 ASE Task

Perform a fluid pressure (hydraulic boost) diagnosis on the integral (high-pressure) antilock system; determine the needed repairs.

Problem Area	Symptoms	Possible Causes	Classroom Manual	Shop Manual
REDUCED POWER ASSIST	Excessive brake pedal effort	Reduced accumulator pressure	379	375

Table 15-9 ASE Task

Remove and install ABS components following the manufacturer's procedures and specifications.

Problem Area	Symptoms	Possible Causes	Classroom Manual	Shop Manual
IMPROPER ABS OPERATION	Amber and/or red brake warning lights on with engine running	Defective ABS components, wiring, or computer	388	375
	Improper pedal feel	Low master cylinder fluid level, worn drums, rotors, linings	381	369

Table 15-10 ASE Task

Service, test, and adjust ABS speed sensors following the manufacturer's recommended procedures.

Problem Area	Symptoms	Possible Causes	Classroom Manual	Shop Manual
IMPROPER OR NO ABS FUNCTION	Amber ABS warning light illuminated with engine running	Defective wheel speed sensors, wiring, or computer	388	382

Computer-Controlled Suspension System Diagnosis and Service

CHAPTER 16

Upon completion and review of this chapter, you should be able to:

❏ Diagnose electronic air suspension systems.

❏ Remove, replace, and inflate air springs.

❏ Adjust trim front and rear trim height on electronic air suspension systems.

❏ Service and repair nylon air lines.

❏ Adjust trim height on a rear load-leveling air suspension system.

❏ Diagnose rear load-leveling air suspension systems.

❏ Diagnose programmed ride control (PRC) systems.

❏ Diagnose automatic air suspension systems.

❏ Diagnose road-sensing suspension systems.

Electronic Air Suspension System Diagnosis and Service

Diagnostic Procedure

If the air suspension warning lamp is illuminated with the engine running, the control module has detected a defect in the electronic air suspension system. The electronic air suspension diagnostic and service procedures vary depending on the vehicle. Always follow the vehicle manufacturer's recommended procedures in the service manual. The following diagnostic and service procedures apply to electronic air suspension systems on some Lincoln Continentals and Mark VIIs. When the air suspension warning lamp indicates a system defect, the diagnostic procedure may be entered as follows:

1. Be sure the air suspension system switch is turned on.
2. Turn on the ignition switch for 5 seconds and then turn it off. Leave the driver's door open, and the other doors closed.
3. Ground the diagnostic lead located near the control module, and close the driver's door with the window down.
4. Turn on the ignition switch. The warning lamp should blink continuously at 1.8 times per second to indicate that the system is in the diagnostic mode.

There are ten tests in the diagnostic procedure. The control module switches from one test to the next when the driver's door is opened and closed. The first three tests in the diagnostic procedure are:

1. Rear suspension
2. Right front suspension
3. Left front suspension

During these three tests, each suspension location should be raised for 30 seconds, lowered for 30 seconds, and raised for 30 seconds. For example, in test 2 this procedure should be followed on the right front suspension. If the expected signal or an illegal signal is received during the test procedure, the test stops and the air suspension warning lamp is illuminated. If all the signals and commands are normal during the first three tests, the warning lamp continues to flash at 1.8 times per second.

While tests 4 through 10 are being performed, the air suspension warning lamp flashes the test number. For example, during test 4 the warning lamp flashes four times followed by a pause

Basic Tools

Basic technician's tool set

Service manual

Floor jack

Jack stands

Utility knife

Machinist's ruler or tape measure

401

and four more flashes. This flash sequence continues while test 4 is completed. The driver's door must be opened and closed to move to the next test. During tests 4 through 10, the technician must listen to or observe various components to detect abnormal operation. The warning lamp only indicates the test number during these tests. Actions performed by the control module during test 4 through 10 are as follows:

4. The compressor is cycled on and off at one-quarter cycle per second. This action is limited to 50 cycles.
5. The vent solenoid is opened and closed at one cycle per second.
6. The left front air valve is opened and closed at one cycle per second, and the vent solenoid is opened. When this occurs, the left front corner of the vehicle should drop slowly.
7. The right front air valve is opened and closed at one cycle per second, and the vent solenoid is opened. This action causes the right front corner of the vehicle to drop slowly.
8. During this test, the right rear air valve is opened and closed at one cycle per second and the vent valve is opened. This action should cause the right rear corner of the vehicle to drop slowly.
9. The left rear solenoid is opened and closed at one cycle per second and the vent valve is opened, which should cause the left rear corner of the vehicle to drop slowly.
10. The module is returned from the diagnostic mode to normal operation by disconnecting the diagnostic lead from ground. This mode change also occurs if the ignition switch is turned off or when the brake pedal is depressed.

If defects are found during the test sequence, specific electrical tests may be performed on air valve or vent valve windings and connecting wires to locate the problem.

Air Spring Removal and Installation

▲ **WARNING:** The system control switch must be in the off position when system components are serviced.

■ **CAUTION:** The system control switch must be turned off before hoisting, jacking, or towing the vehicle. If the front of the chassis is lifted with a bumper jack, the rear suspension moves downward. The electronic air suspension system will attempt to restore the rear trim height to normal, which may cause the front of the chassis to fall off the bumper jack, resulting in personal injury or vehicle damage.

▲ **WARNING:** When air spring valves are being removed, always rotate the valve to first stage until all the air escapes from the air spring. Never turn the valve to the second release stage until all the air is released from the spring.

Many components in an electronic air suspension system, such as control arms, shock absorbers, and stabilizer bars, are diagnosed and serviced in the same way as the components in a conventional suspension system. However, the air spring service procedures are different from coil spring service procedures on a conventional suspension system.

To remove an air spring:

1. Turn off the electronic air suspension switch in the trunk.
2. Hoist the vehicle and allow the suspension to drop downward, or lift the vehicle with a floor jack and place jack stands under the chassis. Lower the vehicle onto the jack stands and allow the suspension to drop downward.
3. Disconnect the nylon air line from the spring solenoid valve, and rotate the valve to the first stage to allow the air to escape from the spring. Never turn the valve to the second stage until all the air is exhausted from the spring.
4. Disconnect the lower spring retainer and remove the spring from the chassis.

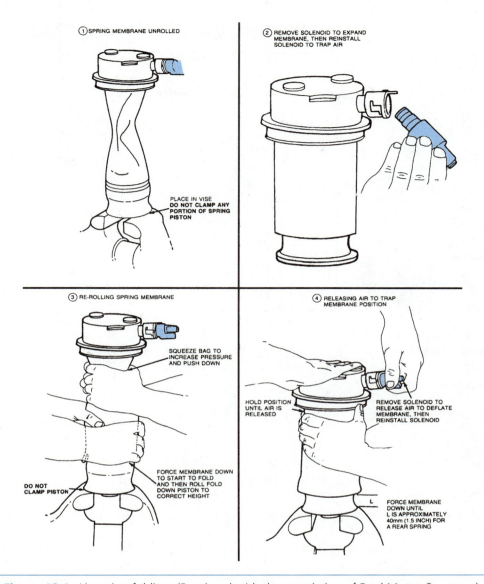

Figure 16-1 Air spring folding (Reprinted with the permission of Ford Motor Company)

5. Before an air spring is installed, it must be properly folded over the piston at the bottom of the membrane (Figure 16-1).
6. Install the spring in the chassis and connect the lower spring retainer. Be sure the top of the spring is properly seated in the spring seat. When an air spring is installed in the front or rear suspension, the spring must be properly positioned to eliminate folds and creases in the membrane (Figure 16-2).

Air Spring Inflation

WARNING: Do not allow the suspension to compress an air spring until the air spring is inflated. This action may damage the air spring.

The weight of the vehicle must not be allowed to compress an uninflated air spring. When an air spring is being inflated, use the following procedure:

1. With the vehicle chassis supported on a hoist, lower the hoist until a slight load is placed on the suspension. Do not lower the hoist until the suspension is heavily loaded.

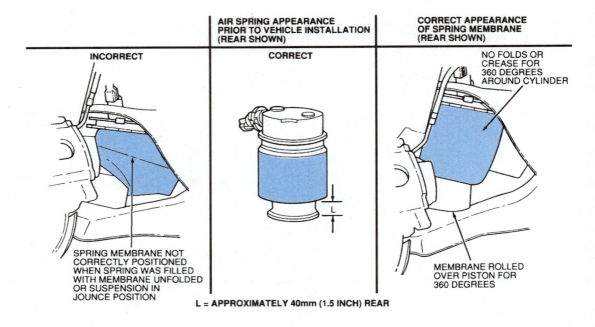

Figure 16-2 When an air spring is installed in the front or rear suspension, the spring must be properly positioned to eliminate folds and creases in the membrane. (Reprinted with the permission of Ford Motor Company)

2. Turn on the air suspension system switch.
3. Turn the ignition switch from off to run for 5 seconds with the driver's door open and the other doors shut. Turn off the ignition switch.
4. Ground the diagnostic lead.
5. Apply the brake pedal and turn the ignition to the run position. The warning lamp will flash every 2 seconds to indicate the fill mode.
6. To fill a rear spring or springs close and open the driver's door once. After a 6-second delay, the rear spring will be filled for 60 seconds.
7. To fill a front spring or springs, close and open the driver's door twice. After a 6-second delay, the front spring will be filled for 60 seconds.
8. When front and rear springs require filling, fill the rear springs first. Once the rear springs are filled, close and open the driver's door once to begin filling the front springs.
9. The spring fill mode is terminated if the diagnostic lead is disconnected from ground. This action is also obtained if the ignition switch is turned off or the brake pedal is applied.

Trim Height Adjustment

> Front and rear trim height is the distance between the vehicle chassis and the road surface measured at locations specified by the vehicle manufacturer.

The trim height should be measured on the front and rear suspension at the locations specified by the vehicle manufacturer (Figure 16-3).

If the rear suspension trim height is not within specifications, it may be adjusted by loosening the attaching bolt on the top height sensor bracket (Figure 16-4). When the bracket is moved one index mark up or down, the ride height is lowered or raised 0.25 in. (6.35 mm).

The front suspension trim height may be adjusted by loosening the lower height sensor-attaching bolt. Three adjustment positions are located in the lower front height sensor bracket (Figure 16-5). If the height sensor-attaching bolt is moved one position up or down, the front suspension height is lowered or raised 0.5 in. (12.7 mm).

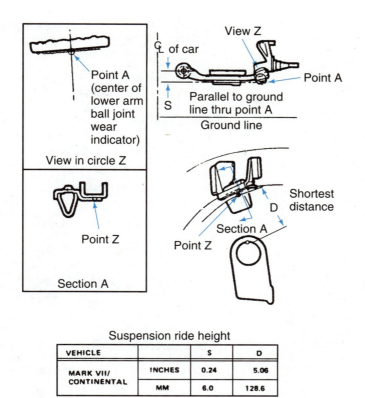

Figure 16-3 Trim height measurement locations (Reprinted with permission of Ford Motor Company)

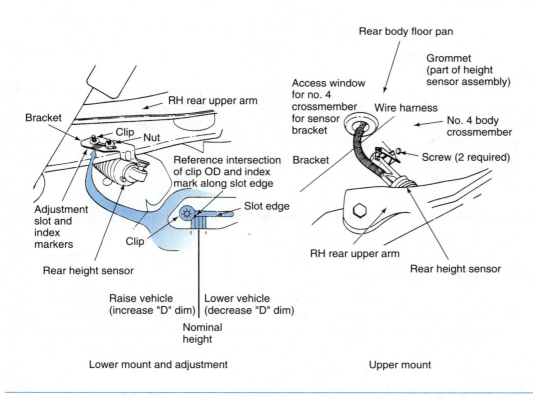

Figure 16-4 Rear trim height adjustment (Reprinted with the permission of Ford Motor Company)

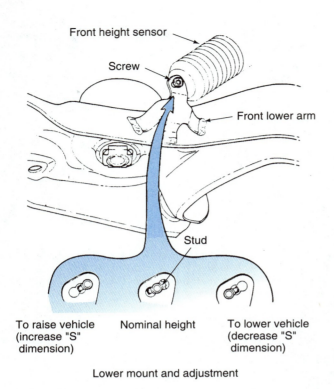

Figure 16-5 Front trim height adjustment (Reprinted with the permission of Ford Motor Company)

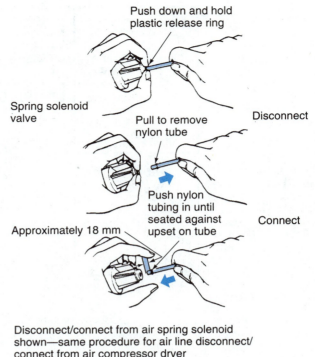

Figure 16-6 Air line removal from air spring valves or compressor outlets (Reprinted with the permission of Ford Motor Company)

Line Service

Nylon lines on the electronic air suspension system have quick-disconnect fittings. These fittings should be released by pushing downward and holding the plastic release ring, and then pulling outward on the nylon line (Figure 16-6). Simply push the nylon line into the fitting until it seats to reconnect an air line.

If a line fitting is damaged, remove it by looping the line around your fingers and pulling on the line without pushing on the release ring (Figure 16-7). When a new collet and release ring is installed, the O-ring under the collet must be replaced. If a leak occurs in a nylon line, a sharp knife may be used to cut the defective area out of the line. A service fitting containing a collet fitting in each end is available for line repairs. After the defective area is cut out of the line, push the two ends of the line into the service fitting.

Classroom Manual
Chapter 16, page 410

Rear Load-Leveling Air Suspension System Service and Diagnosis

Trim Height Adjustment

The trim height is measured between the top of the rear axle housing and the frame (Figure 16-8). The upper height sensor bracket mounting holes have long slots for adjustment purposes. If a trim height adjustment is required, the upper height sensor mounting bolts may be loosened and the bracket moved downward to increase trim height, or upward to decrease trim height (Figure 16-9).

Diagnosis

 SERVICE TIP: It is difficult to diagnose a rear load-leveling suspension system without a scan tester.

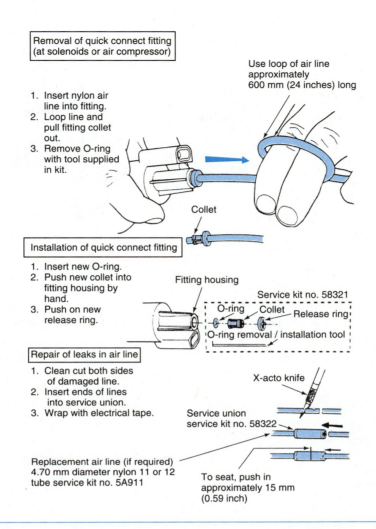

Figure 16-7 Removal of quick-disconnect fittings, and air line repair procedure (Reprinted with the permission of Ford Motor Company)

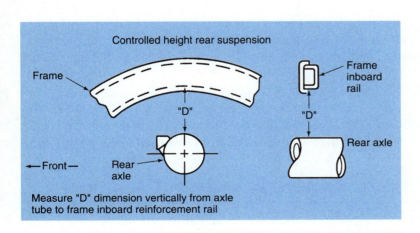

Figure 16-8 Trim height measurement location (Reprinted with the permission of Ford Motor Company)

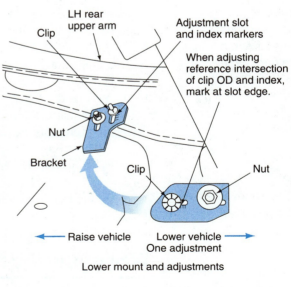

Figure 16-9 Trim height adjustment procedure (Reprinted with the permission of Ford Motor Company)

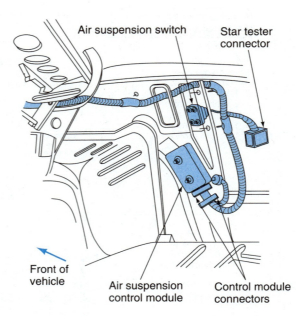

Figure 16-10 Diagnostic connector for rear load-leveling air suspension system (Reprinted with the permission of Ford Motor Company)

Special Tools

Scan tester

A scan tester is a digital tester that may be used to diagnose various on-board computer systems.

The same control module operates the rear load-leveling air suspension system and the electronic variable orifice (EVO) steering system. In the EVO steering system, the control module operates a solenoid in the power steering pump so that less effort is required to turn the steering wheel at low speeds and more steering effort is required at higher speeds.

Classroom Manual Chapter 16, page 418

A scan tester with air suspension test capabilities may be connected to the test connector in the trunk to obtain fault codes from the control module in the combined EVO and rear load-leveling air suspension system (Figure 16-10).

When the scan tester is connected to the diagnostic connector, the following tests can be completed:

1. Auto/manual diagnostic checks
2. Fault code display
3. Pinpoint tests
4. Functional test

During the auto/manual tests, the control module checks the system components for electrical defects. After this electrical check, the control module raises and lowers the vehicle to test all three height sensor states. This height sensor check is followed by a manual test in which the control module monitors the door switches and steering sensor as each door is opened and the steering wheel is turned about one-half turn in each direction.

The fault codes are obtained after the auto/manual test is completed. If there are no codes in the control module, code 11, system pass, is displayed (Figure 16-11).

In the manufacturer's service information, a pinpoint test is provided for each fault code. These pinpoint tests are voltmeter and ohmmeter tests which may be followed to locate the exact cause of fault codes.

The functional test may be used at the end of the auto/manual tests to cycle the component represented by a fault code. Never enter the functional test unless a pinpoint test directs this action. If a functional test is entered without this specified direction, control module damage may result. Detailed test procedures are provided in the manufacturer's service manuals.

Programmed Ride Control System Diagnosis

General Diagnosis

The PRC control module monitors the complete system while the vehicle is driven. If a defect occurs in the PRC system, the mode indicator light starts flashing. When the mode selector switch

AIR SUSPENSION DIAGNOSTICS SERVICE CODES

SERVICE CODE	PINPOINT TEST	DESCRIPTION	SERVICE PRIORITY
10		DIAGNOSTICS ENTERED, AUTO TEST IN PROGRESS	
11		VEHICLE PASSES. IF VEHICLE IS STILL LOW OR HIGH IN REAR, CHECK REAR RIDE HEIGHTS.	
12		AUTO TEST PASSED, PERFORM MANUAL INPUTS	
13		AUTO TEST FAILED, PERFORM MANUAL INPUTS	
16	*	EVO SHORT CIRCUIT	
17	*	EVO OPEN CIRCUIT	
18	*	EVO ACTUATOR VALVE	
39	A	COMPRESSOR RELAY CIRCUIT SHORTED TO BATTERY	2ND
42	B	AIR SPRING SOLENOID CIRCUIT SHORTED TO GROUND	2ND
43	C	AIR SPRING SOLENOID CIRCUIT SHORTED TO BATTERY	2ND
44	D	VENT SOLENOID CIRCUIT SHORTED TO BATTERY	2ND
45	E	AIR COMPRESSOR RELAY CIRCUIT SHORTED TO GROUND OR VENT SOLENOID CIRCUIT SHORTED TO GROUND	2ND
46	F	HEIGHT SENSOR POWER SUPPLY CIRCUIT SHORTED TO GROUND OR BATTERY	2ND
51	G	UNABLE TO DETECT LOWERING OF REAR	3RD
54	H	UNABLE TO DETECT RAISING OF REAR	3RD
68	J	HEIGHT SENSOR OUTPUT CIRCUIT SHORTED TO GROUND	2ND
70	K	REPLACE AIR SUSPENSION/EVO MODULE	1ST
71	L	OPEN HEIGHT SENSOR CIRCUIT	3RD
72	M	FOUR OPEN AND CLOSED DOOR SIGNALS NOT DETECTED	4TH
74	*	STEERING WHEEL ROTATION NOT DETECTED	
80	N	INSUFFICIENT BATTERY VOLTAGE TO RUN DIAGNOSTICS	1ST
23		FUNCTIONAL TEST, VENT REAR	
26		FUNCTIONAL TEST, COMPRESS REAR	
31		FUNCTIONAL TEST, TOGGLE COMPRESSOR	— AIR SUSPENSION FUNCTIONAL TESTS, NEVER PERFORM A FUNCTIONAL TEST UNLESS DIRECTED TO DO SO IN A PINPOINT TEST.
32		FUNCTIONAL TEST, VENT SOLENOID TOGGLE	
33		FUNCTIONAL TEST, SPRING SOLENOID TOGGLE	

*FOR CODES 16, 17, 18 AND 74, REFER TO PINPOINT TEST STEP A1 IN SECTION 13-54. FOR ALL OTHER CODES, REFER TO SECTION 14-40 FOR THE TEST INDICATED ON THIS CHART.

Figure 16-11 Fault codes, rear load-leveling air suspension system and electronic variable orifice (EVO) steering system (Reprinted with the permission of Ford Motor Company)

is moved to the opposite position and returned to the same position, erroneous codes are cleared from the control module memory. If the mode indicator light continues flashing, the diagnostic trouble code (DTC) diagnosis should be performed.

Diagnostic Trouble Code (DTC) Diagnosis

The PRC data link connector (DLC) is located under the ashtray on early model systems (Figure 16-12). This connector has two terminals, which must be connected together during the DTC diagnosis.

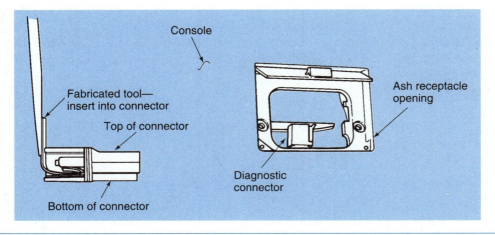

Figure 16-12 Self-test terminals, programmed ride control system (Reprinted with the permission of Ford Motor Company)

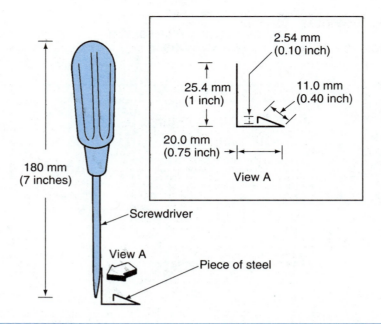

Figure 16-13 Fabricated tool to connect across the PRC data link connector (DLC) (Reprinted with permission of Ford Motor Company)

Fabricate a jumper tool to connect across these terminals, and solder it to the end of a 7-inch slotted screwdriver blade to access the DLC terminals (Figure 16-13).

To perform the DTC diagnosis:

1. Turn off the ignition switch and be sure the headlights and parking lights are off. These lights must remain off during the self-test.
2. Position the mode select switch in the AUTO position.
3. Remove the ashtray and insert the fabricated tool in the DLC connector terminals.
4. Start the engine and leave the mode select switch in the AUTO position.
5. When the engine has been running for 20 seconds or more, remove the tool from the DLC connector.
6. Count the mode indicator light flashes to obtain the trouble codes. For example, if the light flashes six times, code 6 has been provided. The light flashes each code four times at nine-second intervals. The mode indicator light is positioned in the tachometer (Figure 16-14).

Special Tools

Fabricated jumper tool

A light emitting diode (LED) is a special type of diode from which light is emitted when the diode conducts current.

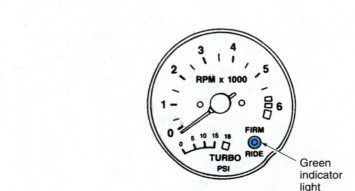

Figure 16-14 The mode indicator light in the tachometer flashes trouble codes. (Reprinted with the permission of Ford Motor Company)

Diagnostic Trouble Codes

 SERVICE TIP: DTCs together with the vehicle manufacturer's service manual diagnostic procedures will usually locate the problem in a PRC system.

The following DTCs are available on the PRC system:

Code	Defect
6	No problem
1	Left rear actuator circuit
2	Right rear actuator circuit
3	Right front actuator circuit
4	Left front actuator circuit
5	Soft relay control circuit shorted
7	PRC control module
13	Firm relay control circuit shorted
14	Relay control circuit

The DTCs vary depending on the vehicle make and year. Always use the fault code list in the manufacturer's service manual. A DTC indicates a defect in a specific area. For example, if code 2 is received, the defect may be in the right front actuator or the connecting wires.

On later model PRC systems, the DLC is located behind the right-hand strut tower (Figure 16-15). A compatable scan tester may be connected to this DLC for system diagnosis. A scan tester diagnostic procedure for a PRC system follows:

1. Be sure the ignition switch and the headlamps and parking lamps are off. Do not turn these lamps on during the test procedure.
2. Set the PRC selector switch to the auto position.
3. Be sure the proper module is installed in the scan tester for the vehicle year and PRC system.
4. Connect the scan tester to the battery terminals with the proper polarity.
5. Enter the vehicle model year and engine vehicle identification (VIN) number, and select PRC system diagnostics.
6. Connect the scan tester to the PRC DLC behind the right-hand strut tower.
7. Start the engine and record the diagnostic trouble codes (DTCs). Do not move the PRC selector switch during the test.
8. Turn off the ignition switch.
9. Disconnect the scan tester diagnostic connector and the power supply wires connected to the battery terminals.

After a PRC system defect has been corrected, the codes should be cleared, and the system should be checked again for codes to make sure there are no other faults in the system.

Photo Sequence 14 shows a typical procedure for diagnosing a programmed ride control system.

Classroom Manual
Chapter 16, page 420

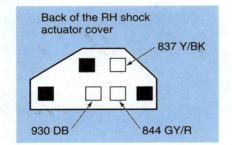

Figure 16-15 Later-model programmed ride control (PRC) data link connector (DLC) positioned behind the right-hand strut tower (Reprinted with the permission of Ford Motor Company)

Photo Sequence 14
Typical Procedure for Diagnosing a Programmed Ride Control System

P14-1 Be sure the ignition switch and the headlamps and parking lamps are off. Do not turn these lamps on during the test procedure.

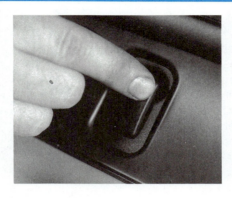

P14-2 Set the PRC selector switch to the auto position.

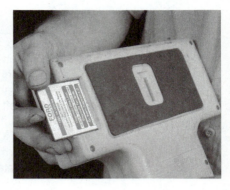

P14-3 Be sure the proper module is installed in the scan tester for the vehicle year and PRC system.

P14-4 Connect the scan tester to the battery terminals with the proper polarity.

P14-5 Enter the vehicle model year and engine vehicle identification (VIN) number, and select PRC system diagnostics.

P14-6 Connect the scan tester to the PRC DLC behind the right-hand strut tower.

P14-7 Start the engine and record the diagnostic trouble codes (DTCs). Do not move the PRC selector switch during the test.

P14-8 Turn off the ignition switch.

P14-9 Disconnect the scan tester diagnostic connector and the power supply wires connected to the battery terminals.

Automatic Air Suspension Diagnosis and Service

Drive Cycle Diagnostics

The drive cycle diagnostic procedure displays any fault codes that have occurred during the last drive cycle. This test procedure detects intermittent faults which may not appear during the other test procedures.

Drive the car for a minimum of five minutes at different speeds and road conditions or until a ride control message is illuminated in the instrument panel. After this driving period, drive the car directly to the service area and park it with the ignition switch off. The fault codes are now stored in the air suspension control module for one hour of ignition off time or until the ignition switch is turned on. If the ignition switch is turned on during this one-hour period, the fault codes are erased in the air suspension module memory. The automatic air suspension on/off switch in the trunk must be on during the test procedure.

Connect a scan tester with air suspension test capabilities to the diagnostic connector near the air suspension module in the trunk (Figure 16-16). Always follow the instructions in the vehicle manufacturer's service manual and in the scan tester manufacturer's manual. The scan tester displays a pass code or one or more of 32 possible fault codes. Codes are displayed continuously for one hour or until the ignition switch is turned on. Write down the fault codes for later reference.

Service Bay Diagnostics

Auto/Manual Diagnostic Check. During the auto/manual test, the air suspension control module verifies itself and checks other system components. After this check, the scan tester displays "12/okay do manual checks" or "13/faults detected do manual checks." The manual tests allow the technician to test the system inputs to the air suspension module.

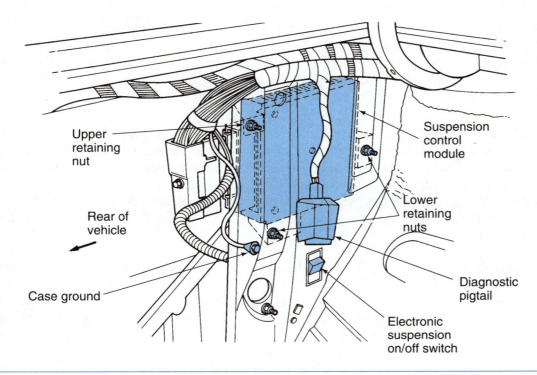

Figure 16-16 Automatic air suspension control module and diagnostic connector (Reprinted with the permission of Ford Motor Company)

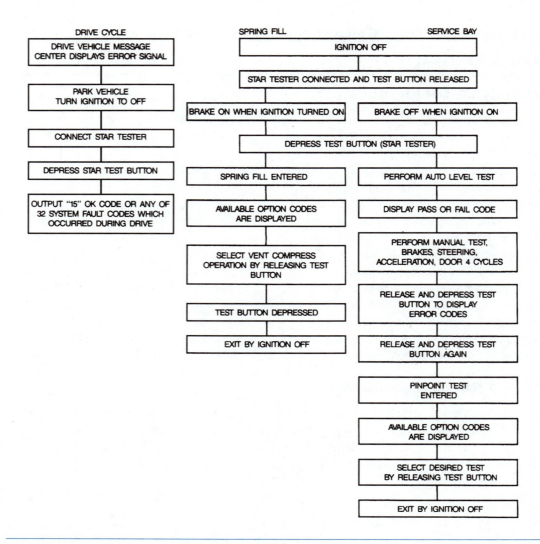

Figure 16-17 Drive cycle, service bay, and spring fill diagnostic procedures (Reprinted with the permission of Ford Motor Company)

Fault Code Display. After the auto/manual test procedure, the fault codes may be displayed by pressing the appropriate scan tester button (Figure 16-17). Each fault code is displayed for approximately 15 seconds, and the fault code display continues until the next test mode is entered. The fault codes obtained during the drive cycle diagnosis should be compared to fault codes that appear during the service bay diagnostics. When a code appears during both test modes, the code represents a hard fault. If the code is only present during the drive cycle diagnosis, the code indicates an intermittent fault.

Pinpoint Tests. Many of the fault codes have a pinpoint procedure represented by two letters. Detailed pinpoint test procedures must be followed in the car manufacturer's service manual. The pinpoint tests are designed to locate the exact cause of a fault code. For example, fault code 40 may be obtained indicating a problem in the left front air spring solenoid valve circuit. Pinpoint test EA will instruct the technician to perform specific tests to prove if the trouble is in the solenoid valve, connecting wires, or module (Figure 16-18).

The pinpoint tests are priorized, with one being the highest priority, and seven being the lowest priority. One fault may result in the display of a second fault. The pinpoint tests must be performed in the order of priority, starting with the highest priority.

STAR CODE	PINPOINT PROCEDURE	DESCRIPTION	SERVICE PRIORITY
10		Service Bay Diagnostics Entered	
11		System Checked Out Okay	
12		Automatic Test Completed — No Faults Detected — Perform Manual Inputs	
13		Automatic Test Completed — Faults Detected — Perform Manual Inputs	
15		No Faults Detected	
21		Vent Right Front Air Spring	
22		Vent Left Front Air Spring	
23		Vent Right Rear Air Spring	
24		Inflate Right Front Air Spring	
25		Inflate Left Front Air Spring	
26		Inflate Right Rear Air Spring	
27		Vent Left Rear Air Spring	
28		Inflate Left Rear Air Spring	
31		Air Compressor Toggle	
32		Vent Solenoid Valve Toggle	
33		Air Spring Solenoid Valve Toggle	
34		Shock Actuator Toggle (Firm/Soft)	
35		Door Open & Door Closed Detection	
40	EA	Short — Left Front Air Spring Solenoid Valve Circuit	2nd
41	EB	Short — Right Front Air Spring Solenoid Valve Circuit	2nd
42	EC	Short — Left Rear Air Spring Solenoid Valve Circuit	2nd
43	ED	Short — Right Rear Air Spring Solenoid Valve Circuit	2nd
44	EE	Short — Vent Solenoid Valve Circuit	2nd
45	EF	Short — Air Compressor Relay Circuit	2nd
46	EG	Short — Height Sensor Power Supply Circuit	2nd
47	EH	Short — Soft Shock Actuator Relay Circuit	2nd
48	EI	Short — Firm Shock Actuator Relay Circuit	2nd
49	HA	Unable to Detect Lowering of Right Front Corner	5th
50	HB	Unable to Detect Lowering of Left Front Corner	5th
51*	HC	Unable to Detect Lowering of Right Rear Corner	5th
51*		Unable to Detect Lowering of Rear of Vehicle	5th
52	IA	Unable to Detect Raising of Right Front Corner	6th
53	IB	Unable to Detect Raising of Left Front Corner	6th
54*	IC	Unable to Detect Raising of Right Rear Corner	6th
54*		Unable to Detect Raising of Rear of Vehicle	6th
55	JA	Speed Greater Than 15 mph Not Detected	7th
56	GA	Soft Not Detected — Left Rear Shock Actuator Circuit	4th
57	GB	Soft Not Detected — Right Front Shock Actuator Circuit	4th
58	GC	Soft Not Detected — Left Front Shock Actuator Circuit	4th
59	GD	Soft Not Detected — Right Rear Shock Actuator Circuit	4th
60	GA	Firm Not Detected — Left Rear Shock Actuator Circuit	4th
61	GB	Firm Not Detected — Right Front Shock Actuator Circuit	4th
62	GC	Firm Not Detected — Left Frt Shock Actuator Circuit	4th
63	GD	Firm Not Detected — Right Rear Shock Actuator Circuit	4th
64	GE	Soft Not Detected — All Shock Actuator Circuits	
65	GE	Firm Not Detected — All Shock Actuator Circuits	4th
66	EJ	Short — Right Front Height Sensor Circuit	2nd
67	EK	Short — Left Front Height Sensor Circuit	2nd
68	EL	Short — Rear Height Sensor Circuit	2nd
69	FA	Open — Rilght Front Height Sensor Circuit	3rd
70	FB	Open — Left Front Height Sensor Circuit	3rd
71	FC	Open — Rear Height Sensor Circuit	3rd
72	JB	At Least Four Open & Closed Door Signals Not Detected	7th
73	JC	Brake Pressure Switch Activation Not Detected	7th
74	JD	Steering Wheel Rotations Not Detected	7th
75	JE	Acceleration Signal Not Detected	7th
78	HD	Unable to Detect Lowering of Left Rear Corner	5th
79	ID	Unable to Detect Raising of Left Rear Corner	6th
80	DA	Insufficient Battery Voltage to Run Diagnostics	1st

NOTE: System faults have been prioritized for repair. Start with those codes identified with a service priority of: 1st, then 2nd, then 3rd, . . . and finally 7th.

Figure 16-18 Fault codes and pinpoint tests (Reprinted with the permission of Ford Motor Company)

SERVICE TIP: When fault codes are provided on a priority basis, always recheck the system for codes after a fault is corrected. Some computers are only capable of storing a specific number of codes in order of priority. If only one code is displayed, there could be another fault of lesser priority in the system.

Spring Fill Diagnostics

The spring fill diagnostics allow the technician to force the automatic air suspension control module to fill and vent each air spring. This procedure indicates inoperative components such as air spring solenoids or the module. During the spring fill procedure, the technician can check for air line leaks.

Classroom Manual
Chapter 16, page 426

Diagnosis of Air Suspension with Speed Leveling Capabilities

Functional Test 211

The functional test is always performed first in the diagnostic procedure. Since the test procedures vary depending on the vehicle model and year, always follow the procedure in the vehicle manufacturer's service manual. The following procedures are typical. A scan tester must be connected to the suspension diagnostic connector located on the right front shock absorber tower. During the functional test, codes stored in the suspension control module memory are displayed. A code 15 indicates there are no codes stored in the module memory. The auto test is performed next to retrieve fault codes representing existing conditions.

Auto Test

The electronic variable orifice (EVO) steering actuator is a solenoid controlled by the suspension control module, which varies power steering-assist in relation to vehicle speed. When the auto test procedure is entered on the scan tester, the suspension control module checks the inputs and outputs for errors. If no errors are found, the module cycles the compressor and system solenoids, including the electronic variable orifice (EVO) steering actuator.

The suspension control module then pumps and vents each air spring to verify component operation. After this procedure, the module performs a vehicle trim height adjustment.

The module then requests manual inputs from the steering wheel and the doors. To provide these inputs, each door must be opened and closed, and the steering wheel must be rotated at least one-quarter turn in each direction. This action checks the door switches and the steering sensor. After all the tests are completed, code 11 is displayed if there are no faults in the system. If the tests are unsuccessful, codes are displayed indicating the system faults. If the vehicle suspension was below trim height when the auto test was started, a pass code 11 may be received even when the correct trim height was not achieved. When this situation occurs, cycle the ignition switch to reset the trim height, and repeat the auto test.

After any air suspension system component is serviced or replaced, perform the auto test to be sure the service has corrected the problem, and erase any stored fault codes.

Pinpoint Tests

The pinpoint tests are grouped by components within the system. The scan tester is used during the pinpoint tests, and the technician is requested to perform additional functional tests during these tests. These functional tests include many system components, including the height sensors. During the pinpoint functional tests, the air suspension switch must be in the off position. Always follow the exact test procedure in the vehicle manufacturer's service manual. Many of these functional tests are audible tests in which the suspension module produces an audible tone through the scan tester speaker. The tone during the audible tests is a steady tone when the height sensor is at trim height and is slower as the sensor moves away from trim height. The last functional test in the pinpoint tests is an erase code function (Figure 16-19).

Classroom Manual
Chapter 16, page 427

Diagnosis of Road-Sensing Suspension System

Trouble Code Display

⚠️ **WARNING:** Never remove or install the wiring connector on a computer or computer system component with the ignition switch on. This action may result in computer damage.

\multicolumn{2}{c}{FUNCTION TEST INDEX}
Code	Description
211	Display All Diagnostic Trouble Codes In Memory
212	LF Pump with Audible Sensor Check
213	LF Vent with Audible Sensor Check
214	RF Pump with Audible Sensor Check
215	RF Vent with Audible Sensor Check
216	LR Pump with Audible Sensor Check
217	LR Vent with Audible Sensor Check
218	RR Pump with Audible Sensor Check
219	RR Vent with Audible Sensor Check
221	Compressor Run
222	Actuator Output Test (Cycles All Solenoids and EVO Actuator)
223	LF Height Sensor Trim Detection, Audible Output
224	RF Height Sensor Trim Detection, Audible Output
225	LR Height Sensor Trim Detection, Audible Output
226	Speed Sensor Detection
227	Pulse EVO Actuator through Duty Cycle
228	Erase All Codes Stored in Module Memory

Figure 16-19 Functional tests with audible tone in the pinpoint tests (Reprinted with the permission of Ford Motor Company)

WARNING: Do not supply voltage to or ground any circuit or component in a computer system unless instructed to do so in the vehicle manufacturer's service manual. This action may damage computer system components.

WARNING: During computer system diagnosis, use only the test equipment recommended in the vehicle manufacturer's service manual to prevent damage to computer system components.

SERVICE TIP: When removing, replacing, or servicing an electronic component on a vehicle, always disconnect the negative battery cable before starting the service procedure. If the vehicle is equipped with an air bag or bags, wait two minutes after the battery negative cable is removed to prevent accidental air bag deployment. Many air bag computers have a backup power supply which is capable of deploying the air bag for a specific length of time after the battery is disconnected.

To enter the diagnostic mode, turn on the ignition switch and press the off and warmer buttons simultaneously on the climate control center (CCC). Hold these buttons until the segment display occurs in the instrument panel cluster (IPC). The segment display verifies that all the segments are operational in the IPC. The turn-signal indicators are not illuminated during this check. Do not proceed with the diagnostics unless all IPC segments are illuminated during the segment display. If any of the IPC segments are not illuminated, erroneous diagnosis may occur. The IPC must be replaced if any segments do not illuminate.

Once the diagnostics are entered, the HI and LO buttons in the CCC may be used to select or reject test displays. Pressing the HI button may be compared to a yes input, while pressing the LO button may be considered as a no input. After the diagnostics are entered, select the diagnostic code displays from the following computers:

1. Powertrain control module (PCM)
2. Instrument panel cluster (IPC)
3. Air conditioning programmer (ACP)
4. Supplemental inflatable restraint (SIR)
5. Traction control system (TCS)
6. Real time damping (RTD)

The powertrain control module (PCM) controls such output functions as electronic fuel injection, spark advance, and emission devices.

The instrument panel cluster (IPC) module controls most instrument panel displays.

The air conditioning programmer (ACP) controls the air conditioning output functions.

The supplemental inflatable restraint (SIR) module controls air bag inflation.

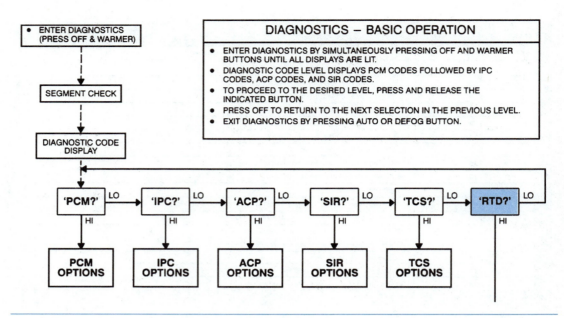

Figure 16-20 Selecting fault code displays from various computers during the diagnostic procedure (Courtesy of Cadillac Motor Car Division, General Motors Corporation)

The traction control system (TCS) module is combined with the antilock brake system (ABS) module, and this module controls wheel lockup during hard braking and wheel spinning during acceleration.

The real time damping (RTD) module controls the road-sensitive suspension (RSS) system. The term *RTD* is used in place of RSS in the system diagnosis.

A hard fault code indicates a fault that is present all the time, including the time of testing.

An intermittent fault code indicates a fault that occurs and then disappears.

The abbreviation for each computer appears in the instrument panel display, and the technician presses the HI button to select fault codes from the computer displayed. If the technician does not want fault codes from a computer, the LO button is pressed, and the display moves to the next computer abbreviation (Figure 16-20).

The fault codes from any computer are three-digit numbers with a one-letter prefix and a one-letter suffix. The fault codes have the following prefixes:

1. PCM—P
2. IPC—I
3. APC—A
4. SIR—R
5. TCS—T
6. RTD—S

Therefore, any fault codes in the RTD module are prefixed by the letter *S*. The suffix is a C or an H. When a code has a C suffix, it is a current code that is present at the time of diagnosis. An H suffix indicates the fault code is a history, or intermittent, code. The fault codes are displayed in numerical order.

If there are no fault codes in the RTD module, NO S CODES appears in the IPC display. If NO S DATA is displayed, the RTD module cannot communicate with the IPC display. When the AUTO button is pressed any time during the fault code display, the system changes from the diagnostic mode to normal operation.

Optional Diagnostic Modes

After the fault code display, the technician may select the following options:

1. RTD data
2. RTD inputs
3. RTD outputs
4. RTD clear codes

When one of these options is displayed in the IPC, the technician selects the option by pressing the HI button. The technician rejects the displayed option by pressing the LO button, and the display moves to the next option.

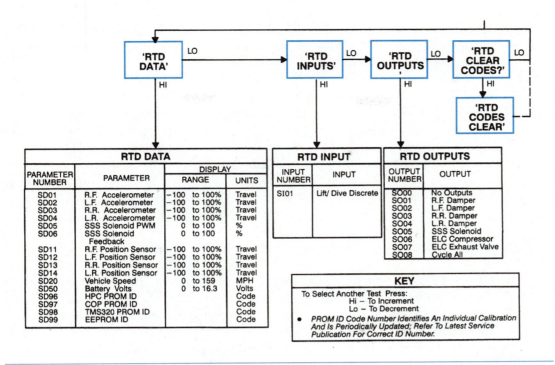

Figure 16-21 Optional test modes, including RTD data, inputs, outputs, and clear codes (Courtesy of Cadillac Motor Car Division, General Motors Corporation)

If RTD data is selected, the data from a number of inputs may be displayed in numerical order beginning with number SD01, R.F. accelerometer (Figure 16-21). The car must be driven to obtain a variable input from this sensor. This sensor input appears as a numerical range, from -100 to 100%.

As the vehicle is accelerated and decelerated, this sensor input should change within the specified range. Always consult the car manufacturer's service manual for more detailed diagnosis and specifications. The next test in the data mode is selected by pressing the HI button. The display goes back to the previous test parameter when the LO button is pressed.

If the technician selects the RTD inputs, the lift/dive discrete input may be checked. When the technician selects RTD outputs, the RTD module cycles the displayed output. The next output in the test sequence is selected by pressing the HI button.

When CLEAR CODES is selected, the codes are erased, and RTD CODES CLEAR is displayed. The AUTO or DEFOG button may be pressed to exit the diagnostics.

● **CUSTOMER CARE:** While discussing computer-controlled suspension systems with customers, remember that the average customer is not familiar with automotive electronics terminology. Always use basic terms that customers can understand when explaining electronic suspension problems. Most customers appreciate a few minutes spent by service personnel to explain their automotive electronic problems. It is not necessary to provide customers with a lesson in electronics, but it is important that customers understand the basic cause of the problem with a car so they feel satisfied the repair expenditures are necessary. A satisfied customer is usually a repeat customer!

Classroom Manual
Chapter 16, page 429

Guidelines for Diagnosing and Servicing Computer-Controlled Suspension Systems

1. When an electronic air suspension system or rear load-leveling air suspension system is serviced, always turn off the air suspension switch in the trunk.

2. When hoisting, jacking, towing, or lifting a vehicle equipped with an electronic air suspension system or a rear load-leveling air suspension system, always turn off the air suspension switch in the trunk.

3. If a defect occurs in an electronic air suspension system, the control module illuminates the suspension warning lamp with the engine running.

4. There are ten steps in the electronic air suspension diagnostic procedure. During the first three tests, the suspension warning lamp flashes at 1.8 times per second if the system is normal. While tests 4 through 10 are performed, the suspension warning lamp flashes the test number. The driver's door is opened and closed to move to the next test in the sequence.

5. To release air from an air spring, the air spring valve should be rotated to the first stage. Never rotate an air spring valve to the second stage until all the air is exhausted from the spring.

6. Prior to installation, an air spring membrane must be folded properly onto the lower piston.

7. After an air spring is installed, there must be no creases in the spring membrane.

8. Never lower the full chassis weight onto an air spring with no air pressure in the spring.

9. In an electronic air suspension system, the front and rear height sensors may be adjusted to obtain the proper trim height.

10. A service fitting is available to splice damaged nylon air lines.

11. The rear height sensor may be adjusted to correct the trim height on a rear load-leveling air suspension system.

12. A scan tester may be connected to a test connector in the trunk to obtain fault codes in a rear load-leveling suspension system.

13. Fault codes may be obtained on a programmed ride control (PRC) system by connecting the two wires in a diagnostic connector behind the ashtray.

14. In a PRC system, the fault codes are flashed by the mode indicator light.

15. The test modes in the automatic air suspension diagnosis are drive cycle diagnosis, service bay diagnosis, and spring fill diagnosis.

16. When diagnosing the air suspension system with speed leveling capabilities, the test modes are functional test 211, auto test, and pinpoint tests.

17. To enter the diagnostics on the road-sensing suspension system, the off and warmer buttons in the climate control center are pressed simultaneously with the ignition switch on.

18. When diagnosing the road-sensing suspension system, the HI and LO buttons in the climate control center are used as yes and no inputs to select fault code displays from various on-board computers.

19. When diagnosing the road-sensing suspension system, the HI and LO buttons are used to select various test options and move ahead or back up within the parameters in a specific test option.

CASE STUDY

A customer complained about excessive body lean while cornering on the left front suspension of a 1993 Thunderbird turbo coupe equipped with a programmed ride control (PRC) system. The customer also said that the programmed ride control lamp in the tachometer started flashing when the problem occurred, but this light would quit flashing after awhile. Further questioning of the customer indicated that the problem was intermittent. Sometimes the problem did not occur during hard cornering. The service writer asked the customer if there was any evidence of noise from the suspension, and the customer indicated that the flashing mode indicator lamp and the body lean on the left front suspension were the only symptoms.

Since the PRC mode indicator lamp was not flashing when the customer brought the car into the shop, the technician took the car for a road test. The technician took with him a digital volt-ohmmeter and a fabricated tool to connect the terminals in the PRC self-test connector. After several attempts at hard cornering, the car displayed the symptoms explained by the customer. While the PRC mode indicator light was still flashing, the technician drove the car into a parking lot and performed a self-test on the PRC system. A code 4 was present during the self-test, indicating a problem in the left front strut actuator circuit. The technician disconnected the wires from the left front actuator and connected an ohmmeter across the strut motor terminals connected to the firm and soft relays. A normal reading was obtained at these terminals. Careful examination of the wiring harness near the strut tower indicated the harness had been punctured with something. The car had a new paint job and closer examination indicated extensive repairs on the front body structure. It was logical to assume that the wiring harness had been punctured during collision damage. The technician removed some of the tape from the wiring harness and found that the insulation was damaged on the wires from the firm and soft relays to the strut actuator, and these wires were touching together at times.

The technician drove the car back to the shop and repaired the wires. Another road test indicated that the PRC system worked perfectly.

Terms to Know

Air conditioning programmer (ACP)
Auto/manual diagnostic check
Auto test
Current code
Data link connector (DLC)
Drive cycle diagnosis
Fault codes
Functional test 211
Hard fault code
History code
Instrument panel cluster (IPC)
Intermittent fault code
Light emitting diode (LED)
Pinpoint tests
Powertrain control module (PCM)
Programmed ride control (PRC) system
Real time damping (RTD)
Road-sensing suspension (RSS)
Scan tester
Service bay diagnostics
Spring fill diagnostics
Supplemental inflatable restraint (SIR)
Traction control system (TCS)
Trim height

ASE Style Review Questions

1. While discussing air spring inflation when a deflated air spring is installed:
   Technician A says the complete vehicle weight should be applied to the spring before spring inflation.
   Technician B says the vehicle weight should be supported on a hoist so that a slight load is placed on the air spring before spring inflation.
   Who is correct?
   A. A only   **C.** Both A and B
   B. B only   **D.** Neither A nor B

2. While discussing electronic air suspension system service on a vehicle with an electronic air suspension system, a bumper jack is used to lift one corner of the vehicle to change a tire.
   Technician A says the electronic air suspension switch should be in the on position.
   Technician B says the air spring should be deflated on the corner of the vehicle being lifted.
   Who is correct?
   A. A only   **C.** Both A and B
   B. B only   **D.** Neither A nor B

3. While discussing electronic air suspension system service:
   Technician A says the suspension warning lamp flashes 1.8 times per second during the first three tests in the test sequence.
   Technician B says the suspension warning lamp flashes the test number during tests 4 through 10 in the test sequence.
   Who is correct?
   A. A only
   B. B only
   C. Both A and B
   D. Neither A nor B

4. While discussing rear load-leveling air suspension system service:
   Technician A says a scan tester may be connected to a diagnostic connector in the trunk to obtain fault codes.
   Technician B says if there are no faults in the control module memory, a code 14 is provided.
   Who is correct?
   A. A only
   B. B only
   C. Both A and B
   D. Neither A nor B

5. While discussing PRC system service:
   Technician A says a glowing mode indicator light indicates a system defect.
   Technician B says a flashing mode indicator light indicates a system defect.
   Who is correct?
   A. A only
   B. B only
   C. Both A and B
   D. Neither A nor B

6. While discussing PRC system diagnosis and service:
   Technician A says the self-test connector is located behind the ashtray.
   Technician B says the self-test connector is mounted beside the fuse panel.
   Who is correct?
   A. A only
   B. B only
   C. Both A and B
   D. Neither A nor B

7. While discussing road-sensing suspension system diagnosis and service:
   Technician A says the diagnostics are entered by pressing the AUTO and DEFOG buttons simultaneously with the ignition switch on.
   Technician B says the HI button in the climate control center may be pressed to move to the next parameter in a test option.
   Who is correct?
   A. A only
   B. B only
   C. Both A and B
   D. Neither A nor B

8. While discussing road-sensing suspension system diagnosis and service:
   Technician A says in the output test option the road-sensing suspension module cycles various outputs.
   Technician B says a history code represents an intermittent fault.
   Who is correct?
   A. A only
   B. B only
   C. Both A and B
   D. Neither A nor B

9. While discussing road-sensing suspension system diagnosis:
   Technician A says all the instrument panel cluster (IPC) segments and the turn signal indicators are illuminated during the segment display when the diagnostic mode is entered.
   Technician B says erroneous diagnosis of various systems may occur if all the IPC segments are not illuminated during the segment display.
   Who is correct?
   A. A only
   B. B only
   C. Both A and B
   D. Neither A nor B

10. While discussing road-sensing suspension system diagnosis:
    Technician A says a diagnostic trouble code (DTC) with an H suffix represents an intermittent fault.
    Technician B says a DTC with a C suffix represents an intermittent fault.
    Who is correct?
    A. A only
    B. B only
    C. Both A and B
    D. Neither A nor B

Table 16-1 ASE Task

Diagnose, inspect, adjust, repair, or replace components of electronically controlled suspension systems.

Problem Area	Symptoms	Possible Causes	Classroom Manual	Shop Manual
RIDE CONTROL	Body sway, dive, or lift	Defective electronic suspension system	420	408
RIDE HEIGHT	Low or high ride height	1. Defective electronic air suspension system	410	401
		2. Trim height adjustment	414	401

Computer-Controlled Air Conditioning Diagnosis and Service

CHAPTER 17

Upon completion and review of this chapter, you should be able to:

- ❏ Perform a preliminary inspection on a computer-controlled air conditioning system.
- ❏ Complete an A/C performance test on a computer-controlled air conditioning system.
- ❏ Perform a refrigerant system charge test on a computer-controlled air conditioning system.
- ❏ Perform self-diagnostic tests to illuminate the A/C control head displays.
- ❏ Complete self-diagnostic tests to obtain diagnostic trouble codes (DTCs) related to input sensors.
- ❏ Perform self-diagnostic tests to obtain DTCs related to mode door position switches.
- ❏ Complete self-diagnostic tests to operate all door actuators, blower motor, and compressor clutch.
- ❏ Perform self-diagnostic tests to display the temperature sensed by each input sensor.
- ❏ Perform control rod adjustments on the door actuator motors.
- ❏ Complete a scan tester diagnosis of computer-controlled air conditioning systems.

Preliminary Inspection

Basic Tools

Basic technician's tool set
Service manuals
Thermometer

In this chapter, we discuss mainly the electronic diagnosing and servicing of computer-controlled air conditioning systems. It is assumed that the student has already studied refrigeration systems and manual air conditioning in a previous class. Before an air conditioning system is diagnosed, a preliminary A/C system inspection must be completed. The procedure for this inspection follows:

1. Visually inspect the air passages in the condenser and radiator for restrictions (Figure 17-1).
2. Visually inspect the cooling system for leaks and proper operation. Tape a thermometer to the top radiator hose and be sure the cooling system is operating at the proper temperature.

Figure 17-1 Visually inspect the condenser air passages for restrictions. (Courtesy of Chrysler Corporation)

3. Check the condition and tension of the compressor and water pump belts.
4. Check all fuses in the A/C system, including the compressor clutch fuse. Some compressor clutches have a thermal fuse.
5. Visually inspect all wiring connections for loose or corroded terminals. Inspect all wiring harnesses for damaged wires.
6. Check for compressor operation when the system is in the A/C mode.
7. Be sure the refrigerant system has the proper refrigerant charge and pressures.

A/C Performance Test

The A/C performance test determines if the A/C system is providing proper in-vehicle cooling. The A/C performance test procedure follows:

High-side pressure in a refrigerant system is the pressure at the compressor outlet side.

Low-side pressure in a refrigerant system is the pressure between the evaporator and the compressor inlet.

Special Tools

Manifold gauge set

 CAUTION: Avoid breathing A/C refrigerant and lubricant vapor or mist. Exposure may irritate eyes, nose, and throat.

 CAUTION: Compressed air should not be used to pressure test R-134a refrigerant systems. Mixtures of R-134a and air are combustible at high pressure, and these mixtures may cause an explosion resulting in personal injury or property damage.

 SERVICE TIP: The manifold gauge sets and service connectors are different for R-12 and R-134a refrigerant systems.

1. Connect a tachometer to the ignition system.
2. Connect a manifold gauge set to the high-side and low-side connectors in the refrigerant system (Figures 17-2 and 17-3).

 SERVICE TIP: The required engine rpm for this test may be obtained with a scan tester in the idle speed control motor mode.

3. Start the engine and set the speed to 1,000 rpm with the A/C off. Leave the vehicle doors closed except to enter and exit the vehicle.

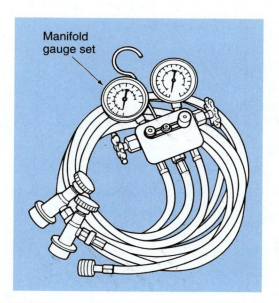

Figure 17-2 Manifold gauge set for R-134a refrigerant systems (Courtesy of Chrysler Corporation)

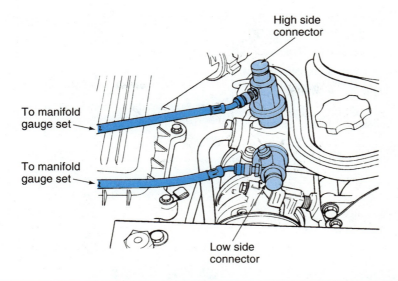

Figure 17-3 Manifold gauge set connections to the high-side and low-side service ports (Courtesy of Chrysler Corporation)

4. On automatic temperature control (ATC) systems, rotate the blower control knob fully clockwise to the high-speed position.
5. Rotate the temperature control knob fully counterclockwise to the fully cold position.
6. Press the recirculation (rec) and A/C buttons. The word *auto* should not appear in the digital display, indicating the system is now manually controlled.
7. Insert a thermometer into the left center A/C panel duct.
8. Operate the system under these conditions for 5 minutes.
9. With the A/C still on and the compressor clutch engaged, compare the discharge air temperature on the thermometer to the vehicle manufacturer's A/C performance temperatures (Figure 17-4). If the thermometer temperature is not within specifications, further refrigerant system diagnosis is required.

Ambient Temperature	21°C (70°F)	26.5°C (80°F)	37.5°C (90°F)	37.5°C (100°F)	43°C (110°F)
Maximum Allowable Air Temperature at Center Left Panel Outlet	7°C (45°F)	9°C (49°F)	12°C (54°F)	13°C (56°F)	15°C (59°F)
Compressor Discharge Pressure	772-1448 kPa (112-210 PSI)	903-1475 kPa (131-214 PSI)	1241-1482 kPa (180-215 PSI)	1400-1986 kPa (203-288 PSI)	1600-2282 kPa (232-331 PSI)
Compressor Suction Pressure	69-255 kPa (10-37 PSI)	117-262 kPa (17-38 PSI)	145-324 kPa (21-47 PSI)	193-352 kPa (28-51 PSI)	207-365 kPa (30-53 PSI)

Figure 17-4 A/C performance temperatures (Courtesy of Chrysler Corporation)

Figure 17-5 A label in the engine compartment identifies R-134a refrigerant. (Courtesy of Chrysler Corporation)

Refrigerant System Charge Test

WARNING: Environmental Protection Agency (EPA) regulations require that all refrigerant must be recovered and recycled.

WARNING: Mixing R-12 and R-134a refrigerant or the lubricating oils required in these refrigerants will cause refrigerant system damage. Separate recovery and recycling stations are required for R-12 and R-134a refrigerants.

SERVICE TIP: When a refrigerant system contains R-134a refrigerant, a label in the engine compartment identifies the type of refrigerant and lubricating oil in the refrigerant (Figure 17-5).

CAUTION: If refrigerant contacts the human body, the refrigerant has a freezing action. Always wear gloves and face protection when servicing refrigerant systems.

CAUTION: If the engine has been running, some refrigerant system components may be very hot. Wear gloves to avoid burns.

Vehicle manufacturers recommend different procedures for checking the refrigerant system state of charge. Always follow the procedure recommended in the vehicle manufacturer's service manual. Some vehicle manufacturers recommend using the compressor high-side discharge pressure and the temperature of the refrigerant leaving the condenser to determine the refrigerant state of charge. The procedure for checking the high-side pressure and the condenser outlet temperature follows:

SERVICE TIP: High-side pressure in the refrigerant system can also be read on a scan tester connected to the data link connector (DLC) under the instrument panel.

1. Connect a manifold gauge set to the low-side and high-side connectors in the refrigerant system.
2. Place a thermocouple pickup clamp on the refrigerant outlet line near the condenser (Figure 17-6).
3. Set the thermocouple gauge to read temperature.
4. Start the engine and select A/C, high blower speed, and outside air on the A/C controls.

WARNING: When a piece of cardboard is placed in front of the condenser for test purposes, carefully watch the engine temperature and high-side refrigerant pressure. If the engine begins to overheat or the high-side refrigerant pressure exceeds the normal range, remove the cardboard immediately to prevent engine overheating or refrigerant system damage.

R-12 refrigerant has been used in automotive A/C systems for many years. This refrigerant depletes the ozone layer when released into the atmosphere.

New vehicle production has gradually changed to R-134a refrigerant during the last three years. This refrigerant does not harm the ozone layer.

Special Tools

Scan tester

Thermocouple with temperature reading

A thermocouple changes heat to a voltage.

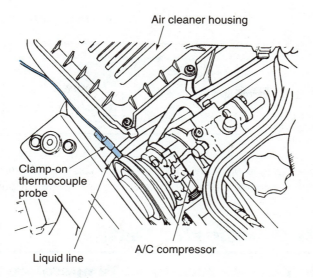

Figure 17-6 Thermocouple clamp on the condenser outlet line (Courtesy of Chrysler Corporation)

5. Install a piece of cardboard in front of the condenser to increase the high-side pressure.
6. Record the high-side pressure on the manifold gauge set or scan tester, and the condenser outlet temperature on the thermocouple meter. Draw a graph of the high-side pressure in relation to the condenser outlet temperature (Figure 17-7).
7. Compare the graph of the high-side pressure and the condenser outlet temperature to the specification graph in the vehicle manufacturer's service manual. If the graph drawn in step 6 is above the specification graph, the refrigerant system is overcharged. When the graph drawn in step 6 is below the specification graph, an undercharged condition is present in the refrigerant system.

Some vehicle manufacturers recommend recovering and recycling the refrigerant, and then installing the proper amount of refrigerant by weight to obtain the proper refrigerant system charge.

Classroom Manual
Chapter 17, page 441

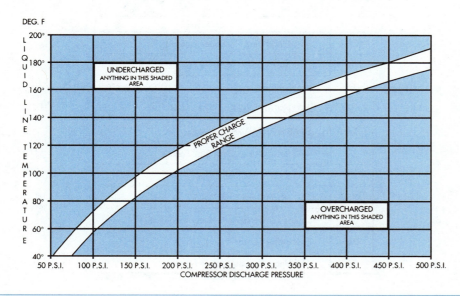

Figure 17-7 Specified high-side pressure in relation to condenser outlet temperature (Courtesy of Chrysler Corporation)

429

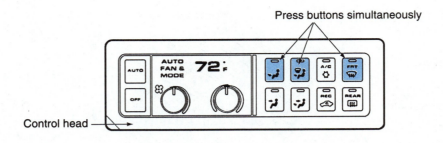

Figure 17-8 The self-diagnostic mode is entered by pressing the floor, mix, and defrost buttons simultaneously with the engine running and the vehicle not moving. (Courtesy of Chrysler Corporation)

Vehicle Self-Diagnostic Tests

Self-Diagnostic Tests on Automatic Temperature Control Systems

Many computer-controlled air conditioning systems have self-diagnostic capabilities. These self-diagnostic tests vary depending on the vehicle make and model year. Always follow the instructions in the vehicle manufacturer's service manual. On some vehicles, such as Chrysler LH and LHS cars, the self-diagnostic mode for the automatic temperature control (ATC) system is entered by pressing the floor, mix, and defrost buttons simultaneously with the engine running and the vehicle not moving (Figure 17-8).

When the diagnostic mode is entered, the digital display in the A/C control head begins blinking. The display continues blinking until self-diagnostic tests are completed. While the display is blinking, the door actuator motors are calibrated to the unit on which they are installed. Any diagnostic trouble codes (DTCs) in the body computer memory are displayed. When one DTC is displayed, the panel button may be pressed to scroll through any other DTCs. DTC numbers range from 23 to 36 (Figure 17-9). If there are no DTCs in the body computer memory, the system returns to normal operation indicated by the temperature display. When DTCs are displayed, voltmeter or

DIAGNOSTIC TROUBLE CODE CHART	
CODE	DESCRIPTION
23	ATC Blend Door Feedback Failure
24	ATC Mode Door Feedback Failure
25	Ambient Sensor
26	ATC In-Car Sensor
27	Sun Sensor Failure
31	ATC Recirculation Door Stall Failure
32	ATC Blend Door Stall Failure
33	ATC Mode Door Stall Failure
34	Engine Temperature Message not Received
35	Evaporator Sensor Failure
36	ATC Head Communication Failure

Figure 17-9 Diagnostic trouble codes (DTCs) related to the automatic temperature control (ATC) system (Courtesy of Chrysler Corporation)

ohmmeter tests are usually required to locate the exact cause of the problem. DTCs may be erased with a scan tester or by disconnecting battery voltage from the body computer for 10 minutes.

Self-Diagnostic Entry and Cancellation

On some cars such as the Nissan Maxima, the self-diagnostic mode in the automatic A/C system is entered by starting the engine and pressing the off button in the A/C control head for 5 seconds. The off button must be pressed within 10 seconds after the engine is started, and the fresh vent lever must be in the off position. The self-diagnostic mode may be cancelled by pressing the auto button or turning off the ignition switch.

The self-diagnostic tests are completed in five steps. The up arrow for temperature setting on the A/C control head is pressed to move to the next step. When the down arrow for temperature setting is pressed, the diagnostic system returns to the previous step (Figure 17-10).

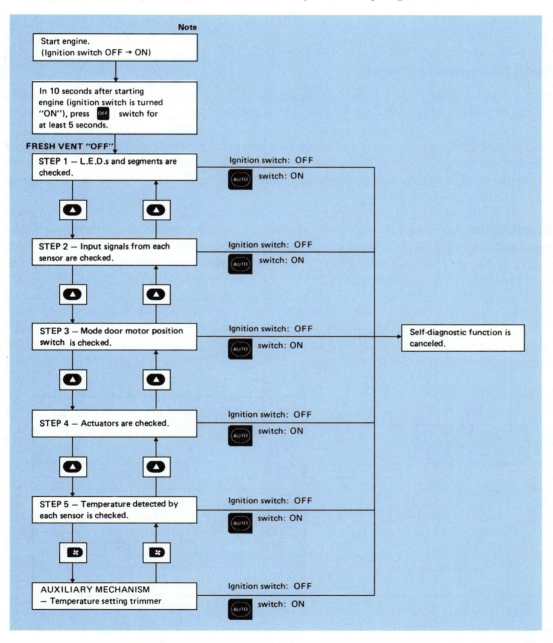

Figure 17-10 Self-diagnostic procedure with five steps (Courtesy of Nissan Motor Co., Ltd.)

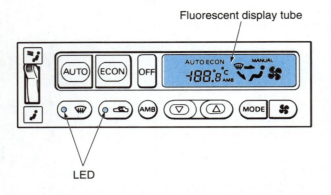

Figure 17-11 In step 1 of the self-diagnostic procedure, all the segments and LEDs are illuminated in the A/C control head. (Courtesy of Nissan Motor Co., Ltd.)

Figure 17-12 When part of a segment does not illuminate, the segment is defective. (Courtesy of Nissan Motor Co., Ltd.)

Self-Diagnostic Steps

Step 1. During step 1, all the LEDs and display segments are illuminated (Figure 17-11). If some segments or LEDs are not illuminated, these components are defective (Figure 17-12).

Step 2. When the up temperature control arrow is pressed to move to step 2, the input sensors are tested by the A/C computer. If there are no defects in the input sensors, 20 is displayed in the A/C control head. If any of the input sensors are defective, a code for that sensor is displayed in the control head (Figure 17-13).

Step 3. When the up temperature arrow is pressed to move to step 3, the mode door position switches are tested. If all these position switches are in satisfactory condition, 30 is displayed in the A/C control head. A defect in one of the position switches results in a code display in the A/C control head. These codes range from 31 to 36 (Figure 17-14).

Step 4. If the up temperature arrow is pressed to move to step 4, the A/C control head displays 41. In this mode, the A/C computer positions the mode door in the vent position. The intake door is in the recirculation (REC) position, and the air mix door is in the full cold position. The defrost (DEF) button is pressed to move to the next mode, which displays 42 in the A/C control head. There are

Code No.	Malfunctioning sensor (including circuits)
21	Ambient sensor
22	In-vehicle sensor
24	Intake sensor
25	Sunload sensor*1
26	P.B.R.

Figure 17-13 In step 2, a defective input sensor causes a code display in the A/C control head. (Courtesy of Nissan Motor Co., Ltd.)

Code No.	Malfunctioning mode door motor position switch (including circuits)
31	VENT
32	B/L
33	B/L
34	FOOT/DEF 1
35	FOOT/DEF 2
36	DEF

Figure 17-14 In step 3, a defective mode door switch causes a code display in the A/C control head. (Courtesy of Nissan Motor Co., Ltd.)

Code No.	Actuators test pattern				
	Mode door	Intake door	Air mix door	Blower motor	Compressor
41	VENT	REC	Full Cold	4 - 5V	ON
42	B/L	REC	Full Cold	9 - 11V	ON
43	B/L	20% FRE	Full Hot	7 - 9V	ON
44	D/F 1	FRE	Full Hot	7 - 9V	OFF
45	D/F 2	FRE	Full Hot	7 - 9V	OFF
46	DEF	FRE	Full Hot	10 - 12V	ON

Figure 17-15 Step 4 contains six modes and in each mode, the A/C computer commands a specific door position, blower motor voltage, and compressor clutch operation. (Courtesy of Nissan Motor Co. Ltd.)

six modes in step 4 and each mode is represented by a number in the A/C control head. These numbers range from 41 to 46. The DEF button is used to select the next mode. In each mode, the A/C computer commands specific door positions, blower motor voltage, and compressor clutch operation (Figure 17-15). Door operation may be checked by the air discharge from the various ducts.

Step 5. When the up temperature arrow is pressed to select step 5, the temperature detected by each input sensor is displayed. After this mode is entered, 5 is displayed in the A/C control head. If the DEF button is pressed, the temperature sensed by the ambient sensor is displayed in the A/C control head. Pressing the DEF button a second time displays the temperature detected by the in-vehicle sensor. If the DEF button is pressed again, the temperature detected by the intake sensor is displayed (Figure 17-16). This sensor is mounted near the evaporator outlet. When the temperature displayed varies significantly from the actual temperature, the sensor and connecting wires should be tested with an ohmmeter and a voltmeter.

Auxiliary Mode. At the end of step 5, the blower speed button may be pressed to enter the auxiliary mode. In this mode, the temperature on the display may be adjusted so it is the same as the in-vehicle temperature felt by the driver. After the auxiliary mode is entered, press the up or down

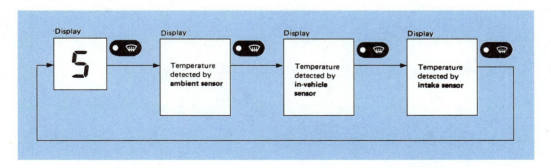

Figure 17-16 Step 5 displays the temperature detected by each sensor. (Courtesy of Nissan Motor Co., Ltd.)

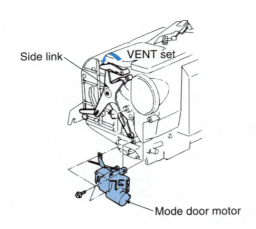

Figure 17-17 Mode door control rod adjustment (Courtesy of Nissan Motor Co., Ltd.)

temperature control buttons until the A/C control head displays the same temperature as the temperature inside the vehicle.

Control Rod Adjustments

Mode Door Control Rod Adjustment. On some systems, the actuator door control rods may be adjusted. These adjustments are necessary after components such as door actuator motors are replaced. The procedure for the mode door control rod adjustment follows:

1. With the mode door installed on the A/C heater case and the wiring connector attached to the actuator, disconnect the door motor rod from the slide link.
2. Enter step 4 in the self-diagnostic mode so 41 is displayed in the A/C control head.
3. Move the slide link by hand until the mode door is in the vent mode (Figure 17-17).
4. Connect the motor rod to the slide link.
5. Continue pressing the DEF button until all six modes have been obtained in step 4, and be sure the door moves to the proper position in each mode.

Air Mix Door Control Rod Adjustment. To adjust the air mix door control rod:

1. Disconnect the air mix door lever from the rod holder. Leave the wiring harness connected to the door motor, and be sure the motor is securely attached to the A/C heater case.
2. Enter step 4 in the self-diagnostic mode so 41 is displayed in the A/C control head.
3. Move the air mix door lever by hand and hold it in the full cold position (Figure 17-18).
4. Connect the air mix door lever to the rod holder.

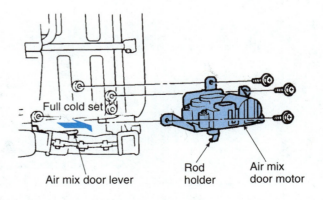

Figure 17-18 Air mix door rod adjustment (Courtesy of Nissan Motor Co., Ltd.)

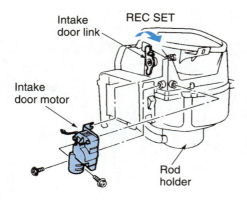

Figure 17-19 Intake door rod adjustment (Courtesy of Nissan Motor Co., Ltd.)

5. Continue pressing the DEF button until all six modes have been obtained in step 4. Be sure the proper door position is provided in each mode.

Intake Door Rod Adjustment. The intake door rod adjustment procedure follows:
1. Be sure the intake air door motor is securely mounted on the A/C heater case and the wiring connector is attached to the motor. Disconnect the intake door link from the rod holder.
2. Enter step 4 in the self-diagnostic procedure so 41 is displayed in the A/C control head.
3. Move the intake door link and hold it in the REC position (Figure 17-19).
4. Connect the intake door link to the rod holder.
5. Continue pressing the DEF button until all six modes are obtained in step 4. Be sure the proper door position is obtained in each mode.

Scan Tester Diagnosis of Computer-Controlled Air Conditioning Systems

 WARNING: Do not disconnect the wiring connector from any computer-controlled A/C system component with the ignition switch on. This action may result in damage to the A/C computer.

 WARNING: Do not connect or disconnect the scan tester to or from the DLC with the ignition switch on. This action may damage the A/C computer.

Initial Menus

The scan tester diagnosis of computer-controlled air conditioning systems varies depending on the vehicle make and model year. Always use the scan tester diagnostic procedures in the vehicle manufacturer's service manual or in the scan tester manufacturer's manual. The proper module for the vehicle make, model year, and air conditioning diagnosis must be installed in the scan tester. The scan tester must be connected to the data link connector under the dashboard (Figure 17-20).

When the initial menu appears on the scan tester, press number 9 on the tester to select climate control. On the first climate control menu, press the proper number to select system test, read faults, state display, actuator tests, or adjustments (Figure 17-21).

System Test

During the system test, pressing number 1 selects the partial charge test. In this mode, the scan tester indicates high-side pressure to check the refrigerant charge as explained previously in this

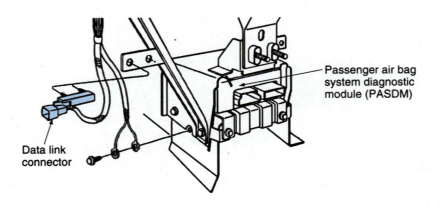

Figure 17-20 Data link connector under the dashboard (Courtesy of Chrysler Corporation)

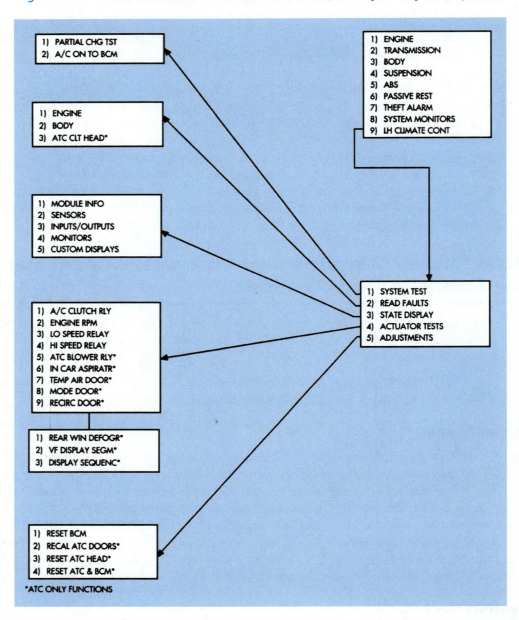

Figure 17-21 Scan tester diagnosis of computer-controlled air conditioning system (Courtesy of Chrysler Corporation)

chapter. If number 2 is pressed in the system test mode, the scan tester may be used to test the A/C switch input to the body computer module (BCM).

Read Faults

When read faults is selected in the initial climate control menu, the technician must then select engine computer faults, body computer faults, or A/C control head faults.

State Display

If the technician selects state display on the initial climate control memory, the scan tester displays the status of each A/C switch and sensor.

Actuator Tests

Selecting actuator tests from the initial climate control menu allows the technician to cycle the relays, operate the door actuators, and illuminate the display segments with the scan tester.

Adjustments

If adjustments is selected on the initial climate control menu, the scan tester may be used to calibrate the door actuator motors to the BCM. Diagnosing an automatic temperature control system with a scan tester is demonstrated in Photo Sequence 15.

Classroom Manual
Chapter 17, page 443

● **CUSTOMER CARE:** Always be honest with customers! One very successful automotive service business used the motto "Honesty is our discount." This motto was even written on their service trucks. Customers soon discovered that the management and staff did business according to their motto. As a result of this motto, plus excellent workmanship at fair prices, this repair shop was always busy. Most customers don't expect you to be perfect, but they do expect honesty and fair treatment.

Guidelines for Diagnosing and Servicing Computer-Controlled Air Conditioning Systems

1. Before computer-controlled air conditioning systems are diagnosed or serviced, a preliminary inspection should be completed.
2. An A/C performance test checks for proper operation of the refrigerant system.
3. Avoid breathing refrigerant and lubricant vapor or mist since this can irritate eyes, nose, or throat.
4. Manifold gauge sets are different for R-12 and R-134a refrigerant.
5. Refrigerant must be recovered and recycled. EPA regulations forbid the release of refrigerant to the atmosphere.
6. Mixing R-12 and R-134a refrigerants or refrigerant oils results in refrigerant system contamination.
7. Separate recovery and recycling stations are required for R-12 and R-134a refrigerant.
8. Some vehicle manufacturers recommend checking the refrigerant charge by checking the high-side pressure and condenser outlet temperature.
9. On some automatic temperature control (ATC) systems, the self-diagnostic mode is entered by pressing the floor, mix, and defrost buttons simultaneously with the engine running and the vehicle not moving.
10. When an ATC system is in the diagnostic mode, the panel button may be used to scroll through the diagnostic trouble codes.

Photo Sequence 15
Typical Procedure for Performing a Scan Tester Diagnosis of an Automatic Temperature Control System

P15-1 Be sure the ignition switch is off.

P15-2 Be sure the proper module is installed in the scan tester for the vehicle make, model year, and A/C diagnosis.

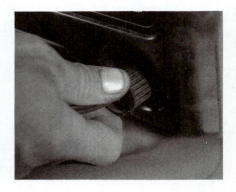

P15-3 Connect the scan tester power lead to the cigarette lighter socket.

P15-4 Program the scan tester for the vehicle make, model year, and VIN number.

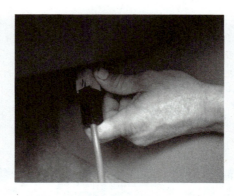

P15-5 Connect the scan tester data lead to the DLC under the dashboard.

P15-6 Select climate control on the initial menu.

P15-7 Select system test, read faults, state display, actuator tests, or adjustments from the climate control menu to diagnose the ATC system.

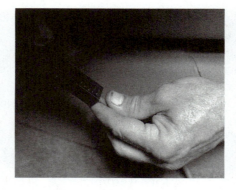

P15-8 Be sure the ignition switch is off, and then disconnect the scan tester data lead from the DLC.

P15-9 Disconnect the scan tester power cord from the cigarette lighter socket.

11. On other computer-controlled air conditioning systems, such as those on the Nissan Maxima, the off button is pressed for 5 seconds to enter the diagnostic mode. This action must be taken within 10 seconds after the engine is started.
12. In some computer-controlled air conditioning systems, the self-diagnosis contains five steps. Step 1 illuminates all the A/C control head displays. Step 2 provides DTCs related to the input sensors. Step 3 indicates DTCs related to the door actuator position switches. During step 4, the DEF button is pressed to move through six modes that check the door actuator motors, blower motor voltage, and compressor clutch operation. In step 5, the temperature sensed by each input sensor is displayed. Step 6 provides a method of adjusting the temperature display so it reads the same as the in-vehicle temperature.
13. On some computer-controlled air conditioning systems, a control rod adjustment is necessary on the door actuator motors.
14. A scan tester diagnosis may be performed on some computer-controlled air conditioning systems to complete a system test, read faults, and display state tests on switches and sensors. During the scan tester diagnosis, the scan tester may be used to actuate relays and door actuators, and illuminate display segments in the A/C control head.

CASE STUDY

A customer complained about no A/C operation on a 1995 Chrysler New Yorker with automatic temperature control. The technician completed a preliminary inspection without finding any problems. During an A/C performance test, the technician found warm air being discharged from the A/C panel ducts. The technician measured the high-side pressure and condenser outlet temperature to check the refrigerant charge and found these readings within specifications.

With the engine running and the car not moving, the technician pressed the floor, mix, and defrost buttons at the same time to enter the self-diagnostics. During the self-diagnostics, code 32 was displayed. This code indicated a blend door stall failure.

The technician connected the scan tester to the DLC and entered the actuator test mode. The next step in the diagnostic procedure was to enter the temp air door test 7 in the actuator tests. In this test, the body computer module (BCM) should move the air blend, or temp air, door actuator in response to commands from the scan tester. However, the blend air door did not move when commanded by the scan tester. The technician realized the problem could be in the BCM, blend air door actuator, or the connecting wires. The technician obtained the manufacturer's Body Diagnostic Procedures manual and located the appropriate test sequence to locate the source of the problem.

While performing the voltmeter and ohmmeter tests at the A/C plenum connector and BCM connector, the technician found the blend drive wire in cavity 20 of the BCM black connector was pushed back in the connector so it did not touch the BCM terminal (Figure 17-22). This wire was pushed firmly into place in the BCM connector. The technician pulled lightly on the wire to be sure it was retained securely in the connector. The black BCM connector was reconnected.

The technician reentered the temp air door mode in the actuator tests with the scan tester. The blend air, or temp air, door actuator now responded to commands from the scan tester.

The technician shut off the ignition switch and disconnected the scan tester. An A/C performance test indicated normal A/C operation.

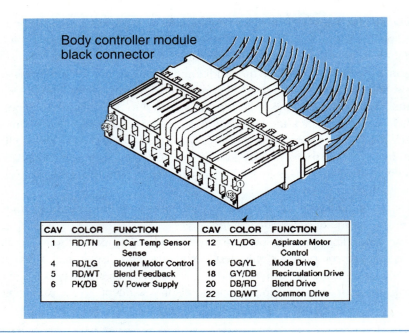

Figure 17-22 Black terminal in the body computer module connector (Courtesy of Chrysler Corporation)

Terms to Know

Air mix door
Automatic temperature control (ATC) system
High side
Intake or recirculation (REC) door
Low side
Manifold gauge set
Mode door
R-12
R-134a
Thermocouple

ASE Style Review Questions

1. While discussing an A/C performance test:
   Technician A says the temperature control knob should be in the full cold position.
   Technician B says the blower speed control knob should be in the low-speed position.
   Who is correct?
   - **A.** A only
   - **B.** B only
   - **C.** Both A and B
   - **D.** Neither A nor B

2. While discussing refrigerant charge testing:
   Technician A says on some A/C systems, the refrigerant charge may be tested by measuring the high-side pressure and condenser outlet temperature.
   Technician B says a thermocouple may be clamped to the condenser outlet pipe to measure the temperature at this location.
   Who is correct?
   - **A.** A only
   - **B.** B only
   - **C.** Both A and B
   - **D.** Neither A nor B

3. While discussing the self-diagnostic mode:
   Technician A says on Chrysler LH and LHS cars, this mode is entered with the ignition switch on and the engine not running.
   Technician B says on these cars, the self-diagnostic mode may be entered while driving the vehicle.
   Who is correct?
   A. A only
   B. B only
   C. Both A and B
   D. Neither A nor B

4. While discussing the self-diagnostic mode:
   Technician A says on some automatic A/C systems, such as those on the Nissan Maxima, the self-diagnostic mode is entered by pressing the A/C off button for 5 seconds within 10 seconds after the engine is started.
   Technician B says on these automatic air conditioning systems, the self-diagnostic mode is entered by pressing the defrost and off buttons at the same time.
   Who is correct?
   A. A only
   B. B only
   C. Both A and B
   D. Neither A nor B

5. While discussing the self-diagnostic mode on the automatic A/C system in a Nissan Maxima:
   Technician A says there are seven steps in the self-diagnostic mode.
   Technician B says the auto button is pressed to move to the next mode in the sequence.
   Who is correct?
   A. A only
   B. B only
   C. Both A and B
   D. Neither A nor B

6. While discussing the self-diagnostic mode on an automatic A/C system in a Nissan Maxima:
   Technician A says step 1 displays faults in the input sensors.
   Technician B says step 2 illuminates all the A/C display segments and LEDs.
   Who is correct?
   A. A only
   B. B only
   C. Both A and B
   D. Neither A nor B

7. While discussing the self-diagnostic mode on an automatic A/C system in a Nissan Maxima:
   Technician A says step 3 checks the door actuator position switches.
   Technician B says step 5 displays the temperature sensed by each input sensor.
   Who is correct?
   A. A only
   B. B only
   C. Both A and B
   D. Neither A nor B

8. While discussing door actuator control rod adjustments:
   Technician A says a control rod adjustment should be performed after a door actuator is replaced.
   Technician B says the door actuator should be securely mounted on the A/C heater case before the control rod adjustment.
   Who is correct?
   A. A only
   B. B only
   C. Both A and B
   D. Neither A nor B

9. While discussing scan tester diagnosis of computer-controlled A/C systems:
   Technician A says in the read faults mode, faults may be read only in the body computer module (BCM).
   Technician B says in the read faults mode, faults may be read in the PCM, BCM, or A/C control head.
   Who is correct?
   A. A only
   B. B only
   C. Both A and B
   D. Neither A nor B

10. While discussing scan tester diagnosis of computer-controlled A/C systems:
    Technician A says the state display checks the status of each A/C switch and sensor.
    Technician B says the state display may be used to cycle the relays in the A/C system.
    Who is correct?
    A. A only
    B. B only
    C. Both A and B
    D. Neither A nor B

Table 17-1 ASE Task

Diagnose temperature control system problems; determine needed repairs.

Problem Area	Symptoms	Possible Causes	Classroom Manual	Shop Manual
IMPROPER IN-VEHICLE TEMPERATURE	Improper air discharge temperature or location	1. Defective refrigerant system 2. Improper door actuator operation 3. Improper cooling or heating system operation	441 451 452	426 430 425

Table 17-2 ASE Task

Diagnose blower system problems; determine needed repairs.

Problem Area	Symptoms	Possible Causes	Classroom Manual	Shop Manual
IMPROPER BLOWER OPERATION	Inadequate or no blower operation	Defective fuse, wiring, relay, power module, or A/C computer	451	436

Table 17-3 ASE Task

Diagnose air distribution problems; determine needed repairs.

Problem Area	Symptoms	Possible Causes	Classroom Manual	Shop Manual
IMPROPER AIR DISCHARGE	Air discharge location wrong in relation to A/C control head setting	Defective actuator, wiring, A/C control head, or A/C computer	453	432

Table 17-4 ASE Task

Diagnose compressor clutch control systems; determined needed repairs.

Problem Area	Symptoms	Possible Causes	Classroom Manual	Shop Manual
IMPROPER CLUTCH OPERATION	Inadequate or no A/C operation; noise when clutch engaged	1. Defective wiring, relay, clutch, or A/C computer	442	436
		2. Defective clutch or compressor components	442	436

Table 17-5 ASE Task

Inspect, test, adjust, and replace blend door/power servo (vacuum or electric) system.

Problem Area	Symptoms	Possible Causes	Classroom Manual	Shop Manual
IN-VEHICLE TEMPERATURE	Improper in-vehicle temperature in relation to A/C control head setting	Defective air blend door actuator, wiring, A/C control head, or A/C computer	450	434

Table 17-6 ASE Task

Inspect, test, and replace A/C control head.

Problem Area	Symptoms	Possible Causes	Classroom Manual	Shop Manual
IMPROPER A/C SYSTEM OPERATION	Improper air discharge temperature or location	Defective A/C control head, wiring, actuators, or A/C computer	453	436

Table 17-7 ASE Task

Inspect, test, adjust, or replace the ATC microprocessor (climate control computer).

Problem Area	Symptoms	Possible Causes	Classroom Manual	Shop Manual
IMPROPER A/C SYSTEM OPERATION	Improper air discharge temperature or location	Defective A/C control head, wiring, actuators, or A/C computer	453	437

Table 17-8 ASE Task

Check and adjust the calibration of the ATC system.

Problem Area	Symptoms	Possible Causes	Classroom Manual	Shop Manual
IMPROPER AIR DISCHARGE	Air discharge wrong in relation to A/C control settings	Improper actuator door adjustment, calibration	451	434, 437

APPENDIX A

Tool and Equipment Suppliers List

Bear Automotive Inc.
12121 West Ferrick Place
Milwaukee WI 53222

Ferret Instruments
1310 Higgins Dr.
Cheboygan MI 49721

Fluke Corporation
P.O. Box 9090
Everett WA 98206

Kent-Moore Automotive Group
Kent-Moore Division
SPX Corporation
29784 Little Mack
Roseville MI 48066-2298

Kleer-Flo Company
15151 Technology Dr.
Eden Prairie MN 55344

Mac Tools, Inc.
South Fayette St.
Washington Court House Ohio 43160

MPSI
6405-19 Mile Road
Sterling Heights MI 48078

OTC A Division of SPX Corporation
655 Eisenhower Dr.
Owatonna MN 55060

SK Hand Tool Corporation
3535 W. 47th St.
Chicago Illinois 60632-2998

Snap-On Tools Corporation
Kenosha WI 53141-1410

Sun Electric Corporation
One Sun Parkway
Crystal Lake Illinois 60014

APPENDIX B

Metric Conversions

	to convert these	to these,	multiply by:
TEMPERATURE	Centigrade Degrees	Fahrenheit Degrees	1.8 then + 32
	Fahrenheit Degrees	Centigrade Degrees	0.556 after − 32
LENGTH	Millimeters	Inches	0.03937
	Inches	Millimeters	25.4
	Meters	Feet	3.28084
	Feet	Meters	0.3048
	Kilometers	Miles	0.62137
	Miles	Kilometers	1.60935
AREA	Square Centimeters	Square Inches	0.155
	Square Inches	Square Centimeters	6.45159
VOLUME	Cubic Centimeters	Cubic Inches	0.06103
	Cubic Inches	Cubic Centimeters	16.38703
	Cubic Centimeters	Liters	0.001
	Liters	Cubic Centimeters	1000
	Liters	Cubic Inches	61.025
	Cubic Inches	Liters	0.01639
	Liters	Quarts	1.05672
	Quarts	Liters	0.94633
	Liters	Pints	2.11344
	Pints	Liters	0.47317
	Liters	Ounces	33.81497
	Ounces	Liters	0.02957
WEIGHT	Grams	Ounces	0.03527
	Ounces	Grams	28.34953
	Kilograms	Pounds	2.20462
	Pounds	Kilograms	0.45359
WORK	Centimeter Kilograms	Inch-Pounds	0.8676
	Inch-Pounds	Centimeter-Kilograms	1.15262
	Meter Kilograms	Foot-Pounds	7.23301
	Foot-Pounds	Newton-Meters	1.3558
PRESSURE	Kilograms/Square Centimeter	Pounds/Square Inch	14.22334
	Pounds/Square Inch	Kilograms/Square Centimeter	0.07031
	Bar	Pounds/Square Inch	14.504
	Pounds/Square Inch	Bar	0.0689

GLOSSARY

Note: Terms are highlighted in color, followed by Spanish translation in bold.

ABS inspection connector A connector to which an ABS tester is connected on some vehicles.
Conector de inspección ABS Un conector al cual se conecta un aparato probador ABS en algunos vehículos.

Actuation test mode A scan tester mode used to cycle the relays and actuators in a computer system.
Modo de prueba de actuación Un modo del explorador que sirve para ciclar los relés y los impulsores en un sistema de computadora.

Advance-type timing light A timing light capable of checking the degrees of spark advance.
Luz de encendido tipo avance Una luz de encendido capaz de verificar los grados de avance de la chispa.

Air bag system diagnostic monitor (ASDM) A computer responsible for deploying the air bag. This computer also provides air bag system monitoring and diagnosing.
Monitor diagnóstico del sistema de bolsa de aire (ASDM) Una computadora que se encarga de desplegar la bolsa de aire. Esta computadora también efectua los diagnósticos y verifica al sistema de bolsa de aire.

AIR bypass (AIRB) solenoid A computer-controlled solenoid that directs air to the atmosphere or to the AIR diverter solenoid.
Solenoide desviación AIR (AIRB) Un solenoide controlado por computadora que dirige el aire a la atmósfera o al solenoide de desviación AIR.

Air charge temperature (ACT) sensor A sensor that sends a signal to the computer in relation to intake air temperature.
Sensor de temperatura de la carga de aire (ACT) Un sensor que envía una señal a la computadora con respeto a la temperatura del aire de entrada.

Air conditioning programmer A control unit that may contain the computer, solenoids, motors, and vacuum diaphragms for air conditioning control.
Programador del acondicionador de aire Una unidad de control que puede contener a la computadora, los solenoides, los motores y los diafragmas de vacío para controlar el acondicionador de aire.

AIR diverter (AIRD) solenoid A computer-controlled solenoid in the secondary air injection system that directs air upstream or downstream.
Solenoide de diversión AIR (AIRD) Un solenoide controlado por computadora en el sistema secundario del inyección de aire que dirije el aire hace arriba o abajo.

Air mix door A door in an air conditioning system that directs air through the heater core or evaporator into the passenger compartment to control the in-car temperature.
Compuerta de mezcla de aire Una puerta en un sistema acondicionador de aire que dirije al aire por el núcleo del calefactor o por el evaporador hacia el compartimiento del pasajera con el fín de controlar la temperatura interior del vehículo.

Analog meter A meter with a movable pointer and a meter scale.
Medidor analógico Un medidor que tiene un indicador móvil y una escala de medir.

Antidieseling adjustment An adjustment that prevents dieseling when the ignition switch is turned off.
Ajusto antiautoencendido Un ajusto que previene el autoencendido al apagarse la llave de encendido.

ASE blue seal of excellence A seal displayed by an automotive repair facility that employs ASE-certified technicians.
Sello azul de excelencia ASE Un sello que se manifiesta en las facilidades de refacción automotriz que emplean los técnicos certificados por la ASE.

ASE technician certification Certification of automotive technicians in various classifications by the National Institute for Automotive Service Excellence (ASE).
Certificación técnico ASE Las varias clasificaciones de certificación de técnicos automotrices ortogadas por el Instituto Nacional de Excelencia en Servicio Automovilístico.

Auto/manual diagnostic check A test procedure in diagnosing a Ford computer-controlled suspension system.
Examinación diagnóstico auto/manual Un procedimiento de examinación en diagnosticar un sistema de suspensión Ford controlado por computadora.

Automatic shutdown (ASD) relay A computer-operated relay that supplies voltage to the fuel pump, coil primary, and other components on Chrysler fuel-injected engines.
Relé de cierre automático (ASD) Un relé controlado por computadora que suministra al voltaje a la bomba de combustible, la bobina primaria, y otros componentes de los motores Chrysler de inyección.

Automatic temperature control system A computer-controlled air conditioning system.
Sistema de control automático de temperatura Un sistema de acondicionador de aire controlado por computadora.

Auto test A test procedure in diagnosing a Ford computer-controlled suspension system.
Prueba autoverificante Un procedimiento de pruebas diagnósticas en un sistema de suspensión controlado por computadora.

Barometric (Baro) pressure sensor A sensor that sends a signal to the computer in relation to barometric pressure.
Sensor de presión barométrica (Baro) Un sensor que envía una señal a la computadora con respeto a la presión barométrica.

Belt tension gauge A gauge designed to measure belt tension.
Calibrador de tensión de la correa Un calibre diseñado para medir la tensión en una correa.

Bimetal sensor A temperature-operated sensor used to control vacuum.
Sensor bimetal Un sensor de operación térmica que sirve para controlar un vacío.

Block learn A chip responsible for fuel control in a General Motors PCM.
Bloque de enseñanza Un chip que se encarga de controlar el combustible en el PCM de los motores General Motors.

Blowgun A device attached to the end of an air hose to control and direct air flow while cleaning components.
Soplete Un dispositivo conectado a la extremidad de un tubo de aire para controlar y dirigir un chorro de aire al limpiar los componentes.

Body computer menu The main menu of body computer test selections displayed on a scan tester.

Menu de la computadora El menu principal de las selecciones de las pruebas del chasis que se presentan en un aparato explorador.

Body computer state menu The body computer test selections in the input/output state test menu displayed on a scan tester.

Menu estatal de la computadora del chasis Las selecciones de las pruebas de la computadora del chasis en el menu para las pruebas del estado de entrada/salida que se presenta en un aparato explorador.

Breakout box A terminal box that is designed to be connected in series at Ford PCM terminals to provide access to these terminals for test purposes.

Caja de desconexión Una caja de terminal diseñada a conectarse en serie con los terminales PCM de Ford para dar acceso a estos terminales con el propósito de efectuar las pruebas.

Calibrator package (CAL-PAK) A removable chip in some computers that usually contains a fuel backup program.

Calibrador encapsulado (CALPAK) Un chip removible en algunas computadoras, suele contener una programa de repuesto del combustible.

Canister purge solenoid A computer-operated solenoid connected in the evaporative emission control system.

Solenoide para purgar el bote Un solenoide controlado por computadora que se conecta en el sistema de control de emisión evaporativa.

Canister-type pressurized injector cleaning container A container filled with unleaded gasoline and injector cleaner and pressurized during the manufacturing process or by the shop air supply.

Recipiente tipo bote para limpiar los inyectores bajo presión Un recipiente lleno de gasolina sin plomo y de limpiador de inyectores presurizado durante el proceso de fabricación o por una provisión de aire comprimido en el taller.

Carbon dioxide (CO_2) A gas formed as a by-product of the combustion process.

Bióxido de carbono (CO_2) Un gas que es un subproducto del proceso de combustión.

Carbon monoxide A gas formed as a by-product of the combustion process in the engine cylinders. This gas is very dangerous or deadly to the human body in high concentrations.

Monóxido de carbono (CO) Un gas formado como un subproducto del proceso de combustión en los cilíndros del motor. Este gas es muy peligroso y hasta fatal al cuerpo humano en concentraciones altas.

Catalytic converter vibrator tool A tool used to remove the pellets from catalytic converters.

Herramienta vibrador del convertidor catalítico Una herramienta que sirve para sacar los gránulos de los convertidores catalíticos.

Chrysler collision detection (C^2D) multiplex system A twisted pair of wires that transmits data between various onboard computers on Chrysler vehicles.

Sistema C^2D multiplex detector de choques Chrysler Un par de alambres trenzados que transmite los datos entre varias de las computadoras a bordo en los vehículos Chrysler.

Clockspring electrical connector A conductive ribbon contained in a plastic case and mounted under the steering wheel. This conductive ribbon maintains positive electrical contact between the air bag electrical system and the inflator module while allowing steering wheel rotation.

Lámina conectora eléctrica Una cinta conductiva en una caja de plástico posicionada bajo el volante de dirección. Esta cinta conductiva mantiene una conexión eléctrica positiva entre el sistema electrónico de la bolsa de aire y el módulo de inflación mientras que permite girar el volante de dirección.

Closed loop A computer operating mode in which the computer uses the oxygen sensor signal to help control the air-fuel ratio.

Bucle cerrado Un modo de operación de un sistema de computadora en el cual la computadora usa la señal del sensor de oxígeno para controlar la relación del aire combustible.

Compass variance Inaccurate compass readings caused by the difference between the location of the magnetic north pole and the geographic north pole.

Variación de brújula La lectura de la brújula incorrecta debido a la variación de posición del polo norte magnético y el polo norte geográfico.

Compression gauge A gauge used to test engine compression.

Compresómetro Un calibrador que sirve para comprobar la presión del motor.

Computed timing test A computer system test mode on Ford products that checks spark advance supplied by the computer.

Prueba de encendido computado Una prueba del sistema de computadora en los productos Ford que verifica el avance del encendido provisionado por la computadora.

Computer-controlled carburetor performance test A test that determines the general condition of a computer-controlled carburetor system.

Prueba de función del carburador controlado por computadora Una prueba que determina la condición general de un sistema de carburación controlado por computadora.

Concealment plug A plug installed over the idle mixture screw to prevent carburetor tampering by inexperienced service personnel.

Tapón de encubrimiento Un tapón instalado sobre la tuerca de ralentí (marcha lenta) para prevenir que los personas sin experiencia en el servicio alteran el ajuste del carburador.

Continuous self-test A computer system test mode on Ford products that provides a method of checking defective wiring connections.

Autoverificación continua Un modo de pruebas del sistema de computadora en los productos Ford que provee un método de verificar las conexiones alámbricas defectuosas.

Control unit Contains the computer in an electronic instrument cluster (EIC).

Unidad de control Contiene la computadora en un conjunto de instrumentos electrónicos (EIC).

Coolant hydrometer A tester designed to measure coolant specific gravity and determine the amount of antifreeze in the coolant.

Hidrómetro del fluido refrigerante Un probador diseñado para medir la gravedad específica del flúido refrigerante y para determinar la cantidad del anticongelante en el refrigerante.

Cooling system pressure tester A tester used to test cooling system leaks and radiator pressure caps.

Probador de presión del sistema refrigerante Un probador que sirve para verificar las fugas del sistema de enfriamiento y las tapas de presión del radiador.

Current diagnostic trouble code (DTC) A DTC that is present during the diagnostic procedure.
Código diagnóstico de la avería actual (DTC) Un DTC presente durante un procedimiento diagnóstico.

Data link connector (DLC) A computer system connector to which the computer supplies data for diagnostic purposes.
Conector de enlace de datos (DLC) Un conector de un sistema de computadora al cual la computadora envía datos para efectuar los diagnósticos.

Diagnostic trouble code (DTC) A code retained in a computer memory representing a fault in a specific area of the computer system.
Código de avería diagnóstico (DTC) Un código almacenado en la memoria de una computadora que representa un fallo en una area específica del sistema de computadora.

Diesel particulates Small carbon particles in diesel exhaust.
Partículos de diesel Pequeños partículos de carbono en el vapor de escape diesel.

Digital EGR valve An EGR valve that contains a computer-operated solenoid or solenoids.
Válvula EGR digital Una válvula EGR que contiene un solenoide o unos solenoides controlados por computadora.

Digital meter A meter with a digital display.
Medidora digital Una medidora con una presentación digital.

Digital storage oscilloscope (DSO) An oscilloscope that senses, stores, and then displays a voltage signal. This type of oscilloscope has a much faster signal sampling speed compared to other automotive test equipment such as an analog scope.
Osciloscopio de almanacenamiento digital (DSO) Un osciloscopio que detecta, almacena, y luego presenta una señal de tensión. Este tipo de osciloscopio toma los muestreos de señales con mucho más rapidez comparado con otros equipos de verificación tal como el osciloscopio análogo.

Display check Illuminates all segments in an electronic instrument cluster (EIC) when certain buttons are pressed.
Verificación del presentación Ilumina todos los segmentos en un conjunto de instrumentos electrónicos (EIC) cuando se oprima ciertos botones.

Display unit Contains the electronic instrument cluster (EIC) digital displays.
Unidad de presentación Contiene los presentaciónes digitales de los conjuntos de instrumentos electrónicos (EIC).

Distributor ignition (DI) system SAE J1930 terminology for any ignition system with a distributor.
Sistema de encendido con distribuidor (DI) La terminología SAE J1920 por cualquier sistema de encendidos que tiene un distribuidor.

Downstream air Air injected into the catalytic converter.
Aire abajo El aire inyectado en el convertidor catalítico.

Drive cycle diagnosis A test procedure when diagnosing computer-controlled suspension systems.
Diagnóstico del ciclo de viaje Un procedimiento de pruebas para diagnosticar los sistemas controlados por computadora.

Dynamic test An ABS system test performed by the ABS system computer when the engine is started and the vehicle is accelerated to 5 to 10 mph.
Prueba dinámica Una prueba del sistema ABS ejecutado por la computadora del sistema ABS cuando el motor esta en marcha y el vehículo es acelerado entre 5 a 10 mph.

EGR pressure transducer (EPT) A vacuum switching device operated by exhaust pressure that opens and closes the vacuum passage to the EGR valve.
Transductor de presión EGR (EPT) Un dispositivo interruptor de vacío operado por la presión del escape que abre y cierre el pasillo del vacío a la válvula EGR.

EGR vacuum regulator (EVR) solenoid A solenoid that is cycled by the computer to provide a specific vacuum to the EGR valve.
Solenoide del regulador del vacío (EVR) Un solenoide ciclado por la computadora para provisionar un vacío específico a la válvula EGR.

Electronic automatic transaxle (EATX) A computer-controlled transaxle.
Transeje electrónico automático (EATX) Un transeje controlado por computadora.

Electronic fuel injection (EFI) A generic term applied to various types of fuel injection systems.
Inyección electrónica de combustible (EFI) Un término que puede indicar varios sistemas de inyección de combustible.

Electronic ignition (EI) system SAE J1930 terminology for any ignition system without a distributor.
Sistema de encendido electrónico La terminología SAE J1930 indicando cualquier sistema de encendido sin distribuidor.

Electronic instrument cluster (EIC) A computer-controlled instrument panel display.
Grupo de instrumentos electrónicos (EIC) Una presentación controlada por computadora en el tablero de instrumentos.

Electronic pressure control (EPC) solenoid A computer-controlled solenoid that controls transmission fluid pressure.
Solenoide de control electrónico de presión (EPC) Un solenoide controlado por computador que controla la presión del flúido de la transmisión.

Engine analyzer A tester designed to test engine systems such as battery, starter, charging, ignition, and fuel plus engine condition.
Analizador del motor Un probador diseñado para probar los sistemas del motor tal como la batería, el arrancador, la ignición, y la combustión y también las condiciones del motor.

Engine coolant temperature (ECT) sensor A sensor that sends a voltage signal to the computer in relation to coolant temperature.
Sensor de temperatura de flúido refrigerante (ECY) Un sensor que envía una señal de tensión a la computadora con respeto a la temperatura del flúido refrigerante.

Engine lift A hydraulically operated piece of equipment used to lift the engine from the chassis.
Grúa de motor (burro) Una pieza de equipo de operación hidráulica que sirve para sacar el motor del chasis.

Evaporative (EVAP) system A system that collects fuel vapors from the fuel tank and directs them into the intake manifold rather than allowing them to escape to the atmosphere.
Sistema evaporativo (EVAP) Un sistema que recoge los vapores del combustible del tanque de gas y los dirige hacia el múltiple de admisión en vez de dejarlos escapar a la atmósfera.

Exhaust gas recirculation (EGR) valve A valve that circulates a specific amount of exhaust gas into the intake manifold to reduce NO_x emissions.
Válvula de recirculación de los vapores del escape (EGR) Una válvula que circula una cierta cantidad del combustible de salida dentro del múltiple de admisión para bajar los emisiones de NOx.

Exhaust gas recirculation valve position (EVP) sensor A sensor that sends a voltage signal to the computer in relation to EGR valve position.

Sensor de posición de la válvula de recirculación de los vapores del escape (EVP) Un sensor que envía una señal de tensión a la computador con respeto a la posición de la válvula EGR.

Exhaust gas temperature sensor A sensor that sends a voltage signal to the computer in relation to exhaust temperature.

Sensor de temperatura de los vapores del escape Un sensor que envía una señal de tensión a la computadora con respeto a la temperatura de los vapores del escape.

Fault codes Numeric codes stored in a computer memory representing specific computer system faults. The term *fault code* is replaced by diagnostic trouble code (DTC) in SAE J1930 terminology.

Códigos de averías Los códigos numéricos almacenados en la memoria de la computadora que representan fallos específicos del sistema de computadora.

Feeler gauge Metal strips with a specific thickness for measuring clearances between components.

Calibrador de laminillas Unas láminas de metal de espesores específicos que sirven para medir las holguras entre los componentes.

Field service mode A computer diagnostic mode that indicates whether the computer is in open or closed loop on General Motors PCMs.

Modo de servicio de campo Un modo diagnóstico de la computadora que indica si la computadora esta en bucle abierto o cerrado en los PCMs de General Motors.

Flash code Reading computer system diagnostic trouble codes (DTCs) from the flashes of an instrument panel warning light connected to the computer system being diagnosed.

Código rápido Analizar los códigos diagnóticos de averías (DTCs) por los parpardeos de la luz de avisos en el panel de instrumento que se ha conectado al sistema de la computadora con fines diagnósticos.

Floor jack A hydraulically operated device mounted on casters and used to raise one end or corner of the chassis.

Gato hidráulico Un dispositivo de operación hidráulica montado en ruedecillas que sirve para elevar una extremidad o una parte del chasis.

Four-gas emissions analyzer An analyzer designed to test carbon monoxide, carbon dioxide, hydrocarbons, and oxygen in the exhaust.

Analizador de emisiones de cuatro gases Un analizador diseñado para examinar el monóxido de carbono, el bióxido de carbono, los hidrocarburos, y el oxígeno en los vapores del escape.

Fuel cut rpm The rpm range in which the computer stops operating the injectors during deceleration.

Rpm de cortar el combustible El rango de rpm en el cual la computador no opera los inyectores durante la deceleración.

Fuel pressure test port A threaded port on the fuel rail to which a pressure gauge may be connected to test fuel pressure.

Puerta de prueba de la presión de combustible Una puerta fileteada en el carríl de combustible al cual se puede conectar un indicador de presíon para probar la presíon del combustible.

Fuel pump volume The amount of fuel the pump delivers in a specific time period.

Volumen de la bomba de combustible La cantidad del combustible que entrega la bomba durante un periodo específico de tiempo.

Fuel tank purging Removing fuel vapors and foreign material from the fuel tank.

Purga del tanque de combustible Sacar los vapores de combustible y la material ajena del tanque de combustible.

Functional test 211 A diagnostic test in a computer-controlled suspension system.

Prueba funciónal 211 Una prueba diagnóstica en un sistema de suspensión controlado por computadora.

Function diagnostic mode Provides diagnostic information in the display readings of an electronic instrument cluster (EIC) when certain defects occur.

Modo diganóstico de funciones Provee la información diagnóstica en las lecturas de las presentaciones de un conjunto de instrumentos electrónicos (EIC) cuando ocurre algún fallo.

Hall Effect pickup A pickup containing a Hall element and a permanent magnet with a rotating blade between these components.

Captador efecto Hall Un captador que contiene un elemento Hall y un imán permanente con una hoja giratoria entre estos componentes.

Hand-held digital pyrometer A tester for measuring component temperature.

Pirómetro digital de mano Un aparato para medir la temperatura de los componentes.

Hand press A hand-operated device for pressing precision-fit components.

Prensa de impresión a mano Un dispositivo operado a mano que sirve para imprimir los componentes de ajuste preciso.

Hard fault code A fault code in a computer that represents a fault that is present at all times.

Código de avería continua Un código de averías de una computadora que representa un fallo que siempre esta presente.

Head-up display A display projected in the driver's view near the end of the hood. This display includes the speedometer reading, turn signal indicators, and other warning indicators.

Presentación de proyección en el parabrisa Una presentación proyectada en la linea de vista del conductor cerca de la extremidad del capó. Esta presentación contiene la lectura del velocímetro, los indicadores de vuelta, y otros avisos.

Heated resistor-type MAF sensor A MAF sensor that uses a heated resistor to sense air intake volume and temperature, and sends a voltage signal to the computer in relation to the total volume of intake air.

Sensor MMAF tipo resistor calentado Un sensor MAF que usa un resistor calentado para detectar el volumen y la temperatura del aire de entrada, y envía una señal de tensión a la computadora con respeto al volumen total del aire de entrada.

High pressure accumulator A sealed metal container with a diaphragm in the center. A nitrogen gas charge is located on one side of the diaphragm. In an ABS system, high-pressure brake fluid is stored on the other side of the diaphragm.

Acumulador de alta presión Un envase de metal sellado que tiene en su centro un diafragma. Tiene una carga de gas nitrógeno en un lado del diafragma. En un sistema ABS, el fluido de freno bajo alta presión se encuentra al otro lado del diafragma.

High side The high-pressure side of an air conditioning refrigerant system.

Lado alto El lado de alta presión de un sistema refrigerante del acondicionador de aire.

History diagnostic trouble code (DTC) A DTC stored in the computer memory representing an intermittent fault.

Narrativo de códigos diagnósticos de averías (DTC) Un DTC almacenado en la memoria de la computadora que representa una avería intermitente.

Hot wire-type MAF sensor A MAF sensor that uses a heated wire to sense air intake volume and temperature, and sends a voltage signal to the computer in relation to the total volume of intake air.

Sensor MAF de hilo caliente Un sensor MAF que usa un alambre calentado para detectar el volumen y la temperatura del aire de entrada, y envía una señal de tensión a la computadora con respeto al volumen total del aire de entrada.

Hydraulic press A hydraulically operated device for pressing precision-fit components.

Prensa hidráulica Un dispositivo de operación hidráulica que sirve para prensar los componentes de ajuste preciso.

Hydrocarbons (HC) Left-over fuel from the combustion process.

Hidrocarburos (HC) Lo que sobra del combustible despues del proceso de combustión.

Idle air control bypass air (IAC BPA) motor An IAC motor that controls idle speed by regulating the amount of air bypassing the throttle.

Motor de control de aire desviado en marcha vacío (IAC BPA) Un motor IAC que controla la velocidad de marcha en vacío por medio de la regulación de la cantidad del aire desviado del acelerador.

Idle air control bypass air (IAC BPA) valve A valve operated by the IAC BPA motor that regulates the air bypassing the throttle to control idle speed.

Válvula de control de aire desviado en marcha vacío (IAC BPA) Una válvula operada por el motor IAC BPA que regula el aire desviado del acelerador con el fin de controlar la velocidad de la marcha en vacío.

Idle air control (IAC) motor A computer-controlled motor that controls idle speed under all conditions.

Motor de control de aire de marcha en vacío (IAC) Un motor controlado por computadora que controla la velocidad de la marcha en vacío bajo todas condiciones.

Idle contact switch A switch in the IAC motor stem that informs the computer when the throttle is in the idle position.

Interruptor de contacto en marcha en vacío Un interruptor en el vástago del motor IAC que informa a la computadora cuando esta en posición de marcha en vacío el acelerador.

Idle stop solenoid An electric solenoid that maintains the throttle in the proper idle speed position, and prevents dieseling when the ignition switch is turned off.

Solenoide de parar la marcha en vacío Un solenoide eléctrico que mantiene al acelerador en la posición correcta para marcha en vacío, y previene el autoencendido al apagar la llave de encendido.

Ignition cross-firing Ignition firing between distributor cap terminals or spark plug wires.

Encendido cruzado El encendido entre los terminales de la tapa del distribuidor o los alambres de las bujías.

Ignition module tester An electronic tester designed to test ignition modules.

Probador del módulo de encendido Un probador electrónico para verificar los módulos del encendido.

IM240 emission testing An enhanced four-second emission test procedure run on a dynamometer at various speeds. The average emission levels during the test are recorded in grams per mile (gpm).

Procedimiento prueba de emisión IM240 El procedimiento para una prueba progressiva de cuatro segundos que se lleva acabo en varias velocidades de operación en un dinamómetro. Los niveles típicos de emisión de la prueba se registran en gramos por milla (gpm).

Inflator module A module located on top of the steering wheel that contains the air bag and the chemicals required for air bag inflation.

Módulo inflador Un módulo ubicado en la parte superior del volante de dirección que contiene la bolsa de aire y los productos químicos requeridos para inflar la bolsa de aire.

Initialization test An ABS system test performed by the ABS computer each time the ignition switch is turned on.

Prueba de inicialización Una prueba del sistema ABS que efectua la computadora ABS cada vez que se prende la llave de arranque.

Injector balance tester A tester designed to test port injectors.

Probador de equilibrio de los inyectores Una prueba diseñado a verificar las puertas de los inyectores.

Inspection, maintenance (I/M) testing Emission inspection and maintenance programs that are usually administered by various states.

Pruebas de inspección y de mantenimiento Las programas de inspección y mantenimiento de los emisiones que suelen administrar los varios estados.

Instrument panel cluster A display in the instrument panel that may include gauges, indicator lights, or digital displays to indicate system functions.

Conjunto de instrumentos del tablero Una presentación en el tablero de instrumentos que puede incluir las indicadoras, las luces de indicadoras o las presentaciones digitales para indicar las funciones de los sistemas.

Intake or recirculation (RECIRC) door A door in an air conditioning system that directs outside air or in-car air into the air conditioning-heater case.

Compuerta de entrada o recirculación (RECIRC) Una puerta en el sistema de aire acondicionado que dirige el aire exterior o el aire interior del vehículo al cárter del acondicionador de aire o del calefactor.

Integrator A chip responsible for fuel control in a General Motors PCM.

Integrador Un chip que se encarga del control del combustible en un PCM de General Motors.

Intermittent fault code A fault code in a computer memory that is caused by a defect that is not always present.

Código de fallo intermitente Un código de fallo en la memoria de una computadora causado por un defecto que no siempre esta presente.

International standards organization (ISO) An organization responsible for establishing symbols that are internationally recognized.

Organización patrón internacional (ISO) Una organización cuyo responsabilidad es de establecer los símbolos que se reconocen internacionalmente.

International System (SI) A system of weights and measures in which each unit may be divided by 10.

Sistema internacional (SI) Un sistema de pesos y medidas en el cual cada unidad puede dividirse por 10.

Jack stand A metal stand used to support one corner of the chassis.

Torre o soporte Un soporte de metal que sirve para sostener una esquina del chasis.

Key on engine off (KOEO) test A computer system test mode on Ford products that displays diagnostic trouble codes (DTCs) with the key on and the engine stopped.

Prueba de llave prendida motor apagado (KOEO) Un modo de prueba del sistema de computadora en los productos Ford que presenta los códigos diagnósticos de averías (DTCs) con la llave prendida y el motor parado.

Key on engine running (KOER) test A computer system test mode on Ford products that displays diagnostic trouble codes (DTCs) with the engine running.

Prueba de llave prendida motor en marcha (KOER) Un modo de prueba del sistema de computadora en los productos Ford que presenta los códigos diagnósticos de averías (DTCs) con la llave prendida y el motor en marcha.

Knock sensor A sensor that sends a voltage signal to the computer in relation to engine detonation.

Sensor de golpeteo Un sensor que envía un señal de tensión a la computadora con respeto al golpeteo del motor.

Knock sensor module An electronic module that changes the analog knock sensor signal to a digital signal and sends it to the PCM.

Módulo de sensor de golpeteo Un módulo electrónico que cambia la señal analoga del sensor de golpeteo a una señal digital y la envía al PCM.

Latching fault An ABS system fault that causes the ABS computer to disable the ABS function and illuminate the ABS and traction control warning lights. The system remains disabled until the ignition switch is cycled.

Falta enclavada Una falta del sistema ABS que causa que la computadora ABS incapacita la función ABS y ilumina las luces de aviso del ABS y del control de tracción. El sistema queda incapacitado hasta que termina un ciclo el interruptor del encendido.

Lift A device used to raise a vehicle.

Elevador de vehículos Un dispositivo que sirve para levantar un vehículo.

Light emitting diode (LED) A special diode that emits light when current flows through the diode.

Diodo emisor de luz (LED) Un diodo especial que produce la luz cuando un corriente eléctrico fluye por el diodo.

Linear EGR valve An EGR valve containing an electric solenoid that is pulsed on and off by the computer to provide a precise EGR flow.

Válvula lineal EGR Una válvula que contiene un solenoide que se prenda o se apaga por medio de impulsos mandados por la computadora con el fin de proveer un flujo preciso de EGR.

Locked fault An ABS system fault that causes the ABS computer to disable the ABS function and illuminate the ABS and traction control warning lights. The system remains disabled and the warning lights are illuminated until the fault is erased from the ABS computer memory.

Falta enclavada Una falta del sistema ABS que causa que la computadora ABS incapacita la función del ABS y ilumina las luces de aviso del ABS y del control de tracción. El sistema queda incapacitado y las luces quedan prendidas hasta que se borra el fallo de la memoria de la computadora ABS.

Low side The low-pressure side of an air conditioning refrigerant system.

Lado de bajo presión El lado del sistema de refrigeración del aire acondicionado que tiene bajo presión.

Magnetic probe-type digital tachometer A digital tachometer that reads engine rpm and uses a magnetic probe pickup.

Tacómetro digital magnético tipo sonda Un tacómetro digital que lee el rpm del motor y usa un captador de sonda magnético.

Magnetic probe-type digital timing meter A digital reading that displays crankshaft degrees and uses a magnetic-type pickup probe mounted in the magnetic timing probe receptacle.

Medidor magnético digital del encendido tipo sonda Una lectura digital que presenta los grados del cigueñal y usa un captador de sonda magnético montado en el receptáculo de la captador magnético de medidor del encendido.

Magnetic sensor A sensor that produces a voltage signal from a rotating element near a winding and a permanent magnet. This voltage signal is often used for ignition triggering.

Sensor magnético Un sensor que produce una señal de tensión de un elemento rotativo cerca de una bobina y un imán permanente. Esta señal de tensión se usa muchas veces para disparar el encendido.

Magnetic timing offset An adjustment to compensate for the position of the magnetic receptacle opening in relation to the TDC mark on the crankshaft pulley.

Compensación magnético del encendido Un ajuste para compensar la posición de la apertura del receptáculo magnético en relación a la marca TDC en la polea del cigueñal.

Magnetic timing probe receptacle An opening in which the magnetic timing probe is installed to check basic timing.

Receptáculo de la sonda magnética del encendido Una apertura en la cual se instala la sonda magnética del encendido para verificar el tiempo de la marcha normal.

Malfunction indicator light (MIL) A light in the instrument panel that is illuminated by the PCM if certain defects occur in the computer system.

Luz indicador de defecto (MIL) Una luz en el tablero de instrumentos que es iluminado por el PCM si ocurren ciertos defectos en el sistema de computadora.

Manifold absolute pressure (MAP) sensor An input sensor that sends a signal to the computer in relation to intake manifold vacuum.

Sensor de presión absoluta en el múltiple (MAP) Un sensor de entrada que envía una señal a la computadora con respeto al vacío del múltiple de admisión.

Manifold gauge set A set of gauges with connecting hoses and valves for diagnosing air conditioning systems.

Conjunto de calibres del múltiple Un conjunto de calibres que tiene conectado uno tubos para efectuar los diagnósticos en los sistemas del acondicionador de aire.

Manual valve lever position sensor (MVLPS) A sensor mounted on the transmission linkage that sends a signal to the transmission control module (TCM) in relation to gear selector position.

Sensor de posición de la válvula de la palanca manual (MVLPS) Un sensor montado sobre la biela de la transmisión que envía una señal de tensión al módulo de control de la transmisión (TCM) con respeto a la posición del selector de velocidades.

Mass air flow sensor (MAF) A sensor that sends an input signal to the engine computer in relation to the total volume of air entering the engine.

Sensor de la masa del aire (MAF) Un sensor que envía una señal de tensión de entrada a la computador del motor con respeto al total del volumen de aire entrando al motor.

Memory calibrator (MEM-CAL) A removable chip in some computers that replaces the PROM and CAL-PAK chips.

Calibrador de memoria (MEM-CAL) Un chip removible en algunas computadoras que reemplaza los chips PROM y CAL-PAK.

Meter impedance The total internal electrical resistance in a meter.

Impedancia del medidor La resistencia interna eléctrica total en un medidor.

Mode door A door in an air conditioning system that directs air to the floor, dash, or defrost outlets in an air conditioning system.

Compuerta de modo Una puerta de un sistema de acondicionador de aire que dirige el aire al suelo, al tablero, o a las compuertas de descongelación en el sistema de acondicionador de aire.

Muffler chisel A chisel that is designed for cutting muffler inlet and outlet pipes.

Cincel de mofles Un cincel diseñado para cortar los tubos de admisión y escape del silenciador.

Multiport fuel injection (MFI) A fuel injection system in which the injectors are grounded in the computer in pairs or groups of three or four.

Inyección de combustible multipuertas (MFI) Un sistema de inyección de combustible en el cual los inyectores se conectan a tierra en la computadora en pares o grupos de tres o cuatro.

National Institute for Automotive Service Excellence (ASE) An organization responsible for certification of automotive technicians in the U.S.

Instituto Nacional de Excelencia en Servicio Automovilístico (ASE) Una organización que se encarga de certificar los técnicos automovilísticos de los Estados Unidos.

Negative backpressure EGR valve An EGR valve containing a vacuum bleed valve that is operated by negative pulses in the exhaust.

Válvula EGR de contrapresión negativa Una válvula EGR que contiene una válvula de purgar a vacío que se opera por medio de los impulsos de contrapresión en el sistema de escape.

Neutral/drive switch (NDS) A switch that sends a signal to the computer in relation to gear selector position.

Interruptor de marcha en neutro (NDS) Un interruptor que envía una señal a la computadora con respeto a la posición del selector de velocidades.

Nonerasable fault An ABS system fault that cannot be erased from the ABS computer memory.

Falta no borrable Una falta del sistema ABS que no se puede borrar de la memoria de la computadora ABS.

Nonintegral ABS system An ABS system with some of the components mounted separately from the master cylinder.

Sistema no integral ABS Un sistema ABS que tiene algunos de sus componentes montados aparte del cilindro maestro.

Nonlatching fault An ABS system fault that causes the ABS computer to disable the system and illuminate the ABS and traction control warning lights. The system is enabled and the warning lights go off if the fault disappears.

Falta no enclavada Una falta del sistema ABS que causa que la computadora ABS incapacita la función del ABS y ilumina las luces de aviso del ABS y del control de tracción. El sistema se habilita y las luces de aviso se apagan si se desaparece la falta.

Nose switch A switch in the IAC motor stem that informs the computer when the throttle is in the idle position.

Interruptor proa Un interruptor en el vástago del motor IAC que informa a la computadora cuando el acelerador esta en la posición de marcha en vacío.

Oil pressure gauge A gauge used to test engine oil pressure.

Indicador de presión del aceite Un indicador que sirve para verficar la presión del aceite.

Open loop A computer operating mode in which the computer controls the air-fuel ratio and ignores the oxygen sensor signal.

Bucle abierto Un modo de operación de computadora en el cual la computadora controla la relación de aire combustible pero ignora a la señal del sensor de oxígeno.

Optical-type pickup A pickup that contains a photo diode and a light emitting diode with a slotted plate between these components.

Captador óptico Un captador que contiene un fotodiodo y un diodo emisor de luz y una placa ranurada que gira entre estos componentes.

Oscilloscope A cathode ray tube (CRT) that displays voltage waveforms from the ignition system.

Osciloscopio Un tubo de rayos catódicos (CRT) que presenta las formas de ondas de voltaje del sistema de encendido.

Output cycling test A test that allows the technician to cycle the computer system outputs.

Prueba del ciclado de salida Una prueba que permite que el técnico hace pasar las salidas del sistema de computadora por un ciclo.

Output speed sensor A magnetic-type sensor positioned near the rear of the transmission that sends a voltage signal to the transmission control module (TCM) in relation to transmission output rotational speed.

Sensor de velocidad de salida Un sensor tipo magnético posicionado cerca del posterior de la transmisión que envía una señal de tensión al módulo de control de transmisión (TCM) con respeto a la velocidad rotativa de la salida de la transmisión.

Output state test A computer system test mode on Ford products that turns the relays and actuators on and off.

Prueba de estado de salida Un modo de prueba del sistema de computadora en los productors Ford que prenda y apaga los relés y los impulsores.

Overhead travel information system (OTIS) An overhead digital display that usually includes travel computer displays such as instant and average fuel economy.

Sistema de información de viaje sobrecabeza (OTIS) Una presentación digital sobrecabeza que suele incluir la información de la computadora tal como el rendimiento actual y promedial del combustible.

Oxygen (O_2) A gaseous element that is present in air.

Oxígeno (O_2) Un elemento gaseoso presente en el aire.

Oxygen (O_2) sensor A sensor mounted in the exhaust system that sends a voltage signal to the computer in relation to the amount of oxygen in the exhaust stream.

Sensor de oxígeno (O_2) Un sensor montado en el sistema de escape que envía una señal de tensión a la computadora con respeto a la cantidad del oxígeno contenido en los vapores de escape.

Park/neutral switch A switch connected in the starter solenoid circuit that prevents starter operation except in park or neutral.

Interruptor neutral Un interruptor conectado al circuito del solenoide del arrancador que previene la operación del arrancador menos en parque o neutral.

Parts per million (ppm) The volume of a gas such as hydrocarbons in ppm in relation to one million parts of the total volume of exhaust gas.

Partes por millón (PPM) El volumen de un gas tal como los hidrocarburos en los ppm en relación a un millón de partes del volumen total del vapor de escape.

Photoelectric tachometer A tachometer that contains an internal light source and a photoelectric cell. This meter senses rpm from reflective tape attached to a rotating component.

Tacómetro fotoeléctrico Un tacómetro que contiene un suministro interior de luz y una célula fotoeléctrica. Este medidor detecta las rpm de una cinta reflectiva conectada a un componente rotativo.

Pinpoint tests Specific test procedures to locate the exact cause of a fault code in a computer system.

Pruebas precisas Los procedimientos de pruebas específicos para localizar la causa exacta de un código de falta en un sistema de computadora.

Pipe expander A tool designed to expand exhaust system pipes.

Mandríl de expansión Una herramienta diseñada para expander los tubos del sistema de escape.

Port EGR valve An EGR valve operated by ported vacuum from above the throttle.

Válvula EGR de compuerta Una válvula EGR operada por un vacío con una puerta arriba de los aceleradores.

Port fuel injection (PFI) A fuel injection system with an injector positioned in each intake port.

Inyección de combustible de puerta (PFI) Un sistema de inyección de combustible que tiene un inyector montado en cada apertura de entrada.

Positive backpressure EGR valve An EGR valve with a vacuum bleed valve that is operated by positive pressure pulses in the exhaust.

Válvula EGR de contrapresión positiva Una válvula EGR con una válvula de purga de vacío que se opera por los impulsos de presión positiva en los vapores del escape.

Positive crankcase ventilation (PCV) valve A valve that delivers crankcase vapors into the intake manifold rather than allowing them to escape to the atmosphere.

Ventilación positiva de la caja de cigueñal (PCV) Una válvula que entrega los vapores de la caja de cigueñal al múltiple de admisión en vez de dejarlos escapar a la atmósfera.

Powertrain control module (PCM) SAE J1930 terminology for an engine control computer.

Módulo control del tren de potencia (PCM) La terminología SAE J1930 indicando una computadora que controla un motor.

Power unit Provides the power supply voltage in some electronic instrument clusters (EICs).

Unidad de potencia Provee el voltaje de un suministro de energía para algunos conjuntos de instrumentos electrónicos.

Preprogrammed signal check Tests for defects in specific electronic instrument cluster (EIC) displays.

Verificación de la señal preprogramada Las pruebas para defectos en las presentaciones específicas de conjuntos de instrumentos electrónicos.

Pressurized injector cleaning container A small, pressurized container filled with unleaded gasoline and injector cleaner for cleaning injectors with the engine running.

Recipiente de limpieza bajo presión para los inyectores Un pequeño recipiente bajo presión lleno de la gasolina sin plomo y el fluido de limpiar inyectores que sirve para limpiar los inyectores mientras que esté en marcha el motor.

Programmable read only memory (PROM) A computer chip containing some of the computer program. This chip is removable in some computers.

Memoria sólo para lectura programable (PROM) Un chip lógico que contiene algo de la programa de la computadora. Este chip es removible en algunas computadoras.

Programmed ride control (PRC) system A computer-controlled suspension system in which the computer operates an actuator in each strut to control strut firmness.

Sistema programado de control de viaje Un sistema de suspensión controlado por computadora en el cual la computadora opera una actuador en cada poste para controlar la firmeza de los postes.

Propane-assisted idle mixture adjustment A method of adjusting idle mixture with fuel supplied from a small propane cylinder.

Ajuste de la mezcla de marcha en vacío asistido por propano Un método de ajustar la mezcla de la marcha en vacío usando el combustible suministrado de un cilindro pequeño de propano.

Prove-out display An initial display when all segments are illuminated in an electronic instrument cluster (EIC). This display occurs when the ignition switch is turned on.

Presentación de muestra Una presentación inicial en la cual todos los segmentos se iluminan en el conjunto de instrumentos electrónicos (EIC). Esta presentación ocurre cuando se da la vuelta a la llave de encendido.

Pulsed secondary air injection system A system that uses negative pressure pulses in the exhaust to move air into the exhaust system.

Inyección de aire pulsado secundario Un sistema que utilisa los impulsos de presión negativos del escape para dirigir al aire al sistema de escape.

Quad driver A group of transistors in a computer that controls specific outputs.

Excitador cuartete Un grupo de transistores en una computadora que controla las salidas específicas.

Quick-disconnect fuel line fittings Fuel line fittings that may be disconnected without using a wrench.

Montaje de tubos de combustible de desconexión rápida Los accesorios de conexión del tubo de combustible que se pueden disconectar sín llave.

R-12 An air conditioning system refrigerant that contains chlorofluorocarbons. When released into the atmosphere, this refrigerant is harmful to the earth's ozone layer.

R-12 Un refrigerante del sistema de acondicionador de aire que contiene los clorofluorocarburos. Al descargar este refrigerante a la atmósfera, causa daños a la capa de ozono de la tierra.

R-134a An air conditioning system refrigerant that replaces R-12. R-134a does not deplete the earth's ozone layer.

R-134a Un refrigerante del sistema de acondicionador de aire que reemplaza el R-12. R-134a no agota la capa de ozono de la tierra.

Radiator shroud A circular component positioned around the cooling fan to concentrate the air flow through the radiator.

Cubierto del radiador Un componente circular colocado alrededor del ventilador de circulación para concentrar el flujo de aire por el radiador.

Radio frequency interference A voltage signal with a very high frequency that may cause radio static.

Interferencia de radiofrecuencia Una señal de tensión con una frecuencia muy alta que puede causar el estático de radio.

Real time damping (RTD) The real time damping module controls the road-sensing suspension system.
Amortiguamiento de tiempo real (RTD) El módulo de amortiguamiento de tiempo real controla el sistema de suspensión sensible al camino.

Reference pickup A pickup assembly that is often used for ignition triggering.
Captador de referencia Una asamblea captador que se suele usar para detonar el encendido.

Reference voltage A constant voltage supplied from the computer to some of the input sensors.
Voltaje de referencia Un voltaje constante proveido por la computadora a algunos de los sensores de entrada.

Road sensing suspension (RSS) A computer-controlled suspension system that senses road conditions and adjusts suspension firmness to match these conditions in a few milliseconds.
Sistema de suspensión sensible al camino Un sistema de suspensión controlado por computadora que detecta las condiciones del camino y ajusta la firmeza de la suspensión en unos cuantos milisegundos en respuesta a estas condiciones.

Room temperature vulcanizing (RTV) sealant A type of sealant that may be used to replace gaskets or help to seal gaskets, in some applications.
Sellador vulcanizante a temperatura ambiente Un tipo de sellador que sirve para reemplazar los empaques o para impermeabilizar los empaques, en algunas aplicaciones.

Scan tester A tester designed to test automotive computer systems.
Aparato explorador Un aparato diseñado para comprobar los sistemas de computadora.

Schrader valve A threaded valve on the fuel rail to which a pressure gauge may be connected to test fuel pressure.
Válvula Schrader Una válvula fileteada en el carril de combustible a la cual se puede conectar un medidor de presión para verificar la presión del combustible.

Secondary air injection (AIR) system A system that injects air into the exhaust system from a belt-driven pump.
Inyección de aire secundario Un sistema que inyecta el aire al sistema de escape por medio de una bomba de aire mandada por correa.

Self-powered test light A test light powered by an internal battery.
Luz de prueba de autopotencia Una luz de prueba cuyo potencia proviene de una batería interna.

Self-test input wire A diagnostic wire located near the diagnostic link connector (DLC) on Ford vehicles.
Alambre (hilo) autodiagnóstico de entrada Un alambre diagnóstico ubicado cerca del conector de enlace diagnóstico (DLC) en los vehículos Ford.

Sequential fuel injection (SFI) A fuel injection system in which the injectors are grounded individually into the computer.
Inyección de combustible secuencial (SFI) Un sistema de inyección de combustible en el cual los inyectores se conectan a tierra individualmente en la computadora.

Serial data Data regarding input sensor voltage signals and output functions. This data is sent from an on-board computer to the data link connector (DLC) and displayed on a scan tester.
Datos en serie Los datos que corresponden a las señales de tensión de los sensores de entrada y las funciones de salida. Estos datos se envían desde la computadora a bordo al conector de enlace de datos (DLC) y se presentan en un aparato explorador.

Service bay diagnostics A diagnostic system supplied by Ford Motor Company to their dealers that diagnoses vehicles and provides communication between the dealer and manufacturer.
Diagnóstico del taller Un sistema diagnóstico proporcionado por la compañía Ford Motor a sus repartidores que efectúa los diagnósticos en los vehículos y provee la comunicación entre el repartidor y el fabricante.

Service check connector A connector with two terminals that may be connected together to obtain diagnostic trouble codes (DTCs) on an ABS system.
Conector de verificación de servicio Un conector de dos terminales que se pueden conectar juntos para obtener los códigos diagnósticos de averías (DTCs) en un sistema ABS.

Shift control cable adjustment Adjustment of the cable connected from the gear selector to the transmission linkage.
Ajuste del cable de cambio de velocidades Un ajuste del cable conectado entre el selector de velocidades y el varillaje de la transmisión.

Shop layout The design of an automotive repair shop.
Esquema del taller El diseño de un taller de reparación automovilístico.

Silicone grease A heat-dissipating grease placed on components such as ignition modules.
Grasa silicona Una grasa disipante del calor que se coloca en los componentes tal como los módulos de encendido.

Snap shot testing The process of freezing computer data into the scan tester memory during a road test, and reading this data later.
Prueba de datos captados El proceso de retener los datos de la computadora en la memoria del aparato explorador, para leerlos después.

Special test mode An electronic instrument cluster (EIC) test mode that displays certain numbers in the speedometer and odometer displays.
Modo especial de prueba Una prueba del conjunto de instrumentos electrónicos (EIC) que presenta ciertos números en las presentaciones del velocímetro y el odómetro.

Spring fill diagnostics A diagnostic procedure in a computer-controlled air suspension system.
Diagnóstico de muelles de suspensión Un procedimiento diagnóstico de un sistema de suspensión neumática controlado por computadora.

Stethoscope A tool used to amplify sound and locate abnormal noises.
Estetoscopio Una herramienta que sirve para amplificar el sonido y localizar los ruidos abnormales.

Sulfuric acid A corrosive acid mixed with water and used in automotive batteries.
Acido sulfúrico Un ácido corrosivo mezclado con el agua y usado en las baterías automotrices.

Supplemental restraint system A term for an air bag system. The word *supplemental* indicates that seat belts must be worn in an air bag-equipped vehicle.
Sistema suplementario de sujeción Un término que indica el sistema de bolsa de aire. La palabra *suplementario* indica que los cinturones de seguridad deben usarse en un vehículo equipado con bolsas de aire.

Switch test A computer system test mode that tests the switch input signals to the computer.
Prueba de interruptor Un modo de prueba del sistema de computadora que verifica las señales de entrada del interruptor a la computadora.

Synchronizer (SYNC) pickup A pickup assembly that produces a voltage signal for ignition triggering or injector sequencing.
Captador sincronizador (SYNC) Una asamblea captadora que produce una señal de tensión para disparar el encendido o para ordenar los inyectores.

Tach-dwell meter A meter that reads engine rpm and ignition dwell.

Medidor del ángulo de leva Un medidor que lee las rpm del motor y el ángulo de leva.

Tachometer (TACH) terminal The negative primary coil terminal.

Terminal del tacómetro (TACH) El terminal negativo de la bobina primaria.

Test spark plug A spark plug with the electrodes removed so it requires a much higher firing voltage for testing such components as the ignition coil.

Bujía de prueba Una bujía de la cual se han quitado los electrodos para que requiere un voltage mucho más fuerte para verificar los componentes tal como la bobina del encendido.

Theft security system An electronic system that blows the horn, flashes the parking and tail lights, and prevents engine starting if the car is not entered in the normal manner.

Sistema de seguridad a prueba de hurto Un sistema electrónico que suena al claxon, hace parpardear las luces de estacionamiento y de cola y incapacita el arranque del motor si se entra al vehículo de una manera incorrecta.

Thermal vacuum valve (TVV) A valve that is opened and closed by a thermo-wax element mounted in the cooling system.

Válvula termovacío (TVV) Una válvula que se abre y se cierre por medio de un elemento de cera térmica colocada en el sistema de enfriamiento.

Thermocouple A device that changes a temperature reading to an electric signal.

Termopar Un dispositivo que interpreta una lectura de temperatura como una señal eléctrica.

Thermostat tester A tester designed to measure thermostat opening temperature.

Probador del termostato Un probador diseñado a medir la temperatura en la cual abre el termostato.

Throttle body injection (TBI) A fuel injection system with the injector or injectors mounted above the throttle.

Inyección cuerpo de acelerador (TBI) Un sistema de inyección de combustible con el inyector o los inyectores montados arriba del acelerador.

Throttle cable adjustment Adjustment of the cable connected from the throttle linkage to a transmission linkage.

Ajuste al cable del acelerador Un ajuste al cable conectado entre el varillaje del acelerador y el varillaje de la transmisión.

Throttle kicker An electric- or vacuum-operated device that holds the throttle open to increase idle rpm under certain engine operating conditions.

Impulsor del acelerador Un dispositivo operado por electricidad o por vacío que mantiene abierto al acelerador para aumentar las rpm de ralentí bajo ciertas condiciones de operación del motor.

Throttle position sensor (TPS) A sensor mounted on the throttle shaft that sends a voltage signal to the computer in relation to throttle opening.

Sensor de posición del acelerador (TPS) Un sensor montado sobre la flecha del acelerador que envía una señal de tensión con respeto a la apertura del acelerador.

Throttle position switch A switch that informs the computer whether the throttle is in the idle position. This switch is usually part of the TPS.

Interruptor de la posición del acelerador Un interruptor que informa a la computador si el acelerador esta en posición de marcha en vacío. Este interruptor suele ser parte del TPS.

Timing connector A wiring connector that must be disconnected while checking basic ignition timing on fuel-injected engines.

Conector del tiempo Un conector alámbrico que se debe disconectar al verificar el tiempo de encendido básico en los motores de inyección.

Traction control system A computer-controlled system that prevents drive wheel spinning. These systems are usually combined with the antilock brake system (ABS).

Sistema de control de tracción Un sistema controlado por computadora que previene que giran las ruedas de impulso. Estos sistemas suelen combinarse con un sistema de frenos antideslizantes (ABS).

Transaxle A mechanical device that transfers engine torque to the drive wheels and provides various gear reductions. The transaxle also contains the differential.

Transeje Un dispositivo mecánico que transfere el par del motor a las ruedas de impulso y proporciona varias desmultiplicaciones de engranaje. El transeje también contiene al diferencial.

Transmission A mechanical device that transfers engine torque to the drive wheels and provides various gear reductions.

Transmisión Un dispositivo mecánico que transfere el par del motor a las ruedas de impulso y provee un desmultiplicación del engranaje.

Transmission control module A computer that controls transmission shifting.

Módulo de control de la transmisión (TCM) Una computadora que controla los cambios de velocidad de la transmisión.

Transmission oil temperature sensor A thermistor-type sensor mounted in the transmission and submerged in transmission fluid. This sensor sends a voltage signal to the transmission control module (TCM) in relation to fluid temperature.

Sensor de la temperatura del aceite de transmisión Un sensor tipo termistor montado en la transmisión y sumergido en el lubricante para transmisiones. Este sensor envía una señal de tensión al módulo de control de la transmisión (TCM) con respeto a la temperatura del lubricante para transmisiones.

Transmission tester A tester that is connected to the transmission electrical connector for diagnostic purposes. This tester allows the technician, rather than the computer, to control the transmission shifts.

Probador de la transmisión Un probador que se conecta al conector eléctrico de la transmisión para efectuar los diagnósticos. Este probador permite que el técnico, y no la computadora, controla los cambios de velocidades.

Transmission tester overlay A plastic overlay with various openings that fit on the transmission tester. Each overlay makes the tester controls and indicators vehicle specific.

Superpuesto transparente del probador de la transmisión Una transparencia de plástico que tiene muchas aperturas que se conformen al probador de la transmisión. Cada superpuesto transparente convierte los controles de prueba y los indicadores para que sirven para un vehículo específico.

Trim height The normal chassis riding height on a computer-controlled suspension system.

Altura de orientación La altura normal de un chasis en un sistema de suspensión controlado por computadora.

Two-gas emissions analyzer An analyzer designed to measure hydrocarbons and carbon monoxide in the exhaust.

Analizador de emisiones de dos gases Un analizador diseñado a medir los hidrocarburos y el monóxido de carbono en el vapor del escape.

United States customary (USC) A system of weights and measures.

Acostumbrado a los Estados Unidos (USC) Un sistema de pesos y medidas.

Upstream air Air injected into the exhaust ports.

Aire arriba El aire inyectado en las puertas del escape.

Vacuum operated decel valve A valve that allows more air into the intake manifold during deceleration to improve emission levels.

Válvula de deceleración operada por vacío Una válvula que permite entrar más aire al múltiple de admisión durante la deceleración para mejorar el nivel de las emisiones.

Vacuum pressure gauge A gauge designed to measure vacuum and pressure.

Medidor de presión de vacío Un medidor diseñado a medir al vacío y a la presión.

Vane-type MAF sensor A MAF sensor containing a pivoted vane that moves a pointer on a variable resistor. This resistor sends a voltage signal to the computer in relation to the total volume of intake air.

Sensor MAF con aletas Un sensor MAF que contiene una aleta articulada que mueva un indicador en un resistor variable. Este resistor envía una señal de tensión a la computador con respeto al volumen total de aire de entrada.

Vehicle speed sensor (VSS) A sensor that is usually mounted in the transmission and sends a voltage signal to the computer in relation to engine speed.

Sensor de velocidad del vehículo (VSS) Un sensor que suele ser montado en la transmisión y envía una señal de tensión a la computadora con respeto a la velocidad del vehículo.

Volt-ampere tester A tester designed to test volts and amperes in such circuits as battery, starter, and charging.

Probador voltampere Un probador diseñado a verificar los voltíos y los amperes en los circuitos tal como los de la batería, el arrancador y el cargador.

Wheel speed sensor gap The air gap between the tip of the wheel speed sensor and the notched ring rotating past the sensor.

Holgura del sensor de la velocidad de la rueda El entrehierro entre la punta del sensor de la velocidad de la rueda y el anillo ranurado girando junto al sensor.

Wiggle test A test recommended on some Ford computer systems to locate defective electrical connections.

Prueba de agitación Una prueba que se recomienda para algunos sistemas de computadora Ford para localizar las conexiones eléctricas defectuosas.

INDEX

Note: Page numbers in bold print reference non-text material.

A

accumulator
    bleeding rear brakes with, 380
    disposal of, 376-378
    high-pressure, 371-381
    pressure relief of, 375-376
A/C idle speed check, 131
A/C performance test, 426-427
actuation test mode (ATM), 144
actuator tests, 297-298
advance-type timing light, 35, 85, 141
air bag system
    clock spring electrical connector, centering, 339
    deployment before vehicle scrapping, 340-341
    diagnostic system check, 329-330
    disabling, 335-336
    flash code diagnosis, 330-332
    guidelines for diagnosing and servicing, 341-342
    inflator module removal and replacement, 336-338, **337**
    scan tester diagnosis, 334
    voltmeter diagnosis, 332-333
    wiring repairs, 339-340
air bag system diagnostic module (ASDM), 329
AIR bypass (AIRB) solenoid, 259-260
air charge temperature (ACT) sensor, 53-54
air conditioning. **See** computer-controlled air conditioning
air conditioning programmer (ACP), 417
AIR diverter (AIRD) solenoid, 259-260
air mix door control rod adjustment, 434-435
air quality, 4-6
air spring valve
    inflation, 403-404
    removal and installation, 402-403
air suspension system
    automated, 413-415
    electronic, 401-406
    rear load-leveling, 406-408
    with speed leveling capabilities, 416
alcohol in fuel test, 113
ambient temperature sensor failure, 295
ammeter, 30
analog meter, 30
analog voltmeter, 54, 139
antidieseling adjustment, 134, 265-266
antilock brake system (ABS)
    Delco Moraine ABS VI diagnosis, 382-388, **385**
    four wheel nonintegral ABS with low-pressure accumulators, 382
    guidelines for diagnosing and servicing, 395-396
    integral ABS with high-pressure accumulator, 379-381
    nonintegral ABS with high-pressure accumulator, 371-379
    rear wheel antilock (RWAL) systems, 369-371
    spark advance reduction and transmission upshift traction ABS diagnosis, 391-392
    throttle control, wheel spin control, spark advance reduction traction control ABS diagnosis, 392-395
    wheel spin traction system diagnosis, 388-391
antilock brake tester, 373-374, 378-379
antilock brake warning light, 379
aqueous parts cleaning tank, 24
arc welding, 220
asbestos dust, 2, 5
ASE blue seal of excellence, 8
ASE technician certification, 8
ATC blend door feedback failure, 296
ATC blend door stall failure, 295
ATC head communications test, 296
ATC mode door feedback failure, 296
ATC mode door stall failure, 296
ATC recirculation door stall failure, 296
auto/manual test, 413
automated air suspension system
    drive cycle diagnosis, 413
    service bay diagnosis, 413-415
    spring fill diagnosis, 415
automatic shutdown (ASD) relay, 90, 93
auto test, 416

B

balance test, 156-157
barometric pressure voltage signal, 57-58, **59 table**
baselining a vehicle, 280
battery boosting, 219-220
baud rate, 220
block learn, 193-194
blow-by, 240
body computer systems
    central timer module diagnosis, 304-305
    generic electronic module diagnosis, 305-306
    guidelines for diagnosing and servicing, 306-307
    individual system and component diagnosis, 299-302, **302-303**
    menu tests, 295-298
    preliminary diagnosis, 293
    scan tester diagnosis, 293-295
    state displays, 298-299
brake bleeding, 379-380
    combined ABS and traction systems, 390
    manual, 387-388
    pressure, 386-387
brake signal test, 364
brake washer, 5
breakout box
    specifications display, 198
    testing, 194-195

C

calibration package (CALPAK), 220
California Rule 66, 23
Canada, 7
canister purge solenoid, 261-262
canister-type injector cleaning container, 28
carbon dioxide (CO_2) emissions, 41, 273
carbon monoxide (CO) emissions, 2, 4, 41, 272-273, 280
carburetor. **See** computer-controlled carburetor
carrying, 16
catalytic converter
    diagnosis and service, 238-239
    efficiency testing, 275
    excessive odor from, 179-180, 249, 269
    guidelines for servicing, 250
central timer module (CTM) system, 304-305
Chrysler
    collision detection (C^2D) multiplex system, 294
    oxygen feedback system fault code diagnosis, 143-144

circuit testers, 28-29
cleaning equipment safety, 23-24
climate control center (CCC), 417
clock spring electrical connector, 339
closed loop, 51
cold-start injector, 163-164
combustion chamber leaks, 275
compass variance adjustment, 299-300
component failures
    diagnosing multiple, 219-220, 268
    diagnosing repeated, 217-219, 268
compressed air equipment safety, 18
computed timing test, 141-142
computer
    ground wire diagnosis, 50
    voltage supply wire diagnosis, 49-50
computer-controlled air conditioning
    A/C performance test, 426-427
    guidelines for diagnosing and servicing, 437, 439
    preliminary inspection, 425-426
    refrigerant system charge test, 428-429
    scan tester diagnosis, 435-437, 438
    vehicle self-diagnostic tests, 430-435
computer-controlled carburetor
    antidieseling adjustment, 134
    flash code diagnosis, 137, 139
    guidelines for diagnosis and service, 145-146
    idle speed adjustment, 131-132
    performance test, 134-137, 138
    preliminary diagnosis, 131
    propane-assisted idle mixture adjustment, 132-134
    scan tester diagnosis, 143-144
    voltmeter diagnosis, 139-143
computer-controlled heat riser valve, 266-267
computer-controlled ignition system
    ignition timing tests, 89-90
    no-start ignition tests, 90
    pickup tests, 91
computer-controlled suspension system
    air with speed leveling capabilities diagnosis, 416
    automatic air diagnosis and service, 413-415
    electronic air diagnosis and service, 401-406
    guidelines for diagnosing and servicing, 419-420
    programmed ride control diagnosis, 408-411, **412**
    rear load-leveling air diagnosis and service, 406-408
    road-sensing diagnosis, 416-419
computer-controlled transmissions
    defining complaints, 345
    diagnosis with pattern select switch, 362-365
    electronic diagnosis, 349-350, **351 table**, **352**
        Ford vehicles, 353-355
    fluid condition and level check, 346
    General Motors 4L80-E diagnosis, 359-361
    guidelines for diagnosing and servicing, 365
    hydraulic and mechanical malfunctions, 348, **348 table**
    improper adjustments, 346-348
    reduced engine performance, 345
    transmission tester diagnosis, 356-359
concealment plug
    removal, 132
continuous self-test, 143
control rod adjustments, 434-435
converter clutch engagement, 359
crossfiring, 73
current DTC, 330, 383
cylinder
    misfiring, 181, 249, 270, 275
    output test, 194

D

data link connector (DLC), 135, 153, 294, 331, 334
data list, 382-383
Delco Moraine ABS VI diagnosis
    manual brake bleeding, 387-388
    pressure brake bleeding, 386-387
    scan tester diagnosis, 382-384, **385**
detonation, 73-74, 180, 249, 262, 269-270
diagnosis by symptom, 203
diagnostic charts, 196, **197**
diagnostic energy reserve module (DERM), 329
diagnostic equipment, 25-32
diagnostic trouble codes (DTCs), 137, 139
    ABS and traction control systems, 370-371, 374-375, 380-382
    central timer module system, 304
    computer-controlled suspension systems, 409-411, 416-419
    EGR systems, 242
    erasing, 143
    generic electronic module system, 305
    OBD II systems, 222-223, **224-229 tables**, 227
    TBI, MFI, SFI systems, 169-176, **174**
    transmissions, 349-350, **351 table**, 362-363
diesel particulates, 4
digital EGR valve, 244-245
digital meter, 30
digital pyrometer, 238-239
digital storage oscilloscope (DSO)
    diagnosing AC superimposed over DC, 217
    guidelines for, 230
    idle air control motor diagnosis, 214-216
    ignition waveform diagnosis, 216
    input sensor signal diagnosis, 207-212
    output actuator diagnosis, 213-214
    spark advance control diagnosis, 216-217
digital tachometer, 33-34
digital volt/ohmmeter, 91
display check, 314
distributor
    assembly, 82
    bushing check, 81
    cap and rotor inspection, 78-79
    disassembly, 81
    guidelines for servicing, 104-105
    inspection, 82
    installing and timing to engine, 82-83, **84**
    removal, 80-81
distributor ignition (DI), 89
DLC terminal, 175
door lock system tests, 301-302
DOT 3 brake fluid, 391
double flaring tool, 121
downshift diagnosis, 358-359
downstream air, 259
drive cycle diagnosis, 413
dwellmeter, 32, 134
dynamic test, 388

E

EATX PRNDL message, 296
EEPROM constant checksum, 296
EGR pressure transducer (EPT), 248
EGR vacuum regulator (EVR), 246, **247**
EGR valve pintle, 242
electric fuel pump testing, 126

electronic air suspension system
  air spring valve
    inflation, 403-404
    removal and installation, 402-403
  diagnosis, 401-402
  line service, 406
  trim height adjustment, 404-405
electronically erasable programmable read only memory (EEPROM), 331
electronic automatic transaxle (EATX), 295
electronic fuel injection (EFI), 114
  computer power and ground wire diagnosis, 178
  diagnosis of specific problems, 178-182
  disconnecting battery cables, 150-151
  DSO diagnosis of, 207-217
  flash code diagnosis, 169-176, **174**
  fuel cut RPM check, 168-169
  fuel pressure testing, 151-156
  fuel pump volume testing, 156
  fuel rail, injector, pressure regulator: removal and replacement, 160-162
  guidelines for servicing TBI, MFI, SFI systems, 182
  injector
    cold-start, diagnosis and service, 163-164
    service and diagnosis, 157-160
    testing, 156-157
  locating service information, 151
  mechanical problems vs. electronic problems, 176-177
  minimum idle speed adjustments, 164-166
  preliminary diagnosis, 150
  preliminary engine control inspection, 149-150
  service bulletin information, 177-178
  service precautions, 150
  throttle body adjustments and service, 164-168
electronic ignition (EI) system
  low data rate or high data rate systems, 95-98
  magnetic sensor
    no-start diagnosis, 103
    tests, 102-103
  no-start ignition diagnosis
    cam and crank sensors, 91-93
    coil and PCM tests, 93-94
  sensor replacement, 94
electronic instrument cluster (EIC), **313**
  diagnosis of typical import, 314-318
  function diagnostic mode, 311, 313
  guidelines for servicing, 324-325
  prove-out display, 311
  special test mode, 313
electronic variable orifice (EVO) steering actuator, 416
electronic variable orifice (EVO) steering system, 408
emission control systems
  catalytic converter diagnosis and service, 238-239
  combination throttle kicker and idle stop solenoid diagnosis and service, 264-266
  computer-controlled heat riser valve diagnosis, 266-267
  EGR pressure transducer diagnosis, 248
  EGR systems diagnosis, 241-245
  EGR vacuum regulator tests, 246, **247**
  emissions analyzer testing, 270-277, **276**
  evaporative (EVAP) system diagnosis and service, 260-262
    thermal vacuum valve, 262
  exhaust gas temperature sensor diagnosis, 248
  guidelines for servicing, 281-282
  guidelines for servicing and diagnosing
    catalytic converters, PCV, EGR systems, 250
  heated air inlet system diagnosis, 267-268
  IM240 emission testing, 277-281
  knock sensor and knock sensor module diagnosis and service, 262-263
  locating service information, 237-238
  multiple failures, diagnosing, 268
  PCV valve diagnosis and service, 239-241
  preliminary inspection, 238, 257-258
  repeated failures, diagnosing, 268
  secondary air injection system diagnosis and service, 258-260
    pulsed, 258
  specific problems and corrections, 269-270
  specific problems related to and corrections, 249-250
  vacuum-operated decel valve diagnosis, 263-264
emissions analyzers, 41-43, 270
  calibrating, 271-272
  causes of excessive emissions, 272-274
  five-gas, **276**, 279
  IM240 standards and testing, 277-281
  infrared, 277
  state inspection maintenance testing with, 274-275, **276**
employer/employee obligations, 6-7
engine
  misfiring, 73, 103-104
  power loss, 181, 249-250, 270
  stalling, 180, 249, 270
  surging, after torque converter clutch lockup, 181, 249
  temperature message test, 297
engine analyzers, 43-44
engine compression, 242
engine coolant temperature (ECT) sensor
  ohmmeter diagnosis, 51-52
  voltage drop specifications, **53 table**
  voltmeter diagnosis, 53
  wiring diagnosis, 52-53
engine detonation, 73-74, 180, 249, 262, 269-270
engine dieseling, 181, 270
engine idle speed check, 132, 265
engine lift safety, 21-22
English system, 13
Environmental Protection Agency (EPA), 428
EPC solenoid dynamic testing, 358
evaporative (EVAP) system, 260-262
  leak testing, 275
  purge flow and pressure test failures, 281
  thermal vacuum valve (TVV), 262
evaporator temperature sensor, 297
exhaust gas recirculation (EGR) system, 238
  digital EGR valve diagnosis, 244-245
  general diagnosis, 241-242
  guidelines for servicing, 250
  linear EGR valve diagnosis, 245
  negative backpressure EGR valve diagnosis, 244
  port EGR valve diagnosis, 243-244
  positive backpressure EGR valve diagnosis, 244
  scan tester data diagnosis, 242-243
  scan tester DTC diagnosis, 242
exhaust gas recirculation valve position (EVP) sensor, 66
exhaust gas temperature sensor, 248

F
failure mode effects management, 215
failure records, 227, 229
fault code display, 414
five-gas emissions analyzer, 275, **276**, 279
flash code diagnosis, 137, 139
floor jack safety, 19
flow testing, 159-160
Ford Motor Company
  breakout box testing, 194-195
  cylinder output test, 194

electronic diagnosis of transmission, 353-355
flash code diagnosis, 171-173, **174**, 330
voltmeter diagnosis, 139-143
four-gas emissions analyzer, 41
four wheel nonintegral ABS with low-pressure accumulators, 382
freeze frame, 227, 229
frequency test, 64
front wheel drive vehicle, 345
fuel cut RPM check, 168-169
fuel filter
    guidelines for service and diagnosis, 126-127
    installation, 123
    removal, 122, **124**
fuel line
    guidelines for service and diagnosis, 126-127
    nylon, inspection and service, 120-121
    rubber, inspection and service, 122
    steel, inspection and service, 121
fuel pressure
    gauge, 26, 114, 152-153
    testing, 151-156
    test port, 114
fuel pump
    electric, testing, 126
    guidelines for service and diagnosis, 126-127
    leakage test, 160
    mechanical, servicing, 123, 125-126
    testing pressure with, 153-155
fuel pump volume, 125
    testing, 156
fuel rail, removing and replacing, 160-162
fuel system pressure relief, 113-115, **124**
fuel tank
    electric: removal, replacement, cleaning, 117
    guidelines for service and diagnosis, 126-127
    inspection, 115
    installation, 120
    purging, 117, 119
    removal, 115-116
    steam cleaning and repairing, 120
functional test 211, 416
function diagnostic mode, 311, 313

G

General Motors
    block learn and integrator diagnosis, 193-194
    fault code diagnosis, 137, 139, 173, 175, 330-331
    mass air flow sensor testing, 193
    PCM service, 220-222
    transmission diagnosis (4L80-E), 359-361
General Motors electronic EI system
    cam sensor timing, 3.8-L turbocharged engines, 102
    coil winding ohmmeter tests, 98
    crankshaft sensor adjustment
        3.0-L, 3300, 3.8-L, and 3800 engines, 98-99
        single slot sensor, 99
    no-start ignition diagnosis
        Type 1 and Type 2 systems, 99-101
        Type 1 fast-start systems, 101-102
generic electronic module (GEM) system, 305-306

H

Hall Effect switch, 91
hand press, 19
hand tool safety, 17
hard fault DTCs, 141, 169, 172, 418
hard starting, 179, 249, 269
hash, 217
head gasket leaks, 275
head-up display (HUD) system, 318-320
    guidelines for servicing, 324-325
heated air inlet system, 267-268
heated resistor MAF sensor, 64
heat riser solenoid, 266-267
heat riser valve
    computer-controlled, 266-267
    vacuum-operated, 264
hesitation on acceleration, 181-182, 250, 270
high idle speed, 179, 249, 269
high-pressure accumulator, 371-381
high-side pressure, 426
history DTC, 330, 383
Honda
    voltmeter diagnosis, 332-333
hot wire MAF sensor, 64
housekeeping, 3-4
hydraulic jack safety, 22-23
hydraulic modulator bleeding, 386-388
hydraulic press safety, 18-19
hydrocarbon (HC) emissions, 41, 272, 280

I

idle
    mixture adjustment, propane-assisted method, 132-134
    speed adjustment, 131-132
idle air control bypass motor
    basic diagnosis, 206
    scan tester diagnosis, 206-207, **208**
idle air control (IAC) motor
    adjusting, 205
    bypass motor diagnosis, 206
    DSO diagnosis, 214-216
    idle contact switch test, 205-206
    improper idle speed diagnosis, 203-204
    scan tester diagnosis of, 206-207, **208**
idle contact switch test, 205-206
idle speed adjustment check, 348
idle stop solenoid, 264-266
ignition coil, inspection and tests, 77-78
ignition system
    computer-controlled, service and diagnosis, 89-91
    diagnosis, 73-74
    distributor
        cap and rotor inspection, 78-79
        service, 80-83, **84**
    DSO waveform diagnosis, 216
    electronic ignition (EI) system
        diagnosis and service, 91-94
        General Motors service and diagnosis, 98-102
        low and high data rate systems, 95-98
        with magnetic sensors, 102-103
    engine misfire diagnosis, 103-104
    guidelines for servicing, 104-105
    ignition coil inspection and tests, 77-78
    ignition module
        removal and replacement, 80
        testers, 37, 75-76
    no-start diagnosis, 74-75
    pickup coil adjustment and tests, 76-77
    secondary ignition wire testing, 79
    spark plug service, 87-89
    test equipment, 32-39
    timing check and adjustment, 83, 85-87
IM240 emission testing, 277-281

inaccurate diagnosis, 219
in-car sensor failure, 297
inductive clamp, 30
inflator module, removing and replacing, 336-338, **337**
infrared emissions analyzer, 277
initialization test, 388
injector
    balance test, 156-157
    cleaning, 157-158
    cold-start, diagnosis and service, 163-164
    DSO diagnosis, 213-214
    flow testing, 159-160
    leakage test, 160
    noid light test, 159
    ohmmeter test, 158
    removing and replacing, 160-162
    sound test, 158
    testing, 156-157
injector balance tester, 27-28
injector pulse width, 278-281
input sensors
    air charge temperature (ACT) sensor diagnosis, 53-54
    DSO diagnosis of, 207-212
    engine coolant temperature diagnosis, 51-53
    exhaust gas recirculation valve position (EVP) sensor diagnosis, 66
    guidelines for diagnosis, 67-68
    knock sensor diagnosis, 65-66
    manifold absolute pressure (MAP) sensor diagnosis, 57-60
    mass air flow (MAF) sensor diagnosis, 61-64
    oxygen sensor diagnosis, 50-51
    park/neutral switch diagnosis, 67
    testing using serial data, 195-198
    throttle position sensor (TPS) diagnosis, 54-57
    vehicle speed sensor (VSS) diagnosis, 66
input test, 305
instrument panel cluster (IPC), 417
intake door rod adjustment, 435
integral ABS with high-pressure accumulator
    antilock brake warning lights, 379
    brake bleeding, 379-380
    DTC diagnosis, 380-381
    master cylinder reservoir, checking and filling, 379
    wheel speed sensor adjustments, 380
integrator diagnosis, 193-194
intermittent DTC, 350, 418
internal module tests, 297
International Standards Organization (ISO), 311

J
jack stand safety, 22-23
job responsibilities, 7

K
Keer-Flo Degreasol 99R, 23
key on engine off (KOEO) test, 139-141, 172, 353-354
key on engine running (KOER) test, 141, 172, 354-355
kilovolt, 39
knock sensor
    diagnosis, 65-66, 262-263
    DSO diagnosis, 212
knock sensor module, 262-263

L
lambda reading, 271-272
latching faults, 388-389
layout (shop), 1-2
leaks, testing for, 160, 275

lean air-fuel mixture, 180, 269
length, 14
lifting, 16
lift safety, 19-21
light emitting diode (LED), 410
linear EGR valve, 245
locked fault, 389
low data rate or high data rate systems
    engine misfire diagnosis, 98
    no-start diagnosis, 95-97
low fuel economy, 179-180, 249, 269
low idle speed, 179, 269
low-side pressure, 426

M
magnetic probe-type digital timing meter, 36
magnetic probe-type tachometer, 34
magnetic sensor
    no-start diagnosis, 103
    tests, 102-103
magnetic timing probe, 36-37
magnetic timing probe receptacle, 86
magnetic timing procedure, 86-87
malfunction indicator light (MIL), 137, 169, 331
manifold absolute pressure (MAP) sensor, 60
    barometric pressure voltage signal diagnosis, 57-58, **59 table**
    DSO diagnosis, 212
    voltage frequency signal diagnosis, 58-59
    voltage signal diagnosis, 58
manifold gauge set, 426
manual brake bleeding, 387-388
manual valve lever position sensor (MVLPS), 347
MAP sensor tester, 58
mass, 14
mass air flow (MAF) sensor
    DSO diagnosis, 211
    heated resistor or hot wire frequency test, 64
    testing, 193
    vane-type MAF ohmmeter tests, 62-63
    vane-type MAF voltmeter diagnosis, 61-62
master cylinder, checking and filling reservoir, 379
measurement systems, 13-15
mechanical fuel pump
    inspection, 123, 125
    removal and replacement, 126
    testing, 125
memory calibrator (MEM-CAL), 220-222
memory DTC, 141, 172, 350
menu tests, 295-298
meter impedance, 31
metric system, 13-15
minimum idle speed adjustment, 164-165
    on TBI systems, 165-166
mode door control rod adjustment, 434
modulator, air bleeding with ALB tester, 378-379
motor rehome, 384, 392
multimeter, 30-32
multiport fuel injection (MFI)
    cold-start injector diagnosis and service, 163-164
    computer power and ground wire diagnosis, 178
    diagnosis of specific problems, 178-182
    disconnecting battery cables, 150-151
    flash code diagnosis, 169-176, **174**
    fuel cut RPM check, 168-169
    fuel pressure testing, 151-156
    fuel pump volume testing, 156
    fuel rail, injector, pressure regulator removal and replacement, 160-162

guidelines for servicing, 182
injector service and diagnosis, 157-160
injector testing, 156-157
locating service information, 151
mechanical problems vs. electronic problems, 176-177
minimum idle speed and throttle position sensor adjustments, 164-165
preliminary diagnostic procedure, 150
preliminary inspection, 149-150
service bulletin information, 177-178
service precautions, 150
throttle body service, 166-168

N

National Institute for Automotive Service Excellence, 8
negative backpressure EGR valve, 244
negative trigger slope, 207
Nissan
    flash code diagnosis, 175-176
noid light test, 159
noneraseable fault, 389
nonintegral ABS with high-pressure accumulator
    accumulator
        disposal of, 376-378
        pressure relief, 375-376
    air bleeding modulator with ALB tester, 378-379
    diagnosis with ALB tester, 373-374
        wheel speed sensor test with, 374
    diagnostic trouble code diagnosis, 374-375
    wheel speed sensor gap measurement, 371-373
nonintegral ABS with low-pressure accumulators, four wheel, 382
nonlatching faults, 389
nonmagnetic feeler gauge, 76
no-start, 178, 249
no-start ignition
    diagnosis, 73
    primary circuit diagnosis, 74-75
    secondary circuit diagnosis, 75
nylon fuel line service, 120-121

O

OBD II diagnosis
    conflicting data, 229-230
    datastream and DTC diagnosis, 223, **224-229 tables**, 227
    DTC interpretation, 222-223
    freeze frame and failure records diagnosis, 227, 229
    guidelines for, 230
    readiness status and O_2 tests, 229
    SAE standards, 222
    scan tester diagnosis, 223
ohmmeter, 77
on-demand self-test, 305
open circuits, intermittent, 218-219
open loop, 51
optical-type pickup, 91
oscilloscope, 38
    scales and tests, 39
OSHA (Occupational safety and Health Act), 5, 18
output actuators
    DSO diagnosis, 213-214
    shorted, 217-218
    testing using serial data, 198-201
output cycling test, 354
output shaft speed sensor
    dynamic test, 359
    static test, 358
output state test, 142
overhead travel information system (OTIS), 299

oxides of nitrogen (NO_x), 273, 281
oxygen (O_2), 41, 273
oxygen sensor
    digital storage oscilloscope diagnosis, 207, 209-210
    heating diagnosis, 51
    signal evaluation, 278-280
    voltage signal diagnosis, 50-51
    wiring diagnosis, 51

P

panel lamps failure, 297
park/neutral switch, 67, 347
parts washers, 23-24
pattern select switch, 362-365
PCV valve
    diagnosis and service, 239-241
    guidelines for servicing, 250
pellet-type catalytic converter, 239
performance test, computer-controlled carburetor, 134-137, **138**
personal safety, 15-16
photoelectric tachometer, 33
pickup coil
    ohmmeter tests, 77
    pickup gap adjustment, 76
pinpoint tests, 414, 416
port EGR valve, 243-244
port fuel injection (PFI), 114
port fuel injection pressure tester, 152
positive backpressure EGR valve, 244
positive crankcase ventilation (PCV) system, 238
    test, 275
positive trigger slope, 207
power door lock
    motor tests, 301-302
    switch test, 301
power loss, 74
power tool safety, 17-18
powertrain control module (PCM), 136, 151, 417
    control circuit testing using serial data, 201-202
    ground and power wire diagnosis, 178
    servicing General Motors vehicles, 220-222
power unit check, 316-317
prefixes, metric, 14
preprogrammed signal check, 314-316
pressure bleeder, 379-380
pressure brake bleeding, 386-387
pressure regulator
    leakage test, 160
    removing and replacing, 160-162
pressure relief, fuel system, 113-115, **124**
pressurized injector cleaning container, 28
programmable read only memory (PROM), 220-221
programmed ride control system
    DTC diagnosis, 409-411, **412**
    general diagnosis, 408-409
propane-assisted idle mixture adjustment, 132-134
prove-out display, 311
pulsed secondary air injection system, 258
purge flow test, 281
pyrometer, 238-239

Q

quad driver, 217, 245
quick-disconnect fuel line fittings, 116, 161

R

radio frequency interference (RFI), 217
    excessive, 219, 268

random access memory (RAM), 331
R-134a refrigerant, 428
RCRA Act, 23
read faults, 295-297
real time damping (RTD) module, 418
rear load-leveling air suspension system
    diagnosis, 406, 408
    trim height adjustment, 406
rear wheel antilock (RWAL) systems, 369-371
receiver module, 300-301
reduced fuel mileage, 74
reference pickup, 91
refrigerant system charge test, 428-429
remote keyless entry transmitter, 300-301
resistance tests, 357
rich air-fuel mixture, 179-180, 249, 269
ripple voltage, 217
road-sensing suspension system
    optional diagnostic modes, 418-419
    trouble code display, 416-418
room temperature vulcanizing (RTV) sealant, 51
rough idle, 179, 249, 269
R-12 refrigerant, 428
rubber fuel line service, 122
rules (shop), 2-3

S
SAE J1930 terminology, 89, 135-137, 151, 169, 331
safety
    cleaning equipment, 23-24
    compressed air equipment, 18
    guidelines for, 45-46
    hand tool, 17
    hydraulic pressing and lifting equipment, 18-23
    personal, 15-16
    power tool, 17-18
    training exercises, 44-45
scan tester, 39-41, 408
scan tester diagnosis, **208**
    ABS and traction control systems, 371, 382, **385**, 391-392, **393**
    actuation test mode (ATM), 144
    additional functions, 202
    air bag systems, 334
    block learn and integrator diagnosis, 193-194
    body computer faults, 293-295
    breakout box
        specifications display, 198
        testing, 194-195
    Chrysler oxygen feedback system fault code diagnosis, 143-144
    computer-controlled air conditioning, 435-437, **438**
    cylinder output test, 194
    guidelines for, 230
    initial entries, 189-190
    initial selections, 190
    MAF sensor testing, 193
    OBD II systems, 223
    precautions for, 189
    serial data interpretation, 191-193
    snapshot testing, 194
    switch test mode, 144
    test selections, 191
    transmissions, 349-350, **351 table**
Schrader valve, 114, 153
secondary air injection system
    component diagnosis, 259-260
    efficiency testing, 275
    general diagnosis, 258-259
    pulsed, 258
secondary ignition wires, testing, 79
self-diagnostic tests
    automatic temperature control systems, 430-431
    entry and cancellation, 431
    steps in, 432-434
self-powered test light, 29
sequential fuel injection (SFI)
    cold-start injector diagnosis and service, 163-164
    diagnosis of specific problems, 178-182
    disconnecting battery cables, 150-151
    flash code diagnosis, 169-176, **174**
    fuel cut RPM check, 168-169
    fuel pressure testing, 151-156
    fuel pump volume testing, 156
    fuel rail, injector, pressure regulator removal and replacement, 160-162
    guidelines for servicing, 182
    injector service and diagnosis, 157-160
    injector testing, 156-157
    locating service information, 151
    mechanical problems vs. electronic problems, 176-177
    minimum idle speed and throttle position sensor adjustments, 164-165
    preliminary diagnostic procedure, 150
    preliminary inspection, 149-150
    service bulletin information, 177-178
    service precautions, 150
    throttle body service, 166-168
serial data
    defined, 191
    interpreting scan tester data, 191-193
    testing input sensors using, 195-198
    testing output actuators using, 198-201
    testing PCM control circuit using, 201-202
service bay diagnosis, 413-415
service bulletins, 177-178
shift control cable adjustment, 347
shop practices
    air quality, 4-6
    ASE certification, 8
    employer/employee obligations, 6-7
    housekeeping, 3-4
    job responsibilities, 7
    layout, 1-2
    rules, 2-3
    vehicle operation, 3
short to ground test, 358
show DTC, 383
silicone grease, 80
snapshot testing, 194, 384
Society of Automotive Engineers (SAE), 222
solenoid voltage test, 358
sound test, 158
spark advance diagnosis, 216-217
spark advance reduction, transmission upshift traction ABS, 391-392
spark knock, 74
spark plug
    normal and abnormal conditions, **88 table**
    service, 87, 89
spark plug wire
    inspection, 79
    installation, 79
    testing, 79
special test mode, 313
speedometer diagnosis, 318
speed sensor check, 317-318
spikes, 220
spring fill diagnostics, 415

...ays, 298-299
...e I/M testing, 274-277
steel fuel line service, 121
stethoscope, 25
sulfuric acid, 4-5
sun load sensor failure, 297
supplemental inflatable restraint (SIR) module, 417
surging at idle, 180, 269
suspension system. **See** computer-controlled suspension system
switch power supply test, 301
switch test mode, 144
synchronizer (SYNC) pickup, 91

T

tach-dwellmeter, 32
tachometer, 32
tachometer (TACH) terminal, 90
thermal vacuum valve (TVV), 262
thermocouple, 428
thermometer, 52
throttle, wheel spin, and spark advance reduction traction control ABS, 392-395
throttle adjuster, throttle body and cruise control cable adjustment, 394
throttle body
    on-vehicle cleaning, 166
    on-vehicle inspection, 166-168
throttle body injection (TBI), 115
    cold-start injector diagnosis and service, 163-164
    computer power and ground wire diagnosis, 178
    diagnosis of specific problems, 178-182
    disconnecting battery cables, 150-151
    flash code diagnosis, 169-176, **174**
    fuel cut RPM check, 168-169
    fuel pressure testing, 151-156
    fuel pump volume testing, 156
    fuel rail, injector, pressure regulator removal and replacement, 160-162
    guidelines for servicing, 182
    injector service and diagnosis, 157-160
    injector testing, 156-157
    locating service information, 151
    mechanical problems vs. electronic problems, 176-177
    minimum idle speed and throttle position sensor adjustments, 164-166
    preliminary diagnostic procedure, 150
    preliminary inspection, 149-150
    service bulletin information, 177-178
    service precautions, 150
    throttle body service, 166-168
throttle cable adjustment, 346-347
throttle kicker, 264-266
throttle position sensor (TPS), 364
    adjustment of, 56-57, 164-165
    DSO diagnosis, 211
    four-wire sensor diagnosis, 56
    three-wire sensor diagnosis, 54-56
timing advance check, 87
timing connector, 90
timing light, 34-36
tone and door ajar light test, 304
torque converter clutch lockup, 206
Toyota
    flash code diagnosis, 170-171, 331-332
traction control systems
    guidelines for diagnosing and servicing, 395-396
    spark advance reduction and transmission upshift traction diagnosis, 391-392
    throttle control, wheel spin control, spark advance reduction traction control, 392-395
    wheel spin traction system diagnosis, 388-391
traction control system (TCS), 418
transaxle, 345
transmission oil temperature sensor, 358
transmissions. **See** computer-controlled transmissions
transmission tester, 356-359
trigger, 207
trim height adjustment, 404-407
tune-up equipment, 25-32
two-gas emissions analyzer, 41

U

United States customary (USC) system, 13
upshift diagnosis, 358-359
upshift voltage test, 364
upstream air, 259

V

vacuum diaphragm, 266
vacuum hand pump, 58, 268
vacuum leaks, 275
vacuum-operated decel valve, 263-264
vacuum-operated heat riser valve, 264
vacuum pressure gauge, 26
vane-type mass air flow sensor, 61-63
vehicle operation, 3
vehicle self-diagnostic tests, 430-435
vehicle speed sensor (VSS) diagnosis, 66
vehicle theft security systems, 320, **321-324 tables**
    guidelines for servicing, 324-325
voltage suppression diodes, defective, 219, 268
volt-ampere tester, 30
voltmeter diagnosis
    computed timing test, 141-142
    continuous self-test, 143
    erase code procedure, 143
    key on engine off (KOEO) test, 139-141
    key on engine running (KOER) test, 141
    output state test, 142
volume, 15

W

wheel brake bleeding, 387-388
wheel speed sensor
    adjustments, 380
    gap measurement, 371-373
    servicing, 390-391
    test with ALB tester, 374
wheel spin traction ABS
    brake bleeding, 390
    diagnostic mode, 388-389
    initialization and dynamic tests, 388
    wheel speed sensor service, 390-391
WHMIS, 7
wiggle test, 354

Z

zero pressure rear wheel antilock (ZPRWAL) systems, 371